Collected Experimental Papers of P. W. Bridgman

Volume I

P. W. BRIDGMAN

Collected Experimental Papers

Volume I

Papers 1–11

Harvard University Press
Cambridge, Massachusetts
1964

© Copyright 1964 by the President and Fellows of Harvard College

All rights reserved

Distributed in Great Britain by Oxford University Press, London

Library of Congress Catalog Card Number 64-16060

Printed in the United States of America

CONTENTS

Volume I

xix. Physical Papers Omitted

xxi. Preface by the Editorial Committee

xxv. Introduction

1-1. "The measurement of high hydrostatic pressure, I. A simple primary gauge," *Proc. Am. Acad. Arts Sci.* 44, 201–217 (1909).

2-19. "The measurement of high hydrostatic pressure, II. A secondary mercury resistance gauge," *Proc. Am. Acad. Arts Sci.* 44, 221–251 (1909).

3-51. "An experimental determination of certain compressibilities," *Proc. Am. Acad. Arts Sci.* 44, 255–279 (1909).

4-77. "The action of mercury on steel at high pressures," *Proc. Am. Acad. Arts Sci.* 46, 325–341 (1911).

5-95. "The measurement of hydrostatic pressures up to 20,000 kilograms per square centimeter," *Proc. Am. Acad. Arts Sci.* 47, 321–343 (1911).

6-119. "Mercury, liquid and solid, under pressure," *Proc. Am. Acad. Arts Sci.* 47, 347–438 (1911).

7-213. "Water, in the liquid and five solid forms, under pressure," *Proc. Am. Acad. Arts Sci.* 47, 441–558 (1911).

8-335. "The collapse of thick cylinders under high hydrostatic pressure," *Phys. Rev.* 34, 1–24 (1912).

9-359. "Breaking tests under hydrostatic pressure and conditions of rupture," *Phil. Mag.* [6] 24, 63–80 (1912).

10-379. "Thermodynamic properties of liquid water to 80° and 12,000 kgm.," *Proc. Am. Acad. Arts Sci.* 48, 309–362 (1912).

11-433. "Thermodynamic properties of twelve liquids between 20° and 80° and up to 12,000 kgm. per sq. cm.," *Proc. Am. Acad. Arts Sci.* **49**, 3-114 (1913).

Volume II

12-593. "The technique of high pressure experimenting," *Proc. Am. Acad. Arts Sci.* **49**, 627-643 (1914).

13-611. "Über Tammanns vier neue Eisarten," *Z. Phys. Chem.* **86**, 513-524 (1914).

14-624. "Change of phase under pressure, I. The phase diagram of eleven substances with especial reference to the melting curve," *Phys. Rev. 3*, 126-141, 153-203 (1914).

15-693. "High pressures and five kinds of ice," *J. Franklin Inst.* **177**, 315-332 (1914).

16-712. "Two new modifications of phosphorus," *J. Am. Chem. Soc.* **36**, 1344-1363 (1914).

17-732. "Nochmals die Frage des unbeständigen Eises," *Z. Phys. Chem.* **89**, 252-253 (1914).

18-735. "The coagulation of albumen by pressure," *J. Biol. Chem.* **19**, 511-512 (1914).

19-737. "Change of phase under pressure, II. New melting curves with a general thermodynamic discussion of melting," *Phys. Rev. 6*, 1-33, 94-112 (1915).

20-789. "Polymorphic transformations of solids under pressure," *Proc. Am. Acad. Sci.* **51**, 55-124 (1915).

21-859. "The effect of pressure on polymorphic transitions of solids," *Proc. Nat. Acad. Sci. U. S.* **1**, 513-516 (1915).

22-863. "On the effect of general mechanical stress on the temperature of transition of two phases, with a discussion of plasticity," *Phys. Rev. 7*, 215-223 (1916).

23-873. "Further note on black phosphorus," *J. Am. Chem. Soc.* **38**, 609-612 (1916).

24-877. "Polymorphic changes under pressure of the univalent nitrates," *Proc. Am. Acad. Arts Sci.* **51**, 581-625 (1916).

25-923. "The velocity of polymorphic changes between solids," *Proc. Am. Acad. Sci.* **52**, 57-88 (1916).

26-955. "Polymorphism at high pressures," *Proc. Am. Acad. Arts Sci.* **52**, 91-187 (1916).

27-1053. "The electrical resistance of metals under pressure," *Proc. Am. Acad. Arts Sci.* 52, 573–646 (1917).

28-1128. "The resistance of metals under pressure," *Proc. Nat. Acad. Sci. U. S.* 3, 10–12 (1917).

29-1132. "Note on the elastic constants of antimony and tellurium wires," *Phys. Rev.* 9, 138–141 (1917).

30-1137. "Theoretical considerations on the nature of metallic resistance, with especial regard to the pressure effects," *Phys. Rev.* 9, 269–289 (1917).

31-1159. "Thermo-electromotive force, Peltier heat, and Thomson heat under pressure," *Proc. Am. Acad. Arts Sci.* 53, 269–386 (1918).

Volume III

32-1277. "The failure of cavities in crystals and rocks under pressure," *Am. J. Sci.* 45, 243–268 (1918).

33-1303. "Stress-strain relations in crystalline cylinders," *Am. J. Sci.* 45, 269–280 (1918).

34-1316. "On equilibrium under non-hydrostatic stress," *Phys. Rev.* 11, 180–183 (1918).

35-1321. "A comparison of certain electrical properties of ordinary and uranium lead," *Proc. Nat. Acad. Sci. U. S.* 5, 351–353 (1919).

36-1325. "An experiment in one-piece gun construction," *Mining and Metallurgy*, February 1920, pp. 1–16.

37-1341. "Further measurements of the effect of pressure on resistance," *Proc. Nat. Acad. Sci. U. S.* 6, 505–508 [1–4] (1920).

38-1345. "Electrical resistance under pressure, including certain liquid metals," *Proc. Am. Acad. Arts Sci.* 56, 61–154 (1921).

39-1439. "Measurements of the deviation from Ohm's law in metals at high current densities," *Proc. Nat. Acad. Sci. U. S.* 7, 299–303 (1921).

40-1445. "The effect of tension on the electrical resistance of certain abnormal metals," *Proc. Am. Acad. Arts Sci.* 57, 41–66 (1922).

41-1471. "The effect of pressure on the thermal conductivity of metals," *Proc. Am. Acad. Arts Sci.* 57, 77–127 (1922).

42-1523. "The failure of Ohm's law in gold and silver at high current densities," *Proc. Am. Acad. Arts Sci.* 57, 131–172 (1922).

43-1565. "The compressibility of metals at high pressures," *Proc. Nat. Acad. Sci. U. S.* 8, 361–365 (1922).

44-1571. "The effect of pressure on the electrical resistance of cobalt, aluminum, nickel, uranium, and caesium," *Proc. Am. Acad. Arts Sci.* 58, 151–161 (1923).

45-1583. "The compressibility of thirty metals as a function of pressure and temperature," *Proc. Am. Acad. Arts Sci.* 58, 165–242 (1923).

46-1661. "The compressibility of hydrogen to high pressures," *Rec. Trav. Chim. Pays-Bas* 42, 568–571 (1923).

47-1665. "The thermal conductivity of liquids," *Proc. Nat. Acad. Sci. U. S.* 9, 341–345 (1923).

48-1670. "The volume changes of five gases under high pressures," *Proc. Nat. Acad. Sci. U. S.* 9, 370–372 (1923).

49-1673. "The compressibility and pressure coefficient of resistance of rhodium and iridium," *Proc. Am. Acad. Arts Sci.* 59, 109–115 (1923).

50-1681. "The effect of tension on the thermal and electrical conductivity of metals," *Proc. Am. Acad. Arts Sci.* 59, 119–137 (1923).

51-1701. "The thermal conductivity of liquids under pressure," *Proc. Am. Acad. Arts Sci.* 59, 141–169 (1923).

52-1733. "The compressibility of five gases to high pressures," *Proc. Am. Acad. Arts Sci.* 59, 173–211 (1924).

53-1773. "The thermal conductivity and compressibility of several rocks under high pressures," *Am. J. Sci.* 7, 81–102 (1924).

54-1795. "Some properties of single metal crystals," *Proc. Nat. Acad. Sci. U. S.* 10, 411–415 (1924).

55-1801. "Properties of matter under high pressure," *Mech. Eng.*, March 1925, pp. 161–169 [1–25].

56-1827. "Certain aspects of high-pressure research," *J. Franklin Inst.* 200, 147–160 (1925).

57-1841. "The compressibility of several artificial and natural glasses," *Am. J. Sci.* 10, 359–367 (1925).

58-1851. "Certain physical properties of single crystals of tungsten, antimony, bismuth, tellurium, cadmium, zinc, and tin," *Proc. Am. Acad. Arts Sci. 60*, 305–383 (1925).

Volume IV

59-1931. "Various physical properties of rubidium and caesium and the resistance of potassium under pressure," *Proc. Am. Acad. Arts Sci. 60*, 385–421 (1925).

60-1969. "The effect of tension on the transverse and longitudinal resistance of metals," *Proc. Am. Acad. Arts Sci. 60*, 423–449 (1925).

61-1997. "The viscosity of liquids under pressure," *Proc. Nat. Acad. Sci. U. S. 11*, 603–606 (1925).

62-2002. "Thermal conductivity and thermo-electromotive force of single metal crystals," *Proc. Nat. Acad. Sci. U.S. 11*, 608–612 (1925).

63-2007. "Linear compressibility of fourteen natural crystals," *Am. J. Sci. 10*, 483–498 (1925).

64-2024. "The five alkali metals under high pressure," *Phys. Rev. 27*, 68–86 (1926).

65-2043. "The effect of pressure on the viscosity of forty-three pure liquids," *Proc. Am. Acad. Arts Sci. 61*, 57–99 (1926).

66-2087. "Thermal conductivity and thermal E.M.F. of single crystals of several non-cubic metals," *Proc. Am. Acad. Arts Sci. 61*, 101–134 (1926).

67-2121. "Dimensional analysis again," *Phil. Mag. 2*, 1263–1266 (1926).

68-2126. "The breakdown of atoms at high pressures," *Phys. Rev. 29*, 188–191 (1927).

69-2130. "The transverse thermo-electric effect in metal crystals," *Proc. Nat. Acad. Sci. U. S. 13*, 46–50 (1927).

70-2135. "Some mechanical properties of matter under high pressure," *Proc. 2nd Intern. Congr. Appl. Mechanics*, Zurich, 1926 (Zurich: Orell Füssli, 1927), pp. 53–61.

71-2145. "Electrical properties of single metal crystals," *Atti Congr. Intern. Fis.*, Como, 1927 (Bologna: Nicola Zanichelli, 1928), pp. 239–248.

72-2155. "The viscosity of mercury under pressure," *Proc. Am. Acad. Arts Sci. 62*, 187–206 (1927).

73-2175. "The compressibility and pressure coefficient of resistance of ten elements," *Proc. Am. Acad. Arts Sci.* 62, 207–226 (1927).

74-2195. "The linear compressibility of thirteen natural crystals," *Am. J. Sci.* 15, 287–296 (1928).

75-2205. "The pressure transitions of the rubidium halides," *Z. Krist.* 67, 363–376 (1928).

76-2219. "Resistance and thermo-electric phenomena in metal crystals," *Proc. Nat. Acad. Sci. U. S.* 14, 943–946 (1928).

77-2223. "The compressibility and pressure coefficient of resistance of zirconium and hafnium," *Proc. Am. Acad. Arts Sci.* 63, 347–350 (1928).

78-2227. "The effect of pressure on the resistance of three series of alloys," *Proc. Am. Acad. Arts Sci.* 63, 329–345 (1928).

79-2245. "Thermo-electric phenomena and electrical resistance in single metal crystals," *Proc. Am. Acad. Arts Sci.* 63, 351–399 (1929).

80-2295. "The effect of pressure on the rigidity of steel and several varieties of glass," *Proc. Am. Acad. Arts Sci.* 63, 401–420 (1929).

81-2315. "General survey of the effects of pressure on the properties of matter," *Proc. Phys. Soc. (London)* 41, 341–360 (The Institute of Physics and The Physical Society, 1929).

82-2336. "Irreversible transformations of organic compounds under high pressures" (with J. B. Conant), *Proc. Nat. Acad. Sci. U. S.* 15, 680–683 (1929).

83-2341. "Die Eigenschaften von Metallen unter hohen hydrostatischen Drucken," *Metallwirtschaft* 8, 229–233 [1–10] (1929).

84-2352. "On the nature of the transverse thermo-magnetic effect and the transverse thermo-electric effect in crystals," *Proc. Nat. Acad. Sci. U. S.* 15, 768–773 (1929).

85-2359. "The elastic moduli of five alkali halides," *Proc. Am. Acad. Arts Sci.* 64, 19–38 (1929).

86-2379. "The effect of pressure on the rigidity of several metals," *Proc. Am. Acad. Arts Sci.* 64, 39–49 (1929).

87-2391. "The compressibility and pressure coefficient of resistance of several elements and single crystals," *Proc. Am. Acad. Arts Sci.* 64, 51–73 (1929).

88-2415. "The minimum of resistance at high pressure," *Proc. Am. Acad. Arts Sci. 64*, 75–90 (1930).

89-2431. "The volume of eighteen liquids as a function of pressure and temperature," *Proc. Am. Acad. Arts Sci. 66*, 185–233 (1931).

90-2481. "Compressibility and pressure coefficient of resistance, including single crystal magnesium," *Proc. Am. Acad. Arts Sci. 66*, 255–271 (1931).

91-2498. "The P-V-T relations of NH_4Cl and NH_4Br, and in particular the effect of pressure on the volume anomalies," *Phys. Rev. 38*, 182–191 (1931).

92-2509. "Recently discovered complexities in the properties of simple substances," *Trans. Am. Inst. Mining Met. Engrs.*, General Volume 1931, pp. 17–37.

93-2531. "Volume-temperature-pressure relations for several non-volatile liquids," *Proc. Am. Acad. Arts Sci. 67*, 1–27 (1932).

Volume V

94-2559. "Physical properties of single crystal magnesium," *Proc. Am. Acad. Arts Sci. 67*, 29–41 (1932).

95-2572. "A new kind of e.m.f. and other effects thermodynamically connected with the four transverse effects," *Phys. Rev. 39*, 702–715 (1932).

96-2587. "A transition of silver oxide under pressure," *Rec. Trav. Chim. Pays-Bas 51*, 627–632 (1932).

97-2593. "The pressure coefficient of resistance of fifteen metals down to liquid oxygen temperatures," *Proc. Am. Acad. Arts Sci. 67*, 305–344 (1932).

98-2633. "The compressibility of eighteen cubic compounds," *Proc. Am. Acad. Arts Sci. 67*, 345–375 (1932).

99-2664. "The effect of homogeneous mechanical stress on the electrical resistance of crystals," *Phys. Rev. 42*, 858–863 (1932).

100-2671. "The pressure-volume-temperature relations of fifteen liquids," *Proc. Am. Acad. Arts Sci. 68*, 1–25 (1933).

101-2697. "Compressibilities and pressure coefficients of resistance of elements, compounds, and alloys, many of them anomalous," *Proc. Am. Acad. Arts Sci. 68*, 27–93 (1933).

102-2765. "The effect of pressure on the electrical resistance of single metal crystals at low temperature," *Proc. Am. Acad. Arts Sci. 68*, 95–123 (1933).

103-2795. "Two new phenomena at very high pressure," *Phys. Rev. 45*, 844–845 (1934).

104-2796. "The melting parameters of nitrogen and argon under pressure, and the nature of the melting curve," *Phys. Rev. 46*, 930–933 (1934).

105-2801. "The compressibility of solutions of three amino acids" (with R. B. Dow), *J. Chem. Phys. 3*, 35–41 (1935).

106-2809. "Theoretically interesting aspects of high pressure phenomena," *Rev. Mod. Phys. 7*, 1–33 (1935).

107-2843. "Electrical resistances and volume changes up to 20,000 kg/cm^2," *Proc. Nat. Acad. Sci. U. S. 21*, 109–113 (1935).

108-2849. "On the effect of slight impurities on the elastic constants, particularly the compressibility of zinc," *Phys. Rev. 47*, 393–397 (1935).

109-2855. "The melting curves and compressibilities of nitrogen and argon," *Proc. Am. Acad. Arts Sci. 70*, 1–32 (1935).

110-2887. "Measurements of certain electrical resistances, compressibilities, and thermal expansions to 20,000 kg/cm^2," *Proc. Am. Acad. Arts Sci. 70*, 71–101 (1935).

111-2919. "The pressure-volume-temperature relations of the liquid, and the phase diagram of heavy water," *J. Chem. Phys. 3*, 597–605 (1935).

112-2929. "Effects of high shearing stress combined with high hydrostatic pressure," *Phys. Rev. 48*, 825–847 (1935).

113-2953. "Polymorphism, principally of the elements, up to 50,000 kg/cm^2," *Phys. Rev. 48*, 893–906 (1935).

114-2967. "Compressibilities and electrical resistance under pressure, with special reference to intermetallic compounds," *Proc. Am. Acad. Arts Sci. 70*, 285–317 (1935).

115-3001. "Shearing phenomena at high pressure of possible importance for geology," *J. Geol. 44*, 653–669 (University of Chicago Press, 1936).

116-3018. "Flow phenomena in heavily stressed metals," *J. Appl. Phys. 8*, 328–336 (1937).

117-3028. "Polymorphic transitions of inorganic compounds to 50,000 kg/cm^2," *Proc. Nat. Acad. Sci. U. S. 23*, 202–205 (1937).

118-3033. "Shearing phenomena at high pressures, particularly in inorganic compounds," *Proc. Am. Acad. Arts Sci.* **71**, 387–460 (1937).

119-3107. "Polymorphic transitions of 35 substances to 50,000 kg/cm^2," *Proc. Am. Acad. Arts Sci.* **72**, 45–136 (1937).

120-3200. "The phase diagram of water to 45,000 kg/cm^2," *J. Chem. Phys.* **5**, 964–966 (1937).

121-3203. "The resistance of nineteen metals to 30,000 kg/cm^2," *Proc. Am. Acad. Arts Sci.* **72**, 157–205 (1938).

Volume VI

122-3253. "Rough compressibilities of fourteen substances to 45,000 kg/cm^2," *Proc. Am. Acad. Arts Sci.* **72**, 207–225 (1938).

123-3273. "Polymorphic transitions up to 50,000 kg/cm^2 of several organic substances," *Proc. Am. Acad. Arts Sci.* **72**, 227–268 (1938).

124-3315. *The nature of metals as shown by their properties under pressure*, Am. Inst. Mining Met. Engrs. Technical Publication No. 922 (1938).

125-3337. "Reflections on rupture," *J. Appl. Phys.* **9**, 517–528 (1938).

126-3349. "Shearing experiments on some selected minerals and mineral combinations" (with Esper S. Larsen), *Am. J. Sci.* **36**, 81–94 (1938).

127-3363. "The high pressure behavior of miscellaneous minerals," *Am. J. Sci.* **237**, 7–18 (1939).

128-3375. "Considerations on rupture under triaxial stress," *Mech. Eng.*, February 1939, pp. 107–111.

129-3381. "Absolute measurements in the pressure range up to 30,000 kg/cm^2," *Phys. Rev.* **57**, 235–237 (1940).

130-3385. "Compressions to 50,000 kg/cm^2," *Phys. Rev.* **57**, 237–239 (1940).

131-3388. "New high pressures reached with multiple apparatus," *Phys. Rev.* **57**, 342–343 (1940).

132-3391. "The measurement of hydrostatic pressure to 30,000 kg/cm^2," *Proc. Am. Acad. Arts Sci.* **74**, 1–10 (1940).

133-3401. "The linear compression of iron to 30,000 kg/cm^2," *Proc. Am. Acad. Arts Sci.* **74**, 11–20 (1940).

CONTENTS

134-3411. "The compression of 46 substances to 50,000 kg/cm^2," *Proc. Am. Acad. Arts Sci.* **74**, 21–51 (1940).

135-3443. "Explorations toward the limit of utilizable pressures," *J. Appl. Phys.* **12**, 461–469 (1941).

136-3453. "Compressions and polymorphic transitions of seventeen elements to 100,000 kg/cm^2," *Phys. Rev.* **60**, 351–354 (1941).

137-3458. "Freezings and compressions to 50,000 kg/cm^2," *J. Chem. Phys.* **9**, 794–797 (1941).

138-3463. "Freezing parameters and compressions of twenty-one substances to 50,000 kg/cm^2," *Proc. Am. Acad. Arts Sci.* **74**, 399–424 (1942).

139-3489. "Pressure-volume relations for seventeen elements to 100,000 kg/cm^2," *Proc. Am. Acad. Arts Sci.* **74**, 425–440 (1942).

140-3505. "Recent work in the field of high pressures," *Am. Scientist* **31**, 1–35 (1943).

141-3541. "On torsion combined with compression," *J. Appl. Phys.* **14**, 273–283 (1943).

142-3553. "Some irreversible effects of high mechanical stress," in *Colloid Chemistry*, ed. Jerome Alexander (New York: Van Nostrand, 1944), V, 327–337.

143-3565. "The stress distribution at the neck of a tension specimen," *Trans. Am. Soc. Metals* **32**, 553–572 (1944).

144-3588. "Flow and fracture," *Metals Technol.* **11**, 32–39 (December 1944).

145-3597. Discussion of "Flow and fracture," *Metals Technol.*, supplement to Technical Publication No. 1782, 2 pp. (April 1945).

146-3599. Discussion of Boyd and Robinson, "The friction properties of various lubricants at high pressures," *Trans. ASME* **67**, 56 (1945).

147-3601. "The compression of twenty-one halogen compounds and eleven other simple substances to 100,000 kg/cm^2," *Proc. Am. Acad. Arts Sci.* **76**, 1–7 (1945).

148-3609. "The compression of sixty-one solid substances to 25,000 kg/cm^2, determined by a new rapid method," *Proc. Am. Acad. Arts Sci.* **76**, 9–24 (1945).

149-3626. "Polymorphic transitions and geological phenomena," *Am. J. Sci.* **243a** (Daly Volume), 90–97 (1945).

150-3635.	"Effects of high hydrostatic pressure on the plastic properties of metals," *Rev. Mod. Phys.* 17, 3–14 (1945).
151-3647.	"Recent work in the field of high pressures," *Rev. Mod. Phys.* 18, 1–93, 291 (1946).
152-3741.	"The tensile properties of several special steels and certain other materials under pressure," *J. Appl. Phys.* 17, 201–212 (1946).
153-3753.	"Studies of plastic flow of steel, especially in two-dimensional compression," *J. Appl. Phys.* 17, 225–243 (1946).
154-3772.	"The effect of hydrostatic pressure on plastic flow under shearing stress," *J. Appl. Phys.* 17, 692–698 (1946).
155-3779.	"On higher order transitions," *Phys. Rev.* 70, 425–428 (1946).
156-3784.	"An experimental contribution to the problem of diamond synthesis," *J. Chem. Phys.* 15, 92–98 (1947).
157-3791.	"The rheological properties of matter under high pressure," *J Colloid Sci.* 2, 7–16 (1947).
158-3802.	"The effect of hydrostatic pressure on the fracture of brittle substances," *J. Appl. Phys.* 18, 246–258 (1947).
159-3815.	"The effect of high mechanical stress on certain solid explosives," *J. Chem. Phys.* 15, 311–313 (1947).
160-3819.	"The compression of 39 substances to 100,000 kg/cm^2," *Proc. Am. Acad. Arts Sci.* 76, 55–70 (1948).
161-3835.	"Rough compressions of 177 substances to 40,000 kg/cm^2," *Proc. Am. Acad. Arts Sci.* 76, 71–87 (1948).
162-3852.	"Large plastic flow and the collapse of hollow cylinders," *J. Appl. Phys.* 19, 302–305 (1948).
163-3856.	"Fracture and hydrostatic pressure," in *Fracturing of metals* (Cleveland: American Society for Metals, 1948), pp. 246–261.
164-3873.	"General survey of certain results in the field of high pressure physics," in *Les Prix Nobel* (Stockholm: P. A. Norstedt, 1948), pp. 149–166. Reprinted in *J. Wash. Acad. Sci. 38* (May 15, 1948).
165-3891.	"The linear compression of various single crystals to 30,000 kg/cm^2," *Proc. Am. Acad. Arts Sci.* 76, 89–99 (1948).
166-3903.	"Viscosities to 30,000 kg/cm^2," *Proc. Am. Acad. Arts Sci.* 77, 117–128 (1949).

167-3915. "Further rough compressions to 40,000 kg/cm^2, especially certain liquids," *Proc. Am. Acad. Arts Sci.* 77, 129–146 (1949).

168-3933. "Linear compressions to 30,000 kg/cm^2, including relatively incompressible substances," *Proc. Am. Acad. Arts Sci.* 77, 189–234 (1949).

Volume VII

169-3980. "Effect of hydrostatic pressure on plasticity and strength," *Research (London)* 2, 550–555 (1949).

170-3987. "Volume changes in the plastic stages of simple compression," *J. Appl. Phys.* 20, 1241–1251 (1949).

171-3999. "Physics above 20,000 kg/cm^2" (Bakerian Lecture of the Royal Society), *Proc. Roy. Soc. (London)* [A] 203, 1–17 (1950).

172-4019. "The effect of pressure on the electrical resistance of certain semi-conductors," *Proc. Am. Acad. Arts Sci.* 79, 127–148 (1951).

173-4041. "The electric resistance to 30,000 kg/cm^2 of twenty nine metals and intermetallic compounds," *Proc. Am. Acad. Arts Sci.* 79, 149–179 (1951).

174-4073. "The effect of pressure on the melting of several methyl siloxanes," *J. Chem. Phys.* 19, 203–207 (1951).

175-4079. "Some implications for geophysics of high-pressure phenomena," *Bull. Geol. Soc. Am.* 62, 533–535 (1951).

176-4083. "Properties of materials under superindustrial stresses" (Charles M. Schwab Memorial Lecture, given at the General Meeting of the American Iron and Steel Institute, New York, 23–24 May 1951).

177-4105. "Some results in the field of high-pressure physics," *Endeavour* 10, 63–69 (1951).

178-4113. "The resistance of 72 elements, alloys and compounds to 100,000 kg/cm^2," *Proc. Am. Acad. Arts Sci.* 81, 167–251 (1952).

179-4198. "Acceptance of the Bingham Medal," *J. Colloid Sci.* 7, 202–203 (1952).

180-4201. "Further measurements of the effect of pressure on the electrical resistance of germanium," *Proc. Am. Acad. Arts Sci.* 82, 71–82 (1953).

181–4213. "Miscellaneous measurements of the effect of pressure on electrical resistance," *Proc. Am. Acad. Arts Sci.* **82**, 83–100 (1953).

182–4231. "The effect of pressure on several properties of the alloys of bismuth-tin and of bismuth-cadmium," *Proc. Am. Acad. Arts Sci.* **82**, 101–156 (1953).

183–4287. "High-pressure instrumentation," *Mech. Eng.*, February 1953, pp. 111–113.

184–4291. "Effects of very high pressure on glass" (with I. Šimon), *J. Appl. Phys.* **24**, 405–413 (1953).

185–4300. "The effect of pressure on the tensile properties of several metals and other materials," *J. Appl. Phys.* **24**, 560–570 (1953).

186–4312. "The use of electrical resistance in high pressure calibration," *Rev. Sci. Instr.* **24**, 400–401 (1953).

187–4314. "The effect of pressure on the bismuth-tin system," *Bull. Soc. Chim. Belges* **62**, 26–33 (1953).

188–4323. "Certain effects of pressure on seven rare earth metals," *Proc. Am. Acad. Arts Sci.* **83**, 3–21 (1954).

189–4343. "Effects of pressure on binary alloys, II. Thirteen alloy systems of low melting monotropic metals," *Proc. Am. Acad. Arts Sci.* **83**, 151–190 (1954).

190–4383. "Certain aspects of plastic flow under high stress," in *Studies in mathematics and mechanics presented to Richard von Mises* (New York: Academic Press, 1954), pp. 227–231.

191–4389. "Effects of pressure on binary alloys, III. Five alloys of thallium, including thallium-bismuth," *Proc. Am. Acad. Arts Sci.* **84**, 1–42 (1955).

192–4431. "Effects of pressure on binary alloys, IV. Six alloys of bismuth," *Proc. Am. Acad. Arts Sci.* **84**, 43–109 (1955).

193–4499. "Miscellaneous effects of pressure on miscellaneous substances," *Proc. Am. Acad. Arts Sci.* **84**, 112–129 (1955).

194–4519. "Synthetic diamonds," *Scientific American* **193**, 42–46 [1–11] (November 1955). Reprinted with permission. Copyright © 1955 by Scientific American, Inc. All rights reserved.

195–4531. "High pressure polymorphism of iron," *J. Appl. Phys.* **27**, 659 (1956).

196-4533. "Effects of pressure on binary alloys, V. Fifteen alloys of metals of moderately high melting point," *Proc. Am. Acad. Arts Sci.* 84, 131–177 (1957).

197-4581. "Effects of pressure on binary alloys, VI. Systems for the most part of dilute alloys of high melting metals," *Proc. Am. Acad. Arts Sci. 84*, 179–216 (1957).

198-4620. "Compression and the α–β phase transition of plutonium," *J. Appl. Phys. 30*, 214–217 (1959).

199-4625. "General outlook on the field of high-pressure research," in *Solids under pressure*, ed. W. Paul and D. M. Warschauer (New York: McGraw-Hill, 1963), pp. 1–13.

4639. Index of Substances

4677. Index of Apparatus

Physical Papers Omitted

1. "The electrostatic field surrounding two special columnar elements," *Proc. Am. Acad. Arts Sci.* **41**, 617–626 (1906).
2. "On a certain development in Bessel's functions," *Phil. Mag.*, December 1908, pp. 947–948.
3. "Verhalten des Wassers als Flüssigkeit und in fünf festen Formen unter Druck," *Z. Anorg. Chem.* **77**, 377–455 (1912).
4. "A complete collection of thermodynamic formulas," *Phys. Rev.* **3**, 273–281 (1914).
5. "Tolman's principle of similitude," *Phys. Rev.* **8**, 423–431 (1916).
6. "A critical thermodynamic discussion of the Volta, thermo-electric and thermionic effects," *Phys. Rev.* **14**, 306–347 (1919).
7. "The electrical resistance of metals," *Phys. Rev.* **17**, 161–194 (1921).
8. "The discontinuity of resistance preceding supraconductivity," *J. Wash. Acad. Sci.* **11**, 455–459 (1921).
9. "The electron theory of metals in the light of new experimental data," *Phys. Rev.* **19**, 114–134 (1922).
10. "A suggestion as to the approximate character of the principle of relativity," *Science* **59**, 16–17 (1924).
11. "Rapport sur les phénomènes de conductibilité dans les métaux et leur explanation théorique," in *Conductibilité électrique des métaux et problèmes connexes*, the report of the Fourth Solvay Congress, Brussels, 24–29 April 1924 (Paris: Gauthier-Villars, 1927), pp. 67–114.
12. "The connections between the four transverse galvanomagnetic and thermomagnetic phenomena," *Phys. Rev.* **24**, 644–651 (1924).
13. "The universal constant of thermionic emission," *Phys. Rev.* **27**, 173–180 (1926).
14. "General considerations on the photo-electric effect," *Phys. Rev.* **31**, 90–100 (1928).
15. "Note on the principle of detailed balancing," *Phys. Rev.* **31**, 101–102 (1928).
16. "Thermoelectric phenomena in crystals and general electrical concepts," *Phys. Rev.* **31**, 221–235 (1928).

17. "The photo-electric effect and thermionic emission: a correction and an extension," *Phys. Rev. 31*, 862–866 (1928).
18. "On the application of thermodynamics to the thermo-electric circuit," *Proc. Nat. Acad. Sci. U.S. 15*, 765–768 (1929).
19. "Thermische Zustandsgrössen bei hohen Drucken und Absorption von Gasen durch Flüssigkeiten unter Druck," *Handbuch der Experimentalphysik*, VIII (Leipzig: Akademische Verlagsgesellschaft 1929), 245–400.
20. "General considerations on the emission of electrons from conductors under intense fields," *Phys. Rev. 34*, 1411–1417 (1929).
21. "Permanent elements in the flux of present-day physics," *Science 71*, 19–23 (1930)
22. "Comments on a note by E. H. Kennard on 'Entropy, reversible processes and thermo-couples,'" *Proc. Nat. Acad. Sci. U.S. 18*, 242–245 (1932).
23. "Statistical mechanics and the second law of thermodynamics," *Bull. Am. Math. Soc.*, April 1932, pp. 225–245.
24. "On the nature and the limitations of cosmical inquiries," *Sci. Monthly 37*, 385–397 (1933).
25. "Energy," *Gamma Alpha Record 24*, 1–6 (1934).
26. "A physicist's second reaction to Mengenlehre," *Scripta Mathematica 2*, 3–29 (1934).
27. "The second law of thermodynamics and irreversible processes," *Phys. Rev. 58*, 845 (1940).
28. "Dimensional analysis," in *Encyclopaedia Britannica*.
29. "The principles of thermodynamics," in *Thermodynamics in physical metallurgy* (Cleveland: American Society for Metals, 1950), pp. 1–15.
30. "The thermodynamics of plastic deformation and generalized entropy," *Rev. Mod. Phys. 22*, 56–63 (1950).
31. "Reflections on thermodynamics," *Proc. Am. Acad. Arts Sci. 82*, 301–309 (1953); reprinted in *Am. Scientist 41*, 549–555 (1953).

PREFACE BY THE EDITORIAL COMMITTEE

The idea of reprinting the collected experimental papers of Professor P. W. Bridgman originated in the spring of 1959 with a proposal to the Harvard University Press from a group of scientists at the National Bureau of Standards led by Dr. C. E. Weir. Professor Bridgman was favorable to the idea and undertook to prepare a brief commentary on some of the papers to bring them up to date in the light of recent advances, and also to prepare an index of substances and an index of apparatus. The proposal eventuated in a grant from the National Science Foundation to the Harvard University Press to defray part of the cost of publication and of the preparation of the indexes. These grants were made available by July of 1961.

Progress on the project was delayed by Professor Bridgman's untimely death in August 1961. Nevertheless, by that time he had prepared an introduction and commentaries on many of the papers, and had nearly completed the index of substances. In fact, he had apparently worked on a draft of the latter the day before his death, and it was received in the mail shortly afterward.

The substance index was completed during the following year by Miss Phyllis Adair and Mr. Steven Groves. The editorial committee appointed to carry on the project was fortunate in securing the services of Dr. Alexander Zeitlin of Barogenics, Inc., to carry forward the difficult task of preparing the apparatus index.

The selection and numbering of the papers was done by Professor Bridgman personally, drawing on his own collection of reprints. In most cases the original reprints have been reproduced photographically, although in a few instances it proved necessary to reset a paper in type. Professor Bridgman's commentaries were set in type, enclosed in square brackets, and incorporated as footnotes or at the ends of the papers to which they pertain. His introduction is included, based on a rough draft he prepared early in 1961. The committee has departed from Professor Bridgman's selection of papers in only one respect. Paper 199, the final contribution, was added. It was originally prepared by him as the first paper for the volume entitled *Solids under Pressure*, edited by William Paul and Douglas Warschauer and published by the McGraw-Hill Book Company in 1962. The committee is indebted to the publisher for permission to reprint this paper—Professor Bridgman's last technical paper—which gives his final perspective on his own work and on the future of high-pressure physics.

In taking over its task after Professor Bridgman's death, the editorial committee wrote to a number of scientists active in various aspects of high-pressure research and requested suggestions as to what should be done to carry out Bridgman's intention of bringing his papers up to date. After reviewing the replies the committee concluded that any extensive attempt at updating would be presumptuous, and decided to confine its editorial task to the mention of one or two major points in this introduction.

The principal point that requires mention has to do with the two pressure scales used by Bridgman in his investigations. The first was appropriate for apparatus in which the pressure was truly hydrostatic because it was transmitted by a liquid or a gas. The pressure was measured usually with a manganin-wire gauge, calibrated by recording its resistance at the freezing pressure of mercury and at the lowest solid-solid transition of bismuth. The former was taken to be 7,640 kg/cm^2 at 0°C and the latter, 25,420 kg/cm^2 at 30°C. Both pressures were referred ultimately to measurements with a free-piston gauge. Their accuracy is probably of the order of 1 percent. Recently determinations of somewhat higher accuracy have been reported by Newhall, Abbot, and Dunn in paper no. 62-WA-283, presented at a meeting of the A.S.M.E. on November 19, 1962, and published in *High Pressure Measurement*, edited by A. A. Giardini and E. C. Lloyd (Butterworth, Washington, 1963). Bridgman's measurements were extended into the pressure region above 30 kb, where strictly hydrostatic conditions are not attainable, by using a piston-and-cylinder method and locating phase changes with the aid of volume discontinuities indicated by piston displacements (papers 139, 147, and 160).

For quasihydrostatic experiments with his opposed-anvil apparatus, Bridgman computed the pressure from the total load and the area of the face of the anvil. This method was used in computing the pressures of the sharp electrical-resistance jumps observed in bismuth, thallium, cesium, and barium (paper 178). As a check on this method he used a comparison of results obtained in the region of pressure overlap between his hydrostatic and quasihydrostatic apparatus. It is regrettable that this second scale was accorded, up to 1960, the same acceptance as the first, even though Bridgman himself made it clear that the quasihydrostatic scale measured the relative forces in his own particular design of apparatus rather than absolute pressures. The discrepancy between the first and second scales was first definitively brought to light by measurements of Kennedy and La Mori, using a piston-and-cylinder method modified to minimize uncertainties due to friction. This work, which was first reported at a conference held at Lake George in 1960 and published in *Progress in Very High Pressure Research* (Wiley, New York, 1961), edited by F. Bundy,

W. R. Hibbard, and H. M. Strong, reestablished a pressure scale above 30 kb, which is in much better agreement with Bridgman's older piston-and-cylinder results than with his more recent opposed-anvil results, which give pressures nearly 30 percent too high for the important transitions.

In summary, the redetermination of the mercury transition pressure at 0°C by Newhall, Abbot, and Dunn, which gave 7,715.20 kg/cm^2, should be regarded as the best present low-pressure calibration. There has been no redetermination of the bismuth calibration point near 25,000 kg/cm^2. For a review of the current situation at higher pressures, which is harder to describe in terms of a few established transitions, the reader is referred to Kennedy and La Mori, *Journal of Geophysical Research 67*, 851–856 (1962), and to Bundy and Strong, "High Temperatures and Pressures," in *Solid State Physics*, vol. 13 (Academic Press, New York, 1962), edited by F. Seitz and D. Turnbull, especially pages 91–94.

There are many cases in which Bridgman's data on the compression of various substances have been superseded by later work, either because Bridgman had available only powders instead of single crystals (for example, paper 98) or because of the use of amorphous materials in place of crystals, as for selenium (paper 134), or because of lack of full knowledge of crystal modifications, as for black phosphorus (paper 16). The committee has made no effort to bring this work up to date, since to do so consistently and systematically would have involved unacceptable delays in publication.

January 31, 1964   *Editorial Committee*
HARVEY BROOKS, *Chairman*
FRANCIS BIRCH
GERALD HOLTON
WILLIAM PAUL

INTRODUCTION

In the following volumes are collected all my experimental papers, with the exception of abstracts of papers read at various times before meetings of scientific societies, chiefly the American Physical Society, and two long papers in German in which no results were given not appearing elsewhere in English. Nearly all these papers deal with some aspect of high-pressure phenomena, but there are a few others. The decision was not always easy as to whether a paper should be included in this collection or not. The decision not to include was easy for a number of papers which would be described as relating to "philosophy of science," but there were a number of others in which the contact with experiment is much closer. With two exceptions, the criterion for inclusion was finally taken to be whether the paper involved any immediate experimental work on my part. In accordance with this criterion a paper dealing with departures from Ohm's law at high current densities and another dealing with new thermoelectric effects in single crystals are included, whereas several dealing with such topics as photoelectric emission and the thermodynamics of irreversible processes are omitted. The two exceptions noted above are a paper of 1926 (No. 61) dealing with Dimensional Analysis, which is included because of the important applications of Dimensional Analysis in experimenting with models, and a review paper of 1946 (No. 151), included because of the survey it gives of the whole high-pressure field. For the sake of completeness, the titles of the omitted papers, excluding the "philosophical" ones, are collected at the end of the bibliography of the included papers.

The arrangement is chronological. There is inevitably some duplication, especially a number of short notes in the *Proceedings of the National Academy of Sciences* which were later published in full detail in the *Proceedings of the American Academy of Arts and Sciences*. I would like to express here my great appreciation of the liberality of the officers of the American Academy in consenting to publish through so many years full experimental details which would have been acceptable in no other place. What duplication there is I do not believe will be found particularly harmful. To completely avoid all duplication would have involved completely rewriting all the articles in the form of a connected series of volumes—an impossible task for me.

Before dealing with the high-pressure papers, comment may be made on the couple of papers dealing with deviations from Ohm's law at high current densities. It was recognized early in the theory of electronic conduction in metals that deviations were to be expected at high current densities, but just what the critical densities were did not at first appear. In these papers a method was developed for measuring the resistance at current densities up to 10^7 amp/cm^2. At these densities deviations were found of a few percent in the expected direction. However, later developments of the theory, particularly a paper by Guth and Mayerhöfer, *Physical Review 57*, 908–915 (1940), indicated pretty definitely that deviations of this magnitude were not to be expected until the current densities were perhaps 10,000 greater. The question then becomes to understand the positive effects found in my papers, which I believe to have been real. It appears that the answer may be found in a hitherto unrecognized and up to the present not experimentally established kind of e.m.f. It was shown in my book (republished by Dover as a paperback) on the *Thermodynamics of Electrical Phenomena in Metals* that there is a new kind of e.m.f. in a metal in which there are simultaneously a current and a variation of temperature with time. These conditions were present to a high degree in my experimental set-up above, and it may well be that this is the explanation of the observed effect. Further experiments in this general field appear desirable; in the above-mentioned book will be found suggestions as to other obscure and small effects not yet established experimentally.

The papers are numbered consecutively through all the volumes. The complete list of papers in numbered order, with full titles and journal references, is given at the beginning of Volume I. In the papers as reproduced photographically the original pagination of each article is kept; at the foot of each page there is added the paper number for reference and the serial number of the page.

The index of substances has demanded much labor and the endeavor has been to make it complete. During all the years of my experimental work I have kept a running card index, by substances, to the original experimental notebooks, to the notebooks of numerical computation, to the graphs, of which there were many and on a large scale, and to the final publication. It is the references on these cards to the final publication which are here assembled in the index under the names of the different substances. The index gives, under the various substances, the reference in these volumes, by number of paper and page of original publication, and also the nature of the physical parameter (as, for example, compressibility or shearing strength) there tabulated.

Several matters require detailed comment. During the course of the experiments several improvements and revisions were made in some of the numerical values, particularly in compressibilities. In general, most of my methods for measuring compressibilities have been differential, determining the difference between the compressibility of the substance in question and iron, taken as fundamental. The absolute compressibility of iron has to be determined by independent methods. During the course of the work the absolute measurements of the compressibility of iron were extended in range and in absolute accuracy. Each revision of the compressibility of iron demands a revision in the other dependent compressibilities. Because of the smallness of the effect, the deviations of the compressibility of iron from linearity in pressure are particularly difficult to obtain accurately. My last measurements of the absolute compression of iron cover the range to 30,000 kg/cm², whereas the earliest were to only 6,000. The accuracy of the second-degree term in pressure increases, other things being equal, as the square of the pressure range. The second-degree term obtained in my last measurements in 1940 was only one-third the value used in most of my earlier work, based on measurements to 12,000. The 1940 (and best) value for the linear compression of iron is:

$$-\Delta l/l_0 = 1.942 \times 10^{-7}p - 0.23 \times 10^{-12}p^2,$$

compared with the value used in practically all the calculations made before 1940:

$$-\Delta l/l_0 = 1.953 \times 10^{-7}p - 0.75 \times 10^{-12}p^2.$$

Before 1940, the compressions of solids were uniformly expressed in the form:

$$(\Delta V/V_0)_{\text{former}} = -ap + bp^2.$$

To correct these values to the 1940 compression of iron, the formula

$$(\Delta V/V_0)_{1940} = (-a + 0.033 \times 10^{-7})p$$
$$+ (b - 1.56 \times 10^{-12} - a + 0.022 \times 10^{-7})p^2$$

is to be used.

In addition to this major correction on compressions (the resultant corrections for compressibilities of liquids are in general below experimental error), other corrections arose during the course of the work which should be summarized here. Paper No. 114 (1935) points out errors, just discovered, in the calculation of the second-degree term of compressions determined before 1925. This applies to 16 substances and is not serious. The corrected values for these substances are given in the 1935 paper. Unfortunately, four of these "corrected" values contained an error in the sign of the second-degree term b, which should be positive throughout. This error in the second-degree term becomes more serious the larger the absolute compressibility, and for the three alkali metals, lithium, sodium, and potassium, is so large as to essentially modify the general picture, particularly for potassium. The error here was sufficient to demand the complete withdrawal in the second (1949) edition of *The Physics of High Pressure* of material for these three substances in the original (1931) edition. In these collected papers the corrections for the three alkali metals will be found in paper No. 110. A warning note is here added to the original paper, calling attention to the error and its place of correction.

Professor Francis Birch has corrected the compressibilities of two minerals in paper No. 74 (1928). The errors are both due to misidentification, probably in the labels originally supplied by the Mineralogical Museum. The mineral described in the original paper as andradite was actually grossularite, and the mineral described as garnet (pyrope) was actually almandite. Fortunately, the original specimens had been preserved, so that Professor Birch could verify the corrections, which he originally made on the basis of unexpected values for the densities. These corrections are noted in the text of paper No. 74.

Cautionary comment is called for with regard to one aspect of the shearing measurements. In my very earliest work in which shearing by rotation was combined with average hydrostatic pressures up to 50,000 kg/cm^2 a number of examples were found in which application of shear was followed by a detonation. At that time it was my opinion that this detonation was evidence of a true chemical instability under the combination of shear and pressure. Later I found, however, that a large number of these apparent chemical detonations were actually only cases of mechanical instability, fundamentally due to the fact that static friction is greater than moving friction. The proof that this was the explanation was given by the fact that the effect entirely disappeared if the thickness of the specimen was made less than some critical value. This aspect of the subject was never satisfactorily cleaned up and in my mind it is uncertain whether there are any cases of genuine chemical instability under high shearing stress combined with high pressure. It is my personal opinion that there are genuine chemical effects (for example, certain changes of chemical valence under shear seem pretty certain), but I did not clean up the subject myself, and I think further work should be done here. These remarks do not apply at all to most of the later work in which numerical values of shearing strength were found for many substances; in all this later work care was taken to work well below the critical thickness for mechanical instability.

November 10, 1960 P. W. BRIDGMAN

Collected Experimental Papers
of P. W. Bridgman

CONTRIBUTIONS FROM THE JEFFERSON PHYSICAL
LABORATORY, HARVARD UNIVERSITY.

THE MEASUREMENT OF HIGH HYDROSTATIC PRESSURE.

I. A SIMPLE PRIMARY GAUGE.

By P. W. BRIDGMAN.

Presented by W. C. Sabine, December 9, 1908. Received December 16, 1908.

INTRODUCTION.

THE classical work of Amagat on various physical effects of high hydrostatic pressure is practically the only work we have in which the pressure has been accurately measured with a direct reading gauge over any considerable pressure range. Amagat's pressure measurements were made with his *manometre à pistons libres*, which is too well known to need description here. The gauge gives consistent results, and throughout the pressure range the indications are proportional to the pressure. In fact, the accuracy of the pressure measurements is limited only by the accuracy with which the dimensions of the pistons can be measured. With this primary gauge Amagat measured a number of secondary pressure effects, principally the compressibilities of various liquids and gases over a pressure range of about 3000 kgm. per sq. cm. The value of the compressibility found in this way has in turn been used by other experimenters as a means of calibrating whatever secondary gauge they found it convenient to use. It is thus possible to avoid the direct measurement of pressure with a *manometre à pistons libres*, which is in most cases inconvenient, because of the unavoidable leak and the time necessary to make the readings. The care with which the ground surfaces of piston and cylinder must be kept free from grit, and the expense of the instrument, are other objections to its common use.

With increasing experience in methods of reaching high pressures, and increasing excellence of commercial steels, it has been found possible, however, to exceed the pressure limit set by Amagat. Thus

Tammann[1] on one occasion reached 5000 kgm. per sq. cm., and Carnazzi,[2] working with Lussana's[3] apparatus, has also attained 5000 kgm. Both of these observers measured the pressure with a secondary gauge involving directly the compressibility of water as found by Amagat. But because Amagat's data run to only 3000 kgm., the pressure measurements of both Tammann and Carnazzi must be uncertain at these higher pressures. Tammann had to content himself with an extrapolation beyond 3000, and Carnazzi does not give any results beyond 3000.

The purpose of this paper is to provide data which shall enable others to work, if they desire, beyond Amagat's pressure range with a reasonable degree of confidence in the accuracy of the pressure measurements. It seems that the most feasible way of doing this is to determine, under conditions that may be reproduced with accuracy, the variation with pressure of some easily measurable physical property. Compressibility does not seem to be the best secondary property for this purpose, for it cannot be measured with much accuracy conveniently. In this paper, advantage has been taken of a suggestion of de Forest Palmer's,[4] and the variation of the electrical resistance of pure mercury under pressure has been determined. The secondary gauge, involving the variation of the resistance of mercury, has proved itself trustworthy and accurate.

This matter of a secondary standard is discussed in the second part of this paper. The first part is occupied with a discussion of the slightly novel form of gauge with which the fundamental direct measurements of pressure were made. Amagat's *manometre à pistons libres* is not well adapted for high pressures. Amagat himself was accustomed to use it to only 3000 kgm. per sq. cm., and others following him have not been so high; thus Barus found that above 2000 kgm. the leak became troublesome. In this paper a gauge is described with which, by modifying the design and decreasing the dimensions, it has been found possible to reach higher pressures than Amagat, frequently without perceptible leak. In fact the primary gauge proved itself so manageable, and is so simple to construct, that if it were not for the greater convenience of the secondary gauge, the primary gauge could be used directly in any high pressure investigation. This paper gives results that have been obtained with this gauge up to 6800 kgm. per

---

[1] Tammann, Kristallisieren und Schmelzen, p. 201 (Leipzig, Barth, 1903).
[2] Carnazzi, Nuov. Cim., (5), **5**, 180–189 (1903).
[3] Lussana, Nuov. Cim., (5), **4**, 371–389 (1902).
[4] de Forest Palmer, Amer. Jour. Sci., **6**, 451–454 (1898).

sq. cm. The first part is occupied with a description of the gauge, calculation of the corrections to be applied, and a comparison of two gauges to determine the accuracy and sensitiveness.

DESCRIPTION OF THE GAUGE.

Besides Amagat's [5] manometer, other forms of direct pressure gauge have been used, examples of which are the pressure balance at Stuckrath,[6] and the differential manometer at the National Physical Laboratory at London.[7] Lisell,[8] in his measurement of the pressure coefficient of resistance of wires, used a gauge much like that at Stuckrath. These gauges differ in the manner in which the pressure exerted on the piston is measured. Amagat measures it by measuring with a mercury column the hydrostatic pressure acting on a larger piston which balances the total thrust exerted by the high unknown pressure on a much smaller piston. The thrust is measured at Stuckrath or by Lisell by hanging weights on the piston either directly or with the aid of a lever. At London the action of weights is used to equilibrate the differential effect of the pressure on two pistons of nearly the size. A common feature of all these gauges is the piston fitting accurately in the cylinder, which is subjected to pressure on the inside. The distortion produced by the pressure is, therefore, a compression of the piston, accompanied by a stretching of the cylinder, the resultant effect being to increase the breadth of the crack between piston and cylinder. The leak, therefore, at higher pressures increases because of the increased pressure expelling the liquid and the increased breadth of the crack.

This effect is avoided in the gauge used in this work by subjecting the cylinder in which the piston plays to pressure on the outside as well as on the inside. It is well known that a cylinder subjected to the same pressure externally and internally shrinks to the same extent as a solid cylinder subjected to the same external pressure. By properly decreasing the external pressure on the hollow cylinder, the shrinkage at the inner surface may be made as small as we please, or may be made an expansion. Practically the same result may be obtained by subjecting only a portion of the external surface of the

---

[5] Amagat, Ann. de Chim. et Phys., (6), **29**, 544 (1893).
[6] Zeit. f. Instrk., **14**, 307 (1894). Manometer für hohe Druck.
[7] Engineering, **75**, 31 (1903). The Estimation of High Pressures.
[8] Lisell. Om Tryckets Imflytande på det Elektriska Ledningmotståndet hos Metaller, samt en ny Metod att Mäta Höga Tryck. Upsala, 1903 (C. J. Lundström).

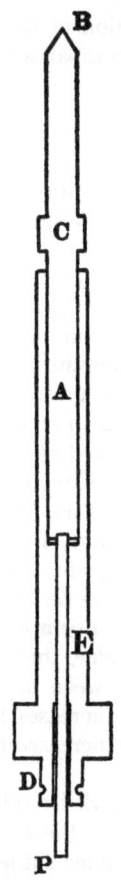

FIGURE 1. The direct reading gauge. P, piston; E, cylinder; A, larger steel rod through which the pressure of the equilibrating weights is transmitted to the piston; B, hardened steel point on which the stirrup carrying the weight pan is hung; C, stop (see Figure 2); D, groove by which the rubber tube containing the viscous mixture of molasses and glycerine is attached.

cylinder to pressure. By suitably changing the area subjected to pressure, the shrinkage of the interior may be controlled. This is the method adopted with the present gauge.

The leak may be further decreased by decreasing as far as possible the dimensions of the piston and cylinder, thus decreasing the circumference of the crack through which leak occurs. Decreasing the size has the additional advantage of making the whole gauge more compact and manageable. In particular, the total thrust becomes small enough to be balanced directly by hanging weights on the free end of the piston. Where the magnitude of the weights is not so great as to make this infeasible, the direct application of weights seems preferable to the usual indirect methods of measuring the thrust. In the gauge adopted in this work, the piston is only $\frac{1}{16}$ in. (0.159 cm.) in diameter, requiring at the maximum pressure of 6800 kgm. an equilibrating weight of about 130 kgm.

The cylinder and piston are shown in Figure 1. In Figure 2 they are shown in place in a large steel block which serves as a reservoir between the gauge and the pressure pump. The dimensions of the important parts are indicated in Figure 3. The thrust on the piston P (Figure 1) is taken by the large cylindrical rod A joined to the piston by a forced fit. A terminates in a hardened point B, on which the weights are hung by a stirrup supporting the scale pan underneath the large steel block. The upper end of the cylinder acts as a guide for the rod A, as does also the attachment screwed onto the top of the cylinder shown in Figure 2. It is essential that fitting here should be accurate, so that the small piston may move freely in a vertical line without danger of any bending of the top end when projecting some distance from the cylinder.

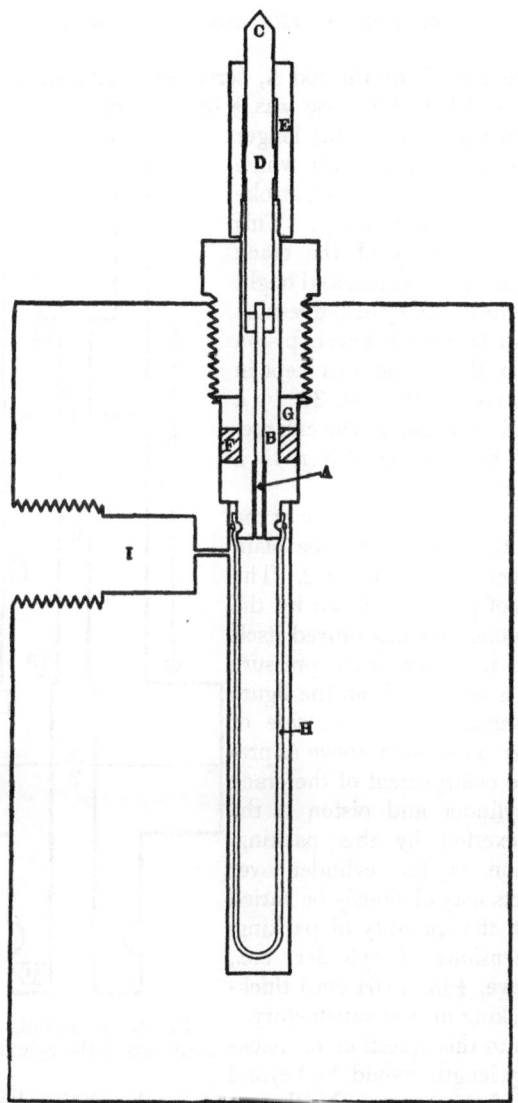

FIGURE 2. A, piston; B, cylinder; C, hardened steel point, on which the equilibrating weights are hung; D, stop, preventing too long a stroke of the piston either up or down. In this stop is placed the rod by which the rotary motion is imparted to the piston to increase sensitiveness. E, guide to insure the upper part of the piston moving rigidly in a straight line. F, rubber packing. G, steel washer, retaining the rubber packing. H, easily collapsible rubber tube, containing the viscous mixture of molasses and glycerine. I, connection to the high pressure pump. The thin mixture of water and glycerine transmitting the pressure is injected through this hole, acts on the outside of the rubber tube H, and so transmits the pressure to the piston A.

The enlargement C, on the rod A, serves as a stop at either end of the stroke, which in this case was ½ in. (1.3 cm.). The piston was made at least ½ in. (1.3 cm.) longer than the hole in the cylinder in which it fits, so that at no part of the stroke is any part of the hole empty. This insures the constancy of the crack through which leak occurs, and ought to increase the accuracy of the results. To diminish friction between piston and cylinder the piston was kept in slow rotary motion through 30° by a rod inserted in a hole in the enlargement C. The rod was driven by a small motor.

The purpose of the shoulder at the bottom of the cylinder will be plain on an inspection of Figure 2. The disposition of packing, shown by the shading, is one that has proved itself serviceable in other high pressure work. It is obvious from the figure that the pressure on the outside of the cylinder mentioned above as preventing the enlargement of the crack between cylinder and piston is the pressure exerted by this packing. The portion of the cylinder over which it acts may obviously be varied by varying the quantity of packing. With dimensions of cylinder, etc., shown above, ¼ in. (0.64 cm.) thickness of packing proved satisfactory.

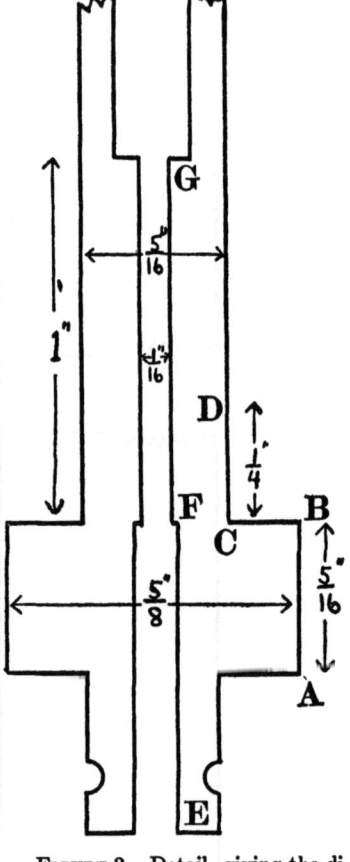

FIGURE 3. Detail, giving the dimensions of the cylinder.

To go into this question of packing at any length would be beyond the scope of this paper. Neither can any description be given here of the apparatus with which the pressure was produced. Briefly, pressure was produced by a small piston pushed by hydrostatic pressure on a larger piston. Pressure was transmitted to various parts of the apparatus by heavy steel tubing. It is hoped that methods of producing high pressures may be made the subject of another paper.

The cylinder (E, Figure 1) was turned in a lathe from a rod of about 1.25 per cent carbon tool steel. The drilling of the hole in which the piston moves demanded care. This was drilled first with a drill about 0.002 in. (0.05 mm.) under $\frac{1}{16}$ in. (1.59 mm.), and then enlarged to full size with a two-lipped $\frac{1}{16}$ in. twist drill. The hole made in this manner proved round, uniform, and satisfactory in every particular. It is a matter of common experience that a two-lipped twist drill hugs the hole very tightly when used as a following drill. For this reason, care is necessary not to push the drill too hard, as otherwise the sharp cutting corners are quickly blunted. After turning and drilling, the cylinder was hardened in water, and the temper drawn below a blue. Drawing the temper is a necessary precaution in the interests of safety, as glass hard steel proved itself very treacherous. The cylinder is so small that with the exercise of a little care in heating and quenching it is not distorted appreciably by the hardening.

The piston was a piece of $\frac{1}{16}$ in. (1.59 mm.) " Crescent " drill rod, hardened in oil, the temper not being drawn further. This drill rod was found to be remarkably round and uniform in diameter, variations of so much as 0.0001 in. (0.0025 mm.) being rare from end to end of the same length. Different pieces, however, of nominally the same size rod may differ by 0.0005 in. (0.0125 mm.) in diameter. It was merely necessary, then, to select from several lengths of drill rod a piece fitting the hole in the cylinder. No grinding whatever was necessary, either on the cylinder or the piston, except rubbing with the finest emery paper to remove the film of oxide after hardening. In fact, it is the salvation of this device that no grinding is necessary, accurate grinding of a piston so small as $\frac{1}{16}$ in. being out of the question, to say nothing of the $\frac{1}{16}$ in. hole in the cylinder. Because of its slenderness, considerable care is necessary in hardening the piston without warping. Several attempts were usually necessary before a perfectly straight piston was obtained. This, however, is a matter of no consequence, because a piston can be made in a few minutes. The writer has himself made two cylinders and pistons complete in one day.

Leak was reduced to a very low value by using a liquid of great viscosity to transmit pressure to the piston. A mixture of molasses and glycerine proved suitable. The viscosity can be given any desired value by boiling away enough water from the molasses before adding the glycerine. Besides increasing the viscosity, the glycerine serves the useful purpose of preventing the molasses from drying where it leaks out between piston and cylinder. The liquid used to transmit pressure from the high pressure pump to the gauge was a mixture of two parts glycerine to one part water. This was prevented from coming into contact with the molasses and glycerine by enclosing the latter in an easily collapsible rubber tube, closed at the lower end, and at the upper end tied over the mouth of the cylinder, as shown in Figure 2.

Molasses was the liquid used by Amagat in his manometer. A heavy mineral oil, such as Barus used in a gauge of Amagat's type, was found to be unsuitable for high pressure work, because it freezes at room temperature under pressure. One grade of heavy oil tried in this experiment froze at 20° under a pressure of 4500 kgm. Presumably vaseline and such soft solids become unsuitable for the same reason, although this point was not tested. For the same reason the glycerine transmitting pressure from the pump had to be diluted with water. The ease with which glycerine subcools, and the difficulty of getting it pure, made any exact determinations impossible; but it was found that commercially pure glycerine was very apt to solidify at 6000 kgm. and 20°.

Corrections to be Applied to the Absolute Gauge

In spite of the simplicity of this gauge, and the directness with which it carries the measurement of pressure back to the fundamental definition, there are two corrections which must be applied in practical use. These corrections are both so small, however, that neither need be determined with much accuracy.

The first correction is introduced by the slow leak, and is in amount equal to the frictional force of the escaping liquid on the piston. The equilibrating force must balance both the hydrostatic pressure on the end of the piston and this frictional force. The effect of the correction, therefore, is to increase slightly the effective area of the piston. If we assume that both cylinder and piston are perfectly cylindrical, and that the crack between them is so narrow that the friction exerted by the escaping liquid is equally divided between cylinder and piston, then we easily see by writing down the equations of steady motion of the escaping liquid that the friction increases the effective diameter of the piston to the mean of the diameter of the piston and cylinder. It appears from the equations that this correction is independent of both the rapidity of leak and pressure. This is usually determined by measuring the diameter of the piston directly, and the diameter of the hole in the cylinder by some such indirect method as weighing the quantity of mercury required to fill it. The dimensions of the gauge used here were so small, however, that direct measurement of even the piston could not be made with the desired percentage accuracy, and accordingly the effective diameter was determined in another way, to be described later.

The second correction is a correction for the distortion of the gauge under pressure, and increases in percentage value directly with the pressure. This correction, of course, varies with the type of gauge, but in the types of gauge described above, and the pressure gauge employed, the correction is practically negligible. A rough calculation showed that at 3000 kgm. the correction in Amagat's manometer is about $\frac{1}{10}$ per cent. Since, however, it was desired in this work to reach an accuracy of $\frac{1}{10}$ per cent, and since the pressure range is 6800 kgm., some approximate evaluation seemed desirable.

No easy experimental method of determining this correction suggested itself, so recourse was had to a calculation, using the theory of elasticity. This was done only as a last resort, because of the doubtful accuracy of the mathematical theory at these pressures, and of the fact that the solution obtained is only an approximation, instead of a rigorous mathematical solution. In fact, the general problem involved has not been solved mathematically, and even if it could be, its application here would be doubtful, because slight irregularities in either cylinder or piston would destroy the ideal boundary conditions of the mathematical problem. In spite of all these objections, however, the magnitude of the approximate correction turned out to be so slight, $\frac{1}{10}$ per cent, that the calculated value can probably be applied with a fair degree of confidence.

The facts used in the following calculation are taken from the most elementary parts of the theory of elasticity, and may be found stated in any book under the calculation of the strains produced in a cylinder by external or internal hydrostatic pressure. It will be noticed that the correction for distortion found below includes the effect of the friction of the escaping liquid.

The strain in the piston can be broken up into two components. The first is that due to the longitudinal compression of the piston by the hydrostatic pressure at one end and the equilibrating weights at the other, and is uniform throughout the piston. The radius increases from this effect by the amount

$$\Delta r = \frac{3\kappa - 2\mu}{18\,\mu\kappa} \times r \times P,$$

where P is pressure in kgm. per sq. cm., κ the compressibility modulus, and μ the shear modulus. These elastic constants vary only slightly in different grades of steel. If we assume as average values that

we find that
$$\mu = 7.8 \times 10^5 \text{ kgm./cm.}^2,$$
$$\kappa = 15 \times 10^5 \text{ kgm./cm.}^2,$$
$$\Delta r = 1.4 \times 10^{-7} \times r \times P.$$

The second component part of the strain is that due to the pressure of the escaping liquid over the curved surface of the piston. Here an approximation must be introduced, for the determination of the strain in a cylinder under a given system of normal stresses on the curved surface seems to be a mathematical problem not yet solved, while in this case the problem is additionally complicated because the stress system is not given but depends in turn on the strain. The approximation made is the assumption that the radial displacement at any point is proportional to the normal pressure at that point, and is the same as that in an infinite cylinder subjected to the same pressure over its entire length. This assumption is probably fairly close to the truth where the extent of the cylinder exposed to the pressure is long compared with the radius, and the pressure varies gradually from point to point, as is the case here.

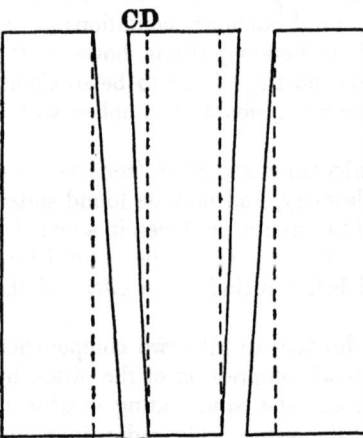

FIGURE 4. Exaggerated effect of the pressure in distorting the cylinder and piston.

The piston then assumes under the external pressure the form of a frustum of cone, as is shown in Figure 4. It will appear in the following that it is absolutely immaterial whether the generating lines of the frustrum of the cone into which the piston has been deformed are straight, as drawn in Figure 4, or not. The displacement at any point due to the external pressure is, on the above assumptions,

$$\Delta r = -\frac{4\mu + 3\kappa}{18\,\mu\kappa} \times r \times p$$
$$= -3.5 \times 10^{-7} \times r \times p.$$

p increases along the piston from its full value, P, at the inner end, E, to zero at the outer end, C. Now by adding these two components of strain, we find that the total radial displacement of the piston consists of a shrinkage of $2.1 \times 10^{-7} \times r \times P$ at the inner end, and a swelling of $1.4 \times 10^{-7} \times r \times P$ at the outer end.

The strain in the cylinder is more difficult to compute because of the uncertainty in the external boundary conditions introduced by the packing. Upon the portion of surface DCB (Figure 3) there is a normal pressure exerted by the packing, equal to 1.32 of the internal hydrostatic pressure. On BAEF there is the normal hydrostatic pressure, and from F to G the same distribution of pressure as on the piston, decreasing from the full value at F to zero at G. The maximum radial displacement due to external pressure may be taken as somewhat less than that from a pressure equal to 1.32 P over the entire external surface, because of the supporting action of the part AB, which is subjected to P only, and of the part beyond D, on which there is no pressure. An upper limit to the distortion is probably set by the distortion of an infinite cylinder subjected to 1.32 P on the outside, and P on the inside. This gives

$$\Delta r = \left(-\frac{0.32\,a^2}{2\,\mu\,(a^2-b^2)} + \frac{4\,\mu+3\,\kappa}{18\,\mu\kappa}\cdot\frac{b^2-1.32\,a^2}{a^2-b^2}\right) \times b \times P$$
$$= -6.9 \times 10^{-7} \times b \times P$$

where a is the external radius, $\tfrac{5}{16}$ in. (0.79 cm.), and b the internal radius, $\tfrac{1}{16}$ in. (0.16 cm.). A value probably nearer the truth is found by assuming for the effective external pressure 1.16 P, i. e., a mean between the maximum and the pressure on AB. This gives

$$\Delta r = \left(-\frac{0.16\,a^2}{2\,\mu(a^2-b^2)} + \frac{4\,\mu+3\,\kappa}{18\,\mu\kappa}\cdot\frac{b^2-1.16\,a^2}{a^2-b^2}\right) b \times P$$
$$= -5.3 \times 10^{-7} \times b \times P$$

and this value will be used in this computation. This represents the maximum radial displacement of the cylinder, which occurs at the inner end; at the outer end there is no pressure either external or internal, and the displacement will be assumed to vanish. Throughout the length of the cylinder the displacement at the inner surface will be assumed proportional to the internal pressure at that point, although the approximation is not so good here as for the piston.

From these displacements of piston and cylinder it is now required to correct for the change in the effective area of the piston. We do this by considering the equilibrium of the escaping liquid. The piston and cylinder each exert on the liquid approximately the same frictional force (F). Furthermore, the cylinder exerts on the escaping liquid a pressure P_1, which is the negative of the component in the direction of the axis of the pressure of the liquid in the crack on the cylinder. P_1 corresponds, therefore, to the axial component of pressure on a ring of breadth AB (Figure 4). Similarly the piston exerts a pressure P_2 equivalent to that on a ring CD. The free liquid at the inner end exerts P_3 on the ring BE. Since the liquid escapes steadily without acceleration, we have

$$2F + P_2 = P_1 + P_3.$$

The effective force on the piston is $F + P_2$

$$F + P_2 = \frac{P_1 + P_2 + P_3}{2}.$$

We now can calculate P_1 and P_2 without any assumption as to the distribution of pressure in the crack if we assume only that at every point the radial displacement is proportional to the pressure at that point. This gives

$$P_1 = 2\pi R \int_{r_1}^{r_2} p\, dr,$$

where r_1 is the value of r at the end ABE of the cylinder, and r_2 at the end CD. R is the average of r_1 and r_2. But

$$r_2 - r = Cp,$$
$$dr = -C dp,$$

$$P_1 = -2\pi C R \int_{P_A}^{P_C} p\, dp$$

$$= 2\pi C \frac{RP^2}{2} = 2\pi \frac{P(r_2 - r_1)}{2} R.$$

That is, P_1 is equal to the pressure exerted by the total internal pressure P on a ring of half the breadth of AB. Similarly, P_2 is the pressure on a ring of half the breadth of CD. If now we put R equal original radius of piston, and $R + \Delta R$ equal original radius of cylinder,

$$AB = 5.3 \times 10^{-7} \times (R + \Delta R) \times P,$$
$$CD = 3.5 \times 10^{-7} \times R \times P,$$
$$BE = \Delta R + (2.1 \times 10^{-7} - 5.3 \times 10^{-7}) \times R \times P,$$
$$= \Delta R - 3.2 \times 10^{-7} \times R \times P.$$

Hence, $$F + P_2 = 2\pi R \frac{(2.6 + 1.8 - 3.2)\, 10^{-7} \times R + \Delta R}{2} \times P$$

$$= 2\pi R \left(\frac{\Delta R}{2} + 1.2 \times 10^{-7} \times R\right) \times P.$$

This force, $F + P_2$, acts in addition to the hydrostatic pressure on the inner end of the piston, which is now decreased in radius by $2.1 \times 10^{-7} \times R \times P$. The new effective radius is therefore

$$R + \frac{\Delta R}{2} - (2.1 - 1.2) \times 10^{-7} \times R \times P,$$

as compared with the original effective radius $R + \Delta R/2$. The correction on the area is therefore $2 \times (2.1 - 1.2) \times 10^{-7} \times P$, or 0.018 per cent per 1000 kgm. The correction turns out, as was to be expected, independent of the size of the crack.

If the maximum value given above for the distortion of the cylinder is used, the effective radius will be found to be

$$R + \frac{\Delta R}{2} - 1.7 \times 10^{-7} \times R \times P,$$

which gives a maximum correction of 0.034 per cent per 1000 kgm. per sq. cm. Experimental reasons will be given later for preferring the lower value for the correction. This value, 0.018 per cent per 1000 kgm., was therefore the correction applied in all the subsequent work.

The Gauge in Practical Use.

The first essential in making an actual measurement with this gauge is a knowledge of the effective area of the piston. As has been intimated above, this could not be determined directly because of the smallness of the parts, and an indirect method was therefore adopted. Briefly, this consisted in subjecting simultaneously to the same hydrostatic pressure the small piston and another piston large enough to be measured accurately, and finding the equilibrating weights required on the two pistons. The effective areas are then in the ratio of the equilibrating weights.

The larger piston was $\frac{1}{4}$ in. (0.635 cm.) in diameter, 2 in. (5.18 cm.) long, ground to fit a reamed $\frac{1}{4}$ in. hole in a large cylinder of Bessemer steel. As this larger gauge was intended for use only to 1000 kgm., the increased breadth of crack produced by exerting the pressure on the interior only of the cylinder was not great enough to give troublesome leak. Also the correction to the effective cross section due to distortion is small enough to be entirely neglected at 1000 kgm. The diameter of the $\frac{1}{4}$ in. piston could be measured certainly to one part in 2500 with a Brown and Sharpe micrometer. The hole in the cylinder was not measured by filling with mercury and weighing, or by any such frequently employed device. It was instead carefully tested against the piston while the latter was in process of being ground to size. The piston was too large to enter the hole except by forcing, when 0.0001 in (0.00025 cm.) larger than the final size. This allowance is probably too much, but still probably not so high as to make the error introduced here in the effective area as much as $\frac{1}{10}$ per cent. This method of measuring the diameter of a hole by testing against plugs of known size is the method used by Brown and Sharpe themselves, and is probably the most accurate that we have, when it is possible to obtain the comparison plugs. The comparison of piston and cylinder was easy in this case because all the work was done in the machine shop of this laboratory.

As preliminary work with this larger gauge, a Bourdon gauge by the Société Genevoise was calibrated to 1000 kgm., and showed a maximum error of 5 kgm. per sq. cm. Various liquids were used to transmit pressure to the $\frac{1}{4}$ in. piston, from vaseline which gave a barely perceptible leak, to a thin mixture of water and glycerine, with which the leak was so rapid that pressure could be maintained only with difficulty. The indications of the gauge, as compared with the Bourdon gauge, proved independent of the rapidity of leak, as they should. In the use of the gauge, sensitiveness was secured as usual, by keeping the piston in continual rotation. Made sensitive in this way, the gauge was very much more sensitive than the Bourdon gauge, responding to about one part in 20,000 at 1000 kgm.

Two high pressure gauges of the type described above were compared with this $\frac{1}{4}$ in. gauge at 1000 kgm. Pressure was kept constant during the comparison by the rise or fall of the $\frac{1}{4}$ in. piston, which had a long enough stroke to accomplish this. As was to be expected, the larger piston proved more sensitive than the smaller ones. The certainty of rise or fall of the small pistons was made greater by observing them with the telescope of a cathetometer. The method of proceeding was to apply a constant weight to the small piston, and then find the two weights on the large piston for which the small piston just began to rise or fall. To accomplish this, the weight on the large piston had to be changed by 0.4 kgm. with a total load of 300 kgm. The mean of these two extreme values gives, therefore, the true equilibrating weight to certainly $\frac{1}{10}$ per cent, and probably much better than this.

From the effective area of either piston found in this way, and the measured diameter, the size of crack between piston and cylinder can be computed. It turned out to be 0.0001 in. (0.00025 cm.) for one gauge, and 0.0003 in. (0.00075 cm.) for the other. This was roughly verified by the more rapid leak shown at higher pressures by the latter gauge. With the former gauge the leak was almost imperceptible after pressure had been kept at 7000 kgm. for an hour. It is a curious fact that the leak around the more loosely fitting piston was distinctly most rapid at 2000 kgm. The decreased leak at higher pressures may probably be taken as proof of the efficiency of the application of pressure to the outside of the cylinder in decreasing the size of the crack, although there is a slight possibility that the effect is due to increased viscosity of the molasses under pressure.

With this calibration, the critical examination of the behavior of the gauges might have been terminated, because the simplicity of the construction is such as to make improbable any error in their use. As a matter of fact, the indications of the various types of gauge described above have usually been accepted at their face value, without comparing with any other absolute gauge. There were means at hand in the present case, however, of so easily comparing the one gauge with the other that it seemed worth while doing. The method adopted was an indirect one, depending on the secondary mercury gauge described in the second part of this paper. It had been found from a great many preliminary comparisons of different mercury gauges that the indications of the mercury gauge were constant, giving a trustworthy measurement of pressure, if once the calibration with a primary gauge could be effected. More detailed proof of this statement will be found in the second part. The two absolute gauges described above were, therefore, compared at different times against the same mercury gauge, and the two sets of readings compared.

The results of the comparisons are shown in Table I. Gauge I was compared twice with the mercury resistance, and Gauge II once. Each number entered in the table is the mean of two or four readings made at increasing or decreasing pressures. The agreement of the two readings under increasing or decreasing pressure, as also of the readings of Guage I on two separate occasions, was as close as it was possible to make the measurements of change of resistance, and, therefore, only averages have been tabulated. The change of resistance could be read to one part in 3000, at the maximum pressure. The average divergence of the readings of either gauge from the mean is well under $\frac{1}{10}$ per cent. The readings of Gauge II are consistently higher than those of Gauge I, a discrepancy which would point to a slight error in determining the effective area of the pistons. The discrepancies also show a tendency to become larger at the higher pressures. This is probably no fault of the gauges themselves, but may be due to the increased difficulty of making fine adjustments of pressure at the higher values. The method of procedure was to apply a known weight to the piston, and then vary the pressure until equilibrium was produced. Setting on this equilibrium pressure was made more difficult by the fact that pressure always showed a tendency to fall after an increase, and to rise after a decrease, a fact that may be explained

TABLE I.

COMPARISON OF TWO ABSOLUTE GAUGES AGAINST THE SAME MERCURY GAUGE.

Gauge I.		Gauge II.		$\frac{\Delta R}{R_0}$ from Gauge I at Gauge II Pressures.	Percentage Divergence from Mean.
Pressure kgm./cm.^{2}	$\frac{\Delta R}{R_0}$.	Pressure kgm./cm.^{2}	$\frac{\Delta R}{R_0}$.		
917	0.002862	929	0.002898	0.002897	−0.015
1501	0.004555	1519	0.004605	0.004604	−0.012
2018	0.005960	2043	0.006032	0.006025	−0.05
2602	0.007491	2634	0.007577	0.007572	−0.03
3196	0.008989	3235	0.009095	0.009083	−0.05
3779	0.010390	3825	0.010530	0.010500	−0.10
4233	0.011420	4285	0.011560	0.011530	−0.15
4816	0.012740	4864	0.012840	0.012840	−0.00
5348	0.013860	5414	0.014020	0.013990	−0.10
5932	0.015030	0005	0.015220	0.015180	−0.13
6452	0.016070	6531	0.016290	0.016230	−0.20
6841	0.016820	....	....	....	....

The absolute gauges were not corrected for distortion, as this is not necessary for the comparison.

by thermal effects of compression, but is more probably due to elastic after effects in the containing steel vessels. It may be concluded, therefore, from the agreement of these comparisons, that even if all the error is in the absolute gauge and none in the mercury resistance, that this type of gauge is good to about $\frac{1}{10}$ per cent.

The comparison with mercury gauges also furnished an estimate of the sensitiveness of the gauge. It was found that throughout the entire pressure range the pistons would respond to differences of pressure that could not be detected by the change of electrical resistance. At 7000 kgm., therefore, the gauges remain sensitive to at least 2 kgm. per sq. cm. The continued sensitiveness of the piston with the crack only 0.0001 in. furnishes an argument against the maximum value set, in the discussion above, on the distortion of the cylinder. For, if we accept the above maximum, we shall find that at 7000 the crack must decrease 0.00018 in., or in this case completely close up. There cannot well be an error of this magnitude in the micrometer measurement of the diameter, and the probable correctness of the average value of the distortion used above is thus increased.

Conclusion.

In this first part of the present paper a description has been given of an absolute gauge, so designed that leak does not become troublesome, at least to 6800 kgm. per sq. cm. The various corrections to be applied have been discussed, and the method by which the dimensions were determined has been described. From a comparison of two gauges of this type with one of another type, the probable accuracy of the gauge is estimated to be at least $\frac{1}{10}$ per cent, and the sensitiveness, 2 kgm. per sq. cm., at 7000 kgm. per sq. cm.

CONTRIBUTIONS FROM THE JEFFERSON PHYSICAL
LABORATORY, HARVARD UNIVERSITY.

THE MEASUREMENT OF HIGH HYDROSTATIC PRESSURE.

II. A SECONDARY MERCURY RESISTANCE GAUGE.

By P. W. Bridgman.

Presented by W. C. Sabine, December 9, 1908. Received December 16, 1908.

IN the introduction to the first part of this paper it was stated that the end sought in designing the primary gauge was the calibration by means of it of some secondary gauge which should be easily reproducible. The secondary gauge that it was proposed to adopt is one involving the variation of mercury resistance with pressure. This is of an entirely different character from the type of secondary gauge in common use, which is usually some form of metallic deformation gauge like that of Bourdon. Undoubtedly the Bourdon is one of the most convenient forms of secondary gauge that it would be possible to devise, being almost immediate in its action, and capable of standing considerable rough handling. If it were applicable over the wide pressure range contemplated for the mercury gauge, its greater convenience would certainly overbalance the fact that every such Bourdon gauge must be initially calibrated against some direct standard.

It seems to be a fact, however, that any elastic deformation gauge becomes unsuitable at high pressures, even when once calibrated, because of the entrance of hysteresis effects. It is true that the existence of elastic hysteresis effects has frequently been doubted, and it has even been stated that proof of their existence would give us knowledge of a new elastic property. It nevertheless seems to be a fact that hysteresis may be inappreciable at low values of the stress, but become increasingly important at higher pressures. This is not the place, however, to enter into a discussion of this point, which will afford the subject for another paper. But in this paper there will be given a somewhat detailed examination of the behavior under pressure of one Bourdon gauge, which will at least show that this type of gauge is irregular at high pressures, whatever the true explanation of this irregularity may be. This paper will be chiefly concerned with a careful examination of the suitability of the proposed mercury standard, and a determination of the constants necessary to its use up to 6800 kgm. per sq. cm. At the end will be found a calculation from the constants of the mercury gauge of the variation of the specific resistance of mercury under pressure. This calculation involves a

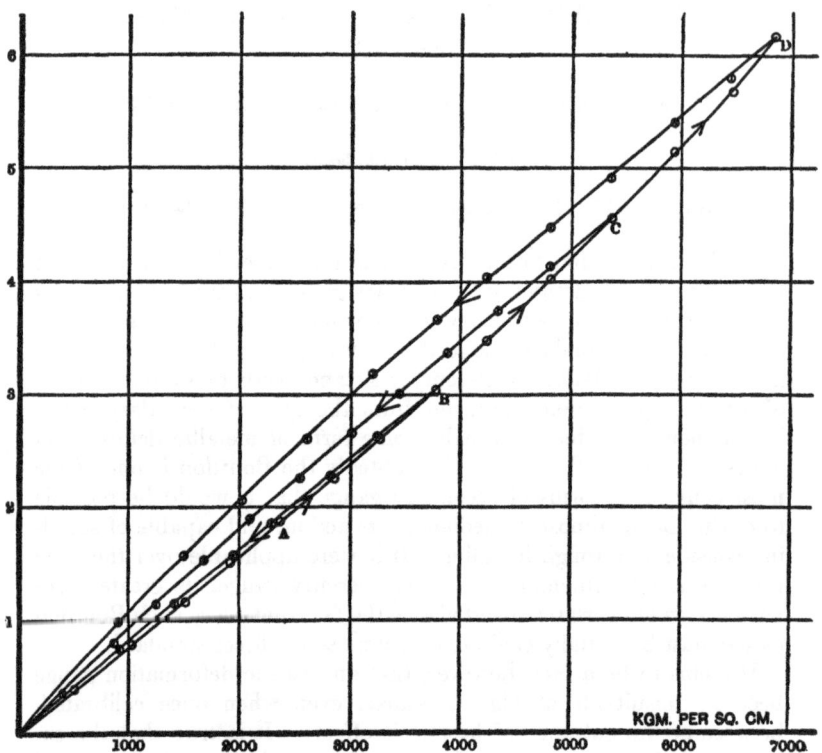

FIGURE 1. Deflection of free end of Bourdon gauge plotted against pressure. Four complete cycles are represented, the points A, B, C, and D being the successive turning points. The figure shows the increasing importance of hysteresis at higher pressures.

knowledge of several compressibilities, which had to be independently determined. In order, however, not to group together in one paper unrelated matter, the determination of compressibilities under high pressures is made the subject of another paper, and only the numerical results there found are used here.

The Bourdon gauge used consisted of hard drawn Shelby steel tubing $\frac{4}{16}$ in. (0.79 cm.) outside diameter and $\frac{1}{16}$ in. (0.159 cm.) inside diameter, wound into a helix of five turns of 5 in. (12.7 cm.) diameter. The tube was not flattened into an elliptic cross section, as in the ordinary Bourdon gauge, since to do this would have too greatly decreased the strength. Even when the cross section is left round, however, the tube unwinds upon the application of pressure, like the ordinary Bourdon. The amount of unwinding was read directly by observing the position of the free end with a microscope, a method of reading which proved more satisfactory than any multiplying mechanism. Thus gauge had been in use for upward of six months before the readings shown in Figure 1 were made. The gauge had been so thoroughly seasoned by the many applications of pressure in this interval that the deflections on many subsequent occasions were found to agree within the errors of reading. Initially, the gauge showed some slight set under the maximum pressure, but after the first few applications of pressure no further set appeared. Elastic after effects, which might be expected to be troublesome over this wide pressure range, could be noticed at every stage of the pressure variations, but were too small to appear on the diagram.

In Figure 1 the deflection of the free end (mm.) is plotted against pressure in kgm., which was measured with a mercury resistance that had been calibrated against an absolute standard, as will be described later. The figure shows the effect of applying four cycles of pressure, from zero by steps to the maximum and by steps back to zero, each subsequent maximum being higher than the preceding. Pressure was first applied in steps from zero to A, and then reduced to zero. The return path coincides so closely with the initial path that the difference cannot be shown on the diagram. Pressure was now increased from zero to B and decreased to zero. The first part of the path zero–B coincides exactly with the path from zero to A. The return path B–zero is sensibly linear, but does not coincide with the path zero–B. We have here, then, the beginning of departure from linearity, and also the beginning of hysteresis. Two more loops, zero–C–zero, and zero–D–zero, reaching to higher pressures, were now described. The essential characteristics are the same, but departure from linearity and hysteresis both increase rapidly with the rise of the range. The return paths for these longer loops do not continue linear, as for zero–B–zero, but they both start as straight lines and run for about the same distance before beginning to curve down to meet the origin. The increasing importance of hysteresis is shown by the fact that the greatest error introduced by hysteresis in the loop zero–B is

4 per cent, while in the loop zero–D it is 40 per cent, an increase of tenfold for a doubling of the pressure range. The return path D–zero was not described at the same time as the part zero–D, because an explosion occurred when the maximum D was reached. It is, however, the return path described on another occasion when the initial path zero–D was identical with the above.

Other types of gauge have shown the same characteristics at high pressures. Whatever the true explanation may be, it has been found in every case that an elastic deformation gauge does show behavior like the above. This type of gauge appears, then, to be unsuitable for the accurate measurement of high pressures, and must be replaced by some form not showing hysteresis; for even if this gauge were readily reproducible, the fact that it shows hysteresis would make its indications such a complicated function of pressure, both present and past, that the meaning of the indications could not be conveniently deciphered.

Any scalar physical property when changed by a strain the same in every direction, such as is produced by hydrostatic pressure in a perfectly homogeneous solid, or a liquid, may be expected to show no hysteresis relative to the stress. Such a property, which has the advantage of being easily measured, is electrical resistance. This has been proposed at least twice as a pressure indicator.

Lisell[1] measured the resistance of a number of metals, drawn out into wires, when subjected to hydrostatic pressures up to 3000 kgm. Pressure was measured on an absolute gauge in which the pressure on the freely moving piston was balanced by weights on a lever. Lisell found no evidence of hysteresis, and proposed the measurement of electrical resistance as a satisfactory means of measuring pressure. The variation of resistance of metallic wires, however, was found by Lisell to have the fatal disadvantage for the present purpose of being so greatly influenced by slight impurities in the metal that specimens of the same metal from different sources gave very different results. This gauge, then, would not be reproducible, but each new specimen of wire would have to be calibrated individually against some absolute standard. In addition, the pressure coefficient is inconveniently small, so that great care must be taken to avoid other effects in measuring the slight change of resistance brought about by pressure. Lisell claims as an advantage of this method that the heat of compression of the metallic wires is smaller than for most substances.

---

[1] Lisell, Om Tryckets Inflytande på det Elektriska Ledningsmotståndet hos Metaller, samt en ny Metod att Mäta Höga Tryck. Upsala, 1903. (C. J. Lundström.)

De Forest Palmer,[2] working with the high pressure apparatus of Barus, made measurements of the electrical resistance of mercury up to 2000 kgm., and suggested it as a suitable secondary standard. He gives data from which the pressure can be calculated if the change of resistance is known. It appears from his work that the pressure coefficient is large enough to make accurate measurements of the change of resistance easy. The additional advantage of presumable reproducibility made it seem worth while to examine with some care its suitability as a secondary standard. The conclusion reached is that with ordinary care the mercury resistance gauge is good to about $\frac{1}{10}$ per cent.

In order to attain this probable degree of accuracy, however, it was necessary to examine several minor points with somewhat greater detail than de Forest Palmer found necessary for the purpose of his work. The probable error in de Forest Palmer's work was $\frac{1}{10}$ per cent on the total resistance, which means an error of 1.5 per cent on the pressure at 2000 kgm. The percentage error at lower pressures is of course proportionally greater. Within these limits of error he found the pressure coefficient to be constant. Furthermore, the mercury was placed in a capillary of some glass not specified, so that the data given will not apply to other mercury gauges with a greater degree of accuracy than the possible error introduced by variations in the compressibility of the glass. It is known that different grades of glass may differ in compressibility by as much as 100 per cent.

In fact, this matter of the glass containing vessel proved to be the chief source of possible error. Pure mercury may with confidence be assumed to be perfectly reproducible, and since internal strains cannot be set up in it, to be also perfectly free from hysteresis. The glass, however, is a solid in which it is particularly difficult to get rid of internal strains. It cannot be assumed, therefore, that a pure hydrostatic pressure will not produce hysteresis, or even set analogous to the volume set shown in thermometers after exposure to changes of temperature. It is an advantage, however, that the total effect of the glass envelope is unusually small, both because of the comparative largeness of the pressure effect on the resistance of the mercury, and because the correction factor is only $\frac{1}{3}$ instead of the whole of the compressibility. This latter fact is due to the simultaneous shortening of the capillary which contains the mercury, and the decrease of the bore, the one resulting in an increase of resistance and the other in a decrease. The total correction on the observed change of resistance

---

[2] de Forest Palmer, Amer. Jour. Sci., **4**, 1–9 (1897), and **6**, 451 (1898).

introduced by the glass envelope is only 2.5 per cent as against 60 per cent in determinations of the compressibility of mercury. Hysteresis and other irregular action will appear, therefore, simply as perturbations of this 2.5 per cent correction. There are a number of smaller sources of error, which, even though very obvious, will be mentioned as occasion presents, because in the justification of a new standard it seems well to record all the sources of error that were considered or guarded against.

The electrical measurements were carried out on a bridge of the Carey Foster type provided with an eight point mercury switch. The variable mercury resistance took the place of one extension coil, and the other was a manganin coil of approximately ten ohms. Measurements were made by setting the slider for no deflection, this being preferable to measuring the current by ballistic or steady throw of the galvanometer. A D'Arsonval galvanometer of low resistance was used, of sensitiveness great enough to indicate changes in the position of the slider of less than $\frac{1}{10}$ millimeter. Extension coils and balancing coils were of seasoned manganin, all approximately ten ohms. In comparing together two mercury resistances the same balancing and extension coils were used, the bridge being provided through leads of $\frac{1}{4}$ in. copper wire of negligible resistance with two slide wires one meter long. The slide wires were interchanged by mercury switches frequently cleaned. The resistance of the extension and balancing coils, as well of the bridge wire, was measured against standard manganin coils known to be correct to 0.01 per cent, which were kindly loaned for the purpose by Professor B. O. Peirce. The bridge wire was calibrated for uniformity by stepping off on it a resistance equivalent to approximately 10 cm. at 3 cm. intervals. The maximum correction of one wire was 0.4 mm., of the other 0.7 mm. The average arithmetic correction of the first was 0.17 mm., of the latter 0.4 mm. Approximately 33 cm. of either wire has a resistance of one ohm. All the connections in the circuit were either soldered without acid for a flux or were through mercury cups, except two connections at the insulating plug leading to the mercury resistance, which were made with nuts. As it was found that induction effects were unnoticeable, the bridge was operated with the galvanometer circuit permanently closed, thus eliminating the principal sources of thermal currents. Two readings of every resistance with the extension coils interchanged were really unnecessary, therefore; but they were always made so as to secure the increased accuracy of two independent settings. Current was supplied by a single Samson cell of about one volt, and was decreased by inserting 100 ohms in the battery circuit. The current through the mercury resistance was therefore about $\frac{1}{350}$ ampere. It was necessary that the current be about as small as this to avoid heating effects in the very fine mercury thread. With this low current, however, the key might be closed indefinitely, with no apparent change in the resistance of the mercury.

In carrying out the measurements, the first and most considerable difficulty that presented itself was the designing of a suitable insulating plug for leading the electrical connections into the pressure chamber. Amagat, and most investigators following him, have used as insulating plug a cone of steel (B, Figure 2) separated from the surrounding walls of the pressure chamber by a thin layer (A) of hard rubber or ivory. Any such arrangement as this proved unsuitable for the pressures dealt with here, the hard rubber flowing completely out of the conical crevice, and exuding in the form of a more or less continuous cylindrical tube. Various modifications of this, using the tougher red fibre instead of hard rubber, were tried with little success. Silk also was used as an insulating material and with better success. The silk was cut out in the form of a number of discs and placed around the shank of the cone, which was then forced into its seat. It was found advisable to make the cone and its shank from one piece of steel, otherwise they were pulled apart by the friction of the silk.

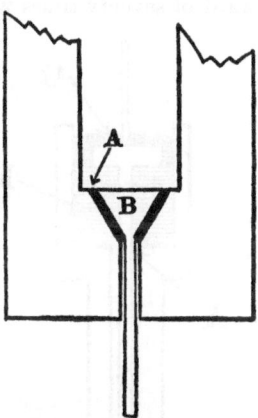

FIGURE 2. Amagat's insulating plug. A, insulating shell of hard rubber or ivory; B, cone of steel. At high pressures the insulating material, A, flows out of the crack.

This form of plug has a high enough insulating resistance and is tight, but has the disadvantage of not being permanent. After ten or twenty applications of pressure the silk loses all semblance of structure, and leaks more and more rapidly with every successive application of pressure.

The cone was now given up and mica tried for insulation, tightness being secured by a layer of marine glue (G, Figure 3). The mica showed no tendency to flow or crumble at the unsupported edge at A. This device was much better than the silk, but it too was not permanent, the marine glue being eventually forced past the mica washers which were a drive fit in the hole. In the form finally adopted (Figure 4) the mica insulation is kept, but tightness was secured by a layer of soft rubber, R, between the mica washers, M. The small steel washer S was necessary to prevent the rubber forcing its way past the mica next the stem, where it is unsupported by the steel at the rear surface. G is an insulating tube of glass. It is well to secure the steel piece B against working loose by the nut and hard rubber washer at A. This plug is the most permanent so far found; one has been subjected to 6500 kgm. upward of seventy times with no sign

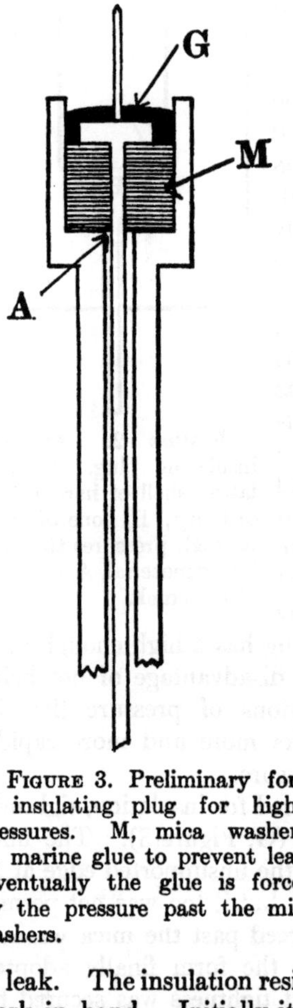

FIGURE 3. Preliminary form of insulating plug for higher pressures. M, mica washers; G, marine glue to prevent leak. Eventually the glue is forced by the pressure past the mica washers.

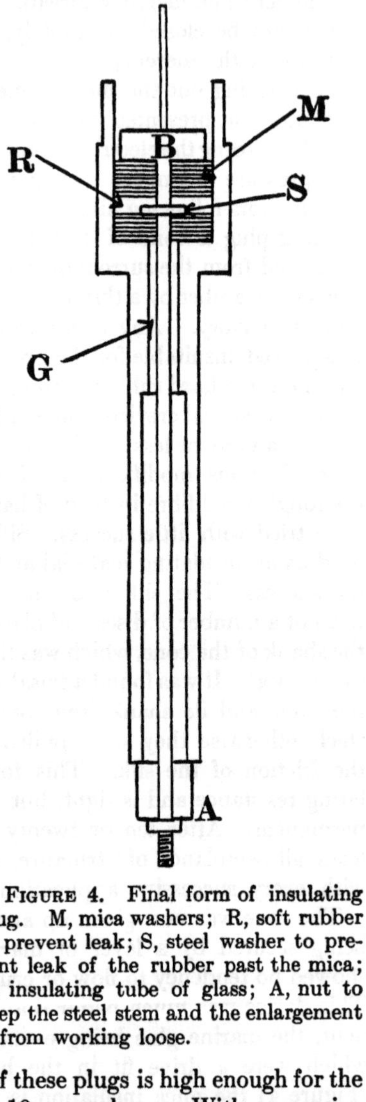

FIGURE 4. Final form of insulating plug. M, mica washers; R, soft rubber to prevent leak; S, steel washer to prevent leak of the rubber past the mica; G, insulating tube of glass; A, nut to keep the steel stem and the enlargement B from working loose.

of leak. The insulation resistance of these plugs is high enough for the work in hand. Initially it is over 10 meg-ohms. With successive applications of pressure the resistance drops considerably, finally reaching a steady value which is of the order of 100,000 ohms. The lowest resistance found in any of these plugs was 30,000 ohms. The resistance of these plugs was measured under pressure, all the conditions of the actual experiment as to position of the electrodes, etc., being reproduced, except for a dummy glass capillary to hold the mercury. When in use, the insulation resistance sometimes increased under pressure, the increase being sometimes as much as 100 per cent. This is still outside the limits of error, the error introduced in the above most unfavorable case being only one part in 6000 on the apparent resistance of the mercury. The performance was usually much better than this. Thus the insulation resistance of one plug which seemed to settle down after several applications of pressure at 150,000 ohms was found to be 220,000 after seven more applications of 7000 kgm.

In devising a form of vessel for holding the mercury, endeavor was made to keep the mercury as much as possible from contact with all sources of contamination by the use of platinum electrodes and a containing vessel entirely of glass. Other experimenters have allowed the mercury to come in contact with the steel of the containing vessel, using the vessel as one electrode, but this seems undesirable in view of the somewhat large effect of minute quantities of impurity. Many forms of glass containing vessel which readily suggest themselves are impractical because of the impossibility of using platinum electrodes sealed into the glass, the difference of compressibility between platinum and glass being sufficiently great to crack the glass around the electrodes. Two forms were finally adopted and used. The form first used was a U capillary (Figure 5), the electrodes dipping into the two cups at the upper end. In the form originally used this was made of thermometer tube of about 6 mm. outside diameter and 0.1 mm. bore. Several times, however, even when carefully annealed, the glass cracked at the bend, apparently because of the unequal strains set up by the hydrostatic pressure within

FIGURE 5. Original and final form of the receptacle for holding the mercury whose resistance is to be measured. If the glass is too thick, it invariably breaks under pressure at the bend A.

the glass, which must have been initially strained. This led to the adoption of a form in which there were no bends in the glass (Figure 6). The glass capillary (A) with the cup on the upper end for an electrode dips into the thin walled tube B containing mercury into which the other electrode dips. This form worked perfectly well, but was somewhat less convenient to handle than the U form. It was finally found that by making the stem of the U capillary very slender, about 1.5 mm., there was no tendency to crack at the bend, and this was the form with which the final determinations were made.

FIGURE 6. Alternative form of containing vessel for the mercury resistance. The resistance of the thin thread of mercury in the capillary A is measured. The containing vessel B must be of thin glass to insure freedom from breakage.

The U capillary (B, Figure 7) is mounted in a split cylindrical piece of steel (A, Figure 7), which is attached to the lower end of the insulating plug. The capillary and plug may thus be connected together and inserted as one piece into the pressure chamber with the certainty that none of the connections will be disarranged in assembling. By making the split steel cylinder containing the U a snug fit, the glass is closely surrounded by metal on all sides, and the quantity of liquid transmitting the pressure is greatly diminished. This has the double advantage of decreasing the total change of volume of liquid necessary to reach a given pressure, and of decreasing the total heat of compression. The heat of compression generated in the small volume

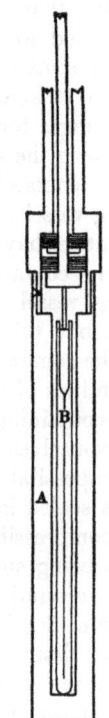

FIGURE 7. Manner of mounting the mercury resistance. The steel envelope A speedily conducts away the heat of compression.

of liquid is so speedily conducted away by the metal that one has to work with inconvenient rapidity after increasing the pressure to find any trace of this effect. This seems to dispose of the only real advantage claimed by Lisell for the solid metallic resistance over the mercury gauge.

The electrodes are of platinum, one soldered to the outside shell of the plug, and the other to the inner stem, which is insulated from contact with the liquid by a layer of marine glue. The electrode leading from this stem is insulated with a soft fine rubber tube, except where it enters the cup of the capillary, where it is covered with a piece of glass tubing, joined continuously to the rubber above it with gutta percha. The electrode from the outer shell of the plug is also protected with glass where it enters the other glass cup. This precaution showed itself necessary, for otherwise if the platinum is not kept from contact with the walls of the cup the liquid above shows an appreciable tendency, with the successive lowerings and raisings of the surface by each application of pressure, to creep down the glass past the mercury.

There are several sources of error here that must be guarded against. Possible short-circuiting from one electrode to the other through the liquid has already been excluded by the measurements of the insulation resistance of the plug with a dummy capillary. In addition, the resistance of the electrodes between the mercury and the plug may change because of (1) lengthening of the free part of the electrode by depression of the mercury surface under pressure or distortions in the containing vessel, (2) pressure effects on the specific resistance of the platinum, (3) and change in resistance at the soldered connection between the electrodes and the plug. The first two sources of error may evidently be eliminated by using heavy enough electrodes. In this work electrodes 0.8 mm. in diameter were large enough. The third effect was found to be troublesome by Lisell, who avoided it by using long metal wires of resistance high in comparison with the resistance of the joint. No trace of this effect could be found, however, in this investigation. The absence of all three effects was tested by measuring the resistance when the terminals were short-circuited by dipping into a large tube of mercury, the resistance of the mercury now being negligible. In this case, the depression of the mercury due to compression is much greater than in the U capillary actually used. In the form tried, this depression may amount to 0.2 mm. Measurements were made up to 7000 kgm., and no change in resistance of the platinum terminal occurred of so much as $\frac{1}{1500}$ ohm, the smallest quantity that could be detected on the bridge. The possible error here, therefore, when the resistance to be measured is 10 ohms, is less than one part in 15000.

During the course of the experiments the steel cylinders containing the mercury resistance were placed in thermostats by which the temperature was usually kept constant within 0.01° during a day's work. Such constancy of temperature as this was not necessary, differences of temperature in the mercury of 0.06° being just perceptible on the bridge wire. Most of this work was carried out at temperatures of about 25°, which was high enough above room temperatures to insure the satisfactory performance of the thermostat. The temperature of the bath was read by a small Goetze thermometer graduated to tenths of a degree and calibrated at the temperature of the bath against a standard Tonnelot thermometer.

Before making the final calibration against the absolute gauge, many preliminary experiments carried out with varying success showed the necessity of observing rather carefully certain apparently insignificant matters of detail.

These preliminary tests were made by comparing together a number of pairs of mercury resistances, there being for this purpose two steel pressure cylinders to contain the resistances, two thermostats, and, as has already been mentioned, two bridge wires, either of which could be connected to the extension and balancing coils. The procedure in comparing two mercury resistances was: read resistance No. 1 on slide wire No. 1; throw in slide wire No. 2 and measure resistance No. 2; interchange the extension coils with the eight point switch and measure resistance No. 2 again; and finally throw in slide wire No. 1 and measure resistance No. 1 again. If these readings were made at equal intervals of time, as they usually were, the average of the two determinations of each resistance gives the value at the same instance of time. In this way the effects of slight changes of pressure due to dissipation of heat of compression and elastic after effects are eliminated. There was no leak. The pressure was roughly measured with the Bourdon gauge described above. These preliminary tests are competent to decide the question of the reproducibility of the mercury resistance gauge. The question of entire freedom from hysteresis, however, cannot be settled merely by a comparison of two gauges, for complete agreement would indicate only that hysteresis in the glass envelope was the same in either gauge. Entire freedom from hysteresis, within the limits of error, can be shown only by a comparison with the absolute gauge.

The results first obtained in the comparison of the two gauges were irregular beyond possibility of experimental error, discrepancies of 1 per cent being not uncommon. This was found to be due principally to three causes: minute impurities in the mercury, the effect of which will be discussed more in detail later; corrections due to air occluded in the mercury; and variations of elastic behavior of the glass envelope under pressure.

With the first few applications of pressure to the glass capillary directly after drawing, the zero value of the mercury resistance undergoes a permanent change, the magnitude of the change decreasing with successive applications of pressure until finally after four or five applications no further change is perceptible. This set is almost certainly due to a change of form of the glass vessel. This initial change has been observed as large as 3 mm. of bridge wire, that is, $\frac{1}{500}$ of the total resistance, and is always in the direction of decreased resistance, that is, toward an increase of cross section of the glass, contrary to what one might expect. If, however, this change of zero is caused by a relieving of the internal strains in the glass, it is in the direction one might expect, because the strains set up by drawing the capillary down from a larger size might decrease upon increasing the size toward its initial value. Not only is there zero change on the first application of pressure, but the elastic behavior over the entire pressure range, as shown by comparison with a well seasoned gauge, is irregular. This irregularity of behavior is shown independently of the resistance measurements by measurements of the compressibility of the glass, which will be given in another paper. The remedy for this defect is to season the glass by gradually applying and relieving the pressure several times. Sudden changes in pressure, such as have sometimes occurred when parts of the apparatus have exploded, are accompanied by large changes in the glass. If the glass has been subjected to considerable temperature changes after being seasoned in this way, it must be seasoned again before its indications are trustworthy.

Occlusion of air in the mercury is likely to cause considerable trouble if present in much quantity. Occluded air, as de Forest Palmer remarks, was doubtless responsible for the surprisingly large pressure coefficient of mercury resistance found by Lenz,[3] 0.0002. The complete removal of the air is difficult and was accomplished only once or twice. Boiling the mercury into the capillary several times is a fairly efficient method, but is open to the objection, as suggested above, that the glass must be seasoned again after each filling. Finally, after several attempts, the following somewhat extravagant method of procedure was found to work satisfactorily: One of the cups of the U capillary was nearly closed by a glass stopper, and the whole U tube

---

[3] Lenz, Wied. Beibl., **6,** 802 (1882).

was then placed in one of the two compartments of a glass vessel which was connected to a mercury pump and exhausted. Heat was then applied to the other compartment of the vessel which was full of mercury, and the mercury slowly distilled over until it covered the capillary, as high a vacuum as possible being maintained all the while by constant operation of the pump. This distillation acts as an additional purification of the mercury, that coming over being dry and presumably free from air. When the capillary was covered with mercury, air was admitted though the pump and mercury forced into the capillary through the open cup, any small possible bubble of air rising to the top of the other cup. In this way nearly all the air can be removed, the slight quantity remaining having probably clung to the inner walls of the capillary throughout the exhaustion. The quantity of air left was usually large enough to introduce an appreciable correction. This correction was determined by measuring the resistance of the mercury at low pressures compared with a calibrated Bourdon gauge of the Société Genevoise, and extrapolating back for the zero from 50 kgm. The tube must be refilled if the correction is large, because it will not remain constant, as obviously the effect of the occluded air on the resistance depends on its position as well as on its quantity. If the correction is small, however, it remains constant apparently indefinitely. In the tubes used, the correction ranged from 0.6 mm. to 0.1 mm. of the bridge wire, that is, a mean correction of about $\frac{1}{5000}$ of the total resistance. That the permanent change of the zero mentioned above was due to set in the glass and not to a curious behavior of the contained air, is proved by the fact that no set was found after filling in the manner above a tube once seasoned, but the correction for air assumed at once its final value.

In addition to showing the necessity of seasoning the glass and removing all air from the mercury, the preliminary comparisons showed that the mercury must be purified with some care. Later, a quantitive determination of the effect of two common impurities will be given. It was found that the mercury could be got sufficiently pure for present purposes by distilling commercial mercury, cleaning with acid, washing and drying, and finally distilling into the U capillary as described above.

When all these precautions are taken, the mercury gauge seems to be reproducible at pleasure. The results of a comparison of two such gauges is shown in Table I. The two mercury resistances compared were each contained in capillaries of the same kind of glass, Jena No. 3880 a. One capillary (R 9), however, was twice the linear dimensions of the other (R 10) because it seemed desirable to eliminate any possible effect of the size of the capillary on its elastic behavior. The smaller capillary, of course, was drawn down farther from the original piece, and so it is conceivable that the internal strains might be enough greater to result in different elastic behavior. In Table I the displacements of the slider of the bridge wire corresponding to the changes of R 9 and R 10, together with the ratio of the displacements, are tabulated against the approximate pressure, which was calculated from the comparison of R 10 later against an absolute gauge. The ratio is constant at 1.007, excepting two values, either of

TABLE I.

COMPARISON OF TWO MERCURY GAUGES TO SHOW REPRODUCIBILITY.

Approximate Pressure kgm./cm.²	Displacement of Slider in Cm.		
	R 10.	R 9.	Ratio.
1040	5.53	5.51	1.003
1930	9.81	9.74	1.007
2870	14.08	13.98	1.007
3750	17.47	17.35	1.007
4390	20.73	20.60	1.007
5650	24.61	24.45	1.007
6600	27.95	27.72	1.008
4250	19.59	19.45	1.007
1990	10.07	10.00	1.007

which could be brought to 1.007 by an error of only 0.1 mm. in the slider settings. The ratio of the resistance R 10 to R 9 multiplied into a constant expressing the different linear resistances of the bridge wires is also 1.007. Within the limits of error of the electrical measurements, therefore, or within $\frac{1}{10}$ per cent, the mercury resistance gauge may be assumed to be reproducible.

There is now left only one point in regard to the suitability of the mercury resistance as a secondary standard to be cleared up by the comparison of the mercury with an absolute gauge, namely complete freedom from hysteresis. The absolute gauge is that described in the first part of this paper. The steel parts of this gauge may of course show hysteresis, but if we assume that the liquid transmitting the pressure shows no hysteresis, which is almost certainly true, it is evident that any hysteresis effects in the steel parts will merely affect the correction for distortion of the gauge. The largest value of this

TABLE II.

COMPARISON OF MERCURY GAUGE AGAINST ABSOLUTE GAUGE AT INCREASING AND DECREASING PRESSURES, TO SHOW FREEDOM FROM HYSTERESIS.

Pressure kgm./cm.^{2}	Slider Displacement Cm.	
	Increasing Pressure.	Decreasing Pressure.
917	4.89	4.89
1501	7.77	7.79
2018	10.17	10.19
2602	12.80	12.79
3196	15.35	15.37
3779	17.73	17.74
4233	19.49	19.51
4816	21.75	21.75
5348	23.70	23.65
5932	25.65	25.67
6452	27.45	27.43
6841	28.71	...

is about $\frac{1}{10}$ per cent. Within the limits of error, therefore, the absolute gauge shows no hysteresis. Freedom of the mercury from hysteresis will be shown by agreement of the resistance measurements under increasing and decreasing pressure.

Comparisons of mercury resistance and absolute gauge were carried out with one mercury resistance (R 9) of soft Jena glass tubing No. 3880 a, and two absolute gauges, as has already been mentioned in the first part of this paper. The results of one of the comparisons under increasing and decreasing pressure, to determine freedom from hysteresis, are given in Table II. Here the displacements of the slider in cm. are tabulated against pressure, calculated from the corrected dimensions of the absolute gauge as described in the first part. The displacements under increasing or decreasing pressure agree within the limits of error of reading the position of the slider. Another comparison of R 9 against the same absolute gauge, as also a comparison against another absolute gauge, led to the same result. These comparisons were taken to afford sufficient proof of freedom from hysteresis of the mercury resistance in the soft Jena glass capillary within errors of $\frac{1}{10}$ per cent.

Having established the reproducibility and freedom from hysteresis of the mercury, we pass to the more important results to be obtained from the comparison with the absolute gauge, namely the final translation of the indications of the mercury gauge into kgm. per cm. The data used for this were those obtained from the two comparisons of R 9 against absolute gauge No. 1, and the one comparison against gauge No. 2. The results of these comparisons have already been given in Part I of this paper, where it appears that the two absolute gauges do not differ on the average so much as $\frac{1}{10}$ per cent from the mean. The average of these two comparisons is taken as the true value and is used in the following computations.

If the change of resistance is to be used as a practical standard of pressure, some empirical formula is desirable connecting the change of proportional resistance with the pressure. In the following, two formulas will be given, the first expressing the change of resistance in terms of the pressure, and the second, which will be more useful in practice, expressing pressure in terms of observed change of resistance. $\frac{\Delta R}{R_0}$ will be abbreviated by ρ, where ΔR is the observed change of the resistance in the soft envelope of Jena glass No. 3880 a, and R_0 the initial resistance measured in this envelope. Then ρ is some function of the pressure, approximately linear. A number of forms of this function were tried, it being desirable for convenience in computation to choose such a form that the number of empirically determined constants is small. It was at once obvious that the ordinary power series representation of the relationship was totally inadequate, at least five and probably more arbitrary constants being necessary to obtain $\frac{1}{10}$ per cent agreement over the entire range. Several other forms of power series tried, with fractional instead of integral exponents, were better, but not sufficiently approximate. Several exponential forms of the type $\rho = ap\, 10^P$, where P is a power series in p, gave still better results. The form finally adopted was, $\rho = ap\,10^{bp^c}$, where

$$\log a = 5.5242 - 10,$$
$$\log(-b) = 6.2486 - 10,$$
$$c = 0.75.$$

This form does not lend itself to computation by least squares, and the best values for a, b, and c were found by trial. Table III shows the

TABLE III.

COMPARISON OF OBSERVED AND CALCULATED CHANGE OF RESISTANCE WITH PRESSURE.

Pressure kgm./cm.²	$\frac{\Delta R}{R_0 p}$		
	Calculated.	Observed.	Difference.
923	0.00003123	0.00003120	+3
1510	0.00003029	0.00003032	−3
2031	0.00002955	0.00002952	+3
2619	0.00002879	0.00002876	+3
3217	0.00002808	0.00002811	−3
3804	0.00002745	0.00002747	−2
4262	0.00002696	0.00002697	−1
4843	0.00002639	0.00002643	−4
5385	0.00002587	0.00002588	−1
5974	0.00002534	0.00002531	+3
6495	0.00002489	0.00002491	−2
6848	0.00002460	0.00002454	+6

$$\frac{\Delta R}{R_0} = -ap\,10^{bp^{\frac{3}{4}}}$$
$$\log a = 5.5242 - 10$$
$$\log(-b) = 6.2486 - 10$$

observed values and the values calculated by the above formula, together with the discrepancies. The divergence is rarely more than $\frac{1}{10}$ per cent and seems irregular in sign. The fairly high discrepancy at 6800 is doubtless because this pressure was reached with only one of the absolute gauges, while all the other values are means of two

TABLE IV.

COMPARISON OF OBSERVED PRESSURE WITH THAT CALCULATED FROM THE CHANGE OF RESISTANCE.

$\frac{\Delta R}{R_0}$	Pressure kgm./cm.²		Difference.	
ρ	Calc.	Obs.	Actual.	Nearest tenth %.
0.02880	925	923	+ 2	+2
0.04578	1512	1510	+ 2	+2
0.05995	2028	2031	− 3	−2
0.07532	2614	2619	− 5	−2
0.09044	3221	3217	+ 4	+1
0.10450	3810	3804	+ 6	+1
0.11490	4266	4264	+ 4	+1
9.12800	4856	4843	+13	+3
0.13940	5393	5385	+ 8	+2
0.15120	5969	5974	− 5	−1
0.16185	6507	6497	+10	+2
0.16810	6831	6848	−17	−3

$p = \rho a \, 10^{\beta \rho^{1.03}}$ $a = \log 4.4871$ $\beta = \log 9.8836 - 10$

determinations. The probable error of the formula itself, calculated by the formula for the error of the mean, is $\frac{1}{30}$ per cent.

The above formula gives the measured change in the resistance of mercury in a specified glass envelope at 25° in terms of the pressure. In practice, it will be necessary to compute the pressure, given the measured change of resistance. The above formula cannot be easily solved for p, and another was set up giving p in terms of ρ. The form of this is exactly the same as for ρ in terms of p, and the procedure in determining the coefficients was the same. It was not found possible to get quite so good an approximation, however, partly because of the shape of the curve itself, which was such that a given percentage error in p produces less percentage error in ρ than the same percentage error in ρ produces in p. In practice, it will be found most convenient to find p graphically from a curve representing the relation between pressure and resistance. The form adopted was

$$p = a\rho\, 10^{\beta \rho^{1.03}},$$

$$a = \log^{-1} 4.4871,$$

$$\beta = \log^{-1} 9.8836 - 10.$$

Table IV shows the observed and computed values for p with the discrepancies. The probable error of a single reading is 0.12 per cent; that of the formula itself much less. This formula holds for mercury in soft Jena glass No. 3880 a at 25°.

At first sight it seems that the two empirical formulas may be combined by eliminating $\dfrac{p}{\rho}$ so as to give a single purely exponential relation between p and ρ which may be readily solved for either. This is not practical, however, because the exponential parts of the above expressions are only slightly affected as to percentage accuracy by relatively large percentage errors in the arguments, and therefore, inversely, small errors in the exponential part may produce large errors in the unknown (p or ρ) calculated from it. Errors of as much as 20 per cent were found to be introduced by the suggested elimination.

The above formulas are only empirical representations of the facts throughout a given pressure range, and their use by extrapolation over any considerably greater range is doubtful. No theoretical value is claimed for them, and it is evident that they cannot represent the actual form of the unknown function. Thus the formula for resistance in terms of pressure predicts a negative minimum of resistance of about -6 at 48,000 kgm. per sq. cm. Neither can extrapolation be carried entirely to the origin of pressure, for the formula demands that $\dfrac{d}{dp}\left(\dfrac{\rho}{p}\right)$ be infinite when $p = 0$, which is almost certainly not the case. The error here is slight, however, and confined to the immediate neighborhood of $p=0$. $\dfrac{\rho}{p}$ at the origin remains finite, with nearly the same values as may be deduced from the formula for p in terms of ρ.

The above formula holds only when the mercury resistance is enclosed in a glass capillary of Jena glass No. 3880 a. If a different glass is used, it will be possible to use the formula by introducing a correction factor. This factor for one other glass, hard Jena combustion tubing No. 3883, was determined by comparing two mercury resistances. The comparison was made not so much with the idea that this hard glass would prove more convenient for practical use, but rather in the hope that these two different kinds of glass, one very infusible and the other very fusible, would show a comparatively large difference of compressibility. Table V shows the ratio of the

TABLE V.
EFFECT OF DIFFERENT GLASS ENVELOPES.

Pressure kgm./cm.²	Ratio $\frac{\Delta R7}{\Delta R9}$	Pressure kgm./cm.²	Ratio $\frac{\Delta R7}{\Delta R9}$
1170	1.025	5800	1.027
1950	1.025	6520	1.026
2960	1.027	4370	1.028
3830	1.028	2100	1.028
4700	1.025	...	...

Mean of ratios of change of resistance weighted according to pressure is 1.0266.
Ratio of initial resistances is 1.0253.
$R7$ is enclosed in hard Jena glass 3883.
$R9$ is enclosed in soft Jena glass 3880 a.

observed changes of resistance in the hard and soft envelopes at different pressure. The mean of the ratios, weighted according to the magnitude of the effect measured, is 1.0266, while the ratio of the initial resistances is 1.0253. The difference between these two numbers is presumably due to the difference of compressibility of the envelopes, which turns out not to be as large as was expected from the character of the glass. The fact that the ratio of the change of resistances is greater than the ratio of the total resistances shows that the hard glass is more compressible than the soft. That the difference is actually due to the difference of compressibility of the glass and is not an experimental error will receive experimental confirmation later by actual measurement of the compressibility of the glass. Resistances in hard as well as in soft glass envelopes may be used as standards, therefore, multiplying, however, the proportional changes of resistance in hard glass by 1.0013 to reduce to soft glass. But it will be noticed from Table V that the ratio of the changes of resistance in the hard and soft glass capillaries varies much more irregularly than the ratio for two capillaries of soft glass (Table I). That this is actually due to irregularities in the deformation of the hard glass will receive confirmation in the paper on compressibility. The hard glass is not so suitable, then, for the capillary as the soft Jena glass.

In practical applications of this gauge it will doubtless be inconvenient to work at the temperature above, 25°, and accordingly the temperature coefficient was determined over a range from 0° to 50°. The determination was made by comparing R 7, which was kept at the standard temperature 25°, with R 9, which was maintained during one set of readings at the given temperature over the entire pressure range. Comparisons were made at six different temperatures, 50.35°, 43.75°, 36.95°, 30.32°, 15.00°, 0.00°. At each temperature seven readings were made with increasing pressure and two with decreasing pressure to avoid all possibility of hysteresis, no evidence of which was found. In making this comparison it appeared necessary after each change of temperature to season the glass by preliminary subjection to the entire pressure range, the irregularities thus eliminated being greater the greater the temperature range. It was found that pressure may be calculated from temperature and the observed proportional change of resistance by the formula:

$$p = a\rho \; 10^{\beta \rho^{1.08}} [1 - a_1 (t - 25°) - b_1 (t - 25°)^2],$$

where a and β have the values previously given, and

$$a_1 = \log^{-1} 7.1253 - 10,$$

$$b_1 = \log^{-1} 4.4487 - 10.$$

a_1 and b_1 were computed by least squares. It was evident on plotting the various points, that a_1 and b_1 are variable with the pressure, becoming less with increasing pressure, but the effect is very slight, and no systematic variation over the entire temperature range could be found. Attempts to introduce such a variation into the general formula would be beyond the accuracy of this work. Table VI shows the value of p computed by the formula for the two extremes of the temperature

TABLE VI.

TEMPERATURE CORRECTION FOR PRESSURE IN TERMS OF RESISTANCE.

51.35°			0°		
Pressure kgm./cm.²			Pressure kgm./cm.²		
Obs.	Calc.	Diff.	Obs.	Calc.	Diff.
1074	1080	+6	1042	1037	− 5
1869	1864	−5	1879	1881	+ 2
2824	2825	+1	2845	2840	− 5
3641	3641	0	3637	3644	+ 7
4478	4478	0	4522	4524	+ 2
5470	5479	+9	5518	5522	+ 4
6527	6528	+1	6560	6573	+13
4249	4243	−6	4262	4256	− 6
1976	1969	−7	0015	2010	− 5

range. The observed pressures tabulated are the pressures computed from the change of R 7 after correction is applied reducing to soft glass. The difference column really contains, therefore, two sources of error. The differences are fairly small and irregular in sign. The irregularity is doubtless due to the incomplete seasoning of the glass by the previous single excursion through the pressure range, and the less regular behavior of the comparison resistance in the hard glass capillary.

During the preliminary comparisons of different mercury resistances, the effect of a known slight quantity of impurity in the mercury was determined. The numerical values thus obtained are given here, as they may be of interest as showing the degree of purity which it is necessary to attain. It was found that metallic impurities have the greatest effect. Impurities that may be absorbed from the glycerine and water unavoidably in contact with the mercury appear to have no effect, as is shown by the constancy of behavior of the gauge over long intervals of time. To test the effect of small metallic impurities, two experiments were made on pure mercury contaminated with known quantities of foreign metal, in the one case 0.1 per cent of zinc, and in the other 0.1 per cent of lead. This is a very large quantity of impurity, much larger than can possibly occur in practice. On standing a short while in the air, the surface of the mercury becomes positively filthy with oxides. The effect of 0.1 per cent zinc is to decrease the resistance by about 1.4 per cent, but the pressure coefficient of resistance by about 5 per cent. Furthermore, the departure from the linear relation between total change of resistance and pressure is less than for pure mercury, being 3 per cent less at 6500 kgm. The results with the lead were not so satisfactory as those with the zinc. It was pretty certain, however, that the effect of the lead is less on the total resistance and greater on the pressure coefficient.

The formulas given above connect the change of resistance of mercury in a capillary of specified glass with the pressure, and are all that is required for use with a secondary standard of pressure. The observed change of resistance, however, is due to a combination of two unrelated effects; the change of dimensions of the glass, and the changed specific resistance of mercury. The results given above will not possess theoretical value, therefore, until the two effects are separated. In the following an experimental determination of these two effects is given.

We may distinguish two specific resistances of mercury, both of which are altered by pressure. The first may be called the specific volume resistance, and is the resistance of a body of mercury of invariable form, but of mass variable with the pressure. The second may be called the specific mass resistance of mercury, and is the specific volume resistance multiplied by the ratio of the masses within the given surface at the variable and standard pressure, i. e., the density. The specific mass resistance seeks to correct for the increased conductivity to be expected at any pressure because of the increased number of conducting particles in a given volume. In order to determine the specific volume resistance, the above results have to be corrected for the compressibility of the glass envelope; to determine the mass resistance, an additional correction must be applied for the compressibility of the mercury. These compressibilities are determined in another paper, to which reference must be made for the methods used. Only the results there found will be used here. It was found that for Jena glass No. 3880 a, $\kappa = 2.17 \times 10^{-6}$, and that the change of volume of mercury is connected with pressure by the relation

$$\frac{\Delta V}{V_0} = bp + cp^2,$$

$$b = \log^{-1} 4.5681 - 10,$$

$$-c = \log^{-1} 9.2977 - 20.$$

Now to find the changed specific volume resistance of mercury we have

$$\frac{\Delta R_s}{R_0} = \rho + a\,p,$$

where ΔR_s is the observed decrease of resistance corrected for changed shape of glass, R_0 is the initial resistance measured in the same glass, a is the linear compressibility of the glass, and ρ has the meaning already given, namely the observed proportional decrease of resistance in the given capillary. But ρ has already been found in terms of p, and a has just been given, so that we have the empirical formula

$$\frac{1}{R_0}\frac{\Delta R_s}{p} = a\,[0.02168 + 10^{bp^2}],$$

where a and b have the values already given, namely,

$$a = \log 5.5242 - 10$$

$$b = -\log 6.2486 - 10$$

The slope of the curve, i. e., the instantaneous pressure coefficient at any point, is:

$$\frac{1}{R_0}\frac{dR_s}{dp} = -a\,[0.02168 + 10^{bp^2}\{1 + \tfrac{3}{4} bp^2 \log_e 10\}],$$

where R_s is the variable resistance corrected for the glass. The instantaneous coefficient per unit resistance is at any point:

$$\frac{1}{R_s}\frac{dR_s}{dp} = -\frac{a\,[0.02168 + 10^{bp^2}\{1 + \tfrac{3}{4} bp^2 \log_e 10\}]}{1 - ap\,[0.02168 + 10^{bp^2}]}.$$

These three quantities were computed by the above formula and are given in Table VII. They are also shown graphically in Figure 8, which indicates the general behavior without, of course, the accuracy of the formula. The general character of all these curves is the same, showing a continually decreasing effect of pressure on change of resistance as the pressure increases, this decrease itself also decreasing.

TABLE VII.

SPECIFIC VOLUME RESISTANCE OF MERCURY.

Pressure kgm./cm.²	$\frac{1}{R_0}\frac{\Delta R_s}{p}$	$\frac{1}{R_0}\frac{dR_s}{dp}$	$\frac{1}{R_s}\frac{dR_s}{dp}$
...	0.00003344	0.00003344	0.00003344
500	0.00003276	0.00003171	0.00003223
1000	0.00003182	0.00003011	0.00003111
1500	0.00003102	0.00002878	0.00003018
2000	0.00003031	0.00002760	0.00002938
2500	0.00002966	0.00002653	0.00002865
3000	0.00002906	0.00002552	0.00002796
3500	0.00002849	0.00002461	0.00002735
4000	0.00002795	0.00002374	0.00002674
4500	0.00002744	0.00002293	0.00002616
5000	0.00002696	0.00002216	0.00002561
5500	0.00002655	0.00002148	0.00002515
6000	0.00002603	0.00002073	0.00002457
6500	0.00002562	0.00002006	0.00002407

The curves do not run to high enough pressures to justify any speculation as to their ultimate behavior.

De Forest Palmer's are the only results with which these can be compared. He found $\frac{1}{R_0}\frac{\Delta R_s}{p}$ to have the constant value 3.224×10^{-5} between 0 and 2000 kgm.[4] There is, however, as already stated, a probable error of 1.5 per cent at 2000 kgm., and proportionally more

---

[4] de Forest Palmer, Amer. Jour. Sci., 4, 8 (1897).

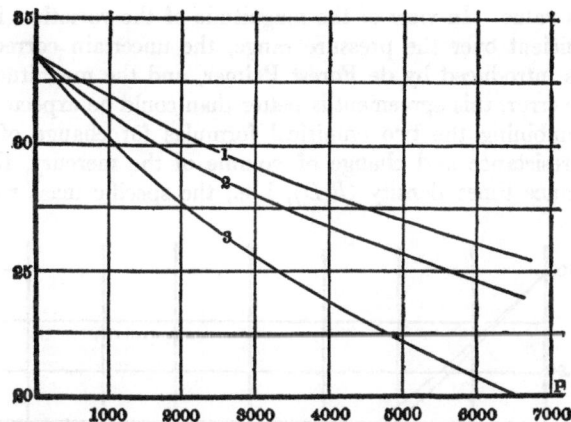

FIGURE 8. Various functions of the specific resistance of mercury plotted against pressure. 1 shows $\frac{1}{R_0}\frac{\Delta R_s}{p}$, 2, $\frac{1}{R_s}\frac{dR_s}{dp}$, and 3, $\frac{dR_s}{dp}$, where R_0 is the initial resistance and R_s is the variable resistance under pressure, corrected for the distortion of the glass containing vessel.

at lower pressures. According to the results above, $\frac{1}{R_0}\frac{\Delta R_s}{p}$ varies from 3.344 to 3.031 × 10^{-5} between 0 and 2000 kgm., giving a mean value of 3.187 × 10^{-5}, which agrees within 1.1 per cent with de Forest

TABLE VIII.

SPECIFIC VOLUME RESISTANCE AND SPECIFIC MASS RESISTANCE OF MERCURY UNDER PRESSURE.

Pressure kgm./cm.²	R_s.	$R_s \times D$.	Pressure kgm./cm.²	R_s.	$R_s \times D$.
0	1.0000	1.0000	3500	0.9003	0.9114
500	0.9836	0.9854	4000	0.8882	0.9010
1000	0.9682	0.9716	4500	0.8765	0.8904
1500	0.9535	0.9588	5000	0.8652	0.8806
2000	0.9394	0.9462	5500	0.8540	0.8708
2500	0.9258	0.9342	6000	0.8438	0.8616
3000	0.9128	0.9228	6500	0.8335	0.8527

Palmer's value. In view of the magnitude of the variation found in the coefficient over the pressure range, the uncertain correction for the glass introduced by de Forest Palmer, and the magnitude of his probable error, this agreement is better than could be expected.

By combining the two empirical formulas for change of specific volume resistance and change of volume of the mercury, the value of resistance times density ($R \cdot D$), i. e., the specific mass resistance,

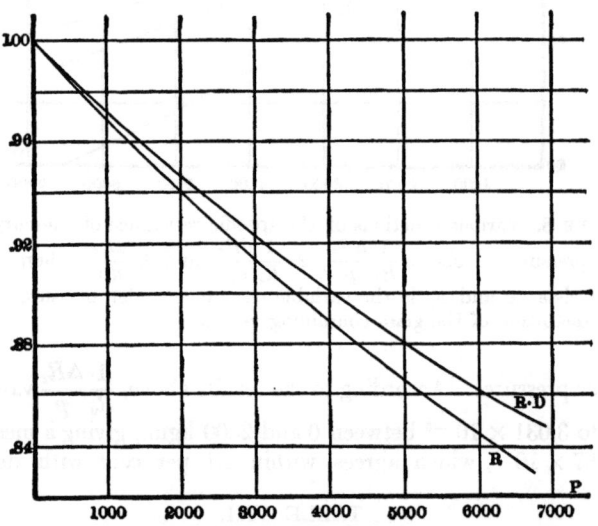

FIGURE 9. The changed resistance of mercury under pressure in terms of the resistance under zero pressure. The curve shows the measured resistance corrected for the distortion of the glass containing vessel. The curve $R \cdot D$ shows the former curve corrected for the changed density of mercury. It shows the pure pressure effect on resistance, that is, the resistance corrected for the increased conductivity due the increased concentration of the molecules. The smallness of the change of resistance due to this concentrating of the molecules is to be noticed.

may be found. The departure of this from constancy may be described as the pure pressure effect on mercury resistance. In Table VIII the specific volume resistance and the specific volume resistance multiplied by the density are given for various pressures. They are also shown graphically in Figure 9. The curves are similar in all respects and show no indications of any remarkable behavior at higher pressures. The comparatively small part played by the change of density in the total change of resistance under pressure is of interest.

Finally, the variation of specific resistance with temperature may be calculated from the formula given for the variation with temperature of p as determined by the measurement of ρ. Retaining only the term of the first degree in p, we have to the degree of experimental accuracy reached in these results:

$$\frac{\Delta R_s(p, t)}{R_s(0, t)} = \frac{\Delta R_s(p, t_0)}{R(0, t_0)} + a^2 a_1 a\, p\, 10^{2bp^{\frac{1}{2}}} (t - t_0),$$

where a, a_1, a, and b have the values already assigned, and t_0 equals 25°. In the deduction of this formula the variation of the compressi-

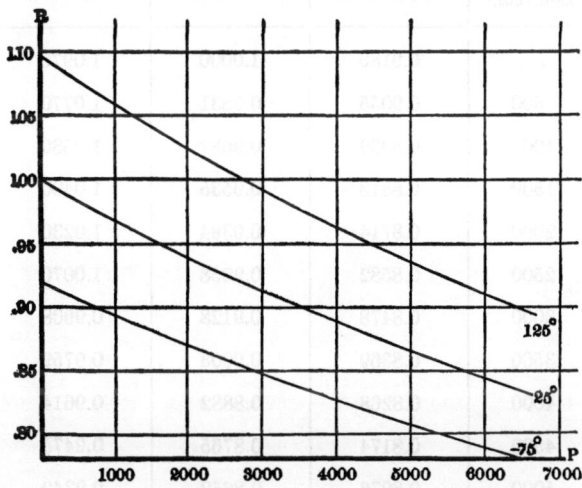

FIGURE 10. The resistance of mercury at various temperatures and pressures in terms of the resistance at zero pressure and 25°.

bility of the glass with the temperature was neglected. This variation is beyond the limits of error if the glass used has a temperature coefficient of the same order as that found by Amagat,[5] who found a change of 10 per cent for 100°. From this formula $R(p, t)$ was calculated for a number of pressures and for the temperatures 125°, 25°, and $-75°$, assuming $R(0, 25°)$ equal to unity, and taking for the temperature coefficient of specific conductivity the value 0.000888. These results are given in Table IX and plotted in Figure 10. This large temperature range was taken merely for convenience in showing

---

[5] Amagat, C. R., **110**, 1248 (1890).

diagrammatically the general tendency of the results. The formula actually does not give results better than $\frac{1}{10}$ per cent beyond the range 0° to 50°. The temperature coefficient found above is nearly ten times de Forest Palmer's value, who, however, worked only at the extremes of a wider temperature range than that used here, namely, 9° to 100°.

TABLE IX.

VARIATION OF MERCURY RESISTANCE WITH PRESSURE AND TEMPERATURE.

Pressure kgm./cm.²	R (p, $-75°$).	R (p, $25°$).	R (p, $125°$).
...	0.9186	1.0000	1.0970
500	0.9055	0.9831	1.0770
1000	0.8930	0.9682	1.0580
1500	0.8818	0.9535	1.0400
2000	0.8714	0.9394	1.0230
2500	0.8582	0.9258	1.0070
3000	0.8478	0.9128	0.9908
3500	0.8369	0.9003	0.9759
4000	0.8268	0.8882	0.9614
4500	0.8174	0.8765	0.9475
5000	0.8076	0.8652	0.9342
5500	0.7982	0.8540	0.9208
6000	0.7896	0.8438	0.9086
6500	0.7807	0.8335	0.8966

No theoretical discussion of the way in which these curves might be expected to behave has been attempted. Only a few points require remark. For instance, it is obvious from the table that temperature has a greater effect on the pressure coefficient of resistance than it does on the resistance itself. The temperature coefficient of the former is 0.00137, and of the latter 0.000888. In other respects the curves behave as one would expect, i. e., at higher pressures the proportionate effect of temperature is reduced. This is shown by the temperature effect both on resistance and on pressures coefficient of resistance. Thus the temperature coefficient of the pressure coefficient has become reduced at 6500 kgm. to 0.7 of its initial value, while the temperature coefficient of resistance is reduced from 0.0009 to 0.0007. This latter effect shows itself in a tendency of the curves for different temperatures to draw together with increasing pressure toward some value of resistance greater than zero. That is, for a large enough value of pressure, the resistance acts as if it might have a definite value independent of temperature.

Conclusion.

In this paper it has been found that the mercury resistance gauge is a reliable secondary standard of pressure if proper precautions are used. The mercury must be pure and free from air. The irregular behavior under pressure of the containing glass capillary is the principal source of error. An easily fusible glass in which the strains left after drawing are presumably small, is better than an infusible glass. The glass must be seasoned by several applications of pressure over the entire range before it becomes regular in behavior. If after this it is exposed to considerable changes of temperature or to sudden changes of pressure, it must be reseasoned. The maximum error that can be introduced by irregularities in the glass is about 2.5 per cent. The dependence of pressure on the measured proportional change of resistance (p) and temperature is given by the equation

$$p = a\rho\, 10^{\beta\rho^{1.08}} [1 - a_1 (t - 25°) - b_1 (t - 25°)^2],$$

where

$$a = \log^{-1} 4.4871;$$
$$\beta = \log^{-1} 9.8836 - 10;$$
$$a_1 = \log^{-1} 7.1253 - 10;$$
$$b_1 = \log^{-1} 4.4487 - 10.$$

This formula, which applies to mercury in a capillary of Jena glass No. 3880 a, gives the pressure correctly to $\frac{1}{10}$ per cent between 500 and 6800 kgm. and 0° and 50° C.

Empirical expressions have also been deduced connecting the specific volume resistance and the specific mass resistance of mercury with the pressure.

CONTRIBUTIONS FROM THE JEFFERSON PHYSICAL
LABORATORY, HARVARD UNIVERSITY.

AN EXPERIMENTAL DETERMINATION OF CERTAIN COMPRESSIBILITIES.

By P. W. Bridgman.

Presented by W. C. Sabine, December 9, 1908. Received December 16, 1908.

In a preceding paper the change of resistance produced by hydrostatic pressure on a fine thread of mercury in a capillary of a specified glass was measured. This change of resistance is the sum of two effects: the change of resistance produced by the changed dimensions of the glass capillary, and the change of resistance due to the changed electrical properties of the mercury under pressure. The change of resistance produced by the distortion of the glass is simultaneously an increase of resistance because of the decreased bore of the capillary, and a decrease because of the decreased length. The total fractional change of resistance is easily seen to be the linear compressibility of the glass. The change of resistance due to the changed electrical properties of the mercury may be further divided into two effects: that due to the change in the conducting power of the separate molecules, and that due to the change in the number of molecules occupying a given space. This latter effect is determined directly by the compressibility of the mercury.

A complete description of the phenomena involved in the measured change of resistance of the mercury involves, therefore, a knowledge of the compressibility of both the glass and the mercury. This paper is occupied with a description of the methods by which these were determined. As the pressure range employed here (6500 kgm.) is somewhat higher than that usually used, modifications of the methods in common use were necessary. It seemed undesirable, however, to bury a description of these methods in a paper on the unrelated subject of the electrical resistance of mercury, and the matter has therefore been made the subject of a separate paper, although the method has been applied to only a few substances, and all the data have been collected solely with a view to the above discussion of the effect of pressure on the resistance of mercury. However, the paper contains an investigation of several minor points that came up in the course of the work, that may be of interest on their own account. Among these is an experimental determination of the difference of linear compressibility of a piece of commercial rolled steel along and perpendicular to the direction of rolling, and some account of the seasoning effect of successive applications of pressure on the elastical behavior of glass. In detail, the paper contains a determination by one method of the compressibility of two kinds of Jena glass, of a piece of commercial aluminum rod, and of several grades of steel; and by another method, the compressibility of mercury, all up to about 6500 kgm. per sq. cm.

In determining the compressibility of a solid the method adopted was to measure the change of length of a rod of the substance produced by hydrostatic pressure applied all over the external surface. This method applies, therefore, only to those solids that can be obtained in the form of a cylindrical rod or tube. The cubic compressibility is found by multiplying the linear compressibility by three. It is a fundamental assumption throughout all the following determinations of the compressibility of solids, therefore, that the substance is so homogeneous and isotropic that the compression under hydrostatic pressure is sensibly the same in all directions. Some experimental proof of the justifiability of this assumption has been attempted in the case of a piece of rolled steel boiler plate.

It is a feature common to all the compressibility methods used in this paper, that the distortion produced by pressure is measured by the displacement of a ring sliding with slight friction on some movable part of the apparatus. The method is not continuous reading, therefore, but the apparatus has to be taken apart and readings made after each application of pressure, the reading obtained corresponding to the maximum pressure. A method of this kind is doubtless inconvenient, but it has the advantage of simplicity and directness over any continuous reading method that would be practical over so wide a pressure range.

In the determination of the compressibility of solids two slightly different methods may be used, according as the solid is of relatively low or high compressibility. The first method, not so accurate as the second, applies to iron and metals of the same order of compressibility. The second applies to substances of higher compressibility, and involves directly the compressibility of iron as determined by the first method.

The first method measures the relative change of length of a rod of the substance and a heavy cylinder of steel. The rod is enclosed in the cylinder, throughout the interior of which hydrostatic pressure is applied. The rod shortens, therefore, under the uniform external pressure, while the cylinder lengthens under the interior pressure. The lengthening of the cylinder is very much less than the shortening of the rod. In the present experiment it was only 5 per cent. The strain in the cylinder is complicated, consisting of a radial displacement away from the centre, and of a longitudinal extension which may produce warping of the originally plain sections. This warping is greatest at the ends, and must vanish at the mid section if the cylinder is symmetrical at the two ends. The warping cannot be easily calculated, and was neglected in the present work. It can in any event constitute only a correction for the above 5 per cent correction term. The method consists, therefore, in subtracting from the

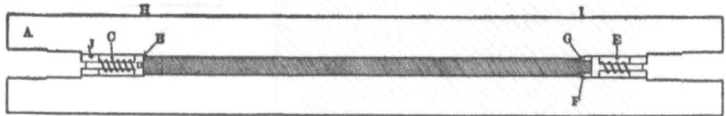

FIGURE 1. Apparatus for measuring the linear compressibility of rods. The rod to be measured is indicated by the shading. The stop D is held permanently against the shoulder B by the spring C, which is kept compressed by the pump connections, not shown. The brass ring F is kept in contact with the shoulder G during increase of pressure by the spring E, which pushes the shortening rod through the ring F, so as always to be in contact with the stop D. When pressure is released the ring comes back with the rod and the displacement is measured. The rod is removed through the end E to make these measurements; the connections at A to the pressure pump are not disturbed during the measurements. The elongation of the cylinder is measured externally at the scratches H and I.

relative change of length of the rod and the cylinder the increase of length of the cylinder as obtained from the measurement of external change of length under pressure. The result is the linear compressibility of the rod, from which the cubic compressibility is calculated.

The cylinder used is shown in Figure 1. It is made of annealed tool steel, 18 in. (45.7 cm.) long, and 2 in. (5.1 cm.) in diameter. It is pierced through the entire length by a reamed ⅜ in. (0.95 cm.) hole, in which the rod to be tested is placed. At either end the ⅜ in. hole is enlarged in several steps in the manner indicated, in order to afford room for the various connections. The enlargements of the holes are precisely alike at the two ends, so as to insure symmetrical warping of the cylinder. The rod to be tested is indicated by the shading. It is carefully turned so as to slide without lateral play into the reamed hole. Three shallow grooves, milled the entire length of this rod, allow the compressing fluid to flow freely from one end of the cylinder to the other. The change of length of the rod is obtained by keeping one end of the rod always fixed opposite the same part of the cylinder, and measuring the relative displacement of the other end, which is free. The fixed end of the rod is kept so by the action of the spring at E, which keeps the rod pressed against the stop at D. This stop D is kept immovable by the spring at C, which keeps D pressed against the shoulder B. This spring C is very much stiffer than the spring E, and is kept permanently compressed by the pump connections (not

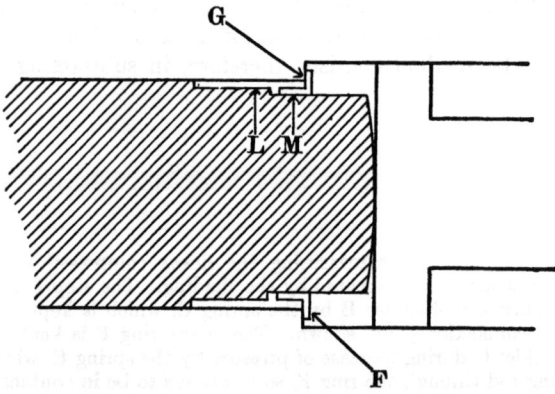

FIGURE 2. Enlarged view of the brass ring, etc., of Figure 1. The displacement of the ring is measured by measuring the distance between the scratches at L and M on the rod and the ring respectively.

shown) which are screwed into the end A, and keep the ring J fast in the position shown. This method of securing the invariable position of the stop seemed preferable to any plug arrangement screwed fast into the cylinder, for the latter might shift slightly, owing to the change produced by the pressure in the dimensions of the thread.

The shift of the free end of the rod relatively to the cylinder was obtained by measuring the displacement on the brass ring F, which is pushed back by the shoulder G. An enlarged view of the ring is shown in Figure 2. The brass ring F is split so as to slide without too great friction on the end of the rod, which is turned down to about $\frac{5}{16}$ in. (8 mm.). There is a fine scratch on the ring at M, and also a scratch on the corresponding ledge L of the rod. The ring and rod are turned in the lathe so that these two scratches are at the same radial distance from the axis of the rod, thus enabling both scratches to be in focus simultaneously under a high power microscope. The effect of an application of pressure is to shorten the rod, pushing up the ring, which stays in its extreme position. The rod is then taken out by unscrewing the plug at the end I and the distance between the scratches L and M measured. The increase of distance over the zero position gives the relative change of length of rod and cylinder. There is here a small source of error in finding the effective length of the rod, which terminates at some unknown place within the brass ring. The effective length used was the length from the fixed end to the middle of the ring when in the zero position. As the breadth of the bearing surface of the ring was only about 2 mm., and the length of the rod was 30 cm., the maximum error here is only 1/300.

It is at once obvious that any slight error in replacing the rod after each measurement in exactly its former position will produce considerable error in the result, since the change of length produced by pressure is small. In the form used, in which the rod is 30 cm. long, the change of length for 1000 kgm. is only 0.05 mm. Slight particles of grit are likely, therefore, to produce considerable irregularities. By working with some care it was found possible, however, to secure fairly uniform results. Particular attention must be given to washing out the cylinder after each application of pressure. The effect of pressure is, of course, to flood the interior of the cylinder with the pump liquid, in this case glycerine and water, which may carry considerable grit in suspension. After each measurement the cylinder was thoroughly washed several times by a jet of water violently expelled from a glass tube reaching into the cylinder as far as the stop D. No cloth or other substance must be used for wiping out the hole. The rod to be tested was also carefully washed under the tap after each measurement, again taking care not to wipe with a cloth or to bring into contact with any possible source of grit. It was found that by decreasing the diameter of the rod for a short distance at the end B, there is less tendency for grit to collect between the end of the rod and the stop D when the rod is replaced in the hole after each measurement.

The change of length of the steel cylinder was not measured at the same time as the relative change of length of rod and cylinder, but was, instead, determined independently as a function of the pressure. Three determinations of this extension were made, one preliminary to, one in the course of, and one after the series of compressibility measurements. The last two agreed within the limits of error; the first was slightly different, as has always been found to be the case when the deformation of a metal is measured on the first application of pressure. In making these measurements, the cylinder was clamped to a heavy comparator bed, which carried two microscopes. The cylinder was clamped at only one point, the middle, so as to avoid any possible distortion of the comparator by the lengthening of the cylinder under pressure. The close contact of cylinder and comparator insured the practical equality of temperature of the two, and the coefficients of expansion of the two pieces proved so close that the few tenths of a degree variation which occurred in the temperature of the room introduced no appreciable error. The microscopes were focussed on fine fortuitous scratches on the cylinder at the points H and I (Figure 1). Change of length was measured by a micrometer eyepiece in either microscope, which had been previously calibrated. Settings on the fine scratches could be made with a maximum error of 0.0003 mm.,

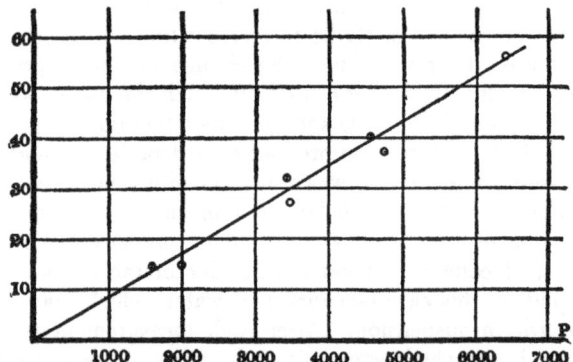

FIGURE 3. The elongation of the cylinder of figure 1, as a function of the pressure. ◯, observations at increasing pressures; ⓛ, at decreasing pressures. The ordinates give the proportional elongation multiplied by 10^6. That is at a pressure of 6400 kgm. per sq. cm. the elongation of the cylinder is 0.000056 per unit length.

thus introducing a possible error of reading of the change of length of 0.0006 mm. The total change of length was found to be 0.02 mm. at 6000 kgm. The maximum error here possible on the extension coefficient of the cylinder is, therefore, 6 parts in 200. The mean of several readings, of course, has a much less probable error.

The results obtained are shown in Figure 3, in which extension of the cylinder is plotted against pressure. The pressure was measured here, as in all subsequent work in this paper, by a secondary gauge depending on the variation of the resistance of mercury under pressure. The justification and calibration of this gauge has been made the subject of another paper. The figure shows distinct evidence of hysteresis, the extension under decreasing pressure being greater than the corresponding extension under increasing pressure. This is the more surprising as the total extension of the cylinder is only $\frac{1}{30}$ of the value of the extension at the elastic limit under pure tension. The departure of the points from a straight line representing the mean is comparatively slight, however, and in applying the corrections determined in this way the relation between extension and pressure was assumed to be linear.

With this apparatus the linear compressibility of a piece of commercial aluminum rod and several specimens of iron and steel were made. In Figure 4 is shown the fractional change of length of the aluminum rod corrected for the extension of the steel cylinder, plotted against pressure. This figure does not include the first observation which was made with a pressure slightly higher than any subsequently reached. The rod took a distinct set on this first application, being permanently shortened by one part in 30,000. No evidence of further set was found on subsequent applications of pressure. This is the first occasion on which a set in any dimension by the application of hydrostatic pressure has been directly observed. No attempt was made to find whether this linear set is accompanied by volume set. The displacement was measured from the mean of several determinations of the position of the ring at zero pressure. But this determination is obviously affected by the same errors as displacement measurements at higher pressures. It is evident from the figure that within the limits of error the points lie on a straight line. This was assumed to be of the form $a + bp$, and a and b determined by least squares, discarding the most discordant results. a is the true zero position and b the pressure coefficient of contraction. In this way every measurement at any pressure contributes to the more accurate determining of

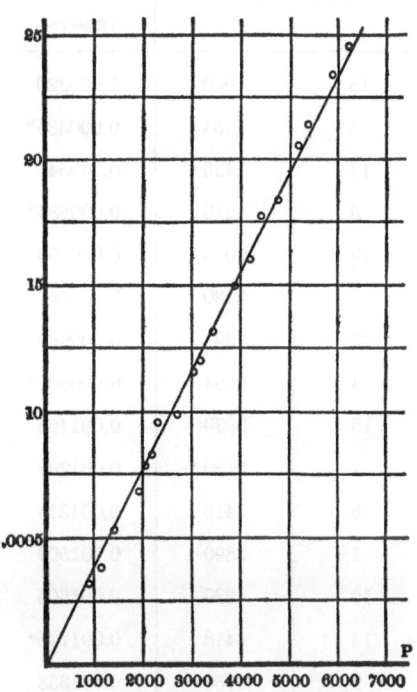

FIGURE 4. The observed proportional change of length of an aluminum rod plotted against pressure.

TABLE I.

Compressibility of Aluminum Rod.

Order of Observation.	Pressure kgm./cm.^{2}	$\frac{\Delta l}{l_0}$		
		Observed.	Calculated.	Difference.
18	900	0.000320	0.000346	+26
6	1154	0.000386*	0.000445	+59
12	1436	0.000545	0.000555	+10
5	1910	0.000684*	0.000741	+57
19	2050	0.000790	0.000796	+ 6
11	2180	0.000867	0.000847	−20
17	2396	0.000960	0.000910	−50
4	2694	0.000990*	0.001048	+58
13	3030	0.001163	0.001179	+16
7	3180	0.001202	0.001238	+36
8	3416	0.001318	0.001330	+12
1	3890	0.001509	0.001515	+ 6
10	4230	0.001605	0.001648	+43
14	4418	0.001775*	0.001722	−53
2	4760	0.001838	0.001855	+17
9	5200	0.002054	0.002027	−27
16	5384	0.002140	0.002099	−41
3	5892	0.002339	0.002297	−42
15	6240	0.002450	0.002434	−16

$$\frac{\Delta l}{l_0} = a + bp. \qquad a = -0.0000056. \qquad b = 0.0000003910.$$

* Discarded in the calculation.

the zero position, the necessity of a large number of determinations of which are therefore avoided. It was found that

$$\frac{\Delta l}{l_0} = -0.0000056 + 0.0000003910\ p.$$

The cubic compressibility is, therefore, 0.000001173 kgm. per sq. cm. In Table I are shown the observed and calculated results. The probable error of a single observation is less than one per cent at the higher pressures. The probable error of b, the compressibility, is about $\frac{1}{3}$ per cent. The value found by Richards [1] for the compressibility of aluminum is 1.28×10^{-6}. He does not state the chemical purity of the aluminum. The specimen used above was commercial aluminum rod, which is usually very pure. No chemical analysis was made, however, and the discrepancy may be due to impurities.

In an exactly similar manner the compressibilities of several samples of iron or steel were determined. The first piece was from a piece of $\frac{1}{2}$ in. (1.27 cm.) Bessemer rod annealed by heating to redness and cooling slowly, and then turned down to $\frac{3}{8}$ in. (0.95 cm.). It was from the same piece of rod as a piezometer for determining the compressibility of mercury, as will be described later. The results obtained for this steel corrected for the extension of the cylinder are plotted in Figure 5, the zero being arbitrary as formerly. The results are better proportionately than for

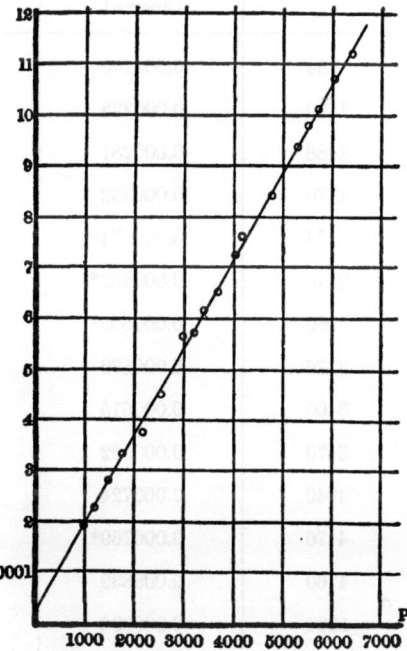

FIGURE 5. The observed proportional change of length of a rod of Bessemer steel plotted against pressure. The zero is here arbitrary.

---

[1] Compressibilities of the Elements and their Periodic Relations. Richards, Carnegie Inst., Washington, p. 61 (1907).

TABLE II.

Compressibility of Bessemer Rod. Same Material as Mercury Piezometer.

Pressure kgm./cm.	$\frac{\Delta l}{l_0}$.		
	Observed.	Calculated.	Difference.
994	0.000195	0.000196	+ 1
1190	0.000228	0.000230	+ 2
1488	0.000281	0.000281	..
1770	0.000332	0.000330	− 2
2174	0.000374*	0.000399	+25
2540	0.000452*	0.000462	+10
2980	0.000565*	0.000538	−27
3176	0.000570	0.000572	+ 2
3400	0.000615	0.000611	− 4
3670	0.000652	0.000657	+ 5
4040	0.000724	0.000721	− 3
4176	0.000769*	0.000744	−25
4760	0.000830	0.000845	+ 0
5294	0.000938	0.000937	− 1
5506	0.000977	0.000973	− 4
5730	0.001013	0.001012	− 1
6060	0.001072	0.001068	− 4
6430	0.001127	0.001133	+ 6

$\frac{\Delta l}{l_0} = a + bp.$ $a = 0.0000249.$ $b = 0.0000001722.$

* Discarded in the calculation.

the aluminum, although, because of the smaller size of the total effect, one would expect greater percentage variation from the particles of grit. Probably the improvement is due to the increased familiarity with the method, which seems capable of giving accurate results. A straight line through the observations, discarding the four worst, was computed by least squares, giving as the linear compressibility 1.722×10^{-7}, and the cubic compressibility 5.166×10^{-7} kgm. per sq. cm. Table II shows the differences between the observed and the computed values. The four starred points are the ones discarded in the computation. The probable error of a single observation, excepting the four discarded ones, is 2.3, less than $\frac{1}{4}$ per cent at the higher pressures. The probable error of the compressibility is $\frac{1}{10}$ per cent, which therefore does not vary more than this from constancy throughout the pressure range. No set was observed in this piece of steel on the first application of pressure, which is perhaps evidence of the freedom from internal strain, and to a less degree evidence for equal compressibility in all directions.

An attempt was made to get some light on the possible magnitude of differences of compressibility in different directions by the following method: Two strips were cut from a very homogeneous piece of $\frac{5}{8}$ in. (1.59 cm.) Bessemer boiler plate, respectively along and perpendicular to the direction of rolling; these were turned down to $\frac{3}{8}$ in. (0.95 cm.) like the other test pieces of steel or aluminum, and the compressibility of each determined. The results are given in Tables III and IV. The compressibility of each was calculated by least squares, discarding only one observation from each set. The probable error of a single observation is approximately the same in either set, $\frac{7}{10}$ per cent at the higher pressures. The probable error of the compressibility in either case is about $\frac{2}{10}$ per cent. The compressibility of the lengthwise piece was 5.298×10^{-7}, and of the transverse 5.303×10^{-7}, agreeing within the limits of error. No claim is made that this settles the question of the equal compressibility of metals in all directions. Doubtless with metals of different character there are internal strains left from working that would produce such a difference.

There are only a few other direct determinations of the compressibility of steel. Amagat[2] measured the change of length by an electric contact device, but does not publish his data. He states that the results agree with a determination by an indirect method involving the theory of elasticity and gives 6.8×10^{-7} as the best value. Richards,[3]

---

[2] Amagat, C. R., **108**, 1199 (1888).
[3] Richards, loc. cit., p. 50.

COMPRESSIBILITY OF BESSEMER BOILER PLATE.

TABLE III. Longitudinal.

Pressure kgm. cm.²	$\frac{\Delta l}{l_0}$ Obs.	$\frac{\Delta l}{l_0}$ Calc.	Diff.
794	0.000120	0.000120	0
984	0.000156	0.000154	− 2
1150	0.000176	0.000186	+10
1396	0.000233	0.000227	− 6
1660	0.000271	0.000273	+ 2
2016	0.000314	0.000336	+22
2228	0.000362	0.000373	+11
2480	0.000431	0.000418	−13
2834	0.000481	0.000480	− 1
3040	0.000500	0.000517	+17
3272	0.000591	0.000558	−33
3540	[0.000663]	0.000595	−68
3646	0.000625	0.000624	− 1
3920	0.000672	0.000672	0
4398	0.000748	0.000757	+49
4400	0.000761	0.000757	− 4
4740	0.000835	0.000817	−18
4920	0.000847	0.000849	+ 2
5340	0.000954	0.000923	−31
5440	0.000938	0.000941	+ 3
5690	0.000988	0.000985	− 3
6164	0.001061	0.001069	+ 8
6430	0.001099	0.001116	+17

$\frac{\Delta l}{l_0} = a + bp.$
$a = -0.000020.$
$b = \log^{-1} 3.2470 - 10.$
Cubic compressibility = $0.0_5 5298.$

TABLE IV. Transverse.

Pressure kgm. cm.²	$\frac{\Delta l}{l_0}$ Obs.	$\frac{\Delta l}{l_0}$ Calc.	Diff.
1000	0.000174	0.000173	− 1
1190	0.000197	0.000206	+ 9
1222	0.000215	0.000212	− 3
1446	[0.000192]	0.000252	+60
1680	0.000280	0.000293	+13
2014	0.000345	0.000353	+ 8
2180	0.000384	0.000381	− 3
2526	0.000439	0.000442	+ 3
2816	0.000518	0.000494	−24
3060	0.000537	0.000537	0
3346	0.000593	0.000586	− 7
3660	0.000635	0.000643	+ 8
3980	0.000703	0.000699	− 4
4186	0.000729	0.000736	+ 7
4472	0.000789	0.000786	− 3
4988	0.000906	0.000877	−29
5294	0.000929	0.000932	+ 3
5456	0.000966	0.000960	− 6
5668	0.000994	0.000998	+ 4
6034	0.001044	0.001063	+19
6210	0.001099	0.001093	− 6
6400	0.001099	0.001128	+29

$\frac{\Delta l}{l_0} = a + bp.$
$a = -0.000004.$
$b = \log^{-1} 3.2474 - 10.$
Cubic compressibility = $0.0_6 5303.$

also observing the change of length by an electrical contact device, finds 3.9×10^{-7}. The iron used by Richards was commercial wrought iron, chemical analysis of which is not given. The mild Bessemer steel used in this investigation is usually as free from carbon as wrought iron, and is very much more likely to be homogeneous. The absence of set is evidence of the closeness of texture, while Richards states that the wrought iron used by him was porous and had to be hammered to give satisfactory results. This possibly may account for some of the difference in the results.

To get some idea of the effect of varying percentage of carbon, the compressibility of a piece of high carbon (1.25 per cent) annealed tool steel was determined with the same probable error as in the other determinations, and was found to be 0.000000525. The discrepancies between Richards' values and the values found in this paper can hardly be explained by impurities of this nature.

It is to be noted that neither the steel nor the aluminum shows any tendency to become decreasingly compressible at higher pressures, in analogy with the behavior of more compressible substances, particularly liquids. In fact, as will be seen from an inspection of either the curves or the table, the aluminum shows a distinct though slight tendency to become more compressible at higher pressures. However, it did not seem that this single example would justify the conclusion that this paradoxical behavior was due to anything except errors of observation, and accordingly the coefficient was calculated by least squares on the assumption that it was constant.

The second modification of the above method for measuring linear compressibility consists in comparing the change of length of a tube of the substance in question with the simultaneous change of length of a piece of steel, both the substance and the steel suffering uniform contraction by the hydrostatic pressure over the whole exterior surface. From the relative change of length the absolute linear compressibility may be found if the linear compressibility of the comparison piece of steel is known. This latter may be found by the first method given above.

The apparatus with which the relative change of length of the tube (in this case of glass) and the steel were determined is shown in Figure 6. The glass tube C was kept pressed against the bottom B of the cylindrical hole in the steel cylinder A, by the spring at G, through the medium of the tie rod H and the nuts E and F. A split brass ring D slides on the glass tube easily, but tightly enough to remain securely in position under moderate jarring. Fine scratches were made on the steel at I and the flange of the brass ring. The whole combination was placed in the pressure chamber, and subjected to hydrostatic pressure all over. Both glass and steel shrink, the glass shrinking the more, and hence the ring D is pushed up on the tube. When pressure is released, D comes back with the tube, and the increased distance between the scratches, measured with two microscopes, gives the relative change of length for the highest pressure reached. The glass tube was taken out of the steel jacket and everything washed carefully after each application of pressure, in order to insure freedom from small particles of grit. It is an advantage of this method over the first, that because of the greater accessibility of the parts, complete freedom from grit is secured by washing after each application of pressure. Repeated measurements of the zero position of the ring gave results agreeing within 0.001 mm., which in this case was about the magnitude of possible errors of reading. The total displacement at 6500 kgm. was about 0.35 mm. in the form above.

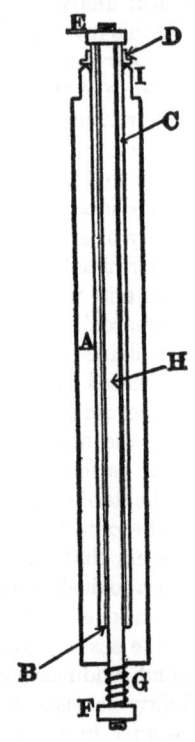

FIGURE 6. Apparatus for comparing the linear compressibility of glass and steel. The glass tube C is compared with the enveloping steel tube A. The relative change of length is measured by measuring the displacement of the ring at D, sliding on the glass tube. The glass tube is kept in contact with the shoulder B by the spring G, acting on the nut F through the tie rod H, which in turn presses on the glass tube by the nut E.

Among possible sources of error we have here again a maximum uncertainty in the effective length of the glass tube of $\frac{1}{2}$ the width of the ring D. In the form used the total length was about 8 cm., and the width of the ring 2 mm. The results may, therefore, be in error by $\frac{1}{80}$, but probably by less than this. This source of error may obviously be decreased at pleasure by increasing the length of the tube. Another possible source of error is temperature change. Error from heat of compression was avoided by operating slowly, applying pressure nearly to the maximum, waiting for the equalization of the temperature, and then applying the last few hundred kilograms. Differences of temperature at different times of measuring the displacement did not prove great enough to introduce perceptible error, since the difference of dilation between the glass and the steel is small. To secure good results, it was found necessary that the glass tube fit closely in the steel cylinder without play sidewise. As it was found difficult to draw a tube accurately enough, this desired freedom from play was secured by wrapping tin foil at either end.

Measurements were made in this way of the change of length of two kinds of Jena glass: a hard combustion tubing No. 3883, and a very fusible glass No. 3880 a. The results at first were discouragingly irregular. After repeated trials, however, they settled down into a fairly regular final form. It became evident on trial with different pieces of glass that there is here, directly observed, the same seasoning effect of successive applications of pressure that was noticed in measurements of electrical resistance. The final behavior never became entirely regular, however. The general effect of frequent applications was to slightly increase, in a totally irregular fashion, however, the observed change in length. In Figure 7 the observed changes of length are plotted against pressures. The irregularity of the results is noticeable, particularly for the hard glass; it approaches, or may sometimes exceed, 5 per cent of the total effect to be expected. The results with the soft Jena glass were only one third as irregular. The explanation suggests itself that the less regularity of the results with the hard infusible glass is because of the greater internal strains set up in this by the long temperature range through which it cools after passing plasticity.

In order to find whether there is any appreciable change in the linear compressibility of glass when it is drawn down from larger sizes, the above form of apparatus was modified by placing the comparison piece of steel inside, instead of outside, the glass tube. The tubes tested were 1 cm. in diameter, which is the original size from which the test pieces mentioned above were drawn down to 0.5 cm. Within the limits of error, no variation of compressibility with absolute size could be detected.

The linear compressibility of the steel against which the glass was compared was determined indirectly by finding the relative change of length in the same manner as for the glass, of the steel and a piece of aluminum cut from the rod whose absolute compressibility was determined above. These readings of the relative change of steel and aluminum are shown in the lower line in Figure 7. The points, except one, lie on a straight line within errors of reading. The one discordant point represents a discrepancy of only 0.0003 mm., and no importance is attached to it. The regularity of these measurements of the aluminum, made with the same apparatus as the measurements of the glass, furnishes additional presumptive evidence, therefore, that the irregularity of the latter is not due to errors of measurement, but is an actual property of the glass. The lower line in Figure 7 was computed by least squares, giving the relative compressibility of the aluminum and the steel. From this and the known absolute compressibility of the aluminum, the cubic compressibility of the steel was

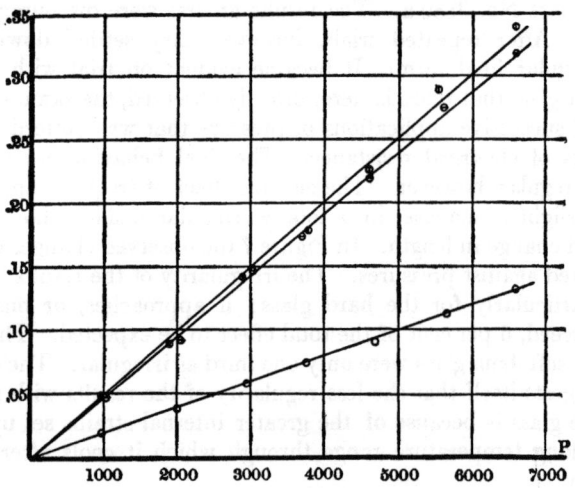

FIGURE 7. Observed relative change of length of steel, and glass or aluminum. The ordinates give the change of length in millimeters, the total length being about 8 cm. ⊕ shows hard Jena glass; ⊖ shows soft Jena glass, and ○ the aluminum.

found to be 4.74×10^{-7}, a value somewhat lower than the values found directly for the other specimens of steel. Similarly, the other lines of Figure 7, connecting relative change of length of glass and steel with the pressure, were computed by least squares. The irregularity of the results is too great to warrant the assumption of any other than a linear relation, although the hard Jena glass in particular shows a tendency toward the paradoxical behavior of higher compressibility at higher pressures already remarked in the aluminum. From these constants calculated by least squares, and the compressibility of the comparison piece of steel determined as above, the compressibility of the glass was found to be:

for Jena glass No. 3880 a 2.17×10^{-6} kgm. per sq. cm.

for Jena glass No. 3883 2.23×10^{-6} kgm. per sq. cm.

The hard glass, contrary to what one might expect, is therefore the more compressible, a result that has already received confirmation by measurements of electrical resistance.

Beside these determinations of the compressibility of glass, it was also necessary to find the compressibility of mercury, in order to find the pressure coefficient of the molecular conductivity of mercury. None of the data at hand reach over a sufficient pressure range for the purpose of this paper, and the data had, therefore, to be extended up to 6800 kgm. The correction introduced by the compressibility of mercury is only 10 per cent of the total change of resistance, so that a highly accurate value of the compressibility was not necessary. The interest of this determination lay rather in finding whether there is any marked decrease of compressibility over the pressure range used. To make this determination, a method was adopted which gives promise of being a considerably better means of determining compressibility even at comparatively low pressures than those methods at present in common use.

The compressibility of mercury at low pressures has been the subject of a number of investigations, and the results which have been obtained recently have been fairly concordant. It is a common feature of all earlier determinations that the mercury has been enclosed in a glass piezometer, the correction for the compressibility of which is 60 per cent of the total effect. The correction for the glass is unusually large in this case because of the comparatively small compressibility of the mercury. For many liquids, the correction for the piezometer is considerably less (6 per cent for water, for example), and the objections urged in the following have proportionally less weight. This correction may be determined in various ways, depending in general on the theory of elasticity, which makes, among others, the assumption of the uniform compressibility of the glass in all directions. Too often, however, the compressibility of the glass has been merely assumed from the work of other investigators on a glass presumably of the same general character as the glass used in the experiments. The correction for compressibility determined by elastic experiments on the same or other pieces of glass seems doubtful in view of the large correction involved. Thus if the behavior of the glass were as irregular as that observed in the case of the hard Jena glass, discrepancies in the compressibility determined with the same piezometer of as much as 3 per cent are to be expected, at least over any considerable pressure range. Doubtless this uncertain correction for the envelope is the cause of the discordant results previously obtained.

The work of Amagat and de Metz along this line seems the most credible. Each gives the mean of the results with several piezometers, where others have used only one. The results of de Metz [4] with four piezometers vary as much as 5 per cent, while those of Amagat [5] with seven piezometers vary 2 per cent. The value of Amagat at 20° is 0.00000380 kgm./cm.^{2} while that of de Metz is 0.00000379 kgm./cm^2. Lately Richards [6] obtained the value 0.00000371, working with a glass piezometer by an electric contact device, but in such a fashion as to eliminate the necessity for calculating the compressibility of the glass, this step being replaced by a calculation from the observed linear compressibility of steel, in which large percentage errors are of much less importance. The values above are for a small pressure range: de Metz and Amagat 50 kgm., and Richards 500 kgm. The results all agree within a unit in the last place, when correction is made for the difference in pressure range.

The only work over a wider range seems to have been done by Carnazzi,[7] who worked between 0 and 200° and went up to 3000 atmos. He used a glass piezometer, assuming Amagat's mean value for the compressibility of the glass, and a manometer depending in a way not entirely free from objection on the compressibility of water as determined by Amagat. Only two significant figures are given in the results, compressibility at 23° being 0.0000038 from 0 to 500 atmos., and 0.0000034 from 2500 to 3000 atmos. These results must be decreased about 3 per cent, becoming 0.0000037 and 0.0000033 respectively, to reduce from atmospheres to kilograms.

In the present determination, a steel instead of a glass envelope was used. The advantages of a steel over a glass piezometer are manifold. The correction for the compressibility of the steel is only 15 per cent of the total effect against 60 per cent when glass is used. Again, the steel is very much more regular in its elastic behavior than the glass; this is obvious at once from an inspection of the curves showing the compressibility of the glass and of the steel. It has been already stated that the irregular behavior of the glass might introduce

---

[4] de Metz, Wied. Ann., **47**, 706 (1892).
[5] Amagat, C. R., **106**, 228 (1888).
[6] Richards, loc. cit., p. 51.
[7] Carnazzi, Nouv. Cim., **5**, 180 (1903).

discrepancies of 3 per cent. Finally, the correction for the glass must be determined from the theory of elasticity, assuming uniform compressibility in all directions. The difficulty of obtaining glass free from internal strain makes the validity of this assumption at least doubtful. Many anomalous results may be explained by this effect. Thus in one case [8] an actual increase of the internal volume of the piezometer under hydrostatic pressure has been recorded. On the other hand, the great homogeneity of steel makes its uniform compression apriori more probable, and here the probability has been greatly increased by an experimental proof of the uniformity of strain in a piece of rolled steel plate, of the same grade of steel as that used in the mercury piezometer.

The method is essentially a revival of one used by Perkins [9] in 1825. Possibly the bad results obtained by Perkins, which were 250 per cent too large, accounts for the subsequent neglect of the method. Several slight modifications were suggested by Professor Sabine, however, so that it has been possible to obtain very satisfactory results. The method consists essentially in observing the extent to which a freely moving piston is pushed into a cylinder containing the liquid to be examined, by the application of hydrostatic pressure all over the exterior of the piston and cylinder. The arrangement used is shown in Figure 8.

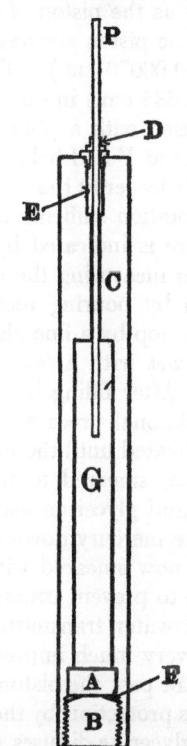

FIGURE 8. Piezometer for determining the compressibility of mercury. C, containing cylinder of steel; G, mercury; P, easily moving piston; D, movable brass ring by which the displacement of the piston is measured. The packing of molasses and glycerine is placed at E. The piezometer is closed at the lower end by the steel plug A, held in place by the screw B. The crack at F is filled with solder.

The containing cylinder C is of

---

[8] M. Schumann, Wied. Ann., **31**, 22 (1887).
[9] Perkins, Phil. Trans. Royal Society, London, p. 324 (1819–1820).

Bessemer steel ½ in. (1.27 cm.) diameter and 3½ in. (8.89 cm.) long. The piston P is $\frac{1}{16}$ in. (0.16 cm.) in diameter, made in exactly the same way as the piston of the absolute gauge described in a previous paper. The piston accurately fits the hole within 0.0002 or 0.0003 in. (0.00051–0.00076 cm.). The cavity G, which is filled with mercury, is ¼ in. (0.635 cm.) in diameter and 2 in. (5.08 cm.) long. The lower end is closed with a plug of steel driven into place and soldered on the outside at F and held additionally by the screw B. The piston P is a slightly looser fit than that used in the absolute gauges, a few ounces without rotation sufficing to displace it. The displacement produced by pressure is indicated by the use of a sliding brass ring at D, exactly as in measuring the change of length of rods. The piezometer was filled by pouring recently distilled mercury through the small hole at the top by a fine glass capillary. The inside of the piezometer was first wet with a few drops of water to insure filling of all the crevices. After filling in this way it was placed under an air pump as an additional precaution against the inclusion of air. The whole was now heated until the mercury rose from the top of the piston hole. The piston, smeared to insure tightness with the same mixture of molasses and glycerine used in the absolute gauge, was inserted and follows the mercury down as it cools. The inside of the enlargement at E was now smeared with molasses, and mercury was poured over the whole to prevent contact of the molasses and the mixture of glycerine and water transmitting the pressure. This packing of viscous molasses very much improved the behavior of the piezometer, reducing the leak past the piston to a minimum. If, however, this packing is used, its protection by the mercury is absolutely necessary, for otherwise the glycerine diffuses through the molasses on each application of pressure, rapidly changing the amount of liquid inside the cylinder.

The method of making the readings was to place the cylinder in the pressure chamber and subject it to hydrostatic pressure all over. By means of the freely moving piston this pressure is transmitted immediately to the interior of the cylinder, the amount of motion of the piston, and so the apparent loss of volume, being indicated by the displacement of the ring D, which is measured after pressure is released and the cylinder removed again from the pressure chamber. This displacement, together with the cross section of the piston and volume of the mercury, gives, therefore, the difference of compressibility between the mercury and the steel of the envelope. The volume of the mercury was obtained by weighing, and the diameter of the piston was measured with a Brown and Sharpe micrometer, the error here not being more than 0.00005 in. on a total of 0.062 in., introducing a possible error of $\frac{1}{800}$ in the area. The determination of the compressibility of the steel, which must be made independently, takes the place of the determination of the compressibility of the glass in previous work.

A variation of temperature of one degree is equivalent in displacement of the piston to about 50 kgm. The pressure chamber in which the cylinder was placed was inserted in a water bath as nearly as possible at room temperature, and the small variations of this temperature were read to 0.01° after every determination. The temperature at the time of measuring the displacement, which was done with a reading microscope, was also recorded and corrections applied for variations. The observations were carried out at temperatures varying only slightly from 20°, and the final results are for this temperature. The error from temperature variations, which were hardly as much as 0.1°, becomes entirely negligible at the higher pressures, in which the principal interest of this work lay. For accurate work at lower pressures it would, of course, be necessary to take more elaborate temperature precautions.

Another correction necessary to apply is a correction on the measured diameter of the piston, because the piston in advancing into the inner cavity draws with it some of the molasses in the crack between piston and cylinder. The effect of this is to increase the effective diameter of the piston. The question has already been discussed in connection with the absolute gauge and a method given for determining the correction, which, however, is not applicable here. In this case the correction was determined by first smearing the hole in the cylinder with a heavy oil, inserting the piston, and then withdrawing it again. A film of oil adheres to the piston equal approximately to one half of the volume of the oil originally in the crack between piston and cylinder. The quantity of oil thus clinging to the piston was determined by weighing, and the crack in this manner found to be 0.0003 in. (0.00076 cm.) wide. The method of course is very inaccurate, but seemed the only practical way of getting any idea of this small quantity. The total correction thus introduced is only 1 per cent, so that fairly large errors in the correction are unimportant.

It seemed necessary to investigate one other source of possible error before confidence could be placed in the results. There has been expressed a feeling that metals might be porous under high pressures, the experience of Amagat in forcing mercury through 8 cm. of cast steel being adduced as evidence on this point. To test this, a piece of steel from the same piece as the piezometer was weighed before and after subjection to pressure, in an endeavor to detect possible increase of weight from the absorbed liquid. No change of weight of more than one part in 400,000 could be detected. On a previous occasion a piece of drawn copper had been found to suffer no increase of weight of one part in 1,800,000. It may be confidently expected, therefore, that ordinary commercial bar metal shows no considerable porosity. Amagat's result was probably due to flaws in the casting.

In Figure 9 the observed proportional changes of volume of mercury measured from an arbitrary zero, as in the case of the determination

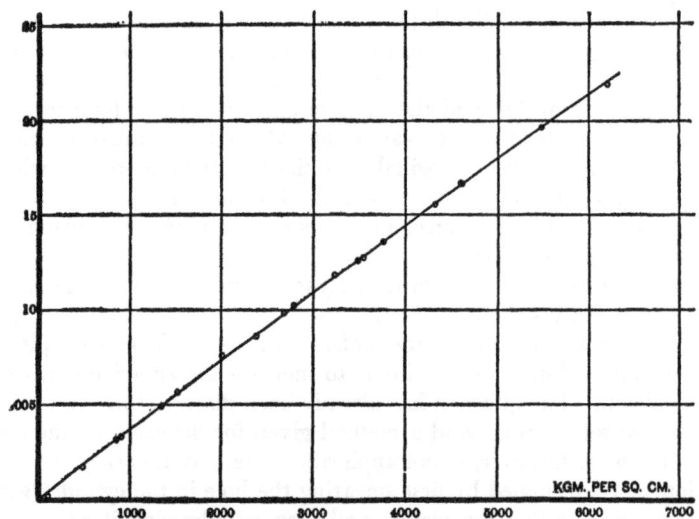

FIGURE 9. The proportional change of volume of mercury, as determined with the piezometer of Figure 8, plotted against pressure.

of the compressibility of rods, are plotted against pressure, measured in the usual way with a mercury resistance. The maximum ordinate corresponds to a displacement of the piston of 1.5 cm. Results obtained with a preliminary piezometer, not so well made as the final one, agree with the curve given within the somewhat larger limits of experimental error. The principal source of error seems to be the inclusion of minute air bubbles. Measurement from an arbitrary zero, determined by backward extrapolation as above, removes this as a consistent source of error, but the measurement of the actual displacement becomes irregular from the lack of certainty with which the piston is returned after release of pressure to the initial position by the comparatively feeble expansive action of the bubble of air. All the precautions described above to remove this bubble appear necessary.

The compressibility of the steel envelope has already been determined, and hence the proportional change of volume of the mercury can be corrected and the true compressibility found. It was assumed that

$$\frac{\Delta V}{V_0} = a + bp + cp^2,$$

and the constants were calculated by least squares. The results are shown in Table V. The constant a has the same significance as in the case of the steel and aluminum rods, the constants b and c alone having significance for the mercury itself. The values found were

$$a = 0.001252,$$
$$b = 3.699 \times 10^{-6},$$
$$c = -1.985 \times 10^{-11}.$$

The compressibility at low pressures is b, 3.70×10^{-6} compared with 3.80×10^{-6}, found by Amagat, de Metz, and Richards, and 3.7×10^{-6} found by Carnazzi. It is to be remarked, however, that the purpose of this investigation was not to find the compressibility at low pressures, only two observations being made at less than 800 kgm. Both the dimensions of the piezometer and the temperature changes make the low pressure values of this determination doubtful. There is, moreover, obvious on inspection of the table a tendency for the low pressure values to lie below the values given by the formula. This would increase the initial compressibility. The experimental error is sufficient, however, to make illusory a more accurate determination of the initial b by passing a curve of the above type through the lower values only. The probable error of a single observation, discarding the first, is $\frac{1}{3}$ per cent at the highest pressure. The probable percentage error of values determined by the formula is 0.25 per cent, discarding the lowest value, or 0.18 per cent, discarding the lowest two.

The departure of the compressibility from constancy is shown by the constant c, which is very small, in fact much smaller than has been found by either Carnazzi or Richards. It may be found from the above formula that the instantaneous compressibility at 2700 kgm. has decreased to 3.58×10^{-6} from its initial value of 3.70×10^{-6}. Carnazzi finds the average compressibility between 2500 and 3000 to be 3.3×10^{-6} against 3.7×10^{-6} between 0 and 500. Richards finds a decrease of compressibility from 3.80×10^{-6} to 3.64×10^{-6} over a pressure range of 500 kgm. However, Richards himself recognized the possibility that his pressure unit might be in error at the higher

TABLE V.

Compressibility of Mercury.

Pressure kgm./cm.^{2}	$\frac{\Delta V}{V_0}$.		
	Observed.	Calculated.	Difference.
116	0.000140	0.000168	+28
496	0.000297	0.000308	+11
850	0.000440	0.000439	− 1
916	0.000458	0.000462	+ 4
1346	0.000619	0.000619	0
1536	0.000691	0.000689	− 2
2050	0.000892	0.000875	−17
2380	0.000990	0.000994	+ 4
2690	0.001117	0.001106	−11
2792	0.001157	0.001142	−15
3224	0.001314	0.001297	−17
3492	0.001393	0.001393	0
3550	0.001408	0.001413	+ 5
3760	0.001497	0.001487	−10
4320	0.001679	0.001486	+.7
4600	0.001796	0.001784	−12
4610	0.001788	0.001787	− 1
5490	0.002097	0.002096	− 1
6216	0.002329	0.002347	+18

$\frac{\Delta V}{V_0} = a + bp + cp^2.$ $\qquad b = \log^{-1} 4.5681 - 10.$

$a = 0.0012523.$ $\qquad -c = \log^{-1} 9.2977 - 20.$

pressures. He finds, e. g., for the compressibility of water at 200 and 400 kgm. 42.5 and 39.6 respectively, against 42.4 and 40.6 as found by Amagat. This points, therefore, to an error in Richards' standard in the right direction, and of approximately the right magnitude to bring his result into agreement with the above. It may also be remarked in this connection that the quantity c is essentially a difference of the second order, and that consequently any increase of the pressure range will give a more than proportionate increase in the probable accuracy of c, other things being equal.

The form of steel piezometer described above may be applied with a few obvious modifications to the determination of the compressibility of other liquids than mercury, and even of liquids that attack the steel. In fact, it seems probable that some such form will prove most useful for high pressure work in general, because the forms of glass piezometer in common use become impracticable at high pressures by the cracking of the glass around any pieces of sealed-in platinum, or even by the cracking of the glass alone, when blown into at all complicated shapes.

Conclusion.

In this paper there have been presented methods applicable over a wide pressure range for finding the compressibility of solids in the form of rods or tubes, and also of liquids. These methods have been applied to the determination of a few compressibilities which were needed for another purpose. The pressure range employed was 6500 kgm. The compressibilities found were as follows: two pieces of Jena glass

No. 3880 a, 2.17×10^{-6} kgm. per sq. cm.

No. 3883, 2.23×10^{-6} kgm. per sq. cm.

Four pieces of steel: two of Bessemer boiler plate, one of Bessemer rod, and one of tool steel, respectively,

5.298×10^{-7}, 5.303×10^{-7}, 5.16×10^{-7}, and 5.25×10^{-7}.

Another piece of Bessemer by an indirect method, not so accurate, gave 4.7×10^{-7}. Compressibility of commercial aluminum rod, 11.7×10^{-7}. The change of volume of mercury is connected with pressure by the relation

$$\frac{\Delta V}{V_0} = bp + cp^2$$
$$b = \log^{-1}(4.5681 - 10);$$
$$-c = \log^{-1}(9.2977 - 20).$$

CONTRIBUTIONS FROM THE JEFFERSON PHYSICAL
LABORATORY, HARVARD UNIVERSITY.

THE ACTION OF MERCURY ON STEEL AT HIGH PRESSURES.

By P. W. Bridgman.

Presented by John Trowbridge, November 9, 1910. Received October 28, 1910.

UNDER ordinary circumstances steel and mercury are inert with respect to each other, as is shown by the possibility of carrying mercury for indefinite periods of time in steel flasks. But there seems to be a widely spread notion that under higher pressures there may be some action not operative at lower pressures. This possibility is usually ascribed to the extraordinary mobility of the mercury molecule. For instance, every one who has had experience in making joints for pressures of a few atmospheres knows that mercury will easily find its way through holes impervious to water or less viscous fluids. It has therefore been thought probable that under higher pressures the easily moving mercury molecule might be forced through the very pores of the solid metal itself, and that in consequence it might be impossible to hold mercury at all in metal receptacles at high pressures. This view has received its highest confirmation from some often cited experiments of Amagat. Amagat[1] has described how in one case mercury was forced by a pressure of 3000 atmospheres in a fine spray through 8 cm. of cast steel, in which no flaw could be afterward detected with the microscope. Amagat explained this effect in the way suggested above by assuming that the mercury was forced by the high pressure through the very intermolecular pores of the solid steel. It is worthy of notice that it was found possible to avoid this difficulty merely by making another apparatus with thicker steel parts. There is also work by Cailletet and Collardeau[2] on the vapor pressure of mercury at high temperatures that seems to demand in explanation

---

[1] Amagat, Ann. de Chim. et Phys. (6), **29**, 87–88 (1893); also Compt. Rend. March 2, 1885.

[2] Cailletet, Colardeau and Riviere, Compt. Rend. **130**, 1585–1591 (1900).

that the mercury was forced through solid steel by comparatively low pressures at sufficiently high temperatures.

In measurements undertaken by the author of various physical constants at high pressures, this question of the action of mercury and steel became of vital importance. For instance, the methods adopted to measure compressibility assumed that there was no penetration of the mercury into the steel containing vessel, as do also the methods used more recently in determining the variation with pressure of the freezing temperature of mercury and its change of volume on freezing. The preliminary work at low pressures made it seem probable that at least over the pressure range used by Amagat the effect described by him does not really exist in the grades of steel used by him, and that the observed effect was due more likely to flaws in the steel. At the same time it was found that at higher pressures there is undoubtedly an effect important enough to demand the redesigning of the apparatus for the measurement of the change of volume on freezing. It is the purpose of this paper to describe the various experiments made to prove the undoubted existence of the effect, and to offer a qualitative explanation. The effect was run across only incidentally, and it was examined only so much as was necessary for the work in hand. No endeavor has been made to make the experimental investigation or the explanation complete, as this would lead too far afield.

The effect was first found during an attempt to measure the change of volume of mercury on freezing by a method similar in many respects to that of Tammann.[3] It was found that cylinders of hardened chrome nickel steel would support very much less internal pressure when this pressure was transmitted by mercury than when the transmitting fluid was some other liquid such as water. The pressure might be less in the ratio of three or four to one; thus cylinders which stood without breaking 24000 atmospheres when the pressure was transmitted to the interior by a mixture of water and glycerine broke on the next application of pressure at 5–8000 atmos. if the transmitting fluid were mercury. These few preliminary experiments under varying conditions made the existence of an effect seem probable, but pointed to nothing conclusively. It might well be that there was a flaw running the entire length of the steel bar from which all these pieces were cut, into which the mercury forced its way in consequence of its greater mobility, in preference to the water, or it might be that there was here a fatigue effect, the steel breaking more readily on the second application of pressure with the mercury because of the exceedingly high pressure to which it had been previously exposed by

[3] Tammann, Kristallisieren und Schmelzen (1903), p. 204.

the water. This explanation, however, was opposed by all previous experience with this steel. In any event, the effect of the mercury was entirely different from that found by Amagat, as there was never any tendency for the mercury to squirt through the steel, but there was always sudden rupture, the cylinder cracking down one side in a plane containing the axis. To show conclusively that a cylinder of hardened nickel steel will really not stand so much pressure when the transmitting liquid is mercury as when it is some other liquid such as water, the following experiment was undertaken. A bar of this special steel (Krupp Special Chrome Nickel Steel E. F. 60.0) was cut into twelve pieces each $8\frac{1}{2}''$ long and $2''$ diameter. (See Figure 1.) The pieces were numbered and their orientation in the original bar carefully noted. They were then each pierced with a $\frac{1}{2}''$ hole reamed to size, turned on the outside true with the hole, and hardened by heating to a bright cherry red and quenching in a heavy tempering oil. Every other cylinder (Nos. 1, 3, 5, 7, 9, 11) was filled with mercury and tested by applying pressure to the mercury by means of a piston actuated by a hydraulic press. The test conditions of these different cylinders were varied somewhat by changing the rapidity with which pressure was applied; in other respects the conditions were the same. The other cylinders were tested in a similar way, except that the fluid transmitting the pressure was not mercury; being in four of the six cases water and glycerine, in the others ether and carbon disulphide respectively. The pressure in the test cylinders was determined from the pressure of the fluid actuating the hydraulic ram, multiplying in the ratio of the areas of the two pistons. An unknown error is introduced here by the friction of the packing, but in other experiments with similar cylinders in which the pressure inside the small cylinders was measured directly it was found that the error so introduced was nearly constant and easy to correct for. The correction so found was used in the results to be given. In any event, the correction is less than the irregularities introduced by other causes.

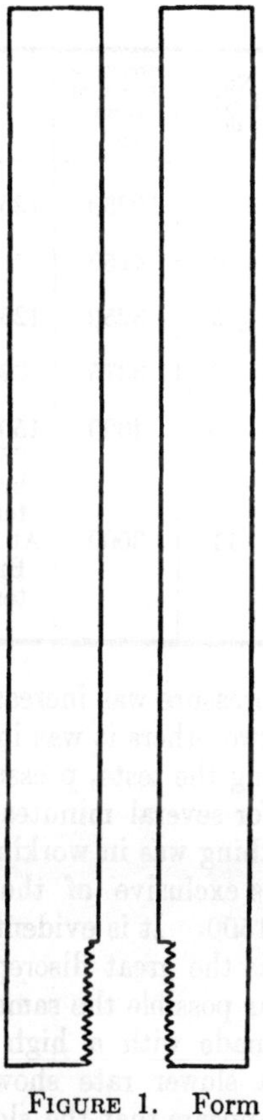

FIGURE 1. Form of the test cylinders broken with mercury.

The accompanying table (see Table I) shows the results found with the six cylinders containing mercury. The pressure was increased more slowly for the higher numbered cylinders; with the four first the

TABLE I.

No. of Cylinder.	Breaking Pressure Kgm $\overline{cm^2}$.	Rate of Increase of Pressure.	Total Duration.	Location of Crack.	Remarks.
1	10250	1250 kgm. in 30 sec.	3½ min.	−123°	Pressure on 1, 3,5, 7, was increased uniformly
3	4750	" " "	1⅓ "	+107°	
5	3250	1250 kgm. in 60 sec.	1⅔ "	+ 64°	Broke on second trial
7	3375	625 kgm. in 60 sec.	4¼ "	+ 11°	Leaked on first at 2750
9	4000	1500 for 3½ min. 2750 for 10 min.; breaks at 4000 after ½ min.	14 "	− 25°	Pressure on 9 and 11 increased discontinuously
11	3000	At 2750 for 46 min. Breaks at 3000 after 5 min.	51 "	−106°	

pressure was increased uniformly at the rate indicated, while with the two others it was increased in discontinuous steps as shown. In making the tests, pressure was first pushed to 1500 atmos. and kept there for several minutes to make sure that there was no leak and everything was in working order. The time given in the duration column is exclusive of the time occupied by the preliminary application of 1500. It is evident that the data admit of no quantitative comparison as the great discrepancy between tests 1 and 3 made under as nearly as possible the same conditions shows. However, these two tests were made with a high rate of increase of pressure. Those made with a slower rate show much more consistent results. In general it appears that the slower the rate of increase of pressure the lower the bursting pressure, which apparently has as its lower limit about 3000 atmos. The data are not inconsistent with the view that there is a critical pressure which will produce rupture if applied for an infinite time; pressures above this produce rupture in constantly less times.

The location of the crack is an important consideration. The crack in each cylinder is in an axial plane, extending the entire length of the cylinder. It was located by measuring the angular distance from a fiducial line marked the length of the bar from which the pieces were cut. The diagram (see Figure 2) showing the location of the crack in the several cylinders makes clear that the rupture does not take place along a flaw in an axial plane extending originally throughout the entire bar. Furthermore, since the manner of fracture of each cylinder demands that the flaw be in an axial plane if the fracture is due to a flaw, and since it seems improbable that a flaw throughout the length of one cylinder should not extend into the neighboring cylinders, the conclusion seems justified that the rupture is not due to a flaw.

Now compare with this the tests for the other set of six cylinders from the same bar. (See Table II.)

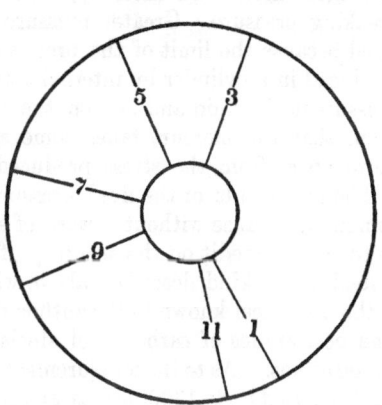

FIGURE 2. Orientation of the crack with respect to the original bar in the six cylinders broken with mercury.

The difference between these two sets of tests is sufficiently striking. Not one of the cylinders filled with a fluid other than mercury was broken during the tests, although in several cases the test was terminated by an irrelevant accident. The

TABLE II.

No. of cylinder.	Maximum Pressure $\frac{Kgm}{cm^2}$.	Manner of applying Pressure.	Remarks.	Fluid.
2	23500	Rate of increase not noted	Packing plug blows out	Glycerine and water
4	19000	From 1500 to 15000 at rate of 1250 in 30 sec. Held at 15000 for 2½ min., 17500 for 3 min.	Packing plug blows out	Glycerine and water
6	15200	From 1500 to 11000 uniformly in 4½ min. Held here for 3 min., then at 13500 for 1 min.	Terminated by leak	Glycerine and water
8	24000	15 min. in reaching max. Held here for 6 min.	Leak after 6 min.	Glycerine and water
10	20250	Held at 16500 for 2 min.; at 20250 for 45 sec.	Plug blows out	Ether
12	24000	1 hour in reaching max. at uniform rate. Held her for 8 min.	No leak	CS_2

maximum pressure reached by three of the cylinders without breaking was 24000 atmos., over 8 times the lowest breaking pressure of the cylinders filled with mercury, and nearly 2½ times the maximum breaking pressure. Greater pressure than 24000 atmos. was not applied because the limit of the press was reached. Of course the stress produced in a cylinder by internal fluid pressure depends only on the pressure in the fluid and not on the nature of the fluid. It must be, then, that the mercury takes some special part in producing rupture quite apart from the stress produced by it.

The magnitude of the fluid pressures mentioned here requires brief comment, because without a word of explanation it may seem so large as to cast discredit on the accuracy of all the data. In the first place, a steel of the kind described above with the indicated heat treatment is the only steel known to the author that will stand the pressures used. The best grades of carbon tool steels will not stand much more than 18000 atmos. As to the measurement of the pressure, similar apparatus has been used up to 13000 atmos. and the pressure measured directly with an absolute gauge with an accuracy of $\frac{1}{10}$ per cent. Absolute measurements have not been made above this, and it is possible that the friction of the packing may become unexpectedly large, although no evidence of this has been found. However, it makes no difference what the friction above 13000 is; the fact stands incontestable, for the breakage produced by mercury is in a region open to easy direct measurement, while the other cylinders have stood a pressure certainly several-fold greater.

All these results so far were obtained with hardened chrome nickel steel. Search was now made for the same effect in other steels. A cylinder of nickel steel of similar dimensions to the above, but left soft, was tried. This was filled with mercury. Pressure was kept at 8000 for 3 hours, then pushed gradually to 15000, where the increasing non-elastic stretch became so great as to let the mercury past the packing. The behavior here was exactly like that of a similar cylinder filled with water and stretched beyond the elastic limit : the elastic limit in the two cases was the same, as also the manner of yield and the shape into which the cylinder was deformed. Apparently, then, mercury exerts no selective action on the soft nickel steel. A similar cylinder of bessemer steel filled with mercury was left exposed to 3500 atmos. for 14 hours, and subsequently the pressure was increased until the mercury blew past the packing, exactly as for similar cylinders filled with water. It should be noted that non-elastic yield occurs for bessemer steel at pressures much lower than 3500, perhaps as low as 2000 atmos. The explanation finally adopted attaches some significance to the value of the elastic limit. The two cylinders above were tested at pressures beyond this limit. To test whether a soft cylinder would be broken by a pressure under the elastic limit if the pressure were sufficiently prolonged, a cylinder of soft nickel steel filled with mercury was exposed for three weeks to 4000 atmos. without rupture. At the end of three weeks, pressure was increased and rupture took place at 12000 atmos. The break was remarkable, as the cylinder showed no preliminary stretch as all soft steels do, but snapped like a piece of hardened tool steel. This may have been due to fatigue from the prolonged application of pressure, or it may be that weakening by the action of mercury had started at the lower pressure and that the cylinder would have finally broken at 4000, had the application been sufficiently prolonged.

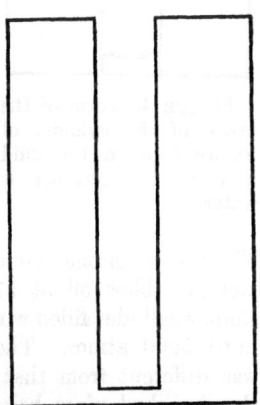

FIGURE 3. Another form of test cylinder for water or mercury, made of hardened nickel steel.

An attempt was made to make a series similar to that on the nickel steel on cylinders cut from a bar of tool steel containing 1.25 per cent carbon. Several of these cylinders cracked in hardening so that a complete set of observations could not be made. The few successful tests made with the hardened tool steel cylinders, however, confirm the conclusions reached with the hardened nickel steel. Thus one cylinder with no hardening flaw burst at 4000 atmos. when filled with mercury, and a similar cylinder, which had a hardening flaw in it, broke at 6500 on the second application of pressure when filled with water.

Another series of tests was now tried which gave the clue to the final explanation. Pressure was applied to cylinders of the form shown (see Figure 3), in which the end opposite the piston is left solid, and may be of varying thickness. In the first tests this bottom was left only $\frac{1}{8}''$ thick. The cylinders were made of the same nickel steel as above, hardened in oil. One, filled with mercury, ruptured by blowing out of the bottom at a pressure of 2400 atmos. The bottom blew out of a second in which the transmitting fluid was water, at a pressure of 12700 atmos. The manner of rupture was very different in the two cases. When water was used, the bottom was blown out in the form of a clean punching, slightly less in diameter than the hole and slightly bulged as one would expect. (See Figure 4.) On the other hand, the piece blown out by the mercury was in the form of a conical cap twice the diameter of the hole, as indicated in Figure 5. The structure of the ruptured metal from A to B was very coarse and granular, remarkably like the structure of brass made rotten by mercury. Between the grains of the steel were minute drops of mercury, and the fractured surface was partially wetted by mercury when plunged beneath a mercury surface. The only clean break was around the edge of the cap (BC), where the rupture took the form of a clean shear as for the piece broken with water.

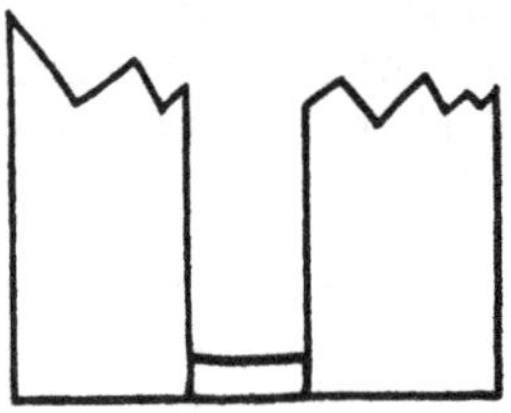

FIGURE 4. Form of the break of the cylinder of Figure 3 when the fluid exerting the pressure is water.

The corresponding tests with tool steel cylinders were hard to carry out because of loss of the cylinders on hardening. However, two successful tests were made. The bottom of one cylinder, when filled with mercury, blew off at 2700 atmos., while a similar cylinder filled with water did not fail until 5200 atmos. The manner of failure was different from that of the nickel steel, the detached piece being in the form of a frustum of a cone, as shown in Figure 6. The form of this cap was the same for both water and mercury, the only difference being that the one fractured with mercury showed unmistakable amalgamation over a limited region (AB). This, together with the curious fracture of the nickel steel piece, suggested the amalgamation of steel by pressure as a possible explanation.

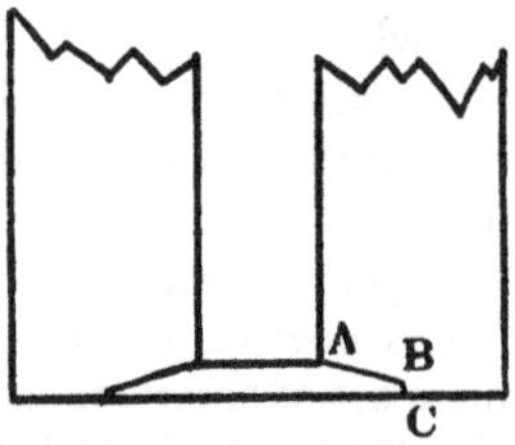

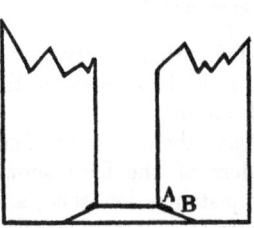

FIGURE 5. Form of the break of the cylinder of Figure 3 when the fluid exerting the pressure is mercury.

FIGURE 6. Form of the break in a cylinder like that of Figure 3 of hardened tool steel when the fluid exerting the pressure is mercury.

Experiment was also made on two cylinders of tool steel similar to the above, except that the bottom was $\frac{1}{2}''$ thick instead of $\frac{1}{8}''$. The cylinder filled with water stretched nonelastically and leaked at a pressure of 8000 atmos., while that filled with mercury failed at 4000 by the blowing out of mercury along a crack. This break was more like that described by Amagat than any other in the course of these experiments, but even here the break was distinctly visible, while Amagat states that he could find no trace of the crack with a microscope. (See Figure 7.)

In all these tests of the soft steel cylinders the punched out pieces were found more or less amalgamated around the sheared edges. In particular, when the cylinder of soft tool steel last mentioned was cut so that the fissure was exposed, it showed beautiful amalgamation over the entire surface of the rupture. This led to a short investigation of the possibility of amalgamating iron or steel. It has been known for some time that mercury is capable of dissolving a small quantity of iron, and that conversely an iron surface may be amalgamated by mercury.[4] This amalgamation is a matter of some difficulty, which may be made to take place by certain chemical or electrical reactions, but under ordinary conditions does not occur at all. It does not seem to have been noticed that under proper conditions the amalgamation of iron is a matter of the greatest ease, the difficulty under ordinary conditions being due apparently to a thin protecting layer of oxide. The following experiment showed strikingly how great the affinity between clean iron and mercury is. A piece of iron was broken underneath a mercury surface so that the freshly ruptured surface came directly in contact with the mercury. The diagram illustrates the form of experiment. (See Figure 8.) The test piece, in the shape of a thin hollow cylinder, is covered with mercury in an iron receptacle, and broken by forcing a wedge into the hole. In every case the broken surface is brightly and completely amalgamated, the mercury wetting it exactly as water wets a surface to which it adheres. The same result was obtained with hard and soft nickel steel, hard and soft tool steel, bessemer steel

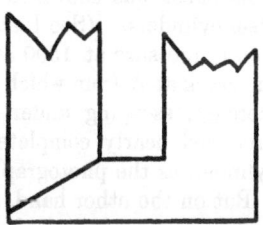

FIGURE 7. Form of the fracture in a cylinder of soft tool steel when the fluid exerting the pressure is mercury.

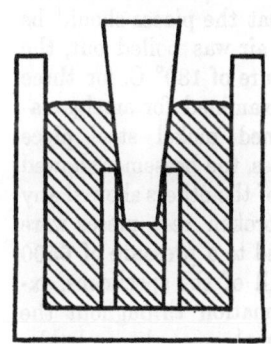

FIGURE 8. Apparatus for fracturing steel under the surface of mercury.

---

[4] See, for example, Richards, Wilson, and Garrod-Thomas, Pub. Carnegie Inst. Wash., No. 118, 54, 1909; also Richards and Garrod-Thomas, ZS. Phys. Chem., **72**, 181 (1910).

both in its natural state and quenched from a red heat, and with cast iron.[5] The same effect is shown by the same grades of steel when broken off by twisting under a mercury surface, showing that the amalgamation does not depend on the nature of the rupturing stress. This effect was shown in a striking way by one of the hardened tool steel cylinders. (See Plate.) This cylinder broke on the first application of pressure at 1500 atmos. Examination showed a small hardening crack at A from which the crack spread into the sound metal. The mercury, escaping under pressure through the freshly opened crack, produced nearly complete amalgamation of the entire section of the cylinder, as the photograph shows.

But on the other hand, all attempts failed to amalgamate the freshly broken surface if it had once come in contact with the air. A test piece like the above was supported directly over a mercury surface so that the fragments fell immediately into the mercury when rupture occurred. No trace of amalgamation was obtained here, as was also no trace with several modified forms of the experiment in which immersion was not so immediate.

Experiments were now made to find whether the amalgamation so produced might be made to work its way through the mass of the metal. Fragments of the above amalgamated test pieces were sealed into a bulb containing mercury so designed that the pieces should be kept below the surface of the mercury. The air was boiled out, the bulb sealed, and the whole kept at a temperature of 180° C. for three hours in an oil bath. The pieces were then examined for amalgamation by breaking them across. The hardened nickel steel piece showed amalgamation throughout its entire mass, the bessemer showed isolated splotches of amalgamation, while none of the others showed any effect. Similar amalgamated bits from the broken test pieces were now submerged in mercury, which was subjected to a pressure of 6500 atmos. at room temperature for six hours. All of the specimens, except the soft tool steel, now showed amalgamation throughout the interior. The hardened nickel steel piece was amalgamated completely, while the amalgamation of the others was not so perfect, being confined to patches near the surface.

Further, pressure by itself is not capable of producing amalgamation

---

[5] Aluminum may also be amalgamated by rupturing under the surface of mercury. Aluminum so treated, when exposed to the air, shows the characteristic tree-like growth of the oxide. The same effect is also shown in nickel and cobalt. It should be remarked that the amalgamated iron surface shows no tendency to oxidize in the air, but keeps its silver luster untarnished at least for months.

on a steel surface or in the mass of the metal unless the amalgamation has been started on the surface by some other means. This was shown by subjecting rods of the three kinds of steel (hard and soft nickel steel, hard and soft tool steel, soft and quenched bessemer) to a pressure of about 6500 atmos. for twelve hours or more under mercury. These rods were scoured bright with fine emery paper immediately before being plunged into the mercury. The fracture after pressure treatment showed not the slightest trace of amalgamation. Reference may be made to some earlier experiments in which no gain of weight could be detected in steel pieces subjected to hydrostatic pressure in mercury. The possibility of amalgamation by pressure was also tried in another form. A hollow cylinder of hardened nickel steel was submerged in mercury, and pressure applied to the outside. The only difference between this and the case of the solid rod is that in the former the stress throughout the metal is not uniform, as it is in the latter. It was thought conceivable that mercury might be forced through metal in which the stress was not the same in every direction, while it might show no tendency to work its way through a mass in which the stress was already hydrostatic. The experiment, however, showed no amalgamation in this case either.

This test for amalgamation by examining the nature of the fracture showed itself so easy to apply and so unmistakable in its indications that it was now applied to the examination of the cylinders which had formed the subject of the first tests. The possibility of the amalgamation of the cylinders as an explanation had at first been discarded because the inner wall of the cylinder, where it was to be most expected, showed no indications of any amalgamation and because attempts to detect the presence of mercury throughout the mass of the metal by microscopic analysis of the polished and etched cross section had given no result. Professor Sauveur had been kind enough to examine four test pieces cut from different cylinders, three of which were ruptured with mercury and one with water. He found martensitic structure in the three former pieces, and only a very fine granular structure in the other. There was no visible trace of mercury in the pores, nor anything to suggest amalgamation. The differences of structure might be due merely to the slight differences in heat treatment occasioned by the separate hardening of the cylinders. More careful regulation of the conditions would be necessary to settle this point.

The fracture test was applied by cutting a scarf about $\frac{1}{4}''$ deep all around the cylinder with a thin emery wheel, and then breaking the cylinder at this scarf with a hammer. All the cylinders broken with the mercury showed the same characteristics. (See Figure 9.) The crack was in a radial plane; surrounding this crack on either side was a band within which the coarser structure and the silver luster showed that the steel had been amalgamated. Besides this band of amalgamation flanking the crack, which was present in every cylinder, there were other irregular splotches of amalgamation growing either from the central hole or from the flanking band. No cases were found of isolated islands of amalgamation in the midst of untouched metal.

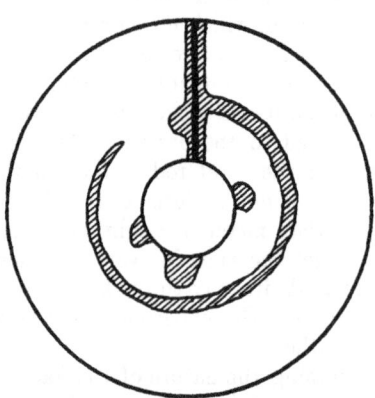

FIGURE 9. The form of the amalgamated region in cylinders like those of Figure 1 when broken with mercury.

The diagram gives an idea of typical forms of this amalgamation, although so many varieties were seldom found in one specimen. In one specimen the amalgamation had grown in the form of a nearly complete ring with no contact with the interior except through the infected region about the crack, in the manner shown in the drawing. The tool steel cylinders showed the same effect, except that the amalgamated band about the crack was not so broad as for the nickel steel, and the other patches of amalgamation were less numerous. The peculiar fracture indicating amalgamation was found in none of the cylinders which were broken in the absence of mercury. Several such cylinders were examined, some of hard or soft tool steel, and some of nickel steel, broken with water after the repeated action of the maximum pressure of 24000 atmos.

Such are the experimental facts which must be explained. It seems evident that the premature breaking of the cylinders filled with mercury was due to the weakening of the steel produced by amalgamation. The fact of amalgamation seems sufficiently proved by the study of the fracture. That amalgamated steel would be weaker than the untouched steel seems obvious enough without the necessity of special experiment to prove it. The fact was proved incidentally several times, however, when parts of the steel packing appliances which had been used with the mercury cylinders were used over again with the cylinders filled with water. The packing plugs in these circumstances always broke at a pressure very much less than the normal breaking pressure and with a fracture showing bright amalgamation. In explanation of the amalgamation of these cylinders we have in the first place the strong natural affinity between iron and mercury. This is prevented from coming into play under ordinary circumstances by a thin layer of dirt on the surface. But it seems reasonable to suppose that amalgamation will start in the mass of the metal if the mercury can be once forced into the pores, since under these circumstances the iron and mercury coming into contact with each other would be clean and the natural chemical affinity come into play. The argument consists in showing that in those cases where amalgamation took place the conditions were such as to favor the introduction of mercury into the pores of the steel, even if the surface were not amalgamated. Then after amalgamation is once started in the mass of the metal it is assisted in the rapidity of its growth from the amalgamated region by the action of hydrostatic pressure.

This discussion demands a slight consideration of the nature of the strain in a hollow cylinder exposed to internal fluid pressure. The stresses in the metal of the cylinder consist of a pressure (negative) across planes perpendicular to the radii, and a tension (positive) across radial planes. These stresses are greatest arithmetically at the interior surface, but the algebraic sum is constant throughout the mass of the cylinder. This has as a consequence that the volume strain in the cylinder is a dilation and is everywhere constant, so that the pores are opened up by the action of the stress and the entrance of mercury facilitated. This holds while the strain remains elastic. But when the internal pressure exceeds a certain value so that at the inner surface the algebraic difference between the radial pressure and the circumferential tension exceeds a critical value depending on the elastic limit, the strain becomes inelastic, the tension changes over to a pressure so that both principle stresses become compressions, the volume strain changes from a dilatation to a compression, and the pores close up. So that with a steel of low elastic limit the type of stress may change, giving a volume compression, at a lower value of the fluid pressure and therefore at a smaller preliminary volume dilatation than in a steel of higher elastic limit. This view as to the nature of the stress in a thick cylinder stressed beyond the elastic limit is supported by many other experiments on the bursting of thick cylinders. An account of these experiments will be published in another paper.

The difference found in the rupture points between soft and hardened steel is to be ascribed to two causes. One is the greater intrinsic ease of driving amalgamation through a mass of hardened steel by hydrostatic pressure. This was proved by the experiments on the broken amalgamated test pieces. It may be due in part to chemical difference between the hard and soft steel, but is almost certainly also due in part to the greater porosity of the hardened steel. It is well known that steel decreases in density on hardening. The other cause is the higher elastic limit of the hardened steel and the consequent wider opening of the pores before non-elastic closing sets in. For facility of comparison the elastic limits of the cylinders of the various grades of steel are given (see Table III), on the usual assumption that the elastic limit is determined by the maximum stretch at the interior. This assumption is probably not very near the truth, so the results can be expected to show only qualitative agreement. This table does not apply to the

TABLE III.

Kind of Steel.	Assumed Elastic Limit in Tension lbs in^2	Internal Pressure required to reach Elastic Limit. Atmos.	Minimum Bursting Pressure with Mercury. Atmos.
Bessemer . . .	40000	2000	Same as with water
Soft tool steel .	75000	4000	Same as with water
Hard tool steel .	150000	8000	4000
Soft nickel steel	90000	4800	Same as with water
Hard nickel steel	225000	12000	3000

cylinders in which the bottoms were blown out, as here the stress is of a different nature and the above values of the elastic limit do not hold. It appears from the table that only those cylinders were broken in which the stress was below the elastic limit. That the cylinders of bessemer and soft nickel steel were not broken may be explained by supposing that the pressure required to open the pores wide enough to force the mercury in is higher than the elastic limit.

It remains to explain the form of the amalgamated region found in the hardened nickel steel cylinders; i. e., bands on either side of the crack. It seems certain that slight inequalities in the structure of the steel will greatly affect the ease of amalgamation. This seems proved by the splotches of amalgamation found throughout the broken test pieces amalgamated by hydrostatic pressure and throughout the broken cylinders. Conceive, then, of a hollow cylinder filled with mercury subjected to pressure. The pressure expands the pores of the metal, and in consequence of the high mobility of the mercury molecule the mercury is forced into these pores through the layer of dirt lining the hole. At certain places where the metal is more susceptible the amalgamation spreads more rapidly than it does at others. But now the metal is weakened at each of these infected places, and the type of strain is modified as it would be by the presence of a flaw. The strain will be redistributed, the brunt of the strain coming at the point farthest removed from the center, where consequently the pores of metal will be still further distended. That is, at this point the amalgamation will proceed most rapidly. It is evident that the continuation of this process will produce a band of amalgamation travelling out along the radius. When the amalgamation reaches the outside, or approaches sufficiently close, the metal gives way, the crack appearing through the midst of the weakest region, that is, through the center of the amalgamated band. It is evident that when the process of amalgamation has once started in this way, it will proceed more and more rapidly as the resisting thickness of sound metal becomes less, thus accounting for the smallness of the other amalgamated regions. One cylinder was found, however, in which an amalgamated patch had worked its way nearly half way to the outside diametrically opposite the crack.

Failure to produce amalgamation in the rods subjected to hydrostatic pressure is to be explained by the fact that neither can the amalgamation begin at the surface, because of the thin layer of dirt, nor can the mercury force its way into the steel to begin amalgamation there because the interstices in the metal are closed up by the hydrostatic pressure. The same argument of course applies still more to the hollow cylinder submerged in mercury and subjected to pressure on the outside.

The experiments with the cylinders in which the bottoms were blown off are to be explained in the same way. The amalgamation grows most rapidly in the direction in which the distension is greatest, which in this case is diagonally from the corner of the hole. As the amalgamation proceeds it carries the hydrostatic pressure with it. When the region over which this pressure acts has extended so far that the sound metal left can no longer support the stress, it gives way as usual by a clean shear. The fact that mercury was forced through the bottom of a cylinder of soft tool steel (Figure 7), while soft tool steel cylinders of the form of Figure 1 were unbroken by the action of pressure, is probably to be explained by the different strain types in the two cases. In the case of Figure 7 distension in a direction diagonally from the corner of the hole is great enough to allow amalgamation before the pores are closed up by viscous yield. The point has not been worked out in greater detail, however. The disconcerting experiments in which a clean punching was blown out of the bottom of a soft cylinder with the sides of the punching amalgamated, are to be explained by the fact previously noted that a freshly fractured surface is amalgamated by mere contact with mercury. Breaking across of the punchings showed no evidence of amalgamation in the mass of the punching, so that all the amalgamation must have taken place after rupture. A similar experiment on the hardened nickel steel punching mentioned above showed more or less complete honeycombing of the metal with mercury.

This is as far as the explanation has been carried. Enough has been done to show that there is here a genuine effect, so that pressure can not be transmitted directly by mercury in hardened steel cylinders, and that the effect is due to amalgamation. One-sided pressure is necessary to start this amalgamation, so that when steel is entirely surrounded by mercury there is no danger of amalgamation or of penetration of the mercury into the pores. This fact was made use of in modifying the design of the apparatus spoken of in the first part of the paper.

SUMMARY.

The fact has been established that cylinders of hardened steel will burst at very much lower than the natural bursting pressure when the fluid exerting the pressure is mercury. Soft steel cylinders show the effect hardly at all, the yield point being reached before the pressure can be raised high enough to produce the effect. The fact that this rupture is due to the amalgamation of the steel is established by the examination of the fracture of such cylinders. The unexpectedly great affinity between steel and mercury was established by the complete amalgamation of surfaces broken under mercury, and the enormous effect of the slightest contact with the air was shown. When this amalgamation is once started, the rapidity with which it spreads through the metal is greatly increased by the action of hydrostatic pressure. The spread of mercury through the mass of the steel and the subsequent destruction of the hollow cylinders is produced by two causes, both of which must act together. One is the natural chemical affinity between mercury and steel, shown by the ready amalgamation of freshly broken surfaces. But the amalgamation is never started by the action of pressure alone. In all those cases in which we have had amalgamation, we have had in addition to the chemical affinity a strain of such a nature as to distend the pores of the metal. This allows the entrance of mercury into the pores so that amalgamation may begin, and also facilitates its further growth, which is most rapid in the direction in which the pores are most distended. It was shown in detail that in all cases in which rupture occurs the strain is of such a type as to distend the metal, and that on the other hand in all those cases in which amalgamation is not produced by pressure, the strain is such as to compress the metal, closing up the pores.

This work was done in the course of an experiment on the thermal properties of mercury and water under high pressure, the expenses of which were partially defrayed by a liberal appropriation from the Rumford Fund of the American Academy of Arts and Sciences.

JEFFERSON PHYSICAL LABORATORY, HARVARD
    UNIVERSITY, CAMBRIDGE, MASS.
        October, 1910.

CONTRIBUTIONS FROM THE JEFFERSON PHYSICAL
LABORATORY, HARVARD UNIVERSITY.

THE MEASUREMENT OF HYDROSTATIC PRESSURES UP TO 20,000 KILOGRAMS PER SQUARE CENTIMETER.

By P. W. Bridgman.

Presented by G. W. Pierce, October 11, 1911. Received October 6, 1911.

In these Proceedings, Vol. XLV, Numbers 8 and 9, 1909, two methods of measuring hydrostatic pressures were described and applied up to 6800 kgm./cm.^{2}. The first method was by means of an absolute gauge of the freely moving piston type; the second method depended ultimately on the first, and utilized the change in the electrical resistance of mercury under pressure. The pressure reached with these two forms of gauge was higher than the highest previous accurately measured pressures, which extended to only 3000 or 4000 kgm./cm.^{2}; but since the earlier paper, the region of attainable and measurable pressures has been still further extended to over 20,000 kgm., so that a re-examination of the gauges there proposed became necessary. Both of the previous methods were found to become inapplicable at pressures much higher than 6800 kgm., the first because of the yielding of the steel of the gauge and the second because of the freezing of the mercury. In this paper the modifications of these two methods are described with which it has been found possible to reach these higher pressures. The freely moving piston gauge has been changed in design so that it has been possible to reach 13,000 kgm., and has been provided with a different reading device. The second method, involving the change in resistance of mercury, has been replaced by another method using the change in resistance of manganin wire. With this an indicated pressure of 20,670 kgm. has been reached. This paper is occupied with a discussion of the calibration, the corrections, and the details of manipulation of these gauges.

The Absolute Gauge.

The novel features of the gauge described in the previous paper which made it possible to more than double the pressure range of

Amagat were the smallness of the piston exposed to pressure (1/16 inch), and the fact that the pressure cylinder in which this piston moved was exposed to pressure on the external as well as on the internal surface, so that the effect of the increasing pressure was to decrease the internal bore of the cylinder and so decrease the leak. The application of pressure to the outside of the cylinder was made through the medium of the packing.[1] The figure referred to shows that the disposition of the packing was such that the cylinder was exposed externally to pressure over the lower end, and to none at all over the upper. At high pressures the effect of such an arrangement is invariably to pinch off the cylinder, the packing forcing its way into the cylinder and separating the upper from the lower end. Two things are necessary to produce this effect: external pressure and non-support of one of the ends. The action is similar to that of a roll of putty which, when squeezed in the hand, will ooze out sidewise if the ends of the roll are left free, but may be prevented by compressing the ends of the roll between the fingers of the other hand.

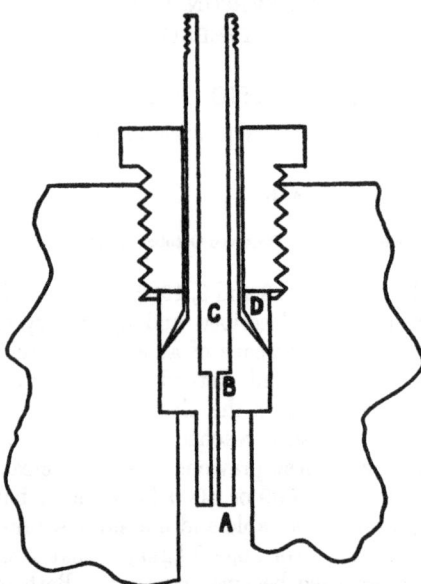

FIGURE 1. The absolute gauge; shows the cylinder of the gauge and the method of mounting it in the containing vessel.

In the new form of gauge the two novel features, smallness of piston and a cylinder exposed to pressure on the outside, have been retained, but the manner of applying the packing has been so changed as to avoid the pinching-off effect. The new gauge is shown in Figure 1. The cylinder AB is placed lower down than in the former gauge, so that now it is in the part of the metal subjected to hydrostatic pressure only. The packing, which now takes the form of a cone of soft steel, D, is placed above the upper end of the cylinder. This new form

---

[1] See Figure 2, p. 205, loc. cit.

gains at three points. The soft steel packing, which takes the place of the rubber, does not transmit pressure hydrostatically and so exerts a smaller pinching-off effect; the cylinder is made of hardened nickel steel instead of tool steel, so that it has a higher yield point and more effectively resists what pinching-off effect there is; and the cylinder itself, which is the only vital part, is placed entirely beyond the reach of this effect. Even in this form, however, the soft steel packing does flow sufficiently to produce the beginning of the effect. After the application of the maximum pressure the bore at C showed a very marked decrease. But since nothing depends on the size of the bore at C, no inaccuracy is introduced.

This form gives up one advantage claimed for the former gauge, namely, that by changing the area over which the packing is distributed it becomes possible to some extent to control the distortion of the cylinder and so the leak. But this question of leak proved of much less importance than was anticipated, it being possible to range from 0 to 13,500 kgm. without inconvenient leak at any point. The leak does, however, as anticipated, become less at higher pressures, partly because of closing up of the crack, so that some sort of control of the leak would become necessary if the crack should ever close up completely. But it has been found possible with the present form of gauge to provide an effective control of leak by providing for the highest pressures a piston slightly smaller than normal. This is evidently easier than to attempt so to design the gauge that the adaptability for different pressures should be secured by changing the packing. The need for even this procedure of changing the piston would probably be slight in practise, for as already stated, one piston sufficed up to 13,500 kgm., and the gauge itself would probably not stand much more.

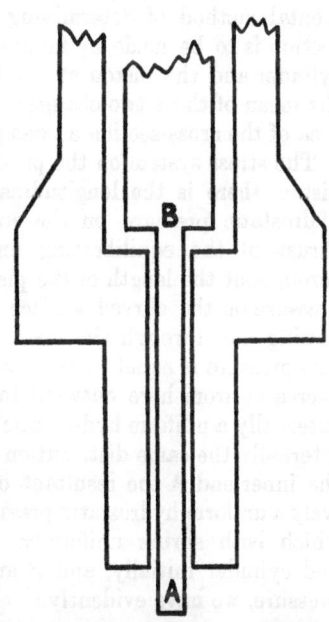

FIGURE 2. The absolute gauge; shows enlarged detail of the cylinder and the piston.

It is evident that the new form given the gauge makes necessary a recomputation of the correction for distortion. The effect of this distortion is to change the effective area of the piston and so to change the total thrust exerted on the piston by a given hydrostatic pressure. As previously explained, this correction is calculated mathematically and is merely a rough approximation. Its justification is that the correction is in any event exceedingly small, and that no easy experimental method of determining it directly presents itself. The correction is to be made by finding the change in the mean area of the cylinder and the piston at the lower and the upper ends, and taking the mean of these two changes. This gives the change in the effective area of the cross-section as was proved in the former paper.

The stress system on the piston and cylinder is as follows. On the piston there is the longitudinal thrust produced by the action of hydrostatic pressure on the end at A, and the equal and opposite thrust of the equilibrating forces at B. This thrust is uniform throughout the length of the piston. In addition there is the normal pressure on the curved surface exerted by the liquid which is slowly flowing out through the crack between piston and cylinder. At A this pressure is equal to the total hydrostatic pressure, and regularly decreases from here outward to zero at B. On the cylinder there is externally a uniform hydrostatic pressure over nearly the entire length; internally the same distribution of pressure as acts on the piston. At the inner end A the resultant of these two systems of stress is effectively a uniform hydrostatic pressure on both piston and cylinder, under which both shrink uniformly. If r and R are the radii of piston and cylinder initially, and r' and R' the corresponding values under pressure, we have evidently :

At the lower end
$$r' = r\left(1 - \frac{pk}{3}\right)$$

$$R' = R\left(1 - \frac{pk}{3}\right).$$

New effective radius
$$= \frac{r' + R'}{2} = \frac{r + R}{2}\left(1 - \frac{pk}{3}\right).$$

Effective area changed by
$$-\frac{2pk}{3}$$

$$= -4 \times 10^{-7} \times p$$

$$\begin{bmatrix} p \text{ is pressure in kgm./cm.}^2 \\ k = \text{compressibility} = 6 \times 10^{-7} \end{bmatrix}$$

At the end B, the piston is exposed merely to the thrust p. It will be assumed that the strain is the same as that under a thrust p uniform throughout the entire length.

Then at B
$$r' = r\left(1 + \frac{p\sigma}{E}\right)$$
$$= r(1 + 1.4 \times 10^{-7} \times p),$$

where σ is Poisson's ratio, assumed $= 0.28$, and $E =$ Young's modulus, taken as 2×10^6 kgm./cm.²

This correction for the distortion of the piston is not open to serious question, because of the smallness of the diameter compared with the length. But the calculation of the distortion of the cylinder at the upper end is open to much more serious question, because the irregular shape and unknown action of the packing produce an unknown stress system in the mass of metal about the upper end. In place, then, of a calculation, the experimental fact was used that in all probability the crack between piston and cylinder would not completely disappear at less than 15,000 kgm., since the piston still possessed some freedom of motion at 13,000. Discussion of this experimental fact is given later. It will be assumed, then, simply that the distortion is proportional to the pressure, and that the combined distortion of cylinder and piston will produce complete closing at 15,000. The initial size of the crack was determined experimentally, by measuring the piston and the effective area, to be 0.00035 inch.

This gives at the upper end $R' = R(1 - ap)$.
When the crack closes $r' = R'$

or $r(1 + 1.4 \times 10^{-7} \times p) = (r + 0.00035)(1 - ap)$.

Substituting $r = 1/16$,

we find $a = 2.3 \times 10^{-7}$.

At B $R' = R(1 - 2.3 \times 10^{-7} \times p)$.

Whence $r' + R' = r + R + \frac{r + R}{2}[(1.4 - 2.3) \times 10^{-7} \times p]$.

Effective radius, $\dfrac{r' + R'}{2} = \dfrac{r + R}{2}(1 - 0.45 \times 10^{-7} \times p)$.

Effective area decreased by $0.90 \times 10^{-7} \times p$.

The average change of effective area, top and bottom, which is the correction desired, is therefore $- 2.4 \times 10^{-7} \times p$.

The correction found in this way is to be regarded as an upper limit, the assumption being made that the crack will close up at 15,000 kgm. The correction found by this assumption is somewhat doubtful. A lower limit to the correction can be found by making the assumption as far removed as possible from that made above, namely, that at the upper end the cylinder suffers no change of internal radius. The change of effective area at the upper end is due to enlarging of the piston alone, therefore, and is evidently $1.4 \times 10^{-7} \times p$. In this case the change in the effective area is $1.3 \times 10^{-7} \times p$. The upper and the lower limits differ only 1/10 per cent at 10,000 kgm. The correction applied in the following work was taken as the mean of these two limits, $1.8 \times 10^{-7} \times p$, and the observed gauge readings have been increased in this ratio. This happens to be the same as the correction used for the former gauge.

The experimental evidence used in part of the above approximations requires brief mention. The calculations given above show a more rapid closing of the crack at the end B, so that the tendency of the piston would be to bind at the upper end. This was verified experimentally by the fact that the upper end of the piston was always more brightly polished after a little use than the lower end, and that sticking could be avoided at the higher pressures by making the piston slightly conical. To accomplish this it was sufficient to make the upper end 0.0002 inch less in diameter than the lower. The figure above for the initial width of the crack at the upper end (0.00035 inch) was obtained by combining with the conicality the measured value for the effective area to be described later.

In actual use care was necessary to be sure that the sticking was really due to closing of the crack, and not to viscosity in the fluid transmitting pressure. In the early experiments, in which the transmitting fluid was molasses and glycerine, almost complete sticking was found at pressures as low as 7500. That this was not due to closing of the crack was shown simply by warming the whole apparatus, thus decreasing the viscosity without materially decreasing the size of the crack. It was thus possible to reach 13,000 with very much less sticking than at 7800 at the lower temperature, and also with much less leak, showing an actual decrease in the size of the crack at high pressures. The liquid finally adopted for use in this work was a mixture of glucose with glycerine and water. Glycerine and water were first mixed in equal parts, and then the glucose thinned with this mixture to a best consistency found by experiment. The advantage is that the pressure effect on viscosity is much less than for the molasses and glycerine mixture, so that a mixture of given consistency will work over twice the pressure range of the molasses mixture.

The functioning of the gauge at high pressures is therefore prevented by two effects, -- increased viscosity of the liquid, and closing of the crack. In view of the fact that the gauge still worked at 13,500 when both these effects were operative, the estimate made above that the functioning would cease at 15,000 by the closing of the crack only would seem to be amply low. The maximum value set on the correction above is probably, therefore, too high.

The calibration of the piston and cylinder at low pressures to determine the effective cross-section was carried out by the method used in the previous paper. Some such indirect method of calibration was made necessary by the fact that the dimensions of the small piston are so small as to make accurate direct measurements of its effective diameter impossible. The method consisted in hanging weights on the piston to be calibrated and on a larger piston of known area, in such a proportion that the pressure produced by the two pistons should be the same. This is done most simply by connecting the two freely moving pistons to the same pressure chamber, keeping the weights on one piston invariable, and changing those on the other until neither rises or falls. The details of the method were the same as that described before, the same comparison piece of apparatus being used.

The results of the comparison showed an effective diameter for the piston of 0.06250 inches. The measured diameter was only 0.0623 inches at the larger end and 0.0622 inches at the center, showing a crack between piston and cylinder 0.0003 inches wide at the center, 0.00025 inches at the lower end, and 0.00035 inches at the upper end, as used in the calculation above.

In the earlier measurement of high pressures, the thrust was found by hanging weights directly on the piston, and determining by trial that weight which produced neither rise nor fall of the piston. This has the advantage of ideal accuracy, but has several serious disadvantages of manipulation. Flexibility of design in the apparatus is sacrificed because the gauge must be kept vertical. The scale pan and weights become increasingly cumbersome at high pressures, so that an assistant is needed. And worst of all, it requires considerable time to make a reading. This is a fatal objection where the pressure must be read instantaneously, as in experiments to be described in a following paper on the freezing of mercury by the method of electrical resistance.

The present gauge was made direct reading and instantaneous by causing the thrust to produce a measurable deflection in a stiff spring. The new process is related to the old exactly as weighing with a spring balance is to weighing with separate weights. The spring balance is less accurate, but very much more convenient. The accuracy obtainable with the new device was, however, sufficient for all requirements. There is an added advantage in that the stroke of the piston over the entire pressure range may be very much decreased. The stroke of 12 mm. in the previous work was reduced to 1.5 mm. in this form. This ensures greater strength at the upper end where the piston pro-

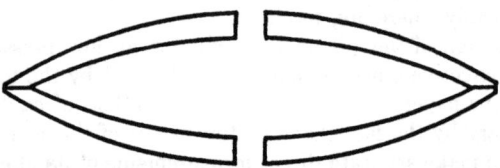

FIGURE 3. The springs with which the thrust on the piston is measured.

jects unsupported from the cylinder, and also ensures greater accuracy, the piston always playing in approximately the same part of the cylinder.

Several forms of spring were tried. The form finally adopted as giving the great stiffness desired without inconvenient bulk was that of a saucer. Two of these saucer-shaped springs were used, placed rim to rim, and the thrust applied at the center (see Figure 3). This arrangement has the advantage of avoiding all friction when the circumference of the saucer increases under pressure, since here the two saucers support each other and move together without slipping.

The choice of steel is of great importance. The most suitable of the many kinds tried was one manufactured by the Halcomb Steel Co. in the form of sheets.

Much experimentation was necessary before the best dimensions and the best heat treatment of the springs was found. The sheet from which the springs were cut was 0.105 inches thick. This was cut into discs 2 1/2 inches in diameter, with a 1/4-inch hole through the center. The disc was then turned on one flat face so as to decrease regularly from 0.105 inches at the center to 0.035 inches at the edge. It was given the saucer shape in a die while hot, the depression at the center being about 1/4 inch. The greatest care was necessary in hardening to prevent warping. This was conveniently done by heating to a bright red heat in a bath of molten lead so as to secure a uniform heat, and then quenching by plunging edgewise into water. The temper was drawn by heating to 370° C. in oil. The temperature was important: 360° C. was too low. Finally, after tempering, the springs were ground flat on the edge to provide a bearing surface against each other. Warping due to hardening was most distinctly shown by variations in the width of the ground strip. A pair of springs so made will support indefinitely at the center a weight of 1350 pounds without permanent set. The deflection under this load is 2 or 3 mm. The actual working pressure did not exceed 650 lbs.

The small motion of the springs was magnified by a simple mirror device, and observed with a telescope and scale rigidly attached to the frame holding the springs. The scale distance was only 30 cm., but it was nevertheless possible to obtain a magnification of over 1500 times with perfect consistency and freedom from back lash or tremor. The magnification was doubled by reflecting twice from the moving mirror. The size of the piston, sensitiveness of the springs, and optical magnification were altogether such that 8 kgm./cm.^{2} on the gauge produced a deflection of 0.1 mm. at the observing telescope. This gives 1/10 per cent as the accuracy of the pressure readings at 8000 kgm./cm.^{2}; proportionally more at the higher pressures.

All of the parts connecting together the springs and the mirrors were made of steel. This has the advantage of avoiding any motion of the mirrors which might be produced by changes of temperature of the surrounding atmosphere. The only temperature effect is that due to the change in the elastic constants of the steel spring with variations of temperature. This was so small that no appreciable error is introduced under the ordinary conditions of use.

Any device for measuring the magnitude of a stress by the deflection of a spring must be subjected to pretty careful scrutiny before the measurements can be accepted as accurate, because there are disturbing effects, such as elastic after-working and hysteresis, which complicate matters. It was hoped to reduce these effects to a negligible value by using as the working stress less than half the stress at the elastic limit as mentioned above. But even with this precaution it seemed desirable to calibrate carefully the springs under working conditions.

In order to facilitate the comparison, the springs, multiplying mechanism, and telescope and scale were rigidly connected in one piece. This could be screwed either to the end of the absolute gauge for the purpose of measuring the thrust on the piston, or to the calibrating device. The calibration was effected at first by hanging weights on a stirrup, but this process, always discontinuous and sometimes as complicated as applying two, removing one, applying two, etc., was so unlike the process of loading during actual use that another method was seen to be necessary. Two freely moving pistons were used, as when finding the area of the 1/16-inch piston, both communicating with a Cailletet pressure pump of the Société Genevoise. One piston, 1/4 inch in diameter, was the same as that used in the previous work. Weights were suspended directly from the upper end of this piston. This piston was kept in constant rotation by a small motor so as to avoid friction. The other freely moving piston, 5/16 inch in diameter, was in direct hydrostatic communication with the 1/4-inch piston, and at its upper end pressed directly against the springs to be calibrated. (This 5/16-inch piston could be rotated by hand as occasion required to destroy friction.) From the weight on the 1/4-inch piston, and the areas of the 1/4-inch and the 5/16-inch pistons, the thrust on the springs could be calculated. The area of the 5/16-inch piston was found in the same way as that of the 1/16-inch piston. The method has all the advantages of the discarded method of the direct application of weights, namely, complete freedom from all elastic effects and hysteresis, and in addition permits very much more convenient and flexible application of pressure.

The procedure of the calibration was to place a weight on the 1/4-inch piston, completely depressing it. The piston was then floated again by raising the pressure to the equilibrium value with the Cailletet pump. The 5/16-inch piston was then rotated to destroy friction, and the deflection of the spring read. Eleven such steps were made with increasing and decreasing pressure, making twenty-two steps in all. The same weights were used in all the calibrations, so that the results were strictly comparable. The pressure exerted by the fluid, as given by the gauge of the Geneva pump, was also recorded as a check. The accuracy of the other readings was so great, however, that these check readings could never be used.

Calibration with this device was first made to find whether the gauge had any error of position, since it was generally calibrated vertically, but used horizontally. This could evidently be done very simply by changing the position of the cylinder with the 5/16-inch piston, a change in the calibrating procedure which could not be made so simply when weights were directly applied. The result of the calibration in the horizontal position showed no detectable error due to this change of position. Of course no such error was to be expected, since all the parts were very stiff in comparison with their weights.

A second result of the calibration was that the springs show no elastic after-effects. By this is meant the gradual creep after application of a load, and gradual recovery after removal. The effect is generally most pronounced at the two ends of the pressure range. These springs, however, showed no tendency to yield viscously under the maximum stress, and never showed any wandering of the zero after release of pressure.

A third result of the calibration was that the springs do not follow the linear law, but are increasingly deflected at the higher pressures. The change was not large, but perfectly distinct, about 6 per cent less weight being required to produce a given deflection at 650 lbs. than initially. This is rather surprising in a substance like this spring steel, which ordinarily follows the linear law to the elastic limit. The

TABLE I.

RELATION BETWEEN LOAD AND DEFLECTION OF SPRINGS WITHOUT DEVICE FOR AVOIDING HYSTERESIS.

Load Kgm.	Mirror Deflection.		Load Kgm.	Mirror Deflection.	
	Increasing Load.	Decreasing Load.		Increasing Load.	Decreasing Load.
00.00	0.00	0.00	95.62	8.52	8.64
16.26	1.42	1.49	112.27	10.06	10.18
33.37	2.94	3.02	128.27	11.55	11.66
49.35	4.35	4.45	143.32	12.97	13.07
65.54	5.81	5.91	160.18	14.57	14.64
81.59	7.24	7.36	173.85	15.90	

effect is evidently due here to the change of geometrical shape, the springs becoming so much flatter under the higher stresses that the geometrical configuration as such has a lower elastic constant, although the elastic constant of the material itself is unaltered. It is customary in the mathematical treatment of the bending of thin rods or plates to assume that the deflection remains proportional to the stress up to the elastic limit. This experiment shows that this approximation may become invalid at considerably less than the elastic limit.

A fourth result was that the gauge does show some hysteresis, a result which was not expected in view of the second result. For as a general rule, hysteresis and elastic after-effects, while not directly related, occur together, both being evidence of some molecular instability. Table I., giving the difference between the reading under increasing and decreasing pressure, shows the usual magnitude of the effect. The first column in the table gives the total load at each step. The effect is much less than that due to departure from linearity mentioned above, so that here we have a hysteresis loop of the unusual shape shown in Figure 4. The lag in similar cases is usually so great that the curvature with decreasing stress is the reverse of that with increasing stress, so that the loop has the general shape of a double convex lens. Furthermore the loop is usually very nearly symmetrical with respect to the line joining the extremities. This is the only example of a loop of the above shape known to the writer. This single example is sufficient to show that there is no necessary connection between hysteresis and departure from the linear relation between stress and strain, as might be supposed if all loops were of the ordinary type.

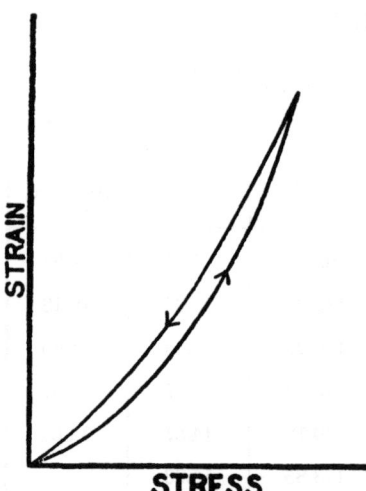

FIGURE 4. Shows the unusual nature of the hysteresis cycles described by the springs.

This hysteresis, while comparatively small, was nevertheless sufficient to reduce the accuracy of measurements made with the gauge far below that desired. A method of avoiding hysteresis was therefore adopted. It depends on the fact, well known for magnetism, that if the stress is varied cyclically by small amounts about any fixed point, a small hysteresis loop is described about this point. The result is that if stress is relieved from the point A (see Figure 5) after increasing pressure, the path AB will be described, while if it is increased from the point C after decreasing pressure, the path CB is described. Suppose that during increasing pressure the point D has been reached, or that during decreasing pressure the same stress, shown at E, has been reached. The difference between the points E and D represents the error due to hysteresis. To make the readings at these two points the same, we may evidently apply a small extra load at D, raising the stress to A, and then remove the extra load, or at E we may remove a slight portion of the load to C and then reapply it. The same point B is finally reached, and the pressure readings have become single valued. The extra load necessary to apply or remove must be determined by experiment, and would be expected to vary at different parts of the hysteresis loop.

This extra load was applied in practise by a very simple lever arrangement, by which the springs could be deflected one way or the other before making readings. The following set of readings, picked at random from a great number, shows how nearly it was possible to avoid hysteresis by the method. The difference of readings is seldom more than the possible error of reading. The small extra load applied or removed before making these readings was constant over the entire range, being sufficient to produce a deflection of about 2.0 divisions. It is evidently not quite enough at the higher pressures and a little too much at the lower pressures. The mean of the two readings nowhere differs from either reading by more than the errors of observation, however, and this simpler procedure was therefore adopted.

The most inconvenient fact disclosed by the calibration was that the constant of the springs varies slowly from time to time. Over two or three days no change whatever is to be noticed, but in a week or a month there are likely to be changes beyond the limits of error. The change is irregular and has no apparent connection with temperature changes. No temperature effect within the limits of working room temperatures was ever found. The change is doubtless due to some slow process of molecular accommodation going on within the metal itself. At one time the springs were permanently deformed by a violent explosion, so that the deflection under the same load was increased in the ratio 16:14. After this deformation for a month or more the change with time was more rapid than usual.

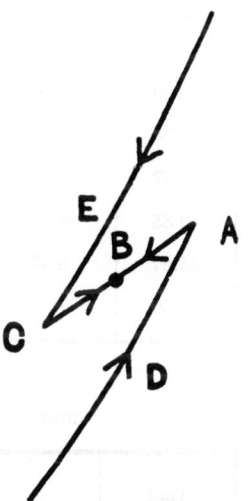

FIGURE 5. Detail of a hysteresis loop, showing the method of avoiding the effects of hysteresis in making the readings with the springs.

The table shows the magnitude of the variation with time. The first set of three, January–April, 1910, was made at intervals during constant use of the gauge. During this time the gauge constant decreased and then increased again. The explosion referred to above took place on December 24, 1909. The last two readings, September 6 and 24, 1910, were made after the springs had been resting for about four months. The constant is in general higher, but the greatest increase has come at the middle of the range, so that the relation between stress and strain is more nearly linear than before. After this prolonged period of rest, the gauge has remained much more nearly constant than before.

TABLE II.

Relation between Load and Deflection of Springs with Device for Avoiding Hysteresis.

Load Kgm.	Mirror Deflection.		Load Kgm.	Mirror Deflection.	
	Increasing Load.	Decreasing Load.		Increasing Load.	Decreasing Load.
00.00	0.00	0.00	95.62	8.50	8.50
16.26	1.43	1.42	112.27	10.02	10.04
33.37	2.94	2.92	128.27	11.51	11.54
49.35	4.35	4.34	143.32	12.98	12.94
65.54	5.80	5.79	160.18	14.54	14.54
81.59	7.23	7.23	173.85	15.85	

TABLE III.

Showing Variation of Springs with Time.

Load Kgm.	Mirror Deflection.				
	Jan. 1.	March 30.	April 27.	Sept. 6.	Sept. 21.
16.26	1.65	1.63	1.65	1.66	1.66
33.37	3.37	3.32	3.36	3.38	3.38
49.35	5.00	4.95	4.98	5.02	5.03
65.54	6.68	6.58	6.65	6.70	6.70
81.59	8.37	8.25	8.35	8.39	8.40
95.62	9.86	9.74	9.83	9.88	9.89
112.27	11.66	11.63	11.64	11.69	11.70
128.27	13.42	13.28	13.40	13.44	13.47
143.32	15.09	14.96	15.09	15.14	15.15
160.18	17.00	16.88	17.02	17.05	17.08
173.85	18.60	18.52	18.63	18.63	18.65

This change of gauge constant with time demands that frequent calibration be made. In all the work of the following papers in which considerable accuracy was desired, such calibrations were made every few days. It is not much trouble to make this calibration. The twenty-two readings with increasing and decreasing pressures can be made in the course of half an hour. When the gauge is so calibrated, and the procedure for avoiding hysteresis is adopted, the readings are consistent to the limit of sensitiveness. This has already been stated to be about 8 kgm./cm.^{2}.

The gauge in actual use has shown itself very convenient. The readings may be made rapidly and immediately after the application of pressure, since there are no thermal effects such as dissipation of heat of compression. The temperature effect is so small that it may be used directly in the room without a thermostat, and the leak is so slight that it may be used in entire comfort in measuring many high-pressure effects. Where applicable, the gauge is more convenient than the electrical resistance gauge and it has been used whenever possible.

The Manganin Resistance Gauge.

In the earlier paper a method was described for measuring high pressures by measuring the electrical resistance of pure mercury under pressure. The method had the advantage of being perfectly reproducible, so that any one could at any time measure pressure without recourse to the then inconvenient fundamental standard of pressure. The advantage has been in large part offset by the devising of the convenient form of absolute gauge described in the first part of this paper. Moreover, the method becomes inapplicable at somewhat higher pressures than those reached formerly because of the freezing of the mercury. Thus at 0°, the freezing pressure is about 7500 kgm./cm.^{2}. Aside from this difficulty, which might be avoided by placing the mercury in a separate vessel, maintained at higher temperature, there are numerous inconveniences of manipulation, as, for instance, that the mercury must always be kept in an upright position. But the greatest inconvenience of all is that the glass capillary containing the mercury is always shattered by an explosion, and explosions become more and more frequent at high pressures.

There are frequently situations, however, where the absolute gauge becomes unavailable, and where a gauge with some of the properties of the mercury gauge becomes desirable. For instance, it is often necessary to secure absolute freedom from leak, and this is obviously impossible with a freely moving piston as in the absolute gauge.

Measurements of compressibility or of change of volume during freezing, as in two following papers, demand this property of freedom from leak.

The method of pressure measurement adopted here, and which secures freedom from leak, has already been described by Lisell.[2] He measured the effect of pressure on the resistance of manganin wire, and proposed that the change of resistance be used as a measure of pressure. Lisell found the effect of pressure between 0 and 4200 atmos to be linear, and showed that there was no appreciable temperature effect between 0° and room temperature. But Lisell also showed that different specimens of manganin show slightly different pressure coefficients, so that the advantage of reproducibility must be given up. Lately, Lafay[3] has also measured the pressure effect on manganin up to 3500 kgm. and has found nearly the same pressure coefficient as did Lisell.

The data of this paper show by direct comparison with the absolute gauge that the manganin is suitable as a pressure gauge over much wider ranges of pressure. Up to 13,000 kgm. the relation is linear within the errors of the absolute gauge. It would not have been surprising if this had not been true, in view of the fact that manganin has a pressure coefficient which is positive instead of negative like that of all the pure metals. Furthermore, over this pressure range the readings are entirely free from hysteresis or creep. This is opposed to the work of Lussana,[4] who found various temporary effects after the application of pressure. His results have not been verified by subsequent observers, however, and the entire absence of the effect here under a very much wider pressure range would seem to make pretty certain that there was some obscure source of error in Lussana's work.

The manganin used in this work was of German manufacture, No. 38, double silk covered, of about 30 ohms to the meter. With change of temperature it shows a maximum resistance at about 27°. The coils were of about 100 ohms resistance, the wire being wound non-inductively on itself in the form of a toroid, about 1 cm. in diameter and 5 mm. thick. To protect the wire, the toroid was covered with a winding of fine silk ribbon, only the ends of the wire being exposed in order to make connections with the insulating plug. This plug was of the same design as that shown in the previous paper, except that it was made of hardened nickel steel instead of tool steel.

---

[2] Lisell, Om Tryckets Inflytande på det Elektriska Ledningsmotståndet hos Metaller samt en ny Metod att Mäta Höga Tryck (Diss. Upsala, 1903)
[3] Lafay, C. R., **149**, 566–569 (1909).
[4] Lussana, Nuov. Cim., **10**, 73–84 (1899); **5**, 305–314 (1903).

The apparatus for producing and measuring pressure consisted of two cylinders connected by a tube of nickel steel. In the lower cylinder the pressure was produced and measured. The absolute gauge of the first part of this paper was screwed directly into the side of this cylinder. The cylinder itself was of Krupp chrome nickel steel, 8 inches outside diameter, 1 1/8 inches inside diameter. It was placed in a hydraulic press of 200 tons capacity, and the pressure produced by a 1 1/8-inch piston forced into the cylinder by the ram of the press. The upper cylinder, also of Krupp chrome nickel steel, 4 1/2 inches outside and 9/16 inch inside diameter, contained the manganin wire to be tested. The connecting tube to the lower cylinder passed through the bottom of the tank of a thermostat, which surrounded the upper cylinder, and with which the temperature could be kept constant to 0.01°. No such temperature precaution was necessary for the absolute gauge. The lower cylinder was filled with the mixture of glucose and glycerine needed to secure tightness of the piston of the absolute gauge, and the upper cylinder was filled initially with either kerosene or gasolene, in which the manganin coil was directly immersed. The action of pressure was to compress the kerosene, glucose passing from the lower cylinder to the lower part of the upper cylinder. The compression was never sufficient, however, to bring the glucose into contact with the manganin. Although kerosene or gasolene are somewhat inconvenient because of their high compressibility, still their use was made necessary by the fact that a heavier oil freezes under pressure, so that it does not transmit pressure hydrostatically to all parts of the wire. The kerosene is also known to become stiff like vaseline at say $10°$ and 8000 kgm., but the viscosity is not so great as to introduce irregularities. When either kerosene or gasolene is used, the insulating qualities of the plug are practically perfect without requiring any special precautions. The insulation resistance was always over 10 megohms, which was the limit of the measuring device conveniently at hand. Formerly, in working with mercury, pressure was transmitted by a mixture of water and glycerine. It was necessary to specially protect the separate parts, and even then the insulation resistance was never greater than several hundred thousand ohms.

The electrical measurements were made by the same null method and on the same Carey Foster bridge as those described in the former paper.

In making the calibration and in using the gauge, there is one fact to be borne in mind which has its analogy in the mercury resistance. This is the seasoning effect of pressure; the gauge does not respond to the first application of pressure in the same way that it does to the second or subsequent applications; there is a gradual settling down to a steady state. In the case of mercury, this was due to the equalization of strains in the containing capillary of glass. In the case of the manganin, some such process of accommodation must be going on within the mass of the metal. This is shown principally by drift of the

TABLE IV.

PRESSURE COEFFICIENT OF RESISTANCE OF MANGANIN.

Pressure Kgm./cm.²	$\frac{\Delta R}{pR_0} \times 10^9$	Pressure Kgm./cm.²	$\frac{\Delta R}{pR_0} \times 10^9$
1260	2272	9290	2302
2610	2282	8000	2304
3810	2299	6450	2302
5000	2301	4790	2304
6180	2299	3250	2291
7210	2300	1740	2308
8230	2302	1000	2272

zero, but the pressure coefficient may change slightly. The seasoning process occupies more time for the manganin than for the glass; it may extend over as much as a month after the first application. It is hastened by frequent applications of pressure, but may apparently run to completion in sufficient time after only one application. To effect the seasoning, it does not seem to be necessary to subject the coil to the maximum pressure under which it is contemplated using it. The coil used to the highest pressure reached in this work, 20,500 kgm., had been seasoned by the application of not more than 12,000 kgm., yet it showed no further change after the application of a pressure 8500 kgm. in excess of the seasoning pressure.

The results obtained with one such well-seasoned coil at 0° are shown in the table. In making the comparison with the absolute gauge all the precautions described in the first part of the paper were observed. It is seen that within the limits of error of the pressure readings, the change of resistance is proportional to pressure.

To show within what limits different pieces from the same spool of wire give the same results, three coils were made from the ends and the middle of a length of wire of 70 m. The constants for the separate coils at 0 were $.0_5 2301$, $.0_5 2307$, and $.0_5 2325$, in the order of the coils, a variation of one per cent. The temperature effect was found by measuring these same coils again at 50°. These same three coils gave 2295, 2319, and 2320 respectively. One of the coils measured at $-12°$ showed no measurable difference between $-12°$ and $0°$.

The maximum pressure to which these calibrations were made varied somewhat with the temperature, because at the lower temperatures the mixture of glucose and glycerine used with the absolute gauge became viscous so rapidly with increasing pressure as to transmit pressure very slowly. At the low temperatures the pressure was increased until this limit was reached. The slow flow of glucose from the lower to the upper cylinder might occupy an hour or more before the equilibrium was complete. The fact that there was such a process of flow was definitely shown by the slow fall of pressure in the lower cylinder as indicated by the absolute gauge, with a simultaneous slow rise of pressure in the upper cylinder, as indicated by the manganin resistance. At 0° the maximum reached was in one case 11,000 kgm. This was probably too high for complete equalization of pressure, for the change of resistance was 1/2 per cent too low at this maximum. The measurements at 0° were not usually carried as far as this, 9500 being the more usual limit. Slight differences in the composition of the glucose mixture made very pronounced differences in the viscosity at high pressure. At 50° the highest reached was 12,000. Even here the viscosity was very considerable. On one occasion pressure was pushed to 13,000 at 50°. There was the same slow equalization of pressure, extending over about half an hour, as was found at 0°. At the end of this time the resistance had nearly acquired the value given by a linear relation, when the experiment was terminated by an explosion. The final reading below this at 11,500, at which there was also some viscous yield, completely satisfied the linear relation.

We may conclude, therefore, that over the temperature range 0°–50° the pressure resistance relation is linear within 1/10 per cent of the change of resistance, up to 13,000 kgm. This was proved by actual experiment at 50° to 12,000, and to 9500 at 0°. The extrapolation to 13,000 at 0° is comparatively slight, and is made all the more probable by the fact that our usual experience would lead us to expect greater departure from linearity at higher temperatures, and no such departure was found.

Although different specimens of manganin do not have the same constant, still it may be worth while comparing the results found here with those of other observers so as to give an idea of the magnitude of the variation of the effect. Lafay[5] found 2.16×10^{-6} per kgm./cm.² as the effect on one specimen. Lisell[6] found for two specimens of annealed wire 2.13×10^{-6} and 2.08×10^{-6}; for three specimens of hard drawn wire results from 2.279×10^{-6} to 2.338×10^{-6}. The data of this paper, which are for hard drawn wire, vary from 2.295×10^{-6} to 2.325×10^{-6}.

Four of these manganin resistance gauges have been in almost constant use for over a year. As compared with the mercury gauge they have had the advantage of greater convenience and ease of manipulation, even over the pressure range within which the mercury remains fluid. During the work explosions have been of frequent occurrence, the shock of any one of which would have broken the glass capillary containing the mercury. It is not necessary to apply elaborate temperature precautions as for the mercury. The only temperature correction necessary to apply is a small one for the shift of the zero. This may be determined accurately enough by hanging a thermometer in the air of the room near the cylinder containing the manganin. At room temperatures of 20° a change of temperature of 1° demands a pressure correction of only 5 kgm.

As compared with the absolute gauge of the first part of the paper, each form has distinct advantages. Where available, the absolute gauge is more convenient, because it is direct reading and immediate. But the absolute gauge has the disadvantage of leak, so that often the manganin gauge becomes absolutely necessary. Futhermore, it has the disadvantage of being more cumbersome. This means that all parts of the apparatus in connection with the gauge must be correspondingly enlarged. This is often a fatal disadvantage, entirely apart from any considerations of expense or convenience, because in order to reach the highest pressures, the steel parts must be hardened, and it is not possible to harden large steel cylinders. The upper limit of pressure attainable will have to be reached with comparatively small apparatus. The reason for not pushing the absolute gauge to its limit was not so much fear of destroying the gauge as the fact that the large 8-inch steel cylinder containing the gauge was of soft nickel steel, and that the yield point would have been reached at 15,000 kgm. This cylinder had been previously seasoned by applying pressures up to 28,000 kgm., which had the effect of increasing the internal diameter

---

[5] Lafay, loc. cit.   [6] Lisell, loc. cit.

from 5/8 inch to 1 1/8 inches, but even this was not sufficient to raise the limit permanently to over 15,000 kgm.

The compactness of the manganin gauge makes it particularly adapted for working with the highest pressures where everything must be in one piece because of the impossibility of making connecting tubes. The gauge has been so used in a number of experiments on the freezing of water under pressure. The gauge was screwed into one end of a large steel cylinder, the plunger was pushed in from the other end, and the water was in the space between. The dimensions were kept down so that the entire block, together with a part of the hydraulic press, could be placed in a thermostat.

The manganin gauge may be used by extrapolation to measure pressures beyond the reach of the absolute gauge. It has been so used in investigating the freezing of water up to an indicated pressure of 20,500 kgm., and this limit could without doubt be exceeded. The limit is not in the manganin itself, but in the hardened steel parts, which have a tendency to stretch too much at pressures as high or higher than 20,000 kgm. Of course the use of any standard by extrapolation is undesirable, but at present any means of measuring these very high pressures with probable accuracy is welcome. In any event, the extrapolation from 12,000 to 20,000 is very much less than the extrapolation from the previous maximum of 4000 to 12,000, which is here shown by actual experiment to be justified. What is more, it will be an easy matter to translate high-pressure readings in terms of a manganin gauge into absolute pressures, if at any time the direct calibration is extended from 13,000 to 20,000, and proves that a linear extrapolation is not sufficiently accurate.

The only serious disadvantage in the manganin gauge as thus far described, when compared with either the mercury gauge or the absolute gauge, is the fact that it is not readily reproducible, so that each new coil of wire must be calibrated against an absolute gauge. This disadvantage may be obviated by the use of fixed pressures of reference analogous to the melting points of the metals used as points of reference in thermometry. The linearity of the relation between resistance and pressure having been established by the work of this paper, it is necessary to know for each coil only the change of resistance corresponding to a single known pressure in order to fix completely the behaviour of the coil.

Such pressures of reference are given very conveniently by the points of transition between the various kinds of ice and water. For low pressures such a pressure of reference is given by the pressure of transition from ice I to ice III, using Tammann's [7] notation. This pressure is very nearly independent of temperature between −22° and −30°. This particular transformation has the advantage that the reaction runs with very great velocity and is accompanied by a comparatively enormous change of volume, 20 per cent, so that in any piece of apparatus, once given the two phases, the equilibrium pressure is automatically set up almost immediately. This pressure has been found to be 2120 kgm. at −23°. For higher pressures, the transition point at 0° from water to a hitherto unknown variety of ice, ice VI may be used. This point has been found by experiment to be 6370 kgm. Equilibrium is reached more slowly than for the transition I–III, but for work at higher pressures the use of this transition point will give a more accurate calibration. This method of calibration has actually been used, with satisfactory results, for two other coils than the four mentioned above.

The calibration of the manganin resistance without the use of an absolute gauge may of course be effected by comparison with any other accurately measured pressure phenomenon. For instance, the calibration may be made by comparing the manganin resistance with the mercury resistance, the data for which have been already published.

Summary of Results.

Two gauges for high pressures are described in this paper. The first is an absolute gauge which has been used up to 13,000 kgm. The construction of the gauge is described, the correction for distortion is determined, and the reading mechanism is discussed. A procedure is given for freeing the deflections of the springs with which the thrust is measured from hysteresis. All of the factors may be determined with sufficient accuracy so that the gauge is accurate to the limit of accuracy of reading, 1/10 per cent at 8000 kgm./cm.^{2}. The second gauge is a manganin resistance. This is shown to be suitable for the purpose, since there is complete freedom from hysteresis and elastic after-effects. The relation between pressure and resistance is shown to be linear up to 12,000 kgm., but the gauge has been used by extrapolation up to 20,500. The accuracy of the readings with this is at least as great as with the absolute gauge. The relative advantages for various kinds of experiment of these two forms of gauge are discussed. Finally, by the use of standard pressures of reference it is

---

[7] Tammann, Kristallisieren und Schmelzen, pp. 315–344 (Barth, Leipzig, 1903).

shown that a manganin resistance gauge may be calibrated without direct reference to an absolute gauge.

This investigation was a necessary preliminary to the measurement of various thermal properties of mercury and water under pressure, the expenses of which have been partially defrayed by several liberal appropriations from the Rumford Fund of the American Academy of Arts and Sciences.

JEFFERSON PHYSICAL LABORATORY,
HARVARD UNIVERSITY, CAMBRIDGE, MASS.,
October, 1911.

CONTRIBUTIONS FROM THE JEFFERSON PHYSICAL
LABORATORY, HARVARD UNIVERSITY.

MERCURY, LIQUID AND SOLID, UNDER PRESSURE.

By P. W. Bridgman.

Presented by G. W. Pierce, Oct. 11, 1911. Received Oct. 6, 1911.

Introduction.

This paper is an attempt to present for the liquid and solid phases of a single simple substance, mercury, data corresponding to the data which have been collected for the gaseous and liquid phases of many substances. These latter data are the relations connecting temperature, pressure, and volume of the gas together with the changes involved when the gas passes to the liquid state. Out of these data have come the theories of the molecular structure of gases and vapors, and the various relations connecting the gaseous with the liquid phase, of which the theory expressed by the equation of van der Waals is the best known. The corresponding work for the liquid and solid phases, that is, the mapping out of the pressure-volume-temperature surface of the liquid and the determination of the quantities involved in the change from the liquid to the solid state, has never been done. As a result, the theory of liquids is entirely undeveloped, and the theory of the relation between liquid and solid hardly begun. For instance, the fundamental question as to the existence of a critical point for the liquid-solid states analogous to that for the vapor-liquid is as yet unsettled.

The apparent reason for this neglect of the thermodynamic data for the liquid and solid is merely experimental difficulty. Whereas for a vapor the critical pressure never exceeds a few hundred atmospheres, the pressures involved in a corresponding study of the liquid-solid phases are of the order of tens of thousands of atmospheres. These pressures are so high as to tax to the utmost the tightness of every joint and the strength of the steel vessels. In this present work it has been found possible by a special packing device, which secures absolute freedom from leak, and by the use of the highest grades of steel manufactured, to reach 30,000 and even 40,000 atmos on the first application of pressure. On subsequent applications rupture may come at half the first maximum. This subsequent early rupture is due to the fact that these pressures are greatly in excess of the theoretical limit of the steel, so that there is a distortion of the inner parts far beyond the elastic limit and consequent rapid fatigue. The pressures reached in this paper are not the greatest attainable because of the questionable scientific economy of pushing pressures to a value where rupture may occur at any moment and destroy valuable apparatus and time.

The previous work in this field is wholly below 5000 kgm. The two principal investigators are Amagat,[1] who studied the compressibility and the thermal dilatation of a number of liquids up to 3000 atmos, without, however, touching on the change of state liquid-solid; and Tammann,[2] who studied the change of state liquid-solid, without determining the behavior of the liquid, over a range of pressure of usually about 3000 kgm., but reaching a few times to 4000 and once to 4800. There are no liquids common to the work of both Amagat and Tammann, so that the complete data are not known for even a single liquid.

The previous work done on mercury is almost negligible for the present purpose. Accurate compressibility determinations run only to 500 kgm. (the highest are by Richards[3]), and the effect of pressure on freezing point has been observed only by Tammann,[4] who found results up to 2000 kgm. that are inconsistent among themselves by 30 per cent.

The data of this paper cover a range of 12,000 kgm., and give both the changes of volume of the liquid with temperature and pressure, and the thermodynamic data required on the freezing curve. The temperature range corresponding to this wide pressure range is comparatively small, since at 12,000 kgm. the freezing temperature has been raised to only 20°, that is, through a range of 60°.

In addition to the data connected more intimately with pressure, several quantities relating to solid mercury at atmospheric pressure have incidentally been determined. It has been a surprise to find how untrustworthy the commonly accepted data for solid mercury are. Thus the only determination of the latent heat of melting dates back to 1838, before the value of the melting temperature itself was known more accurately than within 3°.

---

[1] Amagat, Ann. de Chim. et Phys. (6), **29**, 68–136, 505–574 (1893).
[2] Tammann, Kristallisieren und Schmelzen (Barth, Leipzig, 1903).
[3] Richards, Pub. Carnegie Inst. Wash., No. 7 (1903), and No. 76 (1907).
[4] Tammann, loc. cit., p. 248.

The experimental material of the paper, with respect to both subject matter and experimental method, falls naturally into two parts. The first is concerned with the p-v-t surface of the liquid. To map this the isothermal compressibility of the liquid was found at two temperatures, 0° and 22°, and combined with the known dilatation at atmospheric pressure. This is sufficient to cover the region in question, because the change of dilatation with temperature is slight. The second part is concerned with the changes taking place on the freezing curve. Thermodynamically, the data required are the change of volume on passing from liquid to solid and the variation of freezing temperature with pressure. From these the latent heat may be calculated by Clapeyron's equation. Two independent methods were used; one which gave both the change of volume and the relation between temperature and pressure on the freezing curve, and another which gave only the freezing curve. In addition still another method was used in determining the change of volume on freezing at atmospheric pressure.

In the conclusion, the data collected are discussed from the point of view of their bearing on present theories of the liquid state, and the change solid-liquid. These data for a single substance cannot justify by any means an attempt at a new theory. This is simply a first step, indicating in a general way the nature of the effects to be expected at high pressures. From this point of view there are considerations both for and against the use of mercury. On the one hand, the compressibility and thermal dilatation of mercury are comparatively small, so that the change in these quantities under the pressure range employed is not nearly as great as it would be for many organic liquids; but on the other hand, the internal structure of mercury is probably at least as simple as that of any other known substance, so that in the liquid state the results are not complicated by entrance of polymerization, and on the freezing curve the results are not complicated by the entrance of allotropic forms, as they are in the case of water, for example.

The discussion of the methods and the possibilities of error has necessarily been somewhat minute and painstaking. It is hoped that this will be pardoned when it is realized that the field in which these measurements are made has been hitherto practically unworked, and is one where there are several unusual sources of error to be guarded against. At the same time it is a field in which, by the exercise of proper precautions, an accuracy equal to that of the greater number of physical measurements can be reached. The accuracy of nearly all of the following measurements is of the order of 1/10 per cent.

CONTENTS.

- I. Introduction... 347
- II. The P–V–T Surface of Liquid Mercury................. 351
  - a. Nature of the Experimental Problem — Difficulties in Measuring Compressibility at High Pressures............ 351
  - b. Description of the New Method..................... 354
    - Formulae 357
    - Details of Manipulation.......................... 358
    - Compressibility of the Steel Piezometers......... 362
      - Discussion of the Method..................... 362
        - Hysteresis in the Containing Cylinder..... 364
        - The New Values up to 10,000 kgm.—Temperature Coefficient 366
        - Comparison with Other Results............. 367
  - c. Data on the Compressibility of Mercury............ 368
    - The Method of adjusting the Data................. 369
    - Comparison with Other Results.................... 376
  - d. Discussion of the Results......................... 380
    - Compressibility as a Function of Pressure and Temperature.. 381
      - Dilatation as a Function of the Pressure..... 382
      - Variation with Temperature Imperceptible..... 380
      - The Specific Heats 383
      - The Adiabatic Compressibility 384
      - The Heat of Compression 384
      - Changes of Internal Energy 386
- III. The Change of State, Liquid-Solid................... 387
  - a. The Thermodynamic Data needed — Methods 387
  - b. Co-ordinates of the Freezing Curve by Change of Resistance 388
    - Method and Details of Manipulation 388
    - Data — Method of Adjusting the Values 393
    - Incidental Data during Resistance Measurements .. 394
      - Electrical Resistance of Liquid Mercury under Pressure 395
      - Electrical Resistance of Solid Mercury — Change Solid to Liquid............................... 395
      - Rough Value for Pressure Coefficient of Resistance of Solid 396
      - Comparison with Previous Values for the Resistance of the Solid............................ 397
      - Resistance of an Amalgam under Pressure 398
  - c. The Melting Curve by the Change of Volume Method . 399
    - The Method 399
    - The Apparatus 400
    - Order of Procedure 401
    - Detailed Discussion of Possible Errors 402
      - Leak... 402
      - Elastic After-Effects........................ 402
      - Elastic Deformation — Calculation of Correction 406
      - Temperature Corrections 408
      - Thermal Dilatation of the Transmitting Liquid 410

          Change of State of the Transmitting Liquid 413
          Measurements of Piston Displacement 413
          Pressure Measurements 414
       The Data 414
          Methods of Computation 416
       The Latent Heat — New Value at Atmospheric Pressure . 420
          Recomputed Value for the Specific Heat of the Solid . 422
          Rough Value for the Compressibility of the Solid 423
    d. Change of Volume at Atmospheric Pressure 423
          Probable Error in the Previous Results 424
          The New Method 424
          The New Value 428
          Recomputation of Quantities depending on the Value for
             the Change of Volume 428
          Density of the Solid 428
          Dilatation of the Solid 428
             Direct Experimental Evidence on this Point 429
    e. Subcooling and Superheating 429
IV. Conclusion 431
    Bearing on the Theory of Liquids 432
    Bearing on the Theory of the Change of State, Liquid-Solid . . 433
       Significant Behavior of Latent Heat and Internal Energy. . . 435
V. Summary 437

The P-V-T Surface of Mercury.

The Compressibility of Liquid Mercury.

This first section of the paper is devoted to a determination of the compressibility of mercury at 0° and at 22°. The question of experimental method is important, since the compressibility of mercury is small, and there are various effects which make determinations at high pressures much more difficult than over a smaller pressure range. The section comprises a discussion of the method and the various errors to be avoided, an experimental determination of the compressibility of the steel, which was needed in the computations, the actual data for the compressibility of mercury, with a somewhat detailed discussion of the way of adjusting them so as to give the best results, and finally, a calculation from the data of the change of the various physical properties of mercury under pressure.

One of the chief difficulties in the experimental investigation of the properties of bodies under pressure is the determination of the correction for the distortion of the containing vessel. This is particularly true in the case of compressibility, where the change of shape and volume of the containing vessel is an effect of the same kind as that being measured, and may often constitute a large part of the total change.

To correct for the distortion of the containing vessel only indirect methods are open, as direct observation of the vessel under high pressure is out of the question. All of those indirect methods which are primary, that is, which do not assume the correctness of the work of some other observer, endeavor to determine in one way or another the compressibility of the material of the containing vessel. This may be done by measuring Young's modulus and the bending or the torsion coefficient, and from these calculating the cubic compressibility by the identical relations of the theory of elasticity; or it may be done with the introduction of less questionable assumptions by measuring the linear compressibility and then multiplying by three for the cubic compressibility. This latter method in somewhat different forms has been adopted by Richards [5] and the author. [6] Once the compressibility is known, the change of volume of the vessel is found immediately by multiplying the compressibility by the pressure. This assumes that the containing vessel changes volume without changing shape. This assumption, which is made universally, seems necessary and unavoidable. There can be little question, however, that the assumption is not entirely justified. There are outstanding discrepancies between the determination of the same compressibility by different observers, or by the same observer with different apparatus, or indeed by the same observer with the same apparatus on different occasions, that are beyond the limits of error of the measurements. For instance, four very careful determinations of the compressibility of mercury by de Metz [7] differed among themselves by 5 per cent. The containing vessels were all of the same kind of glass, and the elastic constants had been determined by direct experiment. The only explanation of these discrepancies seems to be that the distortion of the containing vessel is irregular, so that there is change of shape as well as change of volume. The irregularities in the results introduced by irregularities in the piezometer will, of course, be proportional to the total correction introduced by the envelope, and will be greatest in the case of the most incompressible liquids. Thus for mercury in a glass envelope, the correction for the envelope is 60 per cent of the whole effect, while for water it is only 5 per cent.

There can be no doubt that these irregularities are due to lack of homogeneity in the containing vessel, either imperfect annealing or actual variation in the constitution of the material from point to point. It is inconceivable that a perfectly homogeneous body of any shape,

---

[5] Richards, loc. cit., 1907.
[6] Bridgman, These Proceedings, **44**, 255–270 (1909).
[7] De Metz, Wied. Ann., **47**, 706–742 (1892).

when exposed to purely hydrostatic pressure, should suffer anything except mere change of volume without change of shape. If, however, the substance be heterogeneous, so that the elastic constants vary from point to point, then the application of hydrostatic pressure will be followed by change of shape, yielding of one part at the expense of another. If the pressure is high enough, it is conceivable that the elastic limit may be exceeded in some places and not in others, resulting in greatly exaggerated change of shape. Even if the elastic limit is not exceeded, there are other effects which will produce irregular action, such as elastic after-effects and hysteresis, and these effects become rapidly more prominent at high pressures. It follows, therefore, that the irregularities of compressibility determinations over a wide pressure range are going to be more than proportionally greater than the irregularities over a narrow range.

All previous determinations of compressibility have been made under fairly favorable conditions. The work to the highest pressures has been done by Amagat, to 3000 atmos, in which he determined the compressibility of the compressible liquids in a glass envelope. His determinations of the compressibility of the most incompressible liquid, mercury, were only to 50 atmos. The highest accurate work on mercury seems to have been done by Richards to 500 atmos, a comparatively low pressure. Richards also used glass piezometers. The highest previous work of any sort on mercury seems to have been done by Carnazzi,[8] to 3000 atmos, also with a glass piezometer. His accuracy was not such as to justify more than two figures in the final result.

In determining the compressibility of mercury to pressures as high as 12,000 kgm., as it was desired to do in this paper, we have conjunction of the causes most unfavorable to accuracy; low compressibility of the liquid, and greatly exaggerated irregularity in the distortion of the containing vessel. The employment of a glass containing vessel would seem to be out of the question. In a previous paper,[9] the linear compressibility of glass has been determined directly to 6800 kgm. Over this lower range, irregularities were found amounting to 4 per cent. In using a glass piezometer, this irregularity would be magnified by the fact that the compressibility of glass is high compared with that of the mercury. The ideal substance is one which can be obtained in a state of great homogeneity, and which at the same time has as low a compressibility as possible. Steel has these properties in a

---

[8] Carnazzi, Nuov. Cim. (5), 5, 180–189 (1903).
[9] Bridgman, loc. cit.

higher degree than any other substance, the compressibility being as low as any except the platinum metals, and the homogeneity in annealed pieces of soft steel being nearly perfect. The use of a piezometer of steel in determining the compressibility of mercury up to 6800 kgm. has been described in the paper cited above. The advantages of the steel piezometer have been discussed there, and independent experimental evidence has been presented showing that the compressibility of a piece of mild steel is the same in every direction, and that therefore the piezometer does not change its shape. The results obtained with this method were regular, and in so far justify the use of this material; but further work has led to the conviction that there was present in the method a source of error for which it is hopeless to attempt to calculate the correction. This error is concerned with the leak past the freely moving piston; this leak is less on increasing than on decreasing pressure, so that the previous results are all too small. It seems impossible to determine in what way the error so introduced will vary with the pressure. The total magnitude of the error is probably between 2 and 3 per cent.

The method adopted here attempts to keep the advantage of the steel piezometer, while doing away with the uncertainty of the freely moving piston. It was not possible to secure so regular results with the new method as with the old, but on the other hand, any constant source of error seems to be excluded. The method is a modification of one used by Aimé [10] in 1842, and since used in many modified forms by other experimenters. In the original form, the liquid to be investigated was placed in a glass bulb provided with a fine capillary communicating with a vessel containing mercury. The whole affair was then lowered into the sea. Here the increase of pressure forced mercury into the bulb, where it fell to the bottom and remained. On withdrawing from the sea, the maximum volume compression was measured simply by weighing the mercury forced in. Aimé's results were very irregular and incorrect. The most obvious trouble with this method is the tendency of the mercury to collect in drops at the mouth of the capillary. On release of pressure, this drop flows back through the capillary, and the resultant compressibility appears too low. The difficulty may be minimized by making the capillary very small. This is an easy matter when the capillary is made of glass, but when steel is used as here, some special construction must be devised. The form of piezometer adopted is shown in Figure 1. The upper piece B, screws into the shell A. A tight joint between the two is made simply

---

[10] Aimé, Ann. de Chim. et Phys., **8**, 257–280 (1843).

by tightly forcing the narrow edge C on the upper part B into the flat seat at D. It is not desirable to attempt to use any packing material. At the lower part of B there is a very fine channel, E, opening into A. This channel is made by carefully drilling in B a 1/16 inch hole and then plugging the hole with a pin along which a fine scratch has been made. This fine scratch takes the place of the glass capillary when the piezometer is made of glass. It is thus possible to regulate the size of the channel at pleasure. It is a matter of the greatest ease to make such a channel that a pressure of only a few pounds to the square inch will force mercury through it in drops just barely perceptible to the eye, and much too small to be measured by any ordinary balance. In all the work with these piezometers no effect was ever found suggesting in any way the possibility of error introduced by clinging of mercury to the mouth of the channel.

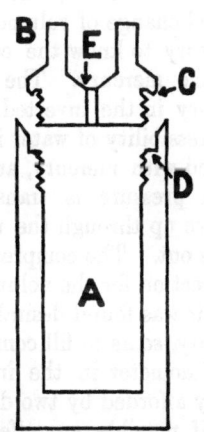

FIGURE 1. The steel piezometer. This is filled with the liquid under investigation and exposed to hydrostatic pressure all over. The pressure forces a measurable quantity of mercury into the chamber A through the fine channel at E, and from this quantity of mercury the change of volume of the liquid originally filling A may be found.

The piezometer may be used in either of two positions, upright or inverted. When used upright, the body A is filled with the liquid to be investigated, water for example, and the cup at B with mercury. The entire piezometer is then placed in a pressure chamber surrounded on all sides by a liquid, to which pressure is applied. This takes the place of lowering into the sea in Aimé's experiment. Mercury flows in through the narrow channel to equalize the pressure within and without, and drops to the bottom. On release of pressure, the water bubbles out through the mercury in B. The piezometer is then unscrewed, the mercury in A weighed, and the total change of volume at the maximum pressure calculated. Into this calculation enter the compressibility of the steel of the piezometer and the compressibility of the mercury. The latter can then be found in terms of the compressibility of water by nearly filling the piezometer with mercury on which floats a little water to fill the piezometer completely. On application of pressure mercury drops through the layer of water, and on release water comes out.

In the inverted position, a sufficient quantity of mercury is placed in the piezometer to cover the bottom, and the fluid with which the piezometer is surrounded in the pressure chamber is chosen the same as that within the piezometer. On the application of pressure, this fluid is forced in, rising through the mercury, and on release the mercury comes out. The quantity of mercury left is weighed, from which the desired change of volume of the fluid is calculated. Here again it is necessary to know the compressibility of the steel of the piezometer and the mercury. The determination of the compressibility of the mercury in the inverted position is as simple as that for finding the compressibility of water in the erect position. The entire piezometer is filled with mercury, and placed inverted in the water by means of which pressure is transmitted. On application of pressure, water bubbles up through the mercury, and on release of pressure, mercury comes out. The compressibility of the water is involved here only as a correction for the volume of the water forced in. In practice, however, it was found desirable to include a little water initially with the mercury, so as to fill completely all the corners. The object of using the piezometer in the inverted position, aside from the check on accuracy afforded by two different arrangements of the apparatus, was to find if possible any effect of the drop clinging to the mouth of the channel. In view of the fact that the surface tension of mercury is greater than that of water, it seemed plausible that the minimum size of a bubble of water that would detach itself from the mouth of the channel and rise through the mercury would be less than the drop of mercury that would fall through water. No such difference could be found, however, in any of the work.

It has been noticed that the method demands the knowledge of two compressibilities besides that of the fluid to be investigated. In the case of mercury these two compressibilities are those of steel and water. The correction for the steel was determined independently by measuring the change of length of a rod under pressure and will be discussed more in detail later in this paper. The effect of the water is small at the low pressures, being simply the change of volume of the slight amount of water forced in, but at higher pressures, where the water may lose as much as 20 per cent in volume, the correction for the change of volume of the water will be 20 per cent for the inverted position and 30 or 40 per cent for the upright position. Given, then, the correction for the steel, it is still necessary to run two sets of determinations to get the compressibility of either water or mercury, and naturally these two determinations give the compressibility of both water and mercury. The data for the water are to be given in a following paper. In calculating the compressibility from these two sets of data, one for the water and the other for mercury, the method of successive approximations was applied. The compressibility of water was first calculated, assuming the compressibility of the mercury to be constant at its initial value at atmospheric pressure. A curve was then plotted giving the volume of water against pressure. From this curve for water, an improved curve for mercury was calculated, from which again a better curve for water was found. In practice it was not necessary to carry the steps further than this.

The relations connecting the compressibility of the water and the mercury with the observed weights of mercury, etc., are given below.

Notation.

V = initial volume of Piezometer,
V_{H_2O} = " " " water,
V_{Hg} = " " " mercury.

Then $V = V_{H_2O} + V_{Hg}$.

Δv_{H_2O} = shrinkage per initial unit vol. of water at pressure under consideration.

Δv_{Hg} = " " " " " " mercury " " " consideration.

Δv_{Fe} = " " " " " " steel " " " consideration;

and in addition:
in the upright position
ΔV_{Hg} = vol. at atmos. pressure of mercury forced in by the pressure in question.
in the inverted position
ΔV_{Hg} = vol. of mercury forced out, etc.

ΔV_{Hg} is the same thing as the volume of the water forced in. The corrected volume of the piezometer at any pressure is evidently $V - V \cdot \Delta v_{Fe}$. It is a simple matter to verify the following equations:

Upright position

$$\Delta v_{H_2O} = \frac{\Delta V_{Hg} + V \Delta v_{Fe} - \Delta v_{Hg}(\Delta V_{Hg} + V_{Hg})}{V_{H_2O}},$$

$$\Delta v_{Hg} = \frac{\Delta V_{Hg} + V \Delta v_{Fe} - V_{H_2O} \Delta v_{H_2O}}{V_{Hg} + \Delta V_{Hg}}.$$

Inverted position

$$\Delta v_{H_2O} = \frac{\Delta V_{Hg} + V\Delta v_{Fe} - V_{Hg}\Delta v_{Hg}}{V_{H_2O} + \Delta V_{Hg}},$$

$$\Delta v_{Hg} = \frac{\Delta V_{Hg} + V\Delta v_{Fe} - \Delta v_{H_2O}(V_{H_2O} + \Delta V_{Hg})}{V_{Hg}}.$$

The following gives a discussion of the minor experimental details that it was necessary to observe.

Considerable care is necessary in filling the piezometer to ensure complete exclusion of air. The procedure in filling was as follows. The piezometer A with the cover B held loosely over it in a suitable frame was placed in a test tube containing distilled water. The test tube was then connected to the air pump and the water boiled under reduced pressure at room temperature, removing the occluded air. The piezometer was then removed from the test tube, the weighed quantity of mercury introduced, and then replaced in the test tube which was again exhausted as before. The mercury had been cleaned with acid and was freshly distilled. Finally, before raising the piezometer above the surface of the water, the cap B was screwed into place with a simple key. Removing from the test tube, the cap B was tightened home by applying a spanner.

This tightening of the cap must be done with especial care, and to irregularities in this most of the irregularities in the data could be traced. Since the piezometer is made of a very mild steel, it is a comparatively easy matter to screw the cap in too far, stripping the thread or shearing off the shoulder. This was avoided by graduating around the top of A, and always screwing B in to the same mark. But even at best, to secure a tight joint, it is necessary to force the cap in pretty tightly, exceeding the elastic limit locally, and thereby introducing probable irregularities as explained before. When the method was first used, a great deal of trouble was found from such irregularities. This was shown most convincingly by the fact that the volume of the piezometer did not remain constant, but changed very slightly after each application of pressure, usually becoming less. This change of volume is not entirely due to the effect of the joint, because it could be mostly removed by annealing. Nevertheless, since the effect of annealing is to remove internal strains, and since internal strains are doubtless introduced at the joint, there is probably some irregular action introduced at the joint. This source of error remains the chief objection to the method, although it can be greatly reduced by proper manipulation. This consists in careful previous annealing, and seasoning by subjecting to pressure several times before the series of measurements is begun. One at least of the previous applications should be to a pressure as high if not higher than that to be reached during the measurements. After treatment like this the piezometers show no further change of volume, and the results with any one are fairly regular. It is a question, however, how good a criterion uniformity of results and permanency of volume are for the uniform compression of the piezometer without change of shape. It must be confessed there seems no valid reason why the piezometer after proper seasoning should not settle down into a steady state in which it gives consistent results with change of shape, and probably it does. The difference of the results with different piezometers is probably to be explained by this effect. The best that can apparently be done is to take the mean of the determinations with several piezometers, all of which individually give consistent results. All of these considerations apply chiefly to determining the compressibility of the incompressible mercury; they have much less weight, and indeed are almost negligible when applied to water.

There is one particular way in which it is very easy to produce irregular results by strains in the piezometer, and one which caused much trouble before the correct explanation was found. If the pressure is pushed too high, the water freezes with sudden decrease of volume. This was not expected when entering on this work, since the freezing curve of water as found by Tammann[11] gives no evidence of the existence of ice above zero. The fact is that there are other forms of ice besides those found by Tammann, one of which is stable at room temperatures. This will be treated in a following paper. The sudden decrease of volume during freezing is accompanied by a rush of liquid through the narrow channel, and, while this rush is taking place, by a momentary excess of pressure outside over that inside. This excess pressure may produce volume set, or if not sufficiently great for that, at least set up internal strains which produce irregularities on subsequent determinations. The irregularities so introduced may be comparatively very high; as much as 10 to 15 per cent in the case of mercury on the next determination. The next value is nearly always too low. The irregularities become smaller with subsequent determinations, but the only satisfactory way is to anneal the piezometer again. Many of the early data had to be discarded because of this freezing effect.

As a consequence of the unforeseen freezing of water at temperatures

---

[11] Tammann, loc. cit., p. 315–344.

above zero under high pressure, it is impossible by this method to push the compressibility determinations of mercury to the freezing curve of mercury, as was desired. The difference between the freezing pressure of water and mercury is not very great, comparatively speaking, being about 1500 kgm./cm.^{2} at 15°. By pushing the pressure into the unstable region for water, it is possible to come closer than this to the freezing curve for mercury. At 0° the approach was made to within 700 kgm., and at 22° to within 1000 kgm. Attempts to come closer than this failed. On several occasions the two piezometers, one filled with water and the other with mercury and water, were subjected to pressure simultaneously beyond the freezing pressure of water. The water alone successfully withstood this subcooling, but the water with the mercury always froze. Even when freezing took place, a rough value for the compressibility could be found by using in the calculations the change of volume of the water on freezing. A compressibility for the mercury was thus found lying on a smooth curve with the values at lower pressures, but of course any such procedure as this is questionable. In view of the strong probability that the compressibility of mercury shows no unusual features in the unreached region, it seemed unnecessary to take the measures that would be needed to explore this region, such as the employment of some auxiliary fluid other than water whose freezing point is higher than that of mercury.

It may be mentioned as confirming these views as to the part played by internal strain in making the results irregular, that the irregularity was always much greater after the freezing had taken place after high super-pressures; that is, after the freezing presumably had been most rapid, and the excess pressure on the outside with its resulting deformation greatest.

It will be noticed that the method is essentially an integrating arrangement for measuring the total increase of pressure during the process; that is, at every increase of pressure mercury is forced into the piezometer whether the previous maximum is thereby exceeded or not. In using the method it is essential, therefore, that pressure should be increased continuously to the maximum with no retrogressions. The form of pressure apparatus used made this particularly easy to accomplish. Pressure was produced by a hydraulic ram, the larger piston of which was 6 inches in diameter and the smaller 1 1/8 inches. The 6-inch piston was actuated by the pressure pump of the Société Genevoise, the pressure being run as high as 600 kgm./cm.^{2}.

The motion of the piston was very slow because of the large volumes involved. About 50 strokes of the pump were necessary for an increase of pressure on the high pressure side of 1000 kgm./cm.^{2}. Furthermore, the friction of the packing material, while not very great, retarded greatly the quickness with which the piston moved in response to an increase of pressure. The piston might continue to move for some minutes in response to an increase of pressure on the low-pressure side. The result was that the step-like advance of pressure with each stroke on the low-pressure side was entirely wiped out and converted into a continuous advance on the high-pressure side. The only way in which a retrogression of pressure was likely to occur was by dissipation of the heat of compression. If pressure is rapidly pushed to a maximum, there is produced an increase of temperature. The final equilibrium pressure is lower than the maximum, which corresponds to an unknown temperature. The difficulty was avoided by compressing so slowly that as the pressure approached the maximum the compression was sensibly isothermal. With the apparatus so designed as to reduce the volume of the transmitting fluid to a minimum, it was easy to attain this condition. By comparing results with slower and more rapid compressions a safe rate was found. The maximum pressure of 12,000, for example, could be reached with entire safety in seven or eight minutes. On releasing pressure there was the reverse effect due to cooling, but it is readily seen that there is no error introduced here unless there is actual retrogression of pressure, which is just as easy to avoid as with increasing pressure.

The temperature was kept constant during the compressibility determinations by surrounding the pressure cylinder with a suitable bath, at 0° with ice and water, and at 22° with an electrically regulated thermostat. The piezometers were always submerged for some minutes in this bath before the final adjustments were made. Thus if the compressibility of water were being determined with the piezometer in the upright position, the piezometer was submerged in a test tube of water placed in the bath without the mercury in the upper cup. After temperature equilibrium had been reached, the upper cup was carefully dried with filter paper and the mercury introduced, the lower part still being in the bath. If the piezometer was to be used in the inverted position the procedure was more simple, merely waiting for temperature equilibrium while submerged in the upright position under water or mercury as the case might be. The variations of temperature in the bath at 22° were never more than a few hundredths of a degree. The actual temperature was read with a standardized thermometer.

The volume of the piezometers was found by weighing when full of water at a known temperature and when empty. The filling was performed in the manner above and was more satisfactory than filling with mercury and weighing, since by the use of water it was possible to entirely fill the cracks. The total volume of the piezometers was 1 or 2 cm.³ for water and 3 or 4 cm.³ for mercury. These small dimensions were necessary to keep the size of the apparatus down within reason, as it was necessary to have the walls of the retaining vessel very heavy in order to withstand the pressures. The cylinder in which the piezometers were placed had a hole 9/16 inches in diameter and was 4 inches on the outside. The small volume of the piezometers places no restriction on the accuracy of the results, however. The weight of the mercury involved in the compression might rise to 3 or 4 gm. for the higher pressures. Weighings were made to 0.0001 gm., so that any error in the weighings is entirely negligible.

Pressure measurements were made with an absolute gauge. This gauge and the various precautions to be observed in its use are described in a previous paper.

Compressibility of the Steel Piezometers.

The experimental determination of the correction for the compressibility of the steel piezometer demands special consideration. The fact must be emphasized that there is no method of determining compressibility in which the compressibility of all the substances concerned can be determined directly, that is, by the application of hydrostatic pressure. There is always a residuum, whatever the method. Thus in the present work, the compressibility of both the water and the mercury involve the compressibility of the steel containing vessels. If an attempt were made to determine by independent experiment the cubic compressibility of the steel, the compressibility of the vessel in which the steel was contained would turn up as a new unknown. The difference of the compressibility of two substances is all that it is possible to get by direct experiment. To get the absolute compressibility of either, the compressibility of the other, the unknown residuum, must be determined by some indirect method. To ensure the greatest accuracy, this residuum should be so chosen as to be as small as possible in comparison with the compressibility in question. Steel, as has been mentioned, is an admirable substance from this point of view.

There are a variety of methods open for the indirect determination of compressibility, all of which depend in some way on the theory of elasticity. A method frequently adopted is to measure two independent elastic constants of a material, such as Young's modulus, or the torsion coefficient, or the bending modulus, and from these to calculate the compressibility by the identical relations of the theory of elasticity. In applying the identical relations, the assumption must be made that the material obeys Hooke's law within the stress range employed, and that it is perfectly homogeneous and isotropic. With these assumptions the identical relations are rigorously exact. The first of these assumptions seems open to but little question. It is the second which is likely to give the most trouble, particularly when the substance experimented on is in the form of hollow tubes, as when the elastic constants have been determined for glass. This method was employed by Amagat [12] and by de Metz,[13] among others.

Another method is to measure the linear compressibilty of the substance in question when subjected to hydrostatic pressure all over and to calculate the cubic compressibility simply by multiplying the linear compressibility by three. This seems preferable to the method of the last paragraph, because the assumptions are reduced to a minimum. The stress, a hydrostatic pressure, is in this method the same during the indirect determination of the cubic compressibility as it is during the actual use of the material in the piezometer. There seems much less chance for complications to be introduced by nonhomogeneity of the material than there is when the compressibility is calculated from two elastic constants, such as Young's modulus and the torsion coefficient. The only assumption made here is that the material is equally compressible in every direction, an assumption which is open to verification by direct experiment. This verification has been made by previous experiments in the paper cited. This method has also been used by several previous experimenters, for example Buchanan,[14] Amagat,[15] and Richards.[16]

In the paper already cited, the linear compressibility of steel was determined at room temperatures up to 6500 kgm./ cm.^{2}. These data were inadequate for the present purpose, however, because they did not run to high enough pressures, and the effect of temperature on compressibility was not determined. But the method there used is entirely satisfactory, and new results have since been obtained by it for the purposes of this paper. Briefly, the method consists in enclosing the metal to be experimented on, which is in the form of a rod, in a long steel cylinder, within which it is exposed to the action of hydrostatic pressure all over. The change produced by the pressure in the length of the rod with respect to that of the cylinder is measured by a ring

---

[12] Amagat, C. R., **107**, 618–620 (1888). Ann. de Chim. et Phys. (6), **22**, 95–141 (1891).
[13] De Metz, loc. cit.
[14] Buchanan, Proc. Roy. Soc. Lon., **73**, 296 (1904).
[15] Amagat, C. R., **108**, 727–730, 1889.
[16] Richards, loc. cit.

slipping on the rod. The change of length of the cylinder itself is measured directly from the outside.

In the previous work the rod was enclosed in a cylinder of soft tool steel. The maximum pressure, 6500 kgm., was beyond the elastic limit of this steel, so that the material did not remain isotropic. That the elastic limit was exceeded is shown by the considerable hysteresis in the observed elongation of the cylinder. The fact that there is hysteresis suggests that possibly there may be also some slight degree of warping. It was stated in the former paper that the only assumption made in the whole work was in regard to this matter of warping, that is, whether the change of length of the cylinder is the same internally as externally. The effect is probably slight, and in the absence of any plausible way of calculating it was neglected altogether.

Apart from the question of the method used in the present work, and not affecting its validity in the least, there was an error made in the previous paper in the calculations from the data. Correction for this error increases the former value for the compressibility of this steel from 5.30×10^{-7} to 5.59×10^{-7} at 20°. This also will change the value for the compressibility of mercury, as will be explained later.

The tool steel cylinder was not available for the present work, because it would not reach the pressure reached here, 10,000 kgm./cm.^{2}. Another cylinder of nickel steel of the same dimensions as the former one of tool steel was made, therefore, and hardened in oil. With this heat treatment, the elastic limit should be somewhere near 15,000 kgm. The pressure was purposely kept at a low value, never exceeding 10,000 kgm., so as to run no risk of producing set in the inner layers of the cylinder. That no such heterogeneity was introduced is made probable by the fact that the cylinder showed no hysteresis, the relation between extension and pressure being linear both for increasing and decreasing pressure. The former measurements of this extension were made directly with a microscope, reading to 0.001 mm. For this work here, a much more sensitive scheme was used, an adaptation of the Maarten's mirror device so fast coming into general use among engineers. The magnification employed here gave a motion of the scale of 5 cm. for an actual elongation of 0.025 mm. The sensitiveness is, therefore, at least twice that of an interference system, there is also the advantage of great simplicity and steadiness, and best of all, the reading is given directly on a scale, there being no necessity for keeping track of fringes. The Maarten's mirror device disclosed no hysteresis in the elongation of the nickel steel cylinder. The freedom from heterogeneity as shown by the absence of hysteresis makes plausible also the absence of warping, although no direct measurements of this warping were possible.

An attempt was made to show directly the presence of warping by measurements of the change of length of the exterior of the cylinder. Measurements of the extension were made over the middle third and the middle two thirds of the cylinder. Any very large warping might be expected to destroy the proportionality of extension to length, since presumably the warping is confined to the neighborhood of the ends. No such effect could be found, however, within the limits of error which were about 1 per cent.

The argument for the probable absence of warping may be stated as follows. The change of internal length of the cylinder with which we are concerned was measured between two points at each of which the internal diameter of the cylinder undergoes an abrupt change. If the elastic limit of the steel were exceeded, the resulting set and heterogeneity would naturally appear first at these places of sudden change in the dimensions. The effect of heterogeneity at these places would be to produce warping, that is, sections originally plane would no longer remain plane. On the other hand, if the material were never strained beyond the elastic limit, the warping would be negligible. Now the cylinder has been shown by direct experiment to show no hysteresis, and therefore probably not to have been strained beyond the elastic limit. The cylinder, therefore, probably also shows no warping. In any event, one would seem justified in accepting the readings of a cylinder showing no hysteresis in preference to one where the hysteresis might amount to 16 per cent. The effect of warping, whatever it is, can be only slight, since it is itself only a correction on a 5 per cent correction term.

The new determinations made for this work with the nickel steel cylinder were made on the same specimens as those formerly used, which were from the same piece of steel as the piezometers. Determinations were made at two temperatures, $10°$ and $50°$, and up to 10,000 kgm./cm.^{2}. There were only a few changes from the experimental procedure of the previous work. Pressure was measured by measuring the change in the electrical resistance of a coil of manganin wire, instead of a capillary of mercury. This is the pressure gauge that was used in the later work of this paper, and it has been described in detail in a previous paper. The fluid transmitting pressure was usually kerosene instead of the former mixture of water and glycerine. At $10°$ and at pressure above 8000, the kerosene becomes so viscous that the ring on the test rod no longer slides freely. The great irregularity of the points first obtained was traced to this effect. It may be avoided by using gasolene instead of kerosene at the higher pressures. At $50°$ the kerosene may be used without trouble over the entire pressure range. The results obtained were not quite so regular as the former results, doubtless because it was impossible to place the cylinder in a position so favorable to the entire removal of all particles of grit after every application of pressure. In this work the cylinder was placed

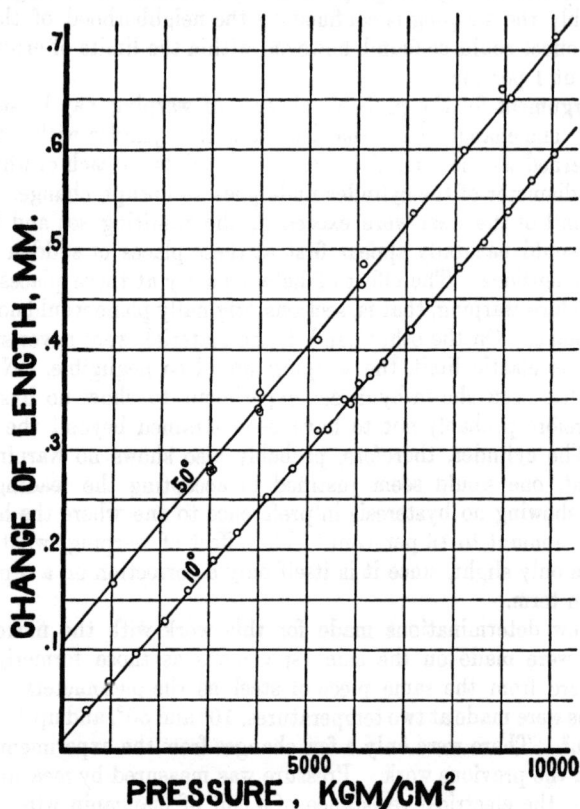

FIGURE 2. Shows the change of length at two temperatures of a bar of bessemer steel 29.95 cm. long under uniform hydrostatic pressure. The zero reading for each of these temperatures is arbitrary.

vertically in a thermostat, whereas in the former work it was placed horizontally.

The results are shown in Figure 2. The values found for the compressibility were 0.0_6583 at 10° and 0.0_6601 at 50°. The result of the previous determination in the tool steel cylinder at 20° was

0.0_6559. A repetition of the former experiment with the tool steel cylinder gave 0.0_6568 at 20°. It is to be noticed that this value is higher than that previously found, and in the direction toward that shown by the nickel steel cylinder. During the repetition of the experiment the tool steel cylinder also showed less hysteresis than in the early determinations, probably due to the gradual disappearance of the heterogeneity introduced by the previous high pressure during the two and one half years for which the cylinder had been resting. During this last set of determinations the pressure was never raised above 4000 kgm., which is below the elastic limit of the steel.

Whether the true explanation of the discrepancies is to be found in the warping of the cross-section of the cylinder or not, the fact seems to be that when the elastic limit of the material is exceeded there is resulting hysteresis in the elongation, and that this hysteresis is accompanied by low values for the compressibility, the greater the hysteresis, the lower the values. It seems to be justifiable therefore, to accept as most probably correct the values given by this new determination with the nickel steel cylinder. The values given above and the temperature coefficient deduced from them were used in the following computations.

It is to be noticed that the elongation of the nickel steel cylinder under internal pressure, which has appeared in the above as a correction factor, is sufficient to give the cubical compressibility of the nickel steel. This is because, as was pointed out by Mallock [17] in 1904, the elongation of a hollow closed cylinder under internal pressure involves only one elastic constant, the cubic compressibility. This gives for the cubic compressibility 6.2×10^{-7} at 20°. The results are not accurate enough to give the temperature coefficient by this method. The high value of this result compared with that by the direct observation of linear compressibility (62 against 58), is probably due to the difference in the materials, the latter value being for an almost pure iron, while the former is for a nickel-chrome alloy with little or no carbon. (Krupp's trade number for this steel is E. F. 60.0.) It was shown in the previous paper that carbon alone makes very little difference in the result.

This same method, the elongation of a hollow cylinder under internal pressure, has been recently used by Grüneisen [18] in determining the cubic compressibility of various metals at different temperatures. His results for two different specimens of iron at 18° varied from $58. \times 10^{-8}$ to $63. \times 10^{-8}$. The temperature effect on this last speci-

---

[17] Mallock, Proc. Roy. Soc. Lon., **74**, 50 (1904).
[18] Grüneisen, Ann. Phys., **33**, 1239–1274 (1910).

men was found to be 3.1×10^{-8} for 100°, against 1.8×10^{-8} for 40° as found above. Grüneisen does not give the temperature coefficient for the iron with the lower compressibility.

In view of the agreement of the results found here with those of Grüneisen, there can seem to be little doubt but that the results found by Richards [19] are affected by some constant error as has been claimed by Grüneisen.[20] The fact that the results here were found by as near an approach to a direct method as possible, that is, by the measurement of the linear compressibility, which also was the method used by Richards, and the fact that these results agree with those of Grüneisen, which were obtained by more indirect methods from the theory of elasticity, would seem to nullify Richards' contention that the results of other observers have been in error because the theory of elasticity gives results 30 per cent out of the way. Richards found for iron 38.5×10^{-8} against 58.3×10^{-8} above. Since his value for the compressibility of mercury depends directly on his value for the compressibility of iron, the effect of the correction will be to change very appreciably the value for the mercury, as will be mentioned later. The source of error in Richards' work is not entirely clear. He used an electrical method, in which contact was made with a drop of mercury resting on the top of the rod of iron to be measured. Change of shape of this drop of mercury under pressure seems the most likely source of error.

The Data for the Compressibility of Mercury and Calculation of Results.

Determinations were carried out at two temperatures, 0° and 22°. The range is sufficient to give the variation of compressibility with temperature. To get the second temperature derivative of compressibility, observations at several other temperatures would have been necessary, as well as a higher degree of accuracy than was possible to attain in this work. The final values given are the mean of a large number of determinations. At 0°, five different piezometers were used, and over 90 determinations made over the pressure range from 0 to 7000 kgm. At 22°, two piezometers were used, and 38 points determined between 0 and 11,000. The points at 0° were obtained first and show considerable irregularities. As familiarity was gained with the method, and the several sources of error mentioned above became recognized, it was found possible to obtain more

---

[19] Richards, loc. cit.
[20] Grüneisen, Ann. Phys., 25, 849 (1908).

regular results. The values at 22° were found after this experience had been gained, and with the two piezometers giving the best results at 0°.

The points obtained with all five piezometers at 0° are shown in Figure 3, and the points with the best two of these piezometers in Figure 4. The points found with the two piezometers at 22° are shown in Figure 5. The greater regularity of these points is manifest. In both of these figures not all the points actually obtained have been represented, for four or five points, obtained directly after the freezing of the water at too high a pressure, have been discarded.

The data were adjusted and the final values computed in the following way. The work was done independently at each temperature. A straight line was first drawn passing through the assemblage of points. It was not necessary that this line pass through the origin. The deviation of the observed points from the points calculated by the linear formula was then determined. The deviations from this line for each piezometer separately were then plotted on a very much enlarged scale against pressure, and a smooth curve drawn through these deviation points. This deviation curve, together with the fundamental linear equation, gave the best smooth curve connecting change of volume with pressure as determined by each piezometer separately. From these best smooth curves (five at 0°), points were determined at even pressure intervals, 500 or 1000 kgm. apart. At each of these evenly distributed pressures, the weighted

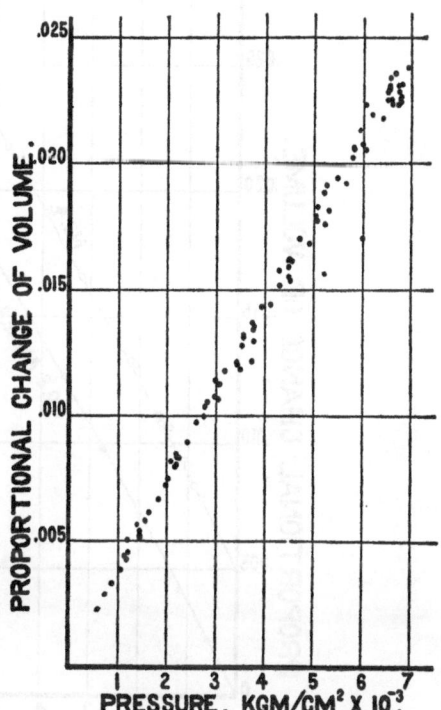

FIGURE 3. Collection of all the results with five piezometers for the compressibility of mercury at 0°.

mean of these five points was taken as the best point at this pressure. Weights were assigned to the smooth curves obtained with each piezometer inversely as the arithmetic difference between the observed deviation points and the smooth deviation curve. The weights at 0°

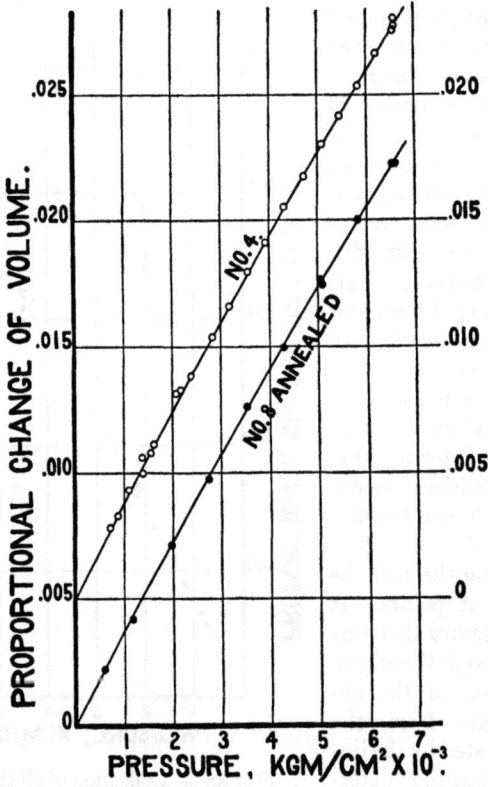

FIGURE 4. The results with the two best piezometers for the compressibility of mercury at 0°. The right-hand vertical scale refers to No. 4, the left-hand scale to No. 8, annealed.

varied from 1 to 7. Finally, a parabola was passed through these best points, the deviations of these weighted best points from the parabola were plotted against pressure, and the smooth curve through these deviations, together with the parabolic formula, was taken as giving the best possible relation between change of volume and pressure.

This was the process used in the final determinations. It has been already explained that the calculation was by means of a method of successive approximations, and that it was necessary to calculate the compressibility of the water as well as that of the mercury. The de-

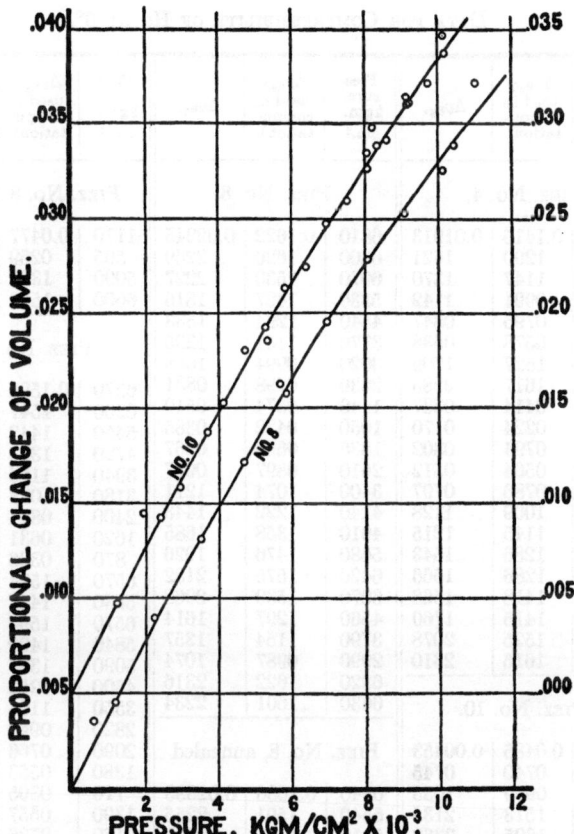

FIGURE 5. The change of volume of liquid mercury at 22° as obtained with two piezometers. The right-hand scale is for No. 10, the left-hand one for No. 8.

tails of this approximation process have been discussed already in the general discussion of the method. Reference is made to a following paper for the values for the compressibility for water. The procedure in the first steps of the approximation process did not need to be so elaborate as in the above final calculation.

The data for 0° and the steps in the final process of adjustment are shown in the tables. Table I. shows the points in the order in which

TABLE I.

Data for Compressibility of Hg at 0°.

Pressure, kgm. cm.²	Δv_{H_2O} (used in computation).	Δv_{Hg}.	Pressure, kgm. cm.²	Δv_{H_2O} (used in computation).	Δv_{Hg}.	Pressure, kgm. cm.²	Δv_{H_2O} (used in computation).	Δv_{Hg}.
Piez. No. 4.			**Piez. No. 8.**			**Piez. No. 8 (cont.).**		
5270	0.1415	0.01913	6810	0.1622	0.02245	1170	0.0477	0.00425
4520	1290	1621	6800	1620	2259	595	0259	0224
3750	1147	1370	6070	1530	2227	5090	1387	1771
3000	0991	1142	5330	1427	1816	6600	1598	2259
2220	0796	0847	4540	1294	1533			
1450	0578	0538	3770	1151	1295	**Piez. No. 11.**		
6050	1527	1709	3020	0994	1068			
6770	1617	2283	2230	0798	0831	6570	0.1594	0.02298
1000	0414	0337	1440	0574	0519	6200	1547	2197
500	0224	0170	1050	0432	0385	5450	1442	1941
2210	0794	0802	1830	0690	0667	4720	1325	1704
1410	0564	0512	2610	0897	0977	3940	1184	1434
2180	0786	0797	3400	1074	1211	3180	1028	1179
3090	1009	1128	4140	1222	1443	2400	0844	0896
3740	1145	1215	4910	1358	1685	1620	0631	0617
4500	1286	1543	5680	1476	1926	870	0366	0335
4500	1286	1566	6420	1575	2182	6570	1594	2310
5290	1418	1568	6070	1530	2068	5830	1497	2064
5270	1415	1760	4560	1297	1614	6570	1594	2340
6030	1525	2078	3790	1154	1357	5840	1499	2064
6760	1616	2310	2990	0987	1074	5090	1387	1829
			6820	1622	2316	4300	1252	1578
Piez. No. 10.			6630	1601	2234	3550	1106	1320
						2820	0948	1054
1190	0.0485	0.00453	**Piez. No. 8, annealed**			2090	0766	0819
2010	0740	0745				1380	0553	0562
2780	0939	1035	6940	0.1636	0.02383	710	0305	0288
5950	1513	2135	6540	1591	2255	1390	0557	0502
6670	1605	2362	5810	1495	2027	2180	0786	0835
694	0298	0297	5050	1380	1790	1550	0610	0582
3570	1110	1310	4300	1251	1521	1110	0455	0441
4490	1284	1620	3540	1104	1279	2190	0788	0836
5220	1407	1881	2750	0931	0998	6540	1590	2289
			1980	0730	0722			

they were determined. The actual computation from the data it is not necessary to show. The formula has already been given by which these computations were made. In Table I., the values for the change of volume of the water, which entered into the computation, are also given, as this effect is rather large. Roughly, the correction introduced by the water was about one third of the total effect due to the change of volume of the mercury. These values for water were taken from the values to be given in greater detail in the following paper.

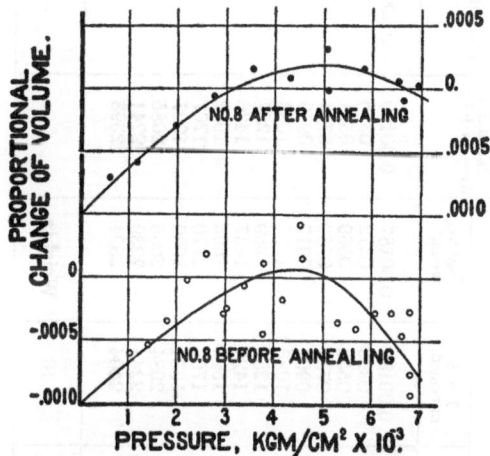

FIGURE 6. The deviations from linearity of the points obtained with one piezometer. This shows the irregularities in the results introduced by internal strains.

Figure 6 gives the deviations of the observed points of two of the sets of readings of Table I. from a straight line. This line, of the form $\Delta V = a + bp$, does not happen to pass through the origin, but gives $\Delta V = 0.001$ at $p = 0$. All the deviation curves, therefore, start from the point $\Delta V = -0.00100$. In Table II. are given the values calculated with the help of the deviation curves for the five piezometers at pressure intervals of 500 kgm., together with the weighted mean of these values. Finally in Table III. are shown the values calculated by the parabolic formula, the weighted experimental values, the deviations, and the finally accepted values, obtained with the smooth deviation curve shown in Figure 7.

The change of volume at any pressure and 0° may be found, either by constructing a curve through the points of Table III., or more accurately by computing at any pressure the change of volume by the formula

$$\Delta v = ap + bp^2,$$

TABLE II.

Adjustment of Data — Final Values of Compressibility of Hg at 0°.

Pressure, kgm. cm.³	Δv_{Hg} at 0°.						Calculated $\Delta v = ap + bp^2$	Deviation.	Final value.
	No. 4.	No. 8.	No. 10.	No. 11.	No. 8 annealed.	Weighted mean.			
500	0.00180	0.00180	0.00196	0.00192	0.00187	0.00187	0.00190	−0.00003	0.00189
1000	0362	0361	0389	0385	0369	0372	0376	−4	0374
1500	0549	0543	0569	0577	0553	0560	0561	−1	0559
2000	0737	0721	0753	0763	0731	0739	0741	−2	0739
2500	0921	0899	0931	0945	0909	0917	0919	−2	0918
3000	1100	1070	1110	1120	1086	1096	1094	+2	1095
3500	1272	1246	1285	1292	1259	1269	1267	+2	1269
4000	1442	1414	1458	1458	1428	1437	1436	+1	1439
4500	1606	1576	1630	1628	1596	1605	1602	+3	1605
5000	1777	1735	1795	1793	1761	1770	1767	+3	1769
5500	1923	1889	1955	1954	1934	1929	1927	+2	1929
6000	2073	2039	2111	2111	2084	2085	2085	0	2085
6500	2220	2192	2244	2268	2244	2239	2241	−2	2240
7000	2366	2344	2390	2422	2404	2394	2393	+1	2393
....	4	2.5	1.5	5	10	Weights	....	....	....

Calculated values are by the formula
$$\Delta v = ap + bp^2$$
$$a = 0.0_3 3818$$
$$b = -\log^{-1} 19.7569 - 10$$

TABLE III.

Data for Compressibility of Hg at 22°.

Pressure, kgm./cm.²	Δv_{H_2O} (used in computation).	Δv_{Hg}.	Pressure, kgm./cm.²	Δv_{H_2O} (used in computation).	Δv_{Hg}.
Piez. No. 8.			Piez. No. 10 (cont.).		
4690	0.1258	0.01720	5840	0.1439	0.02132
5860	1441	2079	4720	1264	1806
6980	1599	2453	3570	1052	1378
1170	0437	0451	2410	0787	0923
2350	0776	0898	1210	0450	0472
3530	1044	1313	1930	0657	0947
8080	1740	2784	2950	0918	1087
9080	1863	3031	4140	1160	1530
10430	2012	3389	5350	1364	1855
*11640	2128	4101	6410	1521	2243
11030	2070	3720	7520	1670	2599
10130	1980	3257	8600	1806	2916
5690	1416	2129	9740	1938	3218
593	0245	0355	10180	1985	3377
			9150	1872	3106
Piez. No. 10.			8080	1766	2765
10150	0.1982	0.03472	8200	1756	2984
9130	1869	3145	5290	1355	1926
8060	1740	2848	8290	1767	2887
6960	1596	2478	9210	1878	3113

* Freezes.

where
$$\log a = 4.5819 - 10,$$
$$\log(-b) = 9.7569 - 20,$$

and applying to this computed value the correction given by the deviation curve of Figure 7.

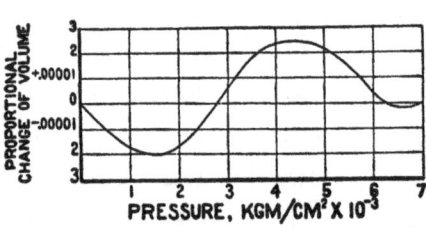

FIGURE 7. The deviation curve for the change of volume at 0°. This is to be used with the formula on page 373.

The data for 22° are shown in a corresponding set of tables and figures. The only difference is that here the parabola was used as the basis for both the first and the final deviation curves. Table IV. corresponds to Table I., Figure 8 to Figure 7, Table V. to Table II., and Table VI. to Table III. The final formula for computation at any pressure is

$$\Delta v = ap + bp^2,$$

where
$$\log a = 4.5911 - 10,$$
$$\log(-b) = 9.7782 - 20,$$

and the final deviation curve from this formula is given in Figure 8.

There are only a few previous determinations of the compressibility of mercury with which these results can be compared, and all of these are at comparatively low pressures, where the percentage accuracy of the above work is naturally less than at the higher pressures. The only way of making the comparison is to find what the compressibility given by the above two expressions would be at atmospheric pressure. The values found in this way from the above data are $0.0_5 380$ at 0° and $0.0_5 395$ at 22°.

The early work, done by Regnault,[21] Grassi,[22] Jamin,[23] Amaury et Descamps,[24] and Tait,[25] may be summarily ruled out as too inaccurate for the comparison. The earlier of these determinations were made

---

[21] Regnault, Mém. Inst. de France, Six. Mem., **21**, 329–428 (1847).
[22] Grassi, Ann. de Chim. et Phys., **31**, 437–478 (1851).
[23] Jamin, C. R., **66**, 1104–1106 (1868).
[24] Amaury et Descamps, C. R., **68**, 1564–1565 (1869).
[25] Tait, Challenger Report (1889), Phys. and Chem., vol. II, p. 1–76.

before the theory of elasticity was well understood, and the work of Jamin, in particular, is a classical instance of how dangerous it is to follow native intuition where the stress and strain system is at all

TABLE IV.

ADJUSTMENT OF DATA — FINAL VALUES OF COMPRESSIBILITY OF HG AT 22°.

Pressure. kgm./cm.²	Δv_{Hg} at 22°.			Calculated, $\Delta v = ap + bp^2$	Deviation.	Final Value.
	No. 8.	No. 10.	Weighted Mean.			
1000	0.00389	0.00389	0.00389	0.00384	+0.00005	0.00389
2000	0766	0766	0766	0756	+10	0766
3000	1129	1131	1130	1116	+14	1130
4000	1482	1485	1483	1465	+18	1483
5000	1818	1823	1820	1800	+20	1820
6000	2141	2149	2144	2124	+20	2144
7000	2451	2461	2454	2436	+18	2454
8000	2746	2761	2751	2736	+15	2751
9000	3026	3044	3030	3024	+06	3030
10000	3290	3313	3297	3300	−03	3297
11000	3540	3568	3549	3564	−15	3549
12000	3776	3809	3787	3816	−29	3787
....	2	1	Wt.	....	....	....

Calculated values are by the formula
$$\Delta v = ap + bp^2$$
$$\log a = 4.5911 - 10$$
$$\log (-b) = 9.7782 - 20$$

complicated. Jamin's results were 50 per cent too small. Of the more accurate work, Amagat[26] gives $0.0_5 379$ for the pressure range 0 to 50 kgm., at a temperature apparently unspecified in the original

---

[26] Amagat, loc. cit.

paper. De Metz,[27] in quoting Amagat, however, gives 0.0_5379 as Amagat's value at 0°. The source from which this knowledge was obtained is not clear. Amagat's results, obtained with seven different piezometers, vary among themselves by 2 per cent. De Metz gives 0.0_5374 at 0° and a pressure range not over 50 kgm., as the mean of results

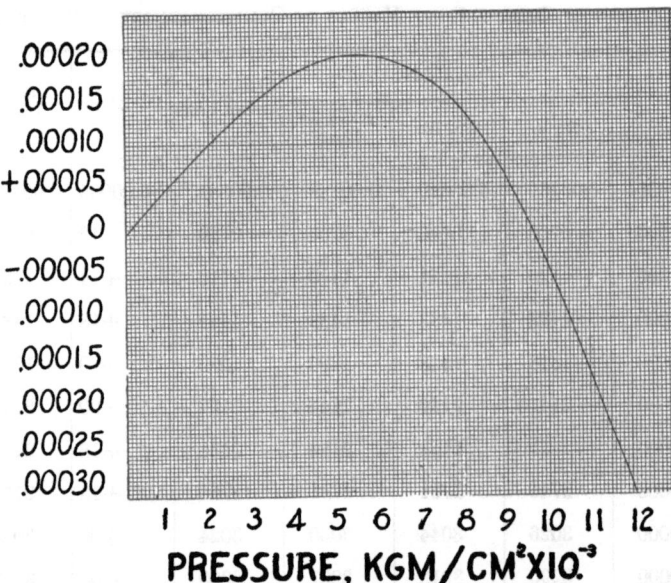

FIGURE 8. The deviation curve for the change of volume at 22°. This is to be used with the formula on page 376.

with four piezometers differing among themselves by 5 per cent. De Metz also gives the temperature coefficient of compressibility. This is higher than the value found above, but is such as to bring de Metz's value at 22° up to 0.0_5394, almost identical with the present determination at that temperature. Lately Richards [28] has determined the compressibility to 500 kgm. He gives 0.0_5371 for the pressure range 0–500, at a temperature not specified, probably room temperature. Previous work by Richards, however, had given a value for the temperature coefficient of the mercury the same as that of the glass piezometer. This temperature coefficient is very much smaller than found above, producing a change from 0.0_53710 at 0° to 0.0_53724 at 20°.

---

[27] De Metz, loc. cit.   [28] Richards, loc. cit.

It is to be noticed, however, that Richards' value involves directly his own value for the compressibility of iron, which is almost certainly much too low. If the more probable value 0.0_638 is used instead of Richards' value 0.0_638, the value of Richards for mercury becomes 0.0_5391 at room temperature. The change of compressibility with pressure found by Richards is higher than that found above, dropping from 3.80 between 0 and 100 kgm. to 3.60 between 500 and 600 kgm. As mentioned in a previous paper, this is probably due to error in the pressure gauge. Carnazzi [29] has published data over a much wider pressure and temperature range than the other previous observers, up to 3000 kgm., and from 23° to 192°, but with less pretense to accuracy, as only two figures are given in the results. Carnazzi gives 0.0_537 as the initial compressibility at 23°. Since Carnazzi's results are the only ones which give the change of compressibility with pressure over an at all wide pressure range, it may be permitted to anticipate a little in order to compare his results with the present ones. Carnazzi finds a drop in the compressibility of 0.0_64 from 250 kgm. to 2750 kgm. at 23°, against 0.0_625 of the present work. His change in the mean dilatation between 23° and 53° is from 181 at 0 pressure to 157 at 3000, against a change in the mean dilatation between 0° and 22° from 181 to 163 for the same pressures as found in the present work.

Finally, the results may be compared with the results of the previous paper.[30] It has been already stated that the method of the paper was open to question, but it should be remembered, as was stated at the time, that the primary object of that work was to find whether the compressibility of mercury showed any marked change over a pressure range from 0 to 6000, and that the data were obtained because of their bearing on another question. The accuracy was sufficient for the purpose in hand, and the employment of the method seems to have been justified in view of the very much greater speed and ease of manipulation. The initial compressibility at 20°, making correction as already explained for the erroneous value used for the compressibility of steel, was 0.0_5375.

Taken altogether, the agreement with the more accurate of the previous determinations is as satisfactory as could be expected: 380 at 0° against 379 of Amagat and 374 of de Metz, and 395 at 22° against 394 of de Metz and 391 of Richards.

---

[29] Carnazzi, loc. cit.
[30] Bridgman, loc. cit.

DISCUSSION OF THE DATA.

These data may be used in mapping out the p-v-t surface, and for giving certain information about the thermodynamic behavior of liquid mercury under pressure. The proportional changes of volume given above were calculated in the usual way, by taking the volume under atmospheric pressure and at the temperature in question as the unit volume. To map the p-v-t surface, correction must be made for the change of volume with temperature at atmospheric pressure. Thus, if the volume when $p = 1$ and $t = 0°$ is taken as unity, the values given by the process just mentioned for Δv at 22° must be corrected by multiplying by the ratio of the volume at 22° and $p = 1$ to the volume at 0° and $p = 1$. The actual volume under different pressures, and at 0° and at 22° is given in Table V. In this the value for the total change of volume between 0° and 22° was taken as 0.00398, as given by the most recent work of Callendar and Moss [31] on the expansion of mercury. It appears from their work that the mean coefficient of expansion at 0° is 0.0001805, while the mean coefficient between 0° and 100° is 0.0001817. This gives for $\left(\dfrac{\partial^2 v}{\partial t^2}\right)_p$ the value 0.0_724, which is evidently beyond the accuracy of this work. That is, for the present purpose, over the temperature and pressure range involved, the isothermals may be assumed to be equally spaced. This is sufficiently accurate at atmospheric pressure, and all our experience with other more compressible liquids would lead us to expect it to be all the more nearly true at higher pressures.

TABLE V.

VOLUME OF HG AT 0° AND 22°.

Pressure. kgm./cm.².	Volume.	
	at 0°.	at 22°.
0	1.00000	1.00398
1000	0.99626	1.00007
2000	99261	0.99627
3000	98905	99264
4000	98561	98909
5000	98231	98571
6000	97914	98246
7000	97607	97934
8000		97637
9000		97356
10000		97088
11000		96835
12000		96596

---

[31] Callendar and Moss, Proc. Roy. Soc. Lon., A, **84**, 595–597 (1911).

Figure 9 shows the compressibility at these two temperatures and different pressures. The compressibility plotted in Figure 9, is the

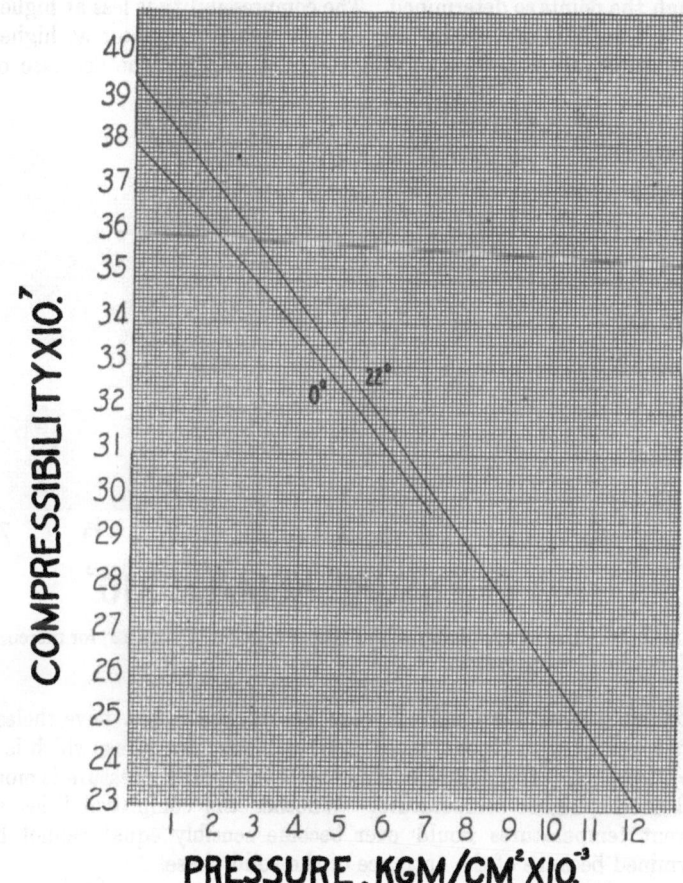

FIGURE 9. The compressibility of mercury at two different temperatures. This compressibility is the instantaneous compressibility $\left(\frac{\partial v}{\partial p}\right)_\tau$.

thermodynamic quantity $\left(\frac{\partial v}{\partial p}\right)_\tau$, which is slightly different from the ordinary meaning of compressibility as used above. It may be obtained at 0° directly from the parabolic formula and the deviation curve. At 22° the results obtained from the parabolic formula and the deviation curve are to be corrected by about 0.4 per cent for the increased initial volume at 22°. The results given in the figure are from smooth curves drawn through the points so determined. The compressibility is less at higher pressures, as it is universally for all liquids, and is greater at higher temperatures, as it is for everything except water. The decrease of

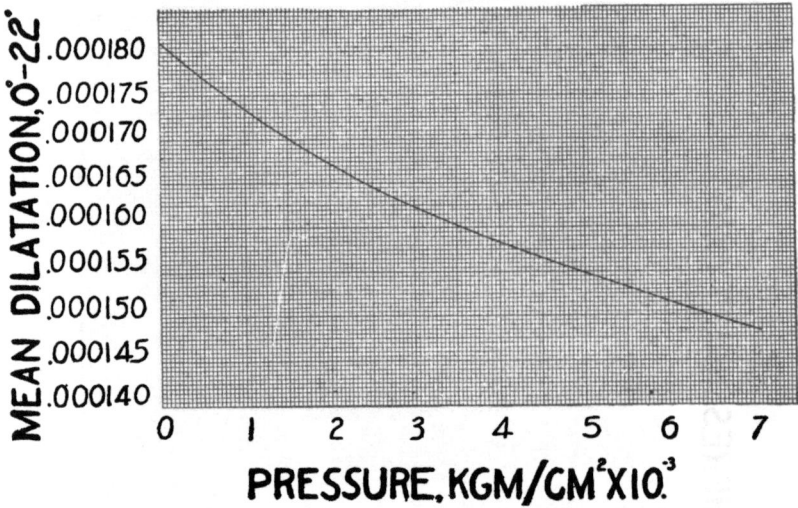

FIGURE 10. The mean dilatation per degree between 0° and 22° for mercury as a function of the pressure.

compressibility with pressure is very nearly linear, but nevertheless shows a tendency to become more rapid at higher pressures, which is a little surprising. The decrease of compressibility with pressure is more rapid at the higher temperature. Whether the compressibilities at different temperatures would ever become sensibly equal cannot be determined because of the entrance of the solid phase.

The dilatation of the mercury, or rather the derivative $\left(\frac{\partial v}{\partial \tau}\right)_p$, may be found from the distance apart of the isothermals at 0° and 22°. The results are shown in Figure 10. The dilatation becomes less at higher pressures, the rate of decrease also becoming less as one would expect.

The data so found, $\left(\frac{\partial v}{\partial \tau}\right)_p$ and $\left(\frac{\partial v}{\partial p}\right)_\tau$, are sufficient to give the difference of the specific heats at different pressures, for we have the relation,

$$C_p - C_v = -\tau \frac{\left(\frac{\partial v}{\partial \tau}\right)_p^2}{\left(\frac{\partial v}{\partial p}\right)_\tau}$$

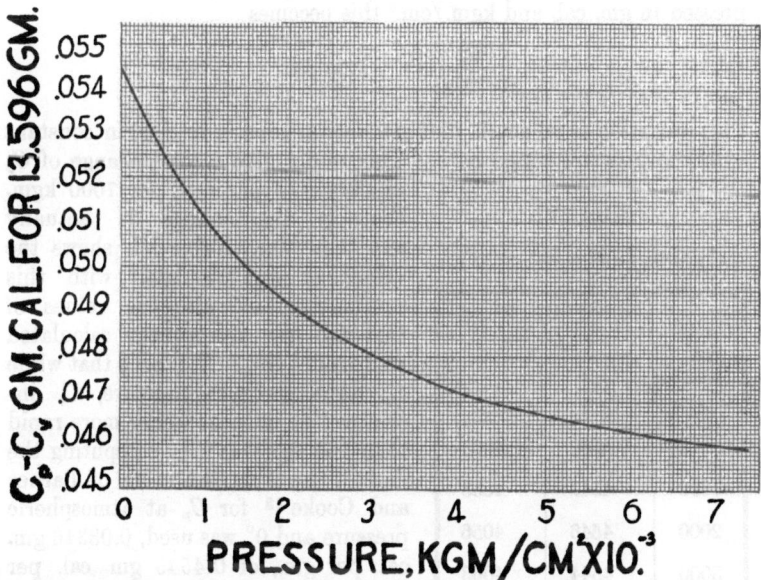

FIGURE 11. The difference of the specific heats as a function of the pressure in gm. cal. for a quantity of mercury occupying 1 cm.³ at 0° and atmospheric pressure.

The results in gm. cal. along the isothermal at 0° are given in Figure 11. The difference becomes less at high pressures, the rate of decrease itself decreasing rapidly. The behavior is thus different from that of water, which shows an increase.

The data are not sufficient to give the actual value of either specific heat along the isothermals, but we can find the initial rate of variation from the thermodynamic formula

$$\left(\frac{\partial C_p}{\partial p}\right)_\tau = -\tau \left(\frac{\partial^2 v}{\partial \tau^2}\right)_p.$$

The work of Callendar and Moss [32] gives at atmospheric pressure and 0°

---

[32] Callendar and Moss, loc. cit.

$$-\tau\left(\frac{\partial^2 v}{\partial \tau^2}\right)_p = -273 \times 0.0_724$$
$$= -0.0_5655,$$

C_p being expressed in ergs per cm.³ and p in dynes per cm.². Expressed in gm. cal. and kgm./cm.² this becomes

$$\left(\frac{\partial C_p}{\partial p}\right)_\tau = -0.0_6154.$$

This variation is so small that we may assume it to remain constant over the pressure range here. This would give a total change of C_p of only 1/4 per cent for 7000 kgm. Probably the change is actually less than this. Table VI. shows the values of C_p calculated with this assumption, and also the values of C_v found from the already calculated values of $C_p - C_v$. It is seen that while C_p decreases with pressure, C_v increases, the increase being more rapid than the decrease. In computing the date of the table, the value of Barnes and Cooke[33] for C_p at atmospheric pressure and 0° was used, 0.03346 gm. cal. per gm., or 0.4549 gm. cal. per cm.³.

There are several other quantities of interest that may be deduced from the given data. One of these is the difference between the isothermal and the adiabatic compressibilities. We have the thermodynamic formula

TABLE VI.

C_p AND C_v FOR HG AT 0°.

Pressure. kgm./cm.²	Specific Heat, gm. cal. for initially Unit Vol. (13.596 gm.)	
	C_p.	C_v.
0	0.4549	0.4003
1000	4547	4035
2000	4546	4056
3000	4544	4063
4000	4543	4073
5000	4541	4073
6000	4540	4077
7000	4538	4079

$$\left(\frac{\partial v}{\partial p}\right)_\tau - \left(\frac{\partial v}{\partial p}\right)_\phi = -\frac{\tau}{C_p}\left(\frac{\partial v}{\partial \tau}\right)_p^2.$$

The results are shown in Figure 12. The adiabatic compressibility, therefore, on the isothermal $t = 0°$, decreases less rapidly than the isothermal compressibility.

Intimately connected with the adiabatic compressibility is the temperature effect of compression. For this we have

---

[33] Barnes and Cooke, Phys. Rev., **16**, 65–71 (1903).

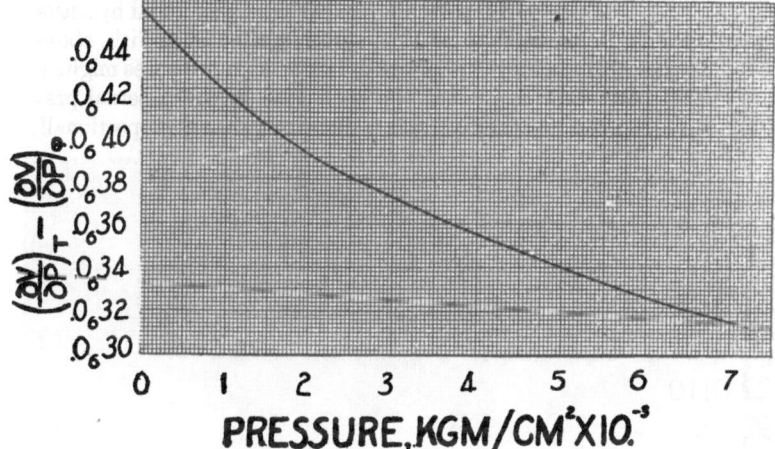

FIGURE 12. The difference between the isothermal and the adiabatic compressibilities as a function of the pressure.

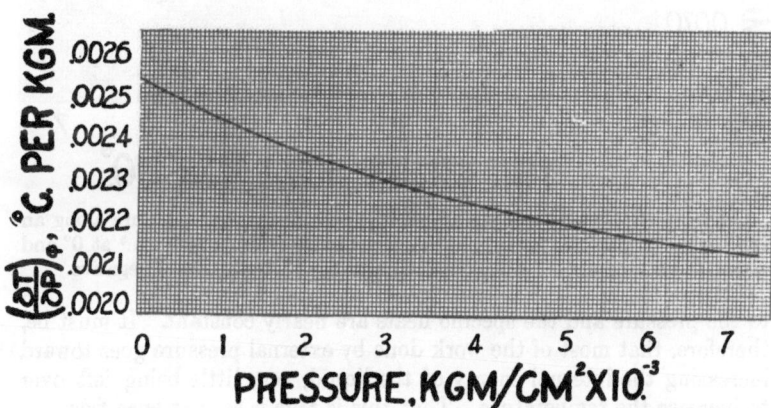

FIGURE 13. The heating effect of compression; that is, the rise of temperature in degrees centigrade for an adiabatic rise of pressure of 1 kgm.

$$\left(\frac{\partial .}{\partial p}\right)_\phi = \frac{\tau \left(\frac{\partial v}{\partial \tau}\right)_p}{C_p}.$$

The numerical values of this quantity are shown in Figure 13. The initial rise of temperature is, therefore, about 2°.5 for 1000 kgm. The rise of temperature becomes less at higher pressures, the rate of decrease also decreasing. The rise of temperature for 500 kgm. was found by actual experiment by Richards [34] to be 1°.2, agreeing essentially with above. The fact that this rise of temperature is less at high pressures might at first sight seem surprising, since the work done by the external pressure during a given pressure increase is increasing nearly proportionally

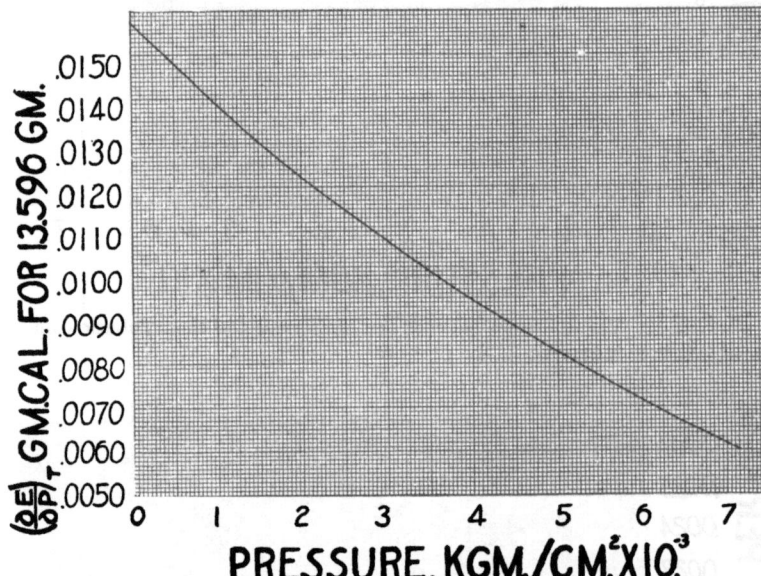

FIGURE 14. The rate of change of internal energy with pressure along an isothermal in gm. cal. for a quantity of mercury occupying 1 cm.³ at 0° and atmospheric pressure. Notice that the sign of the derivative is negative.

to the pressure and the specific heats are nearly constant. It must be, therefore, that most of the work done by external pressure goes toward increasing the internal energy of the liquid, very little being left over to increase the temperature. That this is true is seen at once from the thermodynamic formula

$$\left(\frac{\partial E}{\partial p}\right)_\phi = -p\left(\frac{\partial v}{\partial p}\right)_\phi,$$

which shows that the internal energy is increasing along an adiabatic, and at a rate very nearly proportional to the pressure. The internal energy decreases along an isothermal, however, as may be inferred from the value of the derivative given by

---

[34] Richards, loc. cit. (1903), p. 40.

$$\left(\frac{\partial E}{\partial p}\right)_\tau = -\left[\tau\left(\frac{\partial v}{\partial \tau}\right)_p + p\left(\frac{\partial v}{\partial p}\right)_\tau\right],$$

and shown in Figure 14. The curvature of Figure 14 is so slight, however, that it seems probable that at a high enough pressure the derivative will vanish and then become positive, so that ultimately the internal energy will increase along an isothermal.

The Change of State Liquid–Solid.

The quantities involved thermodynamically in the passage of a substance from one phase to another at a given pressure and temperature are the change of volume and the latent heat. From these quantities the change of freezing temperature with pressure may be calculated by Clapeyron's equation, $\frac{d\tau}{dp} = \frac{\tau \Delta V}{\Delta H}$. But the experimental difficulties of determining the latent heat are almost prohibitive. This is because in order to withstand high pressures the mass of the steel containing vessel must be much greater than the mass of the substance under experiment. Another method of finding these quantities based on Clapeyron's equation was therefore used, as in practically all the work done previously in this field. If the course of the freezing curve is known, $\frac{d\tau}{dp}$ may be found at any point, which, together with the change of volume at this point and the temperature of transformation, gives the latent heat.

The data were obtained by three methods. The first gives simply the course of the freezing curve. This was found by measuring the electrical resistance at constant temperature as a function of the pressure, freezing being indicated by a sudden drop in the resistance to about one third its value for the liquid. The second method, by far the most important, gives both the melting pressure and the change of volume. It is the same as that used by Tammann,[35] and consists in measuring the volume as a function of the pressure at constant temperature. At the freezing pressure, the volume suddenly changes, so that by plotting volume against pressure, the point of discontinuity gives both the equilibrium pressure and the change of volume. The third method is concerned with a single isolated quantity on the freezing curve, the change of volume on freezing at atmospheric pressure. The existing data were not accurate enough, so this quantity had to

---

[35] Tammann, loc. cit., p. 204.

be redetermined. The measurements were made by weighing mercury under CS_2, and plotting weight against temperature. Melting is accompanied by a sudden decrease in the displacement and so by a sudden increase in the apparent weight.

The Freezing Curve by Change of Resistance.

The points on this curve were obtained on two separate occasions; the fall of 1909, and January, 1911, and with two different pieces of apparatus. The data of 1909 are not as accurate as those of 1911, because the gauge was destroyed by an explosion before an entirely satisfactory calibration was made.

The details of the apparatus have been described elsewhere. The absolute gauge, against which the pressure measurements were made directly, has been described in a separate paper. The details of the glass capillary containing the mercury resistance, and the electrical connections with the details of the insulating plug have been described in the paper on the use of mercury as a standard of pressure. The pressure range employed there was only 6800 kgm./cm.^{2}; it has since been found that the same arrangement of apparatus is good to at least 12,000 kgm./cm.^{2}, the range of this paper. The measurements of electrical resistance were made by a null method on a Carey Foster bridge, exactly as in the preceding paper.

One slight change in the details of the measurements as made in 1911 should perhaps be noted, the employment of kerosene as the liquid surrounding the glass capillary instead of a mixture of glycerine and water. This avoids all possibility of short circuiting the terminals of the plug dipping into the mercury cups, and also avoids the necessity for carefully protecting these terminals. A light oil like kerosene is necessary since a heavy oil, such as ordinary lubricating oil, freezes at about 4000 kgm. at room temperature. It was found by independent experiment that at least up to 12,000 kgm. the kerosene does not become so stiff as to refuse to transmit the pressure hydrostatically.

The experimental procedure was the one which would naturally suggest itself, but there are a number of details of manipulation which were necessary to observe. The resistance was measured at appropriate intervals of pressure, the temperature being kept constant. After every increase of pressure it was necessary to wait 10 or 20 minutes for the dissipation of the heat of compression to the surrounding bath, which was kept constant to within 0°.01 with a thermostat. With increasing pressure it was never found possible to reach exactly the freezing pressure, but always the mark was overshot a little. It would have been impossible to land exactly on the mark, in the first place because of the mechanical difficulty of raising the pressure exactly the right amount, in the second place because of the initial effect of heat of compression which would keep the mercury liquid even if the pressure corresponding to the bath temperature were reached, and in the third place, because of the possibility of subcooling the liquid into the region of stability of the solid. With increasing pressure it was always necessary, therefore, to run past the point by from 200 to 400 kgm. Freezing was indicated by a sudden drop in the resistance. This drop was not instantaneous, but might continue for several minutes. Everything points to the spontaneous formation of a crystalline kernal at some point in the capillary, and then the advance of the surface of separation of solid and liquid through the capillary at a rate depending on the amount subcooling, etc.[36] Freezing, when it had once started, usually ran to completion. This was because the volume of the mercury was so small that the change of volume on passing to the solid state was not sufficient to lower the pressure to the equilibrium value. To find the equilibrium pressure, the pressure on the solid phase was lowered until the reaction to the liquid started, and then by rapid work at the pump the pressure was so adjusted that the resistance was in a state of unstable equilibrium, changes of pressure too small to measure on the gauge sufficing to produce a very large increase or decrease of resistance. With a little practice, one could keep the mercury in a partly melted condition for any desired length of time. This was necessary in order to avoid the effect of heat of compression, which might sometimes be so large as to mask the effect sought. The heat of compression of kerosene is so high that it is possible for a decrease of pressure in the kerosene to bring about a freezing of the mercury, the adiabatic drop of temperature with pressure for the kerosene being more rapid than the drop of pressure with temperature on the freezing curve of mercury. To be absolutely certain that there was no such effect the mercury was always kept in the partially melted condition for twenty minutes before recording the equilibrium pressure. To avoid any possibility of error from impurities absorbed by the mercury, pains were taken to find

---

[36] To measure the rate of advance of the surface would have been interesting, but to be of value would have demanded some other disposition of apparatus from that used. Evidently the rate of advance of the surface depends more than anything else on the rate at which the heat of transformation is conducted away, which in turn will depend on the size of the glass capillary, and the thermal conductivity of the surrounding liquid. Any results obtained with the apparatus without modification would have had significance only for this piece of apparatus.

the equilibrium pressure with as little of the solid phase present as possible. This evidently avoids the large depression of the freezing point which would result from the concentration of impurities in the small remaining mass of liquid after the solid had nearly all separated out. The precaution was probably superfluous because the mercury was carefully purified to begin with, and no effect of such a nature was ever detected.

One other source of error was guarded against which was especially prominent in the earlier work of 1909. The fluid transmitting pressure from the lower cylinder, in which was the absolute gauge, to the upper cylinder, in which was the mercury, was a thick mixture of either molasses and glycerine or glucose and glycerine. It was necessary to use such a mixture in connection with the absolute gauge in order to avoid leak at the high pressures. But there is the disadvantage that the viscosity of either of these fluids increases so rapidly with pressure that high pressures are not transmitted immediately through the connecting pipe from the lower to the upper cylinder. This connecting pipe was 15 inches long and 1/16 inch internal diameter. That the pressure was transmitted slowly was shown by the fact that after pressure had been increased in the lower cylinder by advancing the piston, the absolute gauge in the lower cylinder indicated a fall of pressure for some minutes, while the change of mercury resistance in the upper cylinder indicated a corresponding rise, evidently due to the slow flow of liquid from the lower to the upper cylinder. The effect was troublesome only at the higher pressures, the pressure coefficient of the effect being very high, and was very much greater for the mixture of glycerine and molasses used in 1909 than for the glycerine and glucose of 1911. In the extreme case, glycerine and molasses at 12,000 kgm., it was necessary to wait two hours for the equalization of pressure. For most of the pressures used the effect disappeared more rapidly than the heat of compression. To avoid error from this effect, the decrease of resistance indicating freezing must come after an increase of pressure in the lower cylinder, and the converse increase of resistance after a decrease of pressure. The measurements were made with this in mind.

The temperature measurements above zero were made with a Tonnelot thermometer calibrated at the Bureau of Weights and Measures in Paris. Below zero the readings of 1911 were made with a toluol thermometer calibrated at the Reichsanstalt, and in 1909 with a mercury thermometer made by Green of New York, the zero correction for which was the only correction applied. In any case, the errors in the temperature readings were less than the possible error in the pressure. The temperature was kept constant with an ordinary thermostatic device; a stream of ice water, or for the lower temperatures a stream of CaCl$_2$ brine, running constantly into the bath, which was automatically

TABLE VII.

FREEZING PRESSURE AT DIFFERENT TEMPERATURES BY THE METHOD OF CHANGE OF ELECTRICAL RESISTANCE. DATA OF 1909.

Temp. °C.	Equilibrium Pressure, kgm./cm.^{2}.		Difference, kgm./cm.^{2}.
	Observed.	Calculated.	
+14.35	10720	10600	+120
+21.0	11860	11910	− 50
+16.8	10850	11060	−210
+10.06	9620	9730	−110
+16.25	11010	10960	+ 50
+ 4.04	8540	8530	+ 10
0.00	7730	7730	0
0.00	7730	7730	0
− 8.52	6060	6040	+ 20
−15.09	4730	4730	0
+ 9.68	9690	9660	+ 30
+21.42	12070	11990	+ 80

Calculated values by the formula
$$p = a(t + 38.85)$$
where $a = \dfrac{7730}{38.85} = 198.9$

heated with a Simplex heating coil when the temperature fell too low. The temperature did not vary more than 0°.01 during a run, whereas a change of 0°.05 is necessary to produce a change in the equilibrium pressure of 10 kgm. which was about the limit of accuracy of the gauge. The temperature range of the experiment was from −15° to +20°. Since the freezing curve as determined between +20° and −15° extrapolates almost linearly to the ordinary freezing point at −38°.8 and atmospheric pressure, it was not thought necessary to make the special effort demanded to run a thermostat at temperatures below −15°.

One other remote possibility of error was guarded against. The capillaries containing the mercury were fine, the bore being about

TABLE VIII.

FREEZING PRESSURE OF MERCURY AT DIFFERENT TEMPERATURES BY THE METHOD OF CHANGE OF ELECTRICAL RESISTANCE. DATA OF 1911.

Temp. °C.	Equilibrium Pressure. kgm./cm.³.		Difference kgm./cm.³.
	Observed.	Calculated.	
4.95	8570	8660	−90
9.92	9640	9640	0
14.86	10600	10620	−20
19.98	11640	11630	+10
0.00	7680	7680	0
−14.40	4790	4830	−40
−9.70	5630	5760	−130
0.00	7660	7680	−20

Calculated values by the formula
$$p = a(t + 38°.85)$$
where $a = \dfrac{7680}{38.85} = 197.7$

0.1 mm. It was thought barely possible that the curvature of the surface might be sufficient to produce a rise of internal pressure in the mercury sufficient to change slightly the external pressure required to hold liquid and solid in equilibrium. This effect was shown to be negligible by using two capillaries of different sizes, one three times the diameter of the other. The equilibrium pressure found at 0° with the larger capillary was 7680 kgm., and with the smaller 7660. The divergence is in the direction to be expected, but it is of the same order as the experimental discrepancies. The data of 1911 were obtained with the larger of the above mentioned capillaries; those of 1909 with one of intermediate size.

The actual data follow. Those for 1909 are in Table VII. and in Figure 15. The data are given in the order in which they were collected. The two determinations at the higher pressures after the run

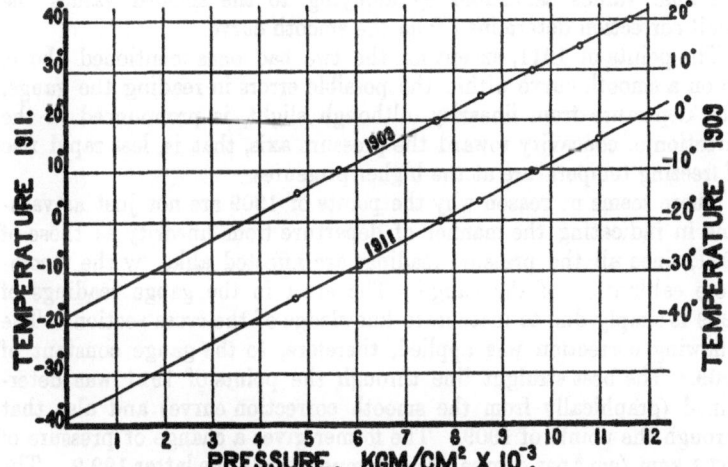

FIGURE 15. The melting curve of mercury by the change of resistance method.

with regularly decreasing pressures were made, the one at 9°.68 to correct the obviously bad point determined first of all, and the one at 21°.4 to make sure that there had been no systematic error introduced after the regular decrease of pressure of the first points. In addition, the three points determined during the initial testing of the apparatus, before the thermostat was installed, are given at the very end. The temperature error here easily accounts for the discrepancies of these last three points. As was stated before, the gauge with which these determinations were made was destroyed before it could be satisfactorily calibrated. The data used in the reduction of the gauge readings were simply the measured dimensions of the gauge, which could not be determined with the requisite accuracy. The column showing the calculated pressure must be taken, therefore, as merely evidence showing the regularity of the results and slight departure from linearity, and not as giving accurately the absolute pressures.

Table VIII. and Figure 15 give the results of 1911. The method by which the results were calculated was the same combination of numerical computation with graphical construction already used in finding the compressibility. A straight line was passed through the point at 9640 and 9°.92, and the departure of the observed pressure values from those calculated by the linear relation plotted on a large scale. A smooth curve was then drawn through the midst of these points, and the final values calculated by applying to the smooth values the small correction determined from the smooth curve.

The points of 1911, except for the two bad ones mentioned above, lie on a smooth curve within the possible errors in reading the gauge. The departure from linearity, although slight, is pronounced in the direction of concavity toward the pressure axis, that is, less rapid rise of freezing temperature at the higher pressures.

There seems no reason why the points of 1909 are not just as valuable in indicating the manner of departure from linearity as those of 1911, since all the pressure readings are affected alike by the incomplete calibration of the gauge. The error in the gauge readings of 1909 is simply due to inaccurate knowledge of the cross-section. The following correction was applied, therefore, to the gauge constant of 1909. The best straight line through the points of 1911 was determined (graphically from the smooth correction curve) and also that through the points of 1909. The former gives a change of pressure of 197.1 kgm./cm.^{2} per degree rise of temperature; the latter 199.2. The observed pressures of 1909 were all reduced in the ratio of 197.1 to 199.2. A smooth curve was then passed through these corrected values by the method described above. Finally, to determine the most probable departure of the freezing curve from linearity, the mean of the two correction curves, those for the data of 1909 and 1911, was taken. These curves differ nowhere by more than the possible errors in reading the pressure gauge. From this curve it is possible to find immediately the tangent to the melting curve at the origin.

Finally, all these results may be collected into the following form. At any given temperature the melting pressure is to be found by adding to the pressure given by the formula $p = 196.5 (t + 38.85)$ the small correction term given graphically in Figure 16. The initial rise of pressure per degree rise of the freezing point is therefore 196.5 kgm./cm.^{2}; at 12,000 kgm. this has become only 1/2 per cent greater, so slight is the departure from linearity. The value of the initial slope, 196.5, has no more certainty than the single series of observations of 1911, but the departure from linearity shown in Figure 16 has the greater weight which should be given to the mean of two independent series of observations with different apparatus.

A number of data collected incidentally in the course of the work may be mentioned here, because they throw some light on certain properties of mercury under pressure, even if these are not the thermodynamic properties which are the particular subject of this paper.

The electrical resistance of the liquid mercury was usually measured over the entire range up to the freezing pressure. The matter of the

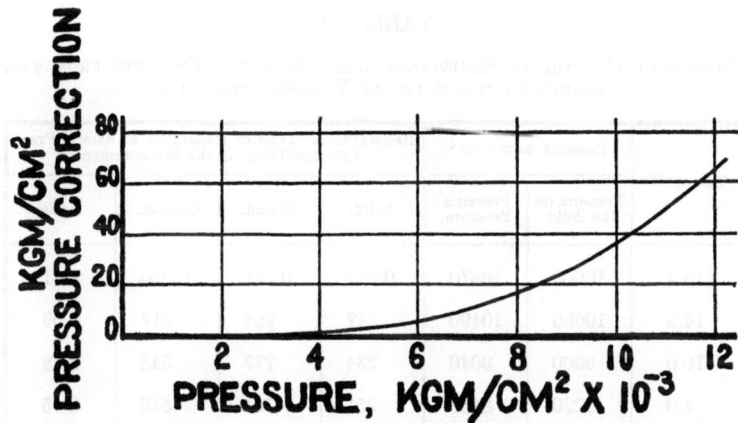

FIGURE 16. Departure of the melting curve of mercury from linearity. The initial slope is 196.5 kgm. for a rise of temperature of 1° C.

electrical resistance of mercury under pressure has already been treated in a paper in which the pressure range was only 6800 kgm./cm.^{2}. The results obtained during the present work were pretty irregular, discrepancies of 1 or 2 per cent not being rare; but within the limits of error, the empirical formula given in the previous paper may be used by extrapolation up to the freezing pressure, at least within the present temperature range from $-15°$ to $+20°$. The irregularities observed in the present work were without doubt due to the strains set up in the glass by the passage of the mercury from one state to the other. To reach the accuracy of which the measurements are susceptible it would be necessary to repeat the above resistance measurements, taking care never to push the pressure beyond the freezing value.

The electrical resistance of the solid mercury was also usually measured. These results cannot be expected to much more than indicate the order of the effect, for in addition to the comparatively small irregularities mentioned above, as due to elastic after-effects and hysteresis, comparatively large temporary alterations in the glass capillary are probably produced by the freezing of the mercury. The results are given in Table IX. R is the resistance of the mercury reduced by taking as unity the resistance of the liquid at atmospheric pressure and the corresponding temperature. R for the liquid was obtained by extrapolating beyond the freezing point to the pressure at which the resistance of the solid was measured. The resistance of the solid shows

TABLE IX.

CHANGE OF ELECTRICAL RESISTANCE WHEN MERCURY CHANGES FROM THE LIQUID TO THE SOLID AT VARIOUS PRESSURES.

Temp. °C.	Pressure, kgm./cm.³.		Resistance in Terms of Resistance at Atmos. Pressure and Temp. of the Measurement.			
	Pressure on the Solid.	Freezing Pressure.	Solid.	Liquid.	Change.	Ratio.
16.2	10960	10870	0.211	0.761	0.550	3.61
14.3	10660	10490	247	764	517	3.09
10.0	9960	9640	234	777	543	3.48
4.0	9220	8450	251	791	540	3.15
0.0	7660	7650	264	817	553	3.09
0.0	7610	7650	241	816	575	3.39
−15.1	5450	4680	258	863	605	3.35

a distinct tendency to become less at the higher pressures and temperatures. The effect of temperature only would be to raise the resistance. The pressure coefficient of resistance must be negative, therefore, as for all other pure metals, and must have a value at least such that 200 kgm. increase of pressure produces a greater fall of resistance than 1° rise of temperature produces rise of resistance. This means that the pressure coefficient of the solid is greater than $0.0004/200 = 0.000002$. The table also makes evident the tendency of the change of resistance liquid-solid to become less at higher pressures and temperatures. This indicates that the effect of pressure on the resistance of the liquid is greater than the effect on the solid, as one would expect, although the reverse is true for the temperature effect. This enables an upper limit to be placed on the pressure coefficient of the solid, since the coefficient of the liquid is known. This upper limit is about ten times the lower limit set above. It hardly seems worth while to attempt to force the above very imperfect data to give more accurate results.

It may be mentioned, however, that the above results are fully as consistent as the results which have been obtained by previous observers for the electrical resistance of solid mercury at atmospheric pressure. Thus Weber [37] found the following values for the ratio of the conductivity of the solid mercury to that of the liquid out of which it is freezing : 3.2, 4.2, 3.9, 4.0, 3.5, 4.0. Taking the reciprocals so as to compare with the above form, these become 0.31, 0.24, 0.26, 0.29, 0.25; different measurements of the same quantity. Cailletet and Bouty [38] give the single number 0.24, while Grunmach in his first paper [39] gives 0.67 and redetermines it in a later paper [40] as 0.40. The high values of Grunmach are doubtless to be explained by the formation of cracks in the mercury when freezing. This error can hardly come in the present work. In explanation of the other outstanding discrepancies, the fact does not seem to have been sufficiently considered that mercury freezes into crystals which may have different conductivities in different directions. It is known that the conductivity of bismuth, for example, may change by 60 per cent in different directions, and that of iron glance by 100 per cent. It is to be expected, therefore, that the particular form of vessel in which the mercury is frozen, by influencing the position in which the crystals tend to separate out, would have an effect on the apparent change of resistance on freezing.[41]

Finally, mention may be made of an unsuccessful attempt to get the heat of transformation from these same resistance measurements. The idea was to measure the depression of the freezing point occasioned by dissolving some metallic impurity in the mercury, the commencing of freezing being indicated by a kink in the curve giving resistance as a function of the pressure. The depression of the freezing point gives the heat of transformation by a well known formula. The difficulty with the method is that there is no metal which gives a depression of

---

[37] Weber, Wied. Ann., **25**, 245–252 (1885), and **36**, 587–591 (1888).
[38] Cailletet et Bouty, C. R., **100**, 1188 (1885).
[39] Grunmach, Wied. Ann., **35**, 764 772 (1888).
[40] Grunmach, Wied. Ann., **37**, 508–515 (1889).
[41] Since the writing of this paper Onnes (Kon. Akad. Wet. Proc., **4**, 113–115 (1911)) has given data from which the ratio of the resistance of solid to liquid mercury may be found to be 0.237. This value was computed by the author from the values given in Science Abstracts for the resistance of a given quantity of mercury (172.7 ohms at 0°C., and 39.7 ohms in the solid state at the melting temperature).

the freezing point sufficiently large to give the desired accuracy. Few metals depress the freezing point at all, and those which do show a eutectic point 1° or 2° below the normal. In a preliminary testing out of this idea, the resistance of a 5 per cent Cd amalgam was measured at 21°.5. Beginning at between 1000 and 2000 kgm. the pressure coefficient was found abnormally large over the pressure range up to

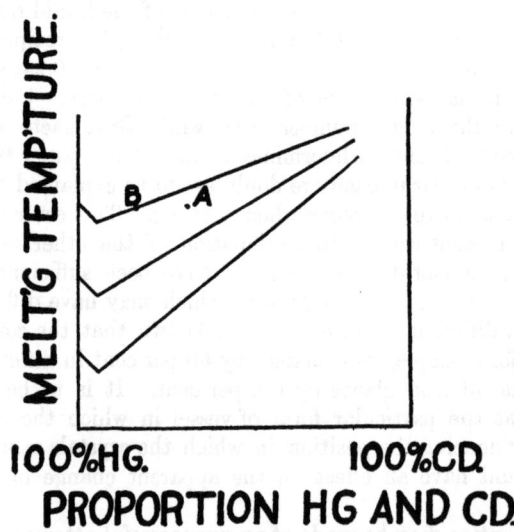

FIGURE 17. Showing the raising of the melting curve of mixtures of mercury and cadmium with increasing pressure.

and including 12,000 kgm., which is the freezing pressure of pure mercury at 21°.5. The time required in reaching equilibrium was also abnormally long. Both these facts indicate the separating out of some component from the amalgam. Between 12,000 and 13,000, this process of separation seemed to have been completed, and from 13,000 to 13,500 the behaviour was entirely normal. The explanation is simple. The effect of pressure is to raise the entire melting diagram Hg—Cd, the Hg end more than the Cd end. (See Figure 17.) This raising is presumably accompanied by a shifting of the proportions of the two metals present in the eutectic mixture. The initial temperature and proportion of the two metals present in the amalgam may be represented by the point A (in this case $t = 21°.5$ and $x = 5$ per cent), but when the pressure has risen sufficiently to bring the melting curve up to A, with further rise of pressure A slides along the rising curve, always at constant temperature, shown by the point B on the highest melting curve of the diagram, until the eutectic point is reached, where complete solidification sets in. This takes place at a pressure slightly higher than the normal freezing pressure for the pure mercury, as shown experimentally above, but the precise amount will depend on the unknown way in which the proportions of the eutectic mixture have changed. To obtain the data desired by this method, A must initially lie to the left of the eutectic point corresponding to the particular freezing pressure instead of to the right, as in the diagram. The results evidently cannot be accurate, and it was therefore not thought worth while to push the investigation further at this time, particularly since the destruction of the gauge mentioned before came as the direct result of the high pressure reached above. An investigation of the change of eutectic proportions under pressure would make an interesting research for its own sake.

THE EQUILIBRIUM DATA BY THE CHANGE OF VOLUME METHOD.

The method was essentially that of Tammann, and gives both the change of volume and the pressure at a given temperature. It consists in measuring the displacement of the piston by which pressure is produced as a function of the pressure. Freezing is indicated by an interruption of the regular rise of pressure with advance of the piston. It is at once obvious that an absolute essential for the success of the method, and at the same time the hardest condition to maintain under high pressures, is the entire freedom from leak of the moving piston and the other parts of the apparatus. Tammann found this trouble at pressures above 2000 kgm./cm.^{2}, and avoided it in part by using a very heavy oil for transmitting pressure. In this present work a new packing device was adopted which secured entire freedom from leak even with so thin a fluid as kerosene or mercury. Data will be given in support of this.

The data needed at any point in addition to the equilibrium pressure and temperature are the amount of the advance of the piston during constant pressure, the quantity of liquid involved in the change, and the cross-section of the advancing piston. The first two may evidently be measured directly, but the last changes with pressure by an amount not susceptible to direct measurement. Here, as in all pressure measurements, the distortion of the containing vessel enters as a disturbing factor. But it should be pointed out that here the effect of the distortion is a minimum. It has been quite impossible up to the present to make direct determinations of compressibility, for example, by this method, because the changes of volume of the liquid are accompanied by unknown changes of volume of the same order in the containing vessel. But here, where the changes take place at constant pressure, the vessel remains constant during the change, and the only distortion which need be taken into account at all is the change of area of the cylinder, a change which is easy to calculate, and which furthermore never in the present work rose above 2 per cent of the total effect. To make this matter clearer, take the example of water. A pressure sufficient to change the volume of water by 20 per cent might introduce a change of about 2 per cent in the volume of the very heavy cylinders of this present work. This change of 2 per cent takes no account of the change of volume introduced by the compression of the packing of the various connecting pipes, etc., which are necessary if the pressure is to be measured at all accurately. This change due to the tightening of the packing is evidently incalculable, but may presumably be as large as the other effect. The change of volume as indicated by the advance of the piston is affected, therefore, by a quantity which may very reasonably be supposed to be 4 per cent on 20 per cent, that is, 1/5 of the total effect measured. Compare this with the change of volume during freezing at constant pressure. Here the change of volume of the rest of the vessel and the packing plays no part whatever, the only correction is 2 per cent for the change in cross-section of the cylinder. That is, a correction of 2 per cent which may be approximately determined, against a wholly unknown correction of 20 per cent. It should be remarked, however, that if once the compressibility of the water has been determined in some other way, it may be used in an apparatus of this form to determine the unknown distortion, and so the compressibility of other liquids may be determined.

Description of Apparatus.

The apparatus consisted essentially of two cylinders, connected by a piece of very heavy nickel steel tubing 12 inches long. The upper cylinder was placed in a hydraulic press, and the pressure was produced in this by a piston forced down by the ram of the press. The amount of motion of the piston was measured with a micrometer screw reading to 0.0001 inch. There were no valves in any part of the apparatus, as it is difficult to make these capable of withstanding high pressures. The apparatus was initially filled entirely full of liquid after the air had been exhausted with an air pump, and the piston was then introduced into the upper cylinder. The dimensions of the various parts were chosen so that the maximum pressure desired could be obtained with a single stroke. This upper cylinder was freely exposed to the air of the room, with no special device to keep the temperature at any definite value.

The lower cylinder contained the mercury, the subject of the experiment, and was surrounded by a bath kept at constant temperature by a thermostat. The mercury was contained in a cylindrical steel shell fitting loosely the hole in the cylinder. The same kerosene which entirely filled all other parts of the apparatus and transmitted pressure from the upper to the lower cylinder also entirely surrounded the walls of the shell containing the mercury. This arrangement was necessary, because if the mercury had been allowed to come in contact with the walls of the cylinder, the steel would have become amalgamated, and the cylinder speedily ruptured. Whereas, the steel shell in which the mercury was contained being surrounded on the outside by kerosene, and so being subjected to a purely hydrostatic pressure equal all over, experienced no amalgamation whatever.[42]

Screwed into the bottom of this lower cylinder, directly immersed in the kerosene and so subjected to the same pressure as the mercury, was an insulating plug carrying a coil of manganin wire with which the pressure was measured.

The procedure was as follows: After filling the apparatus with kerosene, the thermostat was raised around it and adjusted to the temperature desired. When a steady state was reached, the resistance of the manganin coil was read on the Carey Foster bridge, thus giving the pressure zero. The piston was then introduced into the upper cylinder, and pressure increased to a value a few hundred atmospheres short of the freezing pressure at the temperature of the thermostat. After pressure equilibrium had been attained by dissipation of the heat of compression, the pressure was read from the resistance of the manganin coil and the position of the piston was read with the micrometer. Two readings of the resistance of the manganin were made corresponding to the two positions of the eight-point switch of the Carey Foster bridge. These readings were made before and after reading the position of the piston. Since the circuits were non-inductive, the bridge could be used with the galvanometer circuit permanently closed. These two readings are, therefore, really two independent readings of the pressure, and agreement between them is indicative of the attainment of pressure equilibrium. It was not necessary to take any such precautions in measuring the position of the piston, for the

---

[42] See These Proceedings, No. 14, 1911, for a discussion of the amalgamation of steel under high pressures.

friction of the packing prevented all motion after the initial increase of pressure.

The above procedure, reading of pressure and piston position, was repeated at appropriate pressure intervals up to several hundred or perhaps a thousand kgm. beyond the freezing pressure. On plotting piston displacement against pressure, the freezing is shown by a discontinuity in the curve. Similar readings were also made on release of pressure, and so another value for the change of volume found. On releasing pressure the discontinuity is more sudden than with increasing pressure, because it is impossible to superheat the crystalline solid on decreasing pressure, but the liquid may be subcooled on increasing pressure. Because of this, and for another reason to be explained later connected with the distortion of the steel, the value found with decreasing pressure was without exception accepted as the correct value.

In the use of this apparatus and in the calculation of the data, there are numerous corrections or disturbing factors which will now be considered in detail.

Leak. — As already stated, the most obvious and apparently the greatest difficulty with the method, particularly at high pressures, is the question of leak. It might perhaps be possible to correct for this as was done sometimes by Tammann by weighing the quantity of liquid escaping past the piston in a given time, but this correction would always be unsatisfactory in cases where, as here, the total motion of the piston during freezing is small (0.2 inch) and many extend over an hour or more. It has been stated that there was absolutely no leak. This was proved by the fact that the piston and upper end of the packing, when removed from the cylinder after the end of an experiment, were always perfectly dry, even though the total time over which pressure has been applied might reach into several days, as it did on several occasions. That there was no leak in any other part of the apparatus was proved by the self-consistency of the results as well as by special experiment. On one occasion pressure was left at 7500 over night, temperature being kept constant with a bath of ice and water, and in the morning had not dropped by as much as the smallest amount perceptible with the manganin resistance, namely 2 kgm./cm.^{2}.

The effect of leak, if there were one, would be to make the values of Δv during increase larger, and during decrease smaller. The values did actually in the great majority of cases disagree in this way. That this was not due to leak will be evident from a discussion of

Elastic After-Effects. — In the ordinary sense of the word, elastic after-effects, as is indicated by the name, are purely elastic phenomena, complete recovery after any application of stress always occurring, provided time enough is allowed. There are also other effects of somewhat the same general character, such as hysteresis, not usually included in the name. But there will be included in this discussion any effect in virtue of which the strain is not a single valued function of the stress only. The general nature and interpretation of these effects is still obscure, but there are a few facts which are well known. For instance, in all ordinary testing of materials it is first necessary to accommodate the metal to the particular stress conditions of the moment. The strain under a given stress is not the same the first time the stress is applied as the second, and not until the application of the stress a number of times does the strain come to depend always in the same way on the stress. Even after complete accommodation has been attained, the material may continue to show elastic after-effects, that is, continued yield after the application, or continued recovery after the release or stress. Or it may also show hysteresis. These effects become increasingly important as the magnitude of the stress is increased, or as the character of the stress changes from one uniform throughout the mass to one changing from point to point. The effects become most important of all when the material itself has become non-homogeneous by the exceeding of the elastic limit in some places while the remaining parts are without permanent deformation.

Various effects of just this kind are to be expected in these experiments, and may perhaps be fairly large in comparison with the usual magnitude of the effects, because of the wide range of stress here employed. It should be pointed out that one advantage of this method, namely, that the elastic deformation of the vessels does not enter the results, is in part offset by the various effects of the kind mentioned here, because in virtue of them the volume of the containing vessel may change while freezing or melting is going on at constant pressure. Furthermore, these effects, which are themselves intrinsically small, may appear with magnified effect in the result. Thus the actual change of volume of the mercury during freezing is about 1 per cent of the total internal volume of the containing vessel, so that a change of volume of 1/100 per cent of the containing vessel produced by elastic after-effects during freezing will appear in the result magnified to 1 per cent. As a matter of fact, there were found discrepancies between the values under increasing and decreasing pressure amounting in the worst case to 3 per cent.

It was possible, however, by an examination of the discrepancies themselves and by a brief consideration of the method of experiment to select from these the most probably accurate results. During the course of an experiment, the stress history of the steel in its bearing on this effect was as follows. Pressure was pushed to a value somewhat short of the freezing pressure, the readings were made, and then pressure increased again, etc., until the subcooled region had been entered far enough for freezing to begin. This first stage might occupy an hour. The second stage was that of approximately constant pressure during freezing, which might require from an hour to an hour and a half. During the third stage, pressure was increased a few hundred atmospheres beyond the freezing pressure, held at the maximum for a while, and then decreased to the melting point again. This occupied another hour. The fourth stage, melting at constant pressure, ran more rapidly than freezing, but might consume an hour. Finally a few readings below the freezing pressure, rarely more than half an hour. It is to be noticed that during melting the pressure has been held in the immediate neighborhood of the maximum for some hours, and that this melting pressure is only slightly removed from the maximum, whereas during freezing, the stress has just been increased over nearly the entire range and the time for accommodation has been much shorter. All our experience with elastic after-effects would lead us to predict that during melting these effects would be very much less important than during freezing. We would furthermore predict that the change of volume found during freezing would be greater than that during melting, for the piston must advance to compensate for the increasing volume of the steel vessels as well as for the decreased volume of the mercury.

These conclusions were verified by the data. The values found during melting were always self-consistent, lying on a smooth curve, and could be repeated. But those found during freezing might vary wildly within the 3 per cent limit mentioned above. Furthermore, the inconsistent results were always larger than the consistent ones, provided that the procedure was as described above. There seemed to be no connection between the duration of the pressure and the amount of the discrepancy.

We already have evidence enough to make it extremely probable that the discrepancy was not due to leak, although it is of the right sign. But the fact that the discrepancy is reversed in sign so as to indicate a leak in instead of a leak out, if the stress is applied differently from above, renders certain the conclusion that the effect is not due to a leak. The succession of points on the melting curve, the procedure in getting one of which was described above, was in general obtained by starting at low temperatures and pressures and working up to the higher values. Each maximum pressure, then, was higher than the preceding maximum. But if the preceding maximum was higher than the following one, the discrepancy was reversed in sign, the values during freezing being too low, while those during melting still lay on the same smooth curve as before. This certainly cannot be due to leak. Neither is it in accord with the general nature of other elastic after-effects at low pressures. But other experience of this same general nature at these high pressures indicates that this is probably the general nature of the effect at high pressures. It is as if the subsequent application of pressure, and possibly the change of temperature also, changes somewhat the position of the molecules, like shaking a magnet, so that they become free again to resume their natural elastic recovery from the previous greater maximum. One experiment with a hollow steel cylinder subjected to external hydrostatic pressure strikingly suggested the same general explanation. This cylinder had been previously subjected to a pressure many times beyond the elastic limit, so that the metal had flowed toward the center, decreasing the internal volume. On removing the pressure there was some slight elastic recovery of the internal volume, but not nearly as much as was to be expected. But the next increase of external pressure produced in the initial stages a very marked increase of internal volume, instead of a decrease, the natural effect of the pressure. It must be that in some way the pressure loosened up the molecules so that they were free to resume their natural recovery.

Other slight differences suggest also that these discrepancies are of the general nature of elastic after-effects. Because of breakage, several different pieces of apparatus were used. The discrepancy was always greatest with the thickest-walled vessels. It has been observed several times in other connections that elastic after-effects are greater with the greater mass of metal, because this allows greater heterogeneity of strain and greater volume in which to store the effect, with consequent longer time of recovery. At a given pressure the effect in a new piece of apparatus was smaller before it had been subjected to the greatest maximum than it was at this same pressure after being so subjected to the maximum. In general, the effect was likely to increase with time, showing fatigue of the metal. In one case, in the similar measurements on water, to be described elsewhere, rupture of the vessel was foreshadowed by an increase of the effect beyond 3 per cent.

There may be some slight effect of permanent set mixed up with the above effects. An undoubted set was always very distinctly obvious after the first application of the greatest maximum. This application was usually made in a preliminary test in order to season the apparatus before the measurements were undertaken. This applies particularly to those three or four occasions after the lower cylinder had been broken. The seasoning of the upper cylinder was done more carefully, and is described elsewhere. The set must have been nearly all removed by this seasoning, but any subsequent set would still affect only those results for increasing pressure.

In view of all these considerations, the course actually adopted seems justified; namely, to retain only those values found during melting, that is, during decreasing pressure. In consequence, in some two or three of these last determinations, and in many of those on water, the experimental determination with increasing pressure was entirely omitted.

Elastic Deformation. — It has already been stated that the only way in which elastic deformation, that is, a deformation depending only on the pressure, enters, is in the correction for the change of cross-section of the upper cylinder. Although the value finally adopted for this correction is small, only 1.3 per cent at 10,000 kgm./cm.^{2}, still it must be admitted that the greatest uncertainty in the work is precisely in determining this correction. The difficulty comes because of the very high pressure employed, which renders doubtful the calculation by the ordinary theory of elasticity. If the maximum pressure used in the experiment, 12,000 kgm., were the greatest pressure involved, the case would not be so doubtful, because the elastic limit of the steel is not greatly exceeded here. Indeed, if it were possible to harden the steel uniformly throughout its mass, reaching to a depth of 1 1/2 inches, the elastic limit would not be reached until about 15,000 kgm., and there would not be much question in applying the ordinary equations of elasticity. But it is impossible to raise the elastic limit throughout the metal by quenching, and the interior of the cylinder always shows some set at 12,000. Since the accuracy of the whole method depends on accurate knowledge of the cross-section it was necessary to remove every vestige of set in this part of the apparatus. This was done by subjecting the cylinder initially for five or six hours to a pressure of 24,000 kgm. This was sufficient to stretch the inside of the cylinder from its initial value of 1/2 inch to nearly 9/16 inch. The cylinder was then reamed out to 9/16 inch for its entire length, and this was the form in which it was finally used. The effectiveness of the treatment is shown by the fact that the cylinder has shown no further change of so much as 0.0001 inch on the internal diameter.

The present condition of the cylinder is therefore far from one of a state of ease; there are certainly internal strains. But it should be noticed that this fact in itself is not sufficient to destroy the applicability of the equations of elasticity. Provided only that the metal shows no further set after the first application, and that strain remains proportional to stress, any stress will produce the same additional strain as the same stress applied from a state of ease. The long interval of time for which the maximum pressure was exerted, and the large excess of this pressure over any subsequent pressure, make it exceedingly probable that such a state has actually been reached. The residual stress in the cylinder is then of exactly the same nature as the stress in a gun with shrunk-on hoops, and the deformation may be calculated by the ordinary theory. It may be mentioned, in support of this position, that measurements have been made of the external elastic deformation of another such cylinder of the same steel up to 10,000 kgm., showing agreement with the accepted theory to at least 5 per cent, with no trace of hysteresis.

The calculation of the actual deformation of the cylinder according to the elastic theory would be impossible because of the indeterminate nature of the boundary conditions. An approximate solution only can be obtained, in which the end effects are neglected altogether. This solution is nevertheless probably accurate enough, because the end effect in other similar cases in which the rigorous solution is known has been shown to be negligible. It is well to remember in considering the validity of these approximations that the maximum value of the correction at the highest pressure is only 1.6 per cent.

The internal diameter of the cylinder does not expand uniformly under pressure. The expansion may be supposed to have its greatest value at points removed from the position of the piston by at most a few diameters. The expansion throughout the greater length of the cylinder is constant, therefore, and equal to the value it would have in a cylinder infinitely long and of the same radial dimensions. But for a short distance beyond the head of the piston the expansion varies from its maximum value to a value which may be shown to become half this maximum at the piston itself. Now if we suppose that the cylinder is so long that as the piston moves at constant pressure (during freezing or melting) the locality of transition in dimensions travels bodily with the piston, then a moment's consideration shows that the volume swept out by the piston is equal to the length of stroke multiplied by the area of the cylinder at its locality of maximum distortion. The assumption does not seem forced in view of the fact that the length of the cylinder is between ten and fifteen diameters.

The formula for the displacement under these conditions may be found in any book on elasticity. We have

$$\Delta b = \frac{bP}{a^2 - b^2}\left(\frac{a^2}{2\mu} + \frac{2b^2}{3\kappa} \cdot \frac{4\mu + 3\kappa}{12\mu}\right),$$

where b is the internal and a the external radius, μ the shearing modulus, and κ the compressibility modulus.

a was measured directly from the outside of the cylinder and was 3.536 cm.; b was found from a measurement of the plug at the end of the piston after this plug had been forced to conform to the hole by the application of high pressure. It has been already stated that no change in b could be detected during the course of the experiments of so much as 0.00004 cm. b was found to be 0.7183 cm. μ and κ were found by the ordinary equations of elasticity from Young's modulus, taken as 2×10^6 kgm./cm.^{2}, and Poisson's ratio, taken as 0.28. These quantities vary only slightly in different grades of steel. The values of μ and κ calculated in this way were found to be 7.81×10^5 and 15.2×10^5 respectively. Substituting in the formula gives directly

$$\Delta b = 0.00491 \text{ cm. for } 10,000 \text{ kgm./cm.}^2$$

Whence the change of area, $\Delta A = \dfrac{2 \times 0.00491}{.7183}$

$$= 1.35 \text{ per cent for } 10,000 \text{ kgm./cm.}^2$$

Temperature Corrections. — There are a variety of temperature corrections to be considered; those which are to be met in any physical measurement, and those which were peculiar to this particular work. The ordinary corrections concern only the reading of the temperatures and may be dismissed with a few words. Temperature about the lower cylinder was maintained constant to 0°.01 with a thermostat. Above zero this temperature was read with a Tonnelot thermometer calibrated at the Bureau of Weights and Measures in Paris. The correction for the zero was determined, but it was not necessary to apply any of the other corrections. Readings on this are probably accurate to 0°.01. Below zero, readings were made on a toluol thermometer, graduated to 0°.1, and calibrated at the Reichsanstalt. In addition to the corrections of the Reichsanstalt, the zero correction was applied. This is accurate to 0°.1, which corresponds to the accuracy of the pressure measurements.

Of the corrections peculiar to this work, the temperature effect on the pressure readings was the simplest. The manganin resistance coil with which pressure was measured was placed in the lower cylinder with the mercury, and so was exposed to the same changes of temperature. It has already been shown in the paper on the gauge, that within the temperature limits of this work the pressure coefficient of resistance is constant. The only temperature effect is on the total resistance of the coil, which, therefore, merely displaces the zero. This zero was usually redetermined directly at every temperature, but in a few cases where it was inconvenient to release the pressure to zero between two sets of readings at different temperatures a calculated correction was applied. This correction was determined from an independent measurement of the temperature coefficient of the zero of one of the four manganin coils with which pressure was measured. These four coils were made from contiguous pieces of the same piece of wire. The greatest change of the zero from this effect amounts to 400 kgm. between $-20°$ and $0°$. Above zero the correction becomes very much smaller, as this particular kind of manganin shows a maximum resistance at $28°$. It will be noticed that it was necessary to keep the bath temperature constant to only $0°.5$, in order to remain within the errors of the pressure readings.

By far the most serious correction is due to the fact that the lower cylinder with the mercury under experiment is at one temperature, while the upper cylinder where the measurements of the change of volume are made is at another temperature, the temperature of the room. It is readily seen that the change of volume during melting, for example, as measured by the displacement of the piston of the upper cylinder, will be in error by the expansion at constant pressure of the liquid kerosene on passing from the temperature of the lower to that of the upper cylinder. This error has the same sign during both melting and freezing. The change of volume so measured will be too large if the lower cylinder is colder than the upper, and too small in the reverse case. The correction evidently becomes rapidly larger with the increase in the difference of temperature, but because thermal dilatation becomes less with increasing pressure, this correction fortunately becomes smaller at high pressures, where it is harder to determine. The maximum value for the correction in this work was 2 per cent over the range $+20°$ to $-20°$.

This correction seems to have been entirely overlooked by Tammann. It might very possibly amount at the maximum to over 10 per cent for temperature ranges over $100°$ such as he employed. The effect of the correction will not be to change at all his co-ordinates of the melting curve, but might possibly modify somewhat the way in which Δv apparently tends toward zero at high temperatures and pressures. This latter point has some theoretical significance in Tammann's theory of solid-liquid.

The correction had to be determined by direct experiment, as there were no data by which it could be calculated. The discussion, which is somewhat long and involved, will be given in the following separate section.

The Thermal Dilatation of Kerosene at Constant Pressure.

Since the accuracy desired was not very high, the maximum value of the correction being 2 per cent, a procedure was adopted which was rapid, but not accurate enough to employ in determining this quantity as a physical constant interesting in itself. The method consisted essentially in filling the entire apparatus with kerosene, lower as well as upper cylinder, immersing the lower cylinder in the constant temperature bath, and plotting piston displacement against pressure. Then the lower cylinder was allowed to come to room temperature, and again the piston displacement and the corresponding pressures noted. The difference of displacement at the two temperatures gives the change of volume of the kerosene which was in the lower cylinder on expanding from the one temperature to the other. It is to be noticed that this gives the thermal dilatation of a quantity of kerosene of unknown mass but of known volume. To find the dilatation as ordinarily defined, we should require the compressibility, so that we might know the original volume of the kerosene which at the given pressure completely filled the lower cylinder. But the data as given by the experiment are exactly the data needed in making the correction. For suppose the change of volume of the mercury on melting at a certain temperature and pressure is approximately 1 cm.^{3}. We require to know how much this 1 cm.^{3} of displaced kerosene has changed in volume on passing from one temperature to another at constant pressure, and we are not interested at all in the original volume of this 1 cm.^{3} of kerosene.

The procedure as roughly outlined above is evidently liable to a multitude of errors. It is exactly similar to the corresponding method of obtaining compressibilities against which objections were made in the introduction. But it is to be said in extenuation that the correction is small, and that the maximum correction comes at the low pressure where the objections of the introduction have little weight. The correction at $-20°$ of 2 per cent corresponds to only 3500 kgm. At this pressure and with a cylinder seasoned by exposing to 24,000 kgm. there need be no hesitation even in applying the method to accurate determinations of the compressibility. The correction at 0° and 7500 kgm. has already dropped to 0.6 per cent.

The corrections to apply in determining this maximum correction of 2 per cent are the corrections on the volume of the lower cylinder brought about by the changes in temperature and pressure. The volume was found by weighing the lower cylinder when full of mercury. To be subtracted from this volume was the volume of a copper block inserted to reduce to manageableness the total quantity of the kerosene, and the volume of the manganin pressure gauge. The volume occupied by the kerosene increases with pressure both because of the stretching of the cylinder and the shrinking of the copper and the manganin gauge. The stretching of the cylinder was calculated by the theory of elasticity as outlined above. A slight modification of the formula is necessary because the cylinder carries the end thrust of the pressure. The total correction for the stretching of the cylinder is 1.4 per cent at 10,000 kgm. The effect of the copper and the small quantity of mica washers in the insulating plug were assumed to be represented by the compressibility 0.0_51, the compressibility of the copper alone being 0.0_69. This correction amounts to 0.5 per cent at 10,000. Together, the two effects produce a total change on the effective volume of kerosene of 1.00 cm.^{3} per 10,000 kgm., on an initial volume of 28.6 cm.^{3}. The correction for the thermal dilatation of the lower cylinder was found from the known dilatation of steel at ordinary pressures, assuming it to be independent of pressure.

In making the actual measurements, endeavor was made to avoid as far as possible the small effect of mechanical hysteresis. The set of readings at $-15°$, for example, was obtained as follows. The piston displacement was measured with increase of pressure from 3500 to 5800 kgm. with the lower cylinder at $+20°$; the lower cylinder was then cooled to $-15°$, and the displacement measured from 5800 back to 3500; and finally the lower cylinder was warmed again to $+20°$ and the displacement measured from 3500 to 5800. The difference between the second curve and the mean of the first and the third, which differed only very slightly, gives at any one pressure the expansion of the corresponding amount of kerosene on passing from $-15°$ to $+20°$, since the temperature of the upper cylinder was also $+20°$. This difference was found for a number of pressures for the range 3500 to 5800, and a smooth curve drawn through these differences. From this curve was taken the difference corresponding to the freezing pressure of mercury at $-15°$. This process was performed for four temperatures, $-20°$, $15°$, $-7°$, and $0°$, and four points were thus determined on the correction curve. It was not necessary to obtain other points above $0°$ on the curve, because the correction was already small at $0°$ and the other end of the curve is fixed by definition as passing through the value zero at $20°$. Through these five points, $-20°$ to $20°$, another smooth curve was drawn, giving at any temperature the piston displacement when the amount of kerosene in the lower cylinder passes from that temperature to $+20°$ at the pressure corresponding to the freezing pressure of mercury at that temperature. Finally this curve gave the required percentage correction curve by translating piston displacements into volumes and dividing by the volume of the kerosene in the lower cylinder at the corresponding pressure as already determined. The correction curve actually used is shown in Figure 18. It is ob-

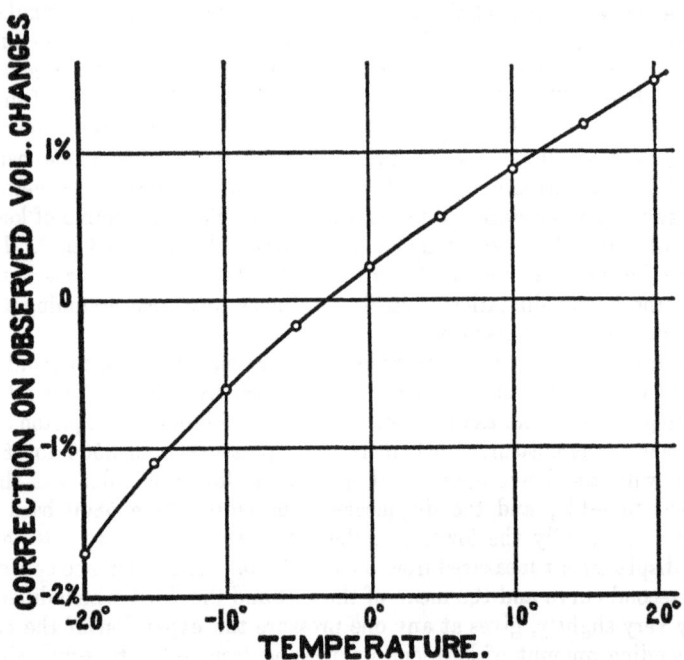

FIGURE 18. The corrections due to the thermal dilatation of the transmitting liquid and the elastic deformation of the cylinder to be applied to the directly measured change of liquid mercury when it passes to the solid state.

tained by combining with this temperature correction the correction already determined for the elastic deformation of the upper cylinder.

The correction curve so determined is not linear, the values being 2.0 per cent at $-20°$, 0.65 per cent at $0°$, and 0 at $+20°$. The departure from linearity is evidently due to the decreasing dilatation of the kerosene under high pressures.

Temperature Correction for the Upper Cylinder. — In the form of apparatus used in this experiment there is one other small correction because the upper cylinder is exposed to the slightly varying room temperature. If the room temperature changes during freezing, the observed change of volume is evidently incorrect by the thermal expansion at the particular pressure of the total quantity of kerosene in the upper cylinder. The temperature changes were usually very slight. They were made still smaller at the cylinder itself by enveloping this in a large mass of cotton batting. The temperature of the cylinder was read with a thermometer directly in contact with it. The correction demands an approximate knowledge of the quantity of kerosene in the upper cylinder, which was obtained by measurement of the position of the piston. The thermal dilatation of the kerosene at any given temperature was found from the correction curve already described, making the justifiable assumption that at constant pressure the dilatation is sensibly constant over the temperature range. The maximum correction was at $-20°$, where the dilatation is a maximum, and where the temperature variation, $0°.7$, happened also to be a maximum. The correction here was 0.5 per cent. Above $10°$ the correction was entirely negligible.

Possible Change of State of the Transmitting Liquid. — The effect of any freezing of the kerosene with contraction during the freezing of the mercury would be to make the change of volume appear too large. This possibility was entirely eliminated, however, incidentally by the measurements determining the thermal dilatation of the kerosene. At any given temperature these measurements extended considerably beyond the freezing pressure of the mercury, and no evidence of a change of state with change of volume was found.

Measurements of the Piston Displacement.—These measurements were made with a Brown and Sharpe micrometer reading to 0.0001 inches. The measurements were made between a point on the head of the advancing ram which drives the piston and a fixed point on the frame of the press against which the cylinder is pressed. There are two possibilities of error here. One, the most serious, is warping of the entire frame of the press. Such an effect was detected and may give rise to discrepancies of 0.002 inch over the entire stroke of 4 inches. It was obviated by making four measurements of every displacement between four pairs of points at the four corners of the frame of the press. The second error may come from elastic deformation of the parts of the press between the head of the ram and the fixed point on the frame, so that the measured change of length does not give the actual piston displacement. The effect, which is in any event small, was obviated by taking advantage of the fact that there is considerable friction in the packings. In virtue of this it is possible to vary the pressure on the low pressure end of the ram over a considerable range without producing motion of the piston, and so without altering the high pressure. Now it is the pressure on the low pressure end which produces distortion in the frame of the press, the cylinder at the high pressure end taking up in itself the effects of the high pressure. The effect of distortion in the frame was avoided, therefore, by adjusting the low pressure before every measurement of displacement to the same constant value for all values of the high pressure in the immediate neighborhood of the freezing point.

The Pressure Measurements. — The method of measuring these high pressures, with a discussion of the various sources of error has been given in another paper. Briefly, the pressures were obtained from the measured change of resistance of a coil of manganin wire immersed directly in the kerosene of the pressure chamber which contained the mercury under investigation. The coils, four of which were used during this investigation, were each calibrated separately against an absolute gauge. These were capable of reproducing the gauge with as much accuracy as the gauge itself could be read, which was about 8 kgm./cm.^{2}.

THE DATA.

The actual determinations by this method extended over about two months, from Oct. 13 to Dec. 3, 1910. In all, eleven points were determined, from $-20°$ to $+18°$, both with increasing and decreasing pressure. The measurements were sandwiched in between similar ones on water which are to be described in another paper. The apparatus was taken apart and set up a great number of times, both in the course of the ordinary manipulations attending the experiment and in making those changes necessary because of the piecemeal explosion of various parts of the apparatus. There were at least five such explosions during the experiment; two lower cylinders being burst, two upper cylinders, and one connecting pipe. The accuracy of the work is spoken for by the fact that determinations with all these different groupings of apparatus, including change of the pressure measuring coils, gave points lying on the same curve. This does not appear so strikingly from the actual data given, but it must be remembered that by far the greater number of measurements during this time were made on water, where the same independence of the particular piece of apparatus is also shown. The first five measurements on mercury were made with different combinations of apparatus; the last six were made with the same set up. The equilibrium pressures found in all eleven sets of readings are equally worthy of acceptance, but the volume measurements of the first five must be discarded because the effect of changes of room temperature was not sufficiently recognized at the time, and no correction was applied for it. The data of the same period for water, however, where the total change of volume is larger and the correction of less account, give consistent values for the change of volume also.

First for the equilibrium pressure. This was ordinarily found at only two points, one with increasing and the other with decreasing

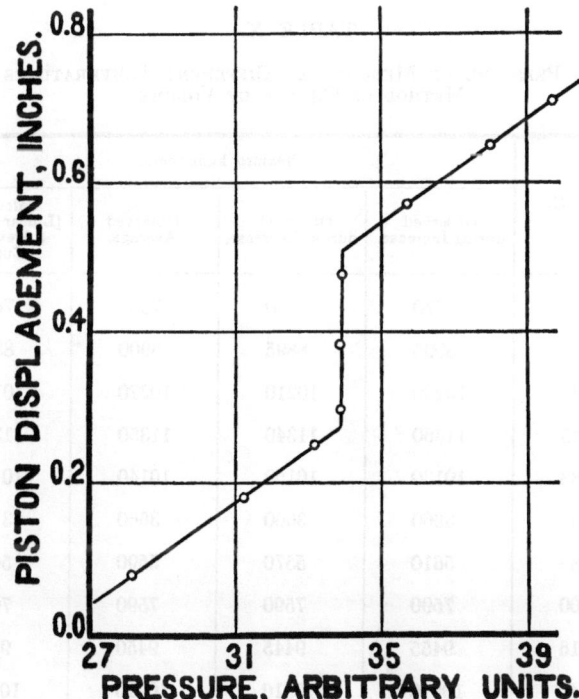

FIGURE 19. Shows the discontinuity in the piston displacement during freezing. (One division on the pressure axis corresponds to 900 kgm.)

pressure, but in order to make sure that this equilibrium pressure was a perfectly definite quantity, independent of the quantity of mercury melted, equilibrium was found, during the first two or three runs, with various proportions of the solid melted. This is shown in Figure 19 (Oct. 13), where the piston displacement is plotted against pressure. The vertical part of the curve shows the period of transition. The equilibrium pressure thus appears to be the same whether the freezing is nearly complete or only just begun. It is possible to get much closer to the upper corner of the curve, as will be shown in the magnified diagram later, but the lower corner cannot be approached very much more closely, because of the subcooling of the liquid mercury.

The data are collected in Table X. and shown in Figure 20. These data can be subjected to a calculation exactly like that used for the equilibrium points found by the change of resistance method. A

TABLE X.

FREEZING PRESSURE OF MERCURY AT DIFFERENT TEMPERATURES BY THE METHOD OF CHANGE OF VOLUME.

Temp. °C.	Pressure, kgm./cm.²			
	Observed during Increase.	Observed during Decrease.	Observed Average.	Calculated [Linear Relation + Deviation Curve].
0.00	7570	7550	7560	7620
6.37	8905	8895	8900	8890
12.87	10230	10210	10220	10180
18.45	11360	11340	11350	11340
12.88	10130	10150	10140	10180
−19.9	3660	3660	3660	3700
− 9.8	5610	5570	5590	5680
0.00	7590	7590	7590	7620
9.16	9455	9445	9450	9440
16.4	10910	10910	10910	10900
5.23	8650	8670	8660	8680

straight line was passed through the midst of the points, the deviations from this line plotted on a large scale, and the best smooth curve drawn through the deviations. The calculated values were obtained from this smooth deviation curve and the original straight line. The points are not so regular here as for the determination by the change of resistance method, and the best smooth curve is open to greater question. The deviations from linearity are nearly twice that found by the resistance method, namely 140 against 60 kgm. at 12,000. This has as a result that the tangent at the origin as determined here gives a rise of pressure of 195.5 kgm./cm.² per degree, against 196.5 found before, a difference of 1/2 per cent. The two series are not really so much different as this would indicate, because the smaller initial slope of this present series is unduly influenced by the very evident greater

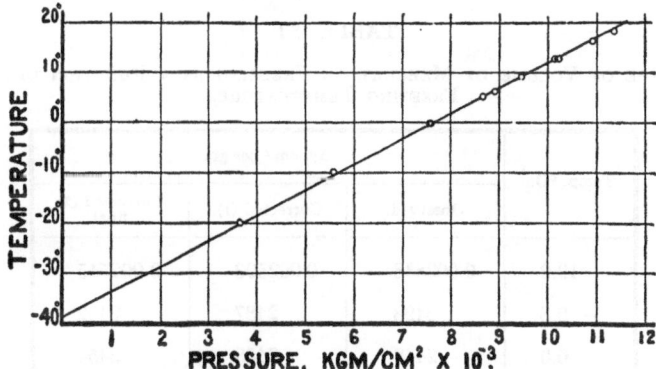

FIGURE 20. The equilibrium curve of liquid and solid mercury obtained by the change of volume method.

irregularity of the low-temperature points, and is compensated for by the greater departure from linearity. If we are willing to admit that the departure from linearity of the resistance measurements is probably more nearly correct, and apply this departure to the seven determinations by the present method above zero, we shall find as the average slope 196.3 kgm./cm.² at the origin, agreeing within 1/10 per cent of the former value. In the rest of the work of this paper the best value for the slope at the origin will be taken as 196.4 kgm./cm.² per degree, and the departure from linearity given by the resistance measurements will be taken as most probably correct.

The changes of volume corresponding to these equilibrium pressures were all determined with the same set up of apparatus as already explained. These results are given in Table XI., and plotted against temperature on an enlarged scale in Figure 21. The observed values were obtained graphically from figures like Figure 19, drawn on a very much enlarged scale. For this purpose, it was not necessary to plot the actual pressures, but the position of the slider on the bridge wire on which the resistance measurements were made was used instead. This enables smaller changes of pressure to be measured with accuracy than could be done in the other way. The absolute gauge against which the calibration of the resistance was made has a limit of sensitiveness of about 8 kgm./cm.², while the resistance measurements are sensitive to about 2 kgm. The scale of the graphical construction was so enlarged that 1 cm. on the bridge wire goes into 5 cm. on the diagram, and 0.01 inch piston displacement into 1 cm. on the diagram.

TABLE XI.

CHANGE OF VOLUME OF MERCURY ON FREEZING AS A FUNCTION OF THE FREEZING TEMPERATURE.

Temp. °C.	Δv, cm.³ per gm.		
	Observed.	Corrected (1).	Corrected (2) final.
−19.9	0.002533	0.002503	0.002515
− 9.8	2498	2487	2492
0.0	2448	2458	2454
5.23	2410	2426	2426
9.16	2376	2399	2399
16.40	2315	2348	2348

Ordinarily four points were found on the curve both above and below the transition point, covering a range of from 500 to 700 kgm. in both directions. Over this range, and within the errors of reading, the curve connecting piston displacement with the position of the slider on the bridge wire was almost without exception linear, and was drawn in with a ruler. There was no tendency whatever for either corner of the melting curve to be rounded off, indicating that the change of state takes place at a single definite temperature and pressure, and affording the best possible evidence of the sufficient purity of the mercury. To this fact of absolutely sharp freezing is to be attributed the self-consistency of the results. With water, where there is some slight rounding off of the corners, the results are not so self-consistent, although the total change of volume is larger.

The process of correction applied to the observed values is shown in the table in two steps so as to give an idea of the magnitude of the corrections. In column (1) the corrections for the thermal dilatation of the kerosene on passing from the lower to the upper cylinder and the correction for the elastic deformation of the upper cylinder are both applied. In the second and finally corrected column, the addi-

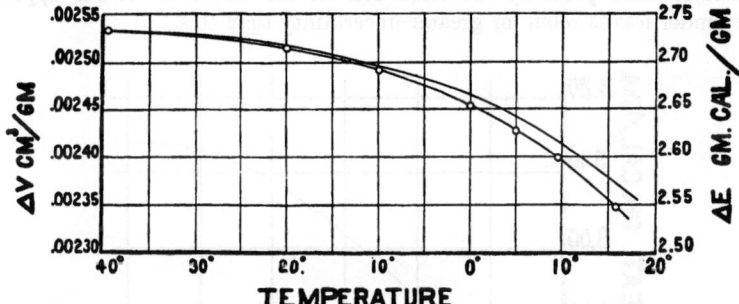

FIGURE 21. Shows the change of volume and of internal energy on passing from the liquid to the solid. The change of volume curve was directly observed, and the experimental points are indicated. The change of energy curve was calculated from the other data.

tional correction for the variation of room temperature is applied. The sign of this correction changes with the direction of variation. As appears from the figure, these points are self-consistent, lying on a

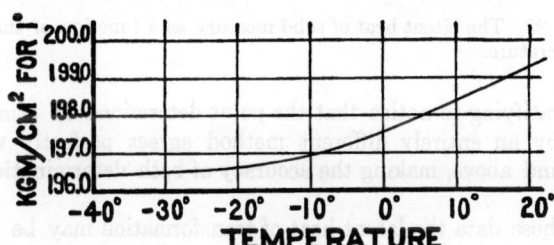

FIGURE 22. Shows the slope of the freezing curve as a function of the equilibrium temperature.

smooth curve to at least the order of accuracy given above, 1 part in 2500, except the point at 9° which is only 1/10 per cent out. The total displacement of the piston during freezing was about 0.2 inch, the quantity of mercury used being in the neighborhood of 350 gm. To obtain the above accuracy, therefore, it was necessary to take advantage of the utmost capacity of the micrometer, making all the readings to 0.0001 inch. That the single measurements do enjoy this degree of accuracy is made evident by a later very much enlarged figure drawn for the purpose of showing the impossibility of subcooling.

It is not claimed, however, that the absolute values at the high pressures have quite so much accuracy as this self-consistency would indicate, since probably the correction for the distortion of the upper cylinder leaves room for greater uncertainty than this.

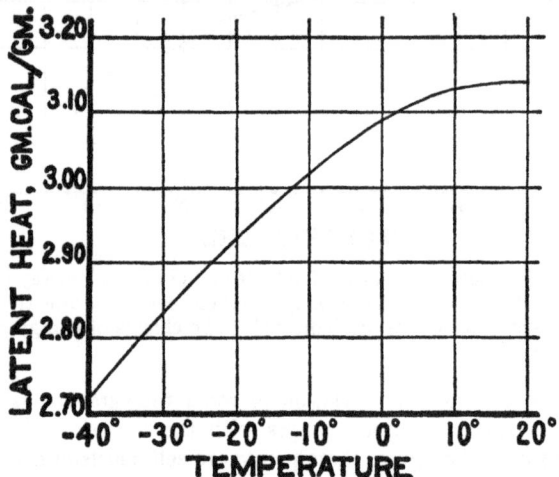

FIGURE 23. The latent heat of solid mercury as a function of the equilibrium temperature.

It is gratifying to notice that the point determined at atmospheric pressure by an entirely different method agrees perfectly with the points found above, making the accuracy of both determinations very probable.

From these data the latent heat of transformation may be found at every point of the melting curve by Clapeyron's equation,

$$\Delta H = \tau \Delta v \frac{dp}{d\tau}.$$

ΔH was calculated from this equation at intervals of 10° from $-40°$ to $+20°$. $\frac{dp}{d\tau}$ was found by adding to the constant value 196.4 kgm. per degree the correction found from the graphically constructed slope of the curve showing the departure of the curve from linearity. The results of this graphical construction are shown in Figure 22. The absolute zero in the calculation was taken as $-273°.1$ C. The results in mean gm. cal. per gm. of mercury are shown in Figure 23 and Table XII.

The rise of ΔH to a probable maximum is the most interesting feature. The course of this curve is to be explained by the increasingly rapid drop of ΔV at high pressures, this drop finally becoming more rapid than the increase in $\frac{dp}{dt}$ and in the absolute temperature.

TABLE XII.

CHANGE OF LATENT HEAT ON THE FREEZING CURVE OF HG.

Temperature.		Δv cm.³ per gm.	$\frac{dp}{dt}$ gm./cm.².	Latent Heat mean gm. cal. per gm.
°C.	Abs.			
−40	233.1	0.002535	196400	2.720
−30	243.1	2526	196500	2.828
−20	253.1	2515	196600	2.939
−10	263.1	2492	196900	3.025
0	273.1	2454	197600	3.103
+10	283.1	2393	198350	3.149
+20	293.1	2311	199300	3.163

It will be noticed here, as for the change of volume, that the curve does not give the usually quoted value for the heat of transformation at $-38°.85$, but here also it seems that there is considerable chance for error in the previous value. The value always given is 2.82, which was obtained by Person[43] in 1847. As far as can be discovered, this is the only determination of this quantity which has ever been published. Person's method consisted essentially in dropping the frozen mercury at a temperature of about $-42°$ into the water or turpentine of a calorimeter, and noting the fall of temperature. The usual corrections for radiation and the heat capacity of the vessel were applied. The actual heat given to the calorimeter was the sum of the heat given out by the solid in cooling to the transformation point, the heat of transformation, and the heat of the liquid in warming to the temperature of the calorimeter. The value $C_p = 0.0333$ was assumed to hold for the liquid over the temperature range. This is probably nearly correct, but the assumption that C_p for the solid is the same as

---

[43] Person, Ann. de Chim. et Phys., **24**, 257–264 (1848).

for the liquid is probably open to greater question. The error introduced by this cannot be very large, however, because of the small temperature range over which the solid cooled. By far the greatest possibility of error seems to be in the simple measurement of temperature. The thermometry of low temperatures must have been in a chaotic state at the time, judging by the values given by Person for known fixed temperatures. He gives the value on the air thermometer of the melting point of mercury as $-41°$. This point is now known to be $-38°.85$. He furthermore states that his alcohol thermometer read higher than a mercury thermometer at $-20°$. No statement whatever is made as to the corrections applied to the readings of this alcohol thermometer. If, however, he did apply the correction of the mercury thermometer to the alcohol thermometer, then there is still the possibility of great variation between $-20°$ and $-40°$. The consistent behavior of three calibrated alcohol thermometers which have been used in this present work indicates a possibility of error of as much as 3 per cent between $-20°$ and $-40°$.

Person made three determinations of the latent heat, giving 2.77, 2.86, and 2.83, mean 2.82, as against 2.735 found above. The use of this latter value seems to be justified, therefore, in preference to that of Person.

Person's value for the latent heat has entered directly into the determination by Regnault[44] in 1849 of the specific heat of solid mercury. Again the value of Regnault is apparently the only value which we have of this quantity. The method of Regnault was exactly the same as that of Person, except that the mercury was initially at $-77°.7$, so that the heat given out by the solid in cooling to the melting point was a very appreciable part of the total heat given out. Incidentally Regnault's temperature measurement was very probably at fault as well as Person's, for he gives $-40°$ as the melting point of mercury. Regnault made two determinations of the mean specific heat from $-77°.7$ to the melting point, giving 0.0314 and 0.325. Evidently the effect of a smaller latent heat of melting would be to give a larger specific heat to the solid. Substituting the value found above produces a change of 6.5 per cent in the values given by Regnault, giving 0.0334 and 0.0346 respectively, mean 0.0340.

The data used in determining the change of volume may also be used to give evidence on one point for which they were not intended. The slope of the curve plotting piston displacement against pressure is evidently going to be different above and below the melting point be-

---

[44] Regnault, Ann. de Chim. et Phys., **26**, 268–278 (1849).

cause of the difference in compressibility between the solid and the liquid mercury. The compressibility of mercury is small in any event, and here the greater part of the piston displacement is used in compressing the kerosene, so that only rough values can be expected from the small difference. There is the further complication of the drift in the temperature of the room. The slope does show a distinct tendency to be less when the solid state is present, but the change is irregular. It can be said, however, that the compressibility of the solid at the freezing point is less than that of the liquid, the difference being in the neighborhood of 10 per cent.

The Change of Volume of Mercury on Freezing at Atmospheric Pressure.

It has already been stated that the necessity for redetermining the change of volume at atmospheric pressure arose from the probable inaccuracy of the previously accepted value (0.00260 cm.^{3} per gm.). The inaccuracy of this number was first suspected when a set of values found for Δv at different pressures, consistent among themselves to 1/25 per cent, did not extrapolate to within 2 per cent of 0.00260. Investigation showed that the value 0.00260 is not a direct experimental determination, but is obtained from the difference between the density of the solid, as found by one observer, and the density of the liquid as given by another. Since the change of volume is only about 3 per cent of the total volume, the density of both solid and liquid must be known with an accuracy of sixty times the accuracy desired in the change of volume. The following short discussion of these previous experimental determinations will make clear that there are here possibilities of error of fully the indicated magnitude.

The value of the density of the solid seems to be open to by far the most serious question. The commonly accepted value, in fact the only value with any claim to accuracy at all, was given by Professor Mallet [45] of the University of Virginia. The method consisted in freezing a weighed quantity of mercury in a glass bulb, and then determining the quantity of alcohol necessary to fill the bulb to a fixed mark. The largest possibility of error seems to be in determining the volume of the bulb at the freezing temperature of mercury. This was found from the quantity of mercury filling it at 0° and 100°, using Regnault's value for the dilatation of mercury, and extrapolating back to the freezing point. No account at all was taken of the possibility of per-

---

[45] Mallet, Proc. Rov. Soc. Lon., **26**, 71 (1877).

manent changes of volume of the glass bulb after subjecting to 100°. That there is such an effect, very noticeable, is shown by the change of zero of glass thermometers after subjecting to a high temperature. No mention is made of the kind of glass; if it were an American lead glass, the effect would be comparatively large. The final value for the density of the solid as given by Mallet was 14.1932. This is the mean of three values, 14.1948, 14.1920, and 14.1929.

The density of liquid mercury at the freezing temperature is given in Landolt and Börnstein's tables as coming from Vicentini and Omodei,[46] but the value given by them is as a matter of fact quoted from a paper of Ayrton and Perry.[47] Their method consisted in reading simultaneously the same temperatures between 0° and —38°.85 with a mercury in glass thermometer and an air thermometer. The readings of one plotted against the other are stated to be "very nearly linear," from which the conclusion is drawn that the dilatation of mercury between 0° and —38°.85 has the same constant value as that found by Regnault to hold between 0° and +100°. The density at —38°.85 is then calculated from the known density at 0°. No statement whatever is made as to the degree of accuracy attained, or even as to the direction in which the plot of the air against the mercury thermometer departs from linearity. However, the fact that one of the conclusions drawn from the results is that mercury shows no maximum of density in the neighborhood of the freezing point analogous to water, suggests that the previous knowledge of this quantity was very imperfect indeed, and that the results of Ayrton and Perry themselves are not so accurate as necessary for the present purpose.

A direct determination of this change of volume with the same apparatus as was used at higher pressures and temperatures would have been possible, but inconvenient for several reasons. Chief of these was the difficulty of running a thermostat of the requisite size with sufficient constancy at this low temperature. Furthermore, if a result concordant with the others could be found by an entirely different method, the results by the two methods would mutually support each other, making any consistent error in either improbable.

The method used consisted in weighing a known quantity of mercury under CS_2 at different temperatures in the neighborhood of the melting point. The weight varies with the temperature. When the temperature passes through the melting temperature, the mercury melts with increase of volume, and the increased displacement causes a sudden

---

[46] Vicentini and Omodei, Atti di Torino, **23**, 38–43 (1887–1888); and **22**, 28–47, 712–726.

[47] Ayrton and Perry, Phil. Mag., **22**, 325–327 (1886).

change in the weight equal in amount to the weight of the displaced CS_2, that is, equal to the weight of CS_2 of volume equal to the change of volume of the mercury. The only quantities directly concerned are the discontinuity of the weights and the density of the CS_2 at the point of discontinuity. It is not necessary to know even the temperature accurately. Any consistent device for indicating changes of temperature, so that the weight may be plotted as some continuous function of the temperature on either side of the melting point, is all that is required.

The temperature was indicated by the resistance of a coil of fine double silk covered iron wire, wound on a glass core 4 inches long and placed in kerosene inside a thin-walled brass tube, which was immersed directly in the CS_2 in which the weighings were made. The coil was not calibrated for absolute temperature, but the resistance could be measured with an accuracy corresponding to changes of temperature of $0°.002$. In addition, the temperature was read on a toluol thermometer, graduated to $0°.1$, which was also immersed in the bath. The readings of the thermometer were taken only to give a check on the readings of the resistance thermometer, since the thermometer readings were inaccurate because of varying parallax and the necessity of removing the thermometer partly from the bath in making the readings.

The cold CS_2 in which the mercury was weighed was contained in a cylindrical Dewar flask, 2 inches in diameter and 12 inches long. The inside of this was lined with a sheet of celluloid to protect the glass from being scratched, and inside this a tube of heavy copper tubing, 5/32 inch thick, to promote rapid equalization of the temperature. The most effective equalization of temperature, however, was provided by a small turbine stirrer, reaching the entire length of the flask. This was run constantly between readings, but was naturally stopped while a weighing was being made. The mercury was contained in a thin steel shell, 9/16 inch diameter, and 4 inches long, suspended by a fine iron wire (0.007 inch diameter), from the beam of a sensitive balance above. The quantity of mercury used was about 140 gm. It was possible to weigh this to 1/4 milligram, giving an accuracy on the discontinuity of weight on freezing, which was about 0.6 gm., of one part in 2000. In the respect of sensitiveness, the CS_2 is probably as convenient a liquid as could be found, because of the high density and very low viscosity and surface tension. Provision was made for warming the bath after each weighing by passing a current through a small heating coil immersed in the bath.

The experimental procedure was the natural one. The Dewar flask was first filled with CS_2 which had been cooled in another dish with

CO_2 snow and ether. In this was suspended the steel shell with the mercury, which had been previously frozen. The temperature was then raised to about $-42°$, and readings made at intervals of $0°.5$. The spontaneous rise of temperature of the bath during a weighing was so small as to introduce no error. During warming, the weight increases linearly, due to the larger thermal expansion of the CS_2 than the mercury. Melting is indicated by a rapid drop in the weight of about 0.6 gm. This drop is not instantaneous, but may extend over $1°.5$, showing temperature lag between the mercury and the bath while the heat of transformation is being absorbed. Completion of melting is indicated by resumption of the linear relation between weight and temperature. To find the discontinuity in the weight, it is merely necessary to extrapolate this straight line back to the point at which melting began.

There are only a few special precautions to observe. The most likely source of error is the formation of minute cracks in the mercury when freezing. To obviate this, the mercury was poured into the steel shell already filled with CS_2, so as to ensure filling of all the corners, and then frozen by placing merely the very bottom of the shell in the freezing mixture. Freezing took place very slowly and uniformly from the bottom, the surface of the finally frozen mercury being as perfectly convex as that of the liquid. Pains had to be taken to scrape away from time to time the minute quantities of ice condensing on the suspending wire. This wire was so fine that no correction whatever had to be applied for the slightly varying depths of immersion during the course of an experiment. Finally, it was necessary to determine the density of the CS_2 at the freezing temperature separately for each experiment. This arose from the fact that during the preliminary work a slight quantity of ether had been accidentally spilled into the CS_2, enough to decrease the density by 1 per cent. This gradually evaporated, producing a slow rise in the density of the CS_2 to a final steady value. The density was given directly by the loss of weight of the mercury and steel in the CS_2 at the freezing temperature. There seems no question whatever but that the density of the liquid mercury at the freezing point is known with the accuracy demanded here, 1 part in 2000, although there may be reasonable question as to whether its accuracy is 60 times greater, as it must be when used to give the change of volume by a subtraction. The steel shell was only a small part of the total weight, and its thermal dilatation is so well known and so small that there is no danger at all in calculating the density of the steel at $-38°.85$ from the experimentally measured density at room temperature. The accuracy of this procedure is vouched for by the fact that the density of the CS_2 when determined by weighing in it a solid steel cylinder agreed within 1 part in 13,000 with the value found by weighing in it the steel shell with the mercury.

Five preliminary determinations of the change of volume were made giving results varying from 0.00253 to 0.00257. The chief cause of

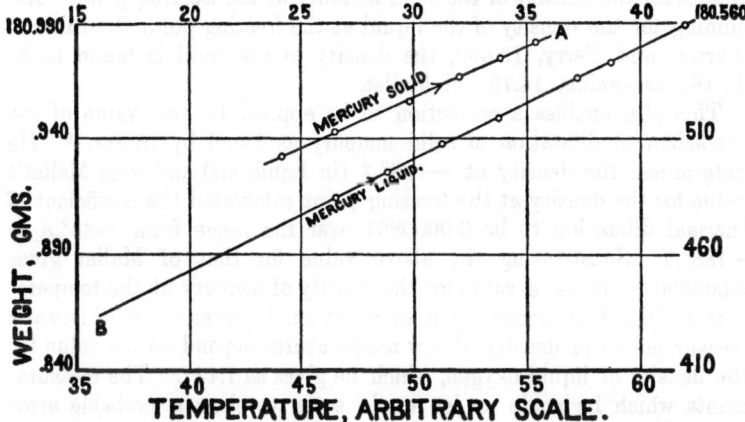

FIGURE 24. Shows the readings from which was found the change of volume when mercury melts at atmospheric pressure. The scale above and to the left refers to the solid mercury, and that below and to the right to the liquid. The point A indicates the beginning of melting, and the point B shows the completion of melting.

inaccuracy in these preliminary results was the irregular behavoir of the resistance thermometer. This was due to the defective insulation of the enamelled wire used in these first experiments. Replacing the enamelled wire by silk-covered wire gave perfectly satisfactory results. Advantage was also taken of this preliminary work to modify slightly various details of apparatus and procedure.

Three independent determinations of the change of volume were made with the final form of apparatus. The experimental points of one of these determinations are shown in Figure 24. The mercury was changed for each of these determinations, the total quantity being varied by 10 per cent. The three values found, using the iron resistance as the temperature indicator, were 0.002533, 0.002536, and 0.002532 cm.³ per gm. The same weighings plotted against the thermometer readings (it is to be remembered that this was done merely as a check) gave 0.002533, 0.002537, and 0.00252 respectively. The thermometer points of the last determination were very irregular.

The resistance points are to be taken as the correct ones, and the mean of the three determinations, 0.002534, is accepted as the best value. The departure from the previous value, 0.00260, is thus nearly 3 per cent.

This result may be used to give a more accurate value than that of Mallet for the density of the solid mercury at the freezing point. Assuming for the density of the liquid at the freezing point the value of Ayrton and Perry, 13.690, the density of the solid is found to be 14.182 as against 14.193 of Mallet.

This also enables a correction to be applied to the value of the coefficient of dilatation of solid mercury as found by Dewar.[48] He determined the density at $-188°.7$ (in liquid air) and from Mallet's value for the density at the freezing point calculated the coefficient of thermal dilatation to be 0.0000887 over the range from $-38°.8$ to $-188°.7$. Substituting the above value for that of Mallet gives 0.0000907. Dewar's value for the density of mercury at the temperature of liquid air seems open to question, however. All of Dewar's measurements of density at low temperatures depend on his value for the density of liquid oxygen, which he gives as 1.137. The measurements which he made on ice would seem to show a probable error here. The value for the thermal dilatation of ice over the range $0°$ to $-190°$ was found to be only half the known value from $0°$ to $-20°$. So large a decrease seems hardly probable. A change in the fundamental constant so as to give a higher value to this dilatation would also give a higher value to the thermal dilatation of mercury. Furthermore, a change in this direction would bring Dewar's value for the dilatation of mercury into agreement with that of Grunmach,[49] namely, 0.000123 from $-38°$ to $-80°$. This is apparently the only other value we have. But Grunmach used a dilatometer method in which he made no correction whatever for the effect of the glass envelope. The value obtained by him for the change of volume on freezing with the same apparatus is certainly wrong, namely, 5 per cent as against 3 per cent found above, and this one fact is enough to cast discredit on the rest of the work. There seems to be no value, therefore, which can be accepted at present with confidence for the thermal dilatation of the solid.

In view of the apparent uncertainty as to the value of the thermal dilatation of the solid, it may perhaps be allowed to force the above weighings to give what information they can on this point. It is not

---

[48] Dewar, Proc. Roy. Soc. Lon., **70**, 237–246 (1902).
[49] Grunmach, Phys. ZS., **3**, 133–136 (1902).

expected that the value so obtained can do more than roughly indicate the probable value, since this use of the data was not contemplated in the original experiment. The information comes by observing at the melting point the change in the slope of the curve plotting weights against temperature. The slope is evidently connected with the various thermal dilatations.

The exact expression is found to be

$$\left(\frac{d\overline{W}}{dt}\right)_1 - \left(\frac{d\overline{W}}{dt}\right)_2 = \frac{dD_{CS_2}}{dt}[V_{Hg,2} - V_{Hg,1}]$$
$$- D_{CS_2}\left[\left(\frac{dV_{Hg}}{dt}\right)_1 - \left(\frac{dV_{Hg}}{dt}\right)_2\right].$$

The subscript 1 refers to the liquid and 2 to the solid. D_{CS_2} is the density of the CS_2. The value $\frac{D_{CS_2}}{dt}$ was taken to be 0.0014. This amounts to assuming the value of the dilatation of CS_2 to be 0.001, which is the value given at $0°$ by Pierre's formula. There is room for considerable question here, but the term into which $\frac{D_{CS_2}}{dt}$ enters is only 10 per cent of the other term involving the dilatation of the liquid mercury.

Only the last two of the sets of weighings are available for this calculation, the CS_2 having then assumed an unvarying value for the density. The dilatation of the solid calculated from these two sets was 0.000165 and 0.000125, mean 0.00014. The value is extremely rough, but probably as good as either the value of Dewar or Grunmach.

These data also give a rough incidental determination of the density of CS_2 at $-38°.85$. This was found to be 1.3460. It has been already mentioned that this is probably slightly in error due to the presence of a slight impurity of ether, but the above value was the final value, and remained constant over two days, showing that what little ether there was had nearly all evaporated. The density as calculated by the third degree formula of Isid. Pierre is 1.3647.[50] The formula of Pierre is intended to hold only for the range $0°$ to $100°$. The discrepancy suggests at any rate the danger of using the formula by extrapolation to considerable negative temperatures.

SUBCOOLING AND SUPERHEATING.

All these observations on the freezing and melting of mercury under pressure also show that one property which apparently holds without

---

[50] Pierre, Ann. de Chim. et Phys. (3), **15**, 325 (1845).

exception at atmospheric pressure also holds under high pressure. This is the fact that it is impossible to superheat a crystalline phase with respect to the liquid phase, although the liquid may be subcooled,

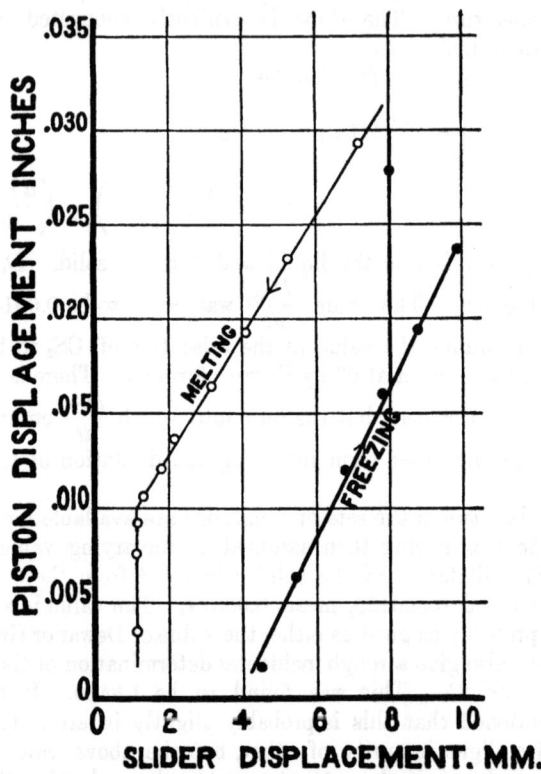

FIGURE 25. Shows on an enlarged scale the possibility of subcooling the liquid but the impossibility of superheating the solid. The two curves have been displaced, the one with respect to the other; the melting and freezing actually takes place at the same pressure. One division on the pressure axis corresponds to 40 kgm.

sometimes very considerably, with respect to the solid. Figure 25, drawn on a very large scale, shows this fact. At the lower corner the subcooling of the liquid with respect to the solid is shown, while at the upper corner, where the pressure is decreased on the solid phase, is shown the sharp change of state when the equilibrium pressure is overstepped. The slight irregularities in the points are due to errors of reading, the several points on the vertical part of the curve at decreasing pressure differing among themselves by only 0.05 mm. of bridge wire, better than the limit of reading, corresponding to pressure differences of 1 kgm./cm.^{2}. The points plotted are the mean of two readings, each to 0.1 mm. The same possibility of subcooling was found during measurements by the change of resistance method.

Attempts to measure the amount of subcooling would have had no absolute value, for its amount depends on many extraneous factors, such as the size and shape of the glass capillary, the presence of small particles of grit in the fluid transmitting pressure, and most important of all, the element of time. The greatest amount of subcooling found from the resistance measurements was 230 kgm. at 10°.2, equivalent to about 1°.1. The mercury remained liquid under these conditions for about 30 minutes, and did not freeze until the pressure was still further increased.

This impossibility of superheating a crystal has been shown to be so universally true that it is coming to be regarded in the light of a natural law. It holds whether the liquid melts with increase or decrease of volume. No satisfactory explanation has yet been given of the fact, however. Nearly all the explanations suggested, applicable at atmospheric pressure, have attempted to make it in some way a surface phenomenon, the melting beginning at the free surface and running in toward the interior. In view of the fact shown above that superheating is also impossible under very high pressures, it would seem somewhat doubtful whether this is the correct method of attack on the problem. Under high pressures the surface of separation between the solid and the surrounding liquid (mercury and kerosene, for example) cannot be thought of properly as a free surface. The molecules of mercury would seem no more free to assume their natural positions at this surface than they would in the interior of the mercury itself. This is particularly probable when it is remembered that the mercury melts with increase of volume.

CONCLUSION.

In this conclusion an attempt will be made to show what bearing the data may have on the theories of the liquid state and on our knowledge of the change of state liquid-solid. As giving the best general survey of all the data, a diagram is presented showing the isothermal lines for the liquid and the sudden change on passing to the solid state. (Plate.) The data from which this figure was constructed are shown in Table XIII. Within the accuracy of this work the isothermals may be taken as equi-spaced, as already explained. The first thing to strike one on looking at the diagram is that the effects are nearly constant over the entire range; the compressibility does not change markedly, nor does the dilatation, and the lines bounding the

TABLE XIII.

Points on the Isothermals of Liquid Hg.
(Computed from Experimental Curves at 0° and 22°.)

Pressure, kgm./cm.².	Volume, cm.³.					
	−30°.	−20°.	−10°.	0°.	+10°.	+20°.
0	0.99457	0.99638	0.99819	1.00000	1.00181	1.00362
1000	99207	99280	99553	0.99626	0.99799	0.99972
2000	98760	98927	99094	99261	99428	0.99593
3000	...	98581	98743	98905	99067	99232
4000	...	98245	98403	98561	98719	98877
5000	...	...	98077	98231	98385	98540
6000	...	...	97763	97914	98065	98216
7000	...	...	...	97607	97756	97904
8000	...	...	...	...	97461	97608
9000	...	...	...	...	97182	97327
10000	...	...	...	...	96915	97059
11000	...	...	...	...	...	96806
12000	...	...	...	...	...	96567
Change of vol. on freezing	0.03435	0.03418	0.03388	0.03337	0.03243	0.03142
Freezing pressure	1740	3710	5670	7640	9620	11600

domain liquid-solid are nearly parallel. It is evident, therefore, that any reversal of the ordinary effects or the appearance of critical points must be far beyond the reach of the diagram. The diagram is competent to show only the initial trend of the various effects.

For the liquid state the results are the same in general as those with which we are familiar for the more compressible liquids under lower pressures. The compressibility and the dilatation both decrease with rising pressure. This has as a consequence that the effect of rising temperature is as usual to increase the compressibility. The behavior of the specific heats is rather unusual, approximate constancy for C_p and increase of C_v with pressure. For water, $C_p - C_v$ increases instead of decreasing. As a little surprising, but not unusual, it may be mentioned that the internal energy decreases along an isothermal with increasing pressure, but increases, as is normal, along an adiabatic. This means that along an isothermal the attractive force between the molecules is still the dominant factor in the situation, even up to 12,000 kgm. At higher pressures a reversal of the effect is indicated, where the repulsion between the molecules would become the important part.

The data are more suggestive in the light they throw on the change from the liquid to the solid state. In this connection it may be useful to outline briefly the present state of the theory, and the points at issue.

For the change of state liquid-vapor, the physical facts have been worked out pretty fully until we now have a fairly definite understanding of the nature of the process. This increase of knowledge has come about almost entirely from the experimental side, the one thermodynamic relation not being of much assistance. In particular, the existence of the critical phenomena, and the possibility of the continuous passage from the liquid to the vapor were facts which were unexpected before the actual experimental proof. There is no thermodynamics which would predict the existence of such a point. Our present knowledge of the nature of the equilibrium between liquid and solid is in much the same state as the knowledge of the transformation liquid-vapor before the discovery of the critical point. The fundamental question as to whether a critical point exists or not is therefore of greatest immediate interest.

At present, both the most important theory of the equilibrium between solid and liquid and the most far reaching experimental work are due to Tammann.[51] The essential part of his theory, as opposed to Ostwald,[52] Poynting,[53] and others, is that for the transition solid-liquid there is no critical point like that for the transition liquid-vapor. The reason for this fundamental distinction is to be found in the essential similarity between a liquid and a vapor on the one hand, and the essen-

---

[51] Tammann, loc. cit.
[52] Ostwald, Lehrbuch der Allgemeinen Chemie, II^{II}, Verwandtschaftslehre, p. 389 (Engelmann, Leipzig, 1902).
[53] Poynting, Phil. Mag. (5), **12**, 32 (1881).

tial dissimilarity between a liquid and a crystal on the other hand. The molecules in both the liquid and the vapor are thought of as arranged at random, "molekular-ungeordnet," the only difference between liquid and vapor lying in the average distance apart of the molecules. Once let the average distance apart of the molecules appropriate to the vapor and liquid become equal, as they do at the critical point, and all further distinction between the two phases vanishes. In particular, the heat of transformation vanishes as a consequence of the equality of the volumes. But for the two phases liquid-crystal, there is besides the distinction of density the further distinction that in the crystal the molecules are arranged in some sort of a framework in space. Equality of density of the two phases liquid and solid does not mean identity, because the crystal still preserves the regular arrangement of the molecules which cannot hold in the liquid. In consequence, the energy difference between the two phases need not become zero when the volume difference vanishes; that is, the heat of transformation does not vanish at the same time with the change of volume, and we do not have a critical point. Tammann goes further than simply to deny the existence of a critical point. He concludes from an examination of all the known data, chiefly his own experiments, that for substances of the ordinary type of freezing with decrease of volume, not only do the change of latent heat and the change of volume fail to pass through the value zero at the same time, but that the change of volume always becomes zero before the heat of transformation. It is then an immediate deduction from Clapeyron's equation that the melting curve has a maximum. That is, there exists for every substance a temperature so high that no pressure, however intense, will bring the point of transition as high as this temperature. Above this temperature the substance is always liquid. This is really the essential part of Tammann's theory of the shape of the equilibrium curve. It has been immediately seized upon by geologists, and made the basis of many speculations as to the state of the interior of the earth, the matter there being supposed fluid because the temperature is higher than the maximum.

The experimental evidence in support of this theory is meagre, the pressure reached by Tammann being sufficient only to show a curvature in the direction demanded by the presence of a maximum, and an initial decrease toward zero of the change of volume more rapid than the decrease of the latent heat. The only case in which Tammann claims to have found a maximum, for $Na_2SO_4 10H_2O$, must be ruled out because of the complication introduced in the equilibrium conditions by the water of crystallization.

The evidence of the present data of significance for this question as to the existence of either a critical point or a maximum are the slope of the freezing curve, the variation of volume, and the variation of the latent heat.

The freezing curve is concave toward the pressure axis, which is curvature in the direction demanded by the presence of a maximum. This direction of curvature is the same as that shown by all the melting curves of Tammann, who uses it as an argument tending to show the existence of a maximum. But the curves liquid-vapor also invariably show the same curvature, and here there is a critical point. It is also easy for a curve to run to infinite temperature and pressure, still being concave toward the pressure axis. The direction of curvature would seem to give absolutely no presumptive evidence, therefore, one way or the other.

The change of volume and the latent heat together may be considered. The change of volume alone invariably decreases with rising temperature. This merely means that the phase with the smaller volume is less compressible. Tammann's argument consists in showing that the change of volume decreases along the equilibrium curve, while the latent heat usually increases, or if it shows a decrease, the decrease is slight in comparison with the decrease in the change of volume. The change of volume becomes zero first, therefore, and we have a maximum freezing temperature.

The curves for ΔV and the latent heat of this present work on mercury have been already shown. ΔV decreases with rising temperature, as it does universally. The latent heat curve is the significant curve, because it increases apparently to a maximum and seems about to descend. Tammann's argument would be perfectly valid for the first 6000 or 7000 kgm., but at higher pressures this reversal of the latent heat curve invalidates the whole thing. There is no reason to anticipate that the decrease in ΔH beyond 11,000 might not be rapid enough to make it vanish with ΔV.

It is interesting in this connection to plot the difference of internal energy between the solid and the liquid against temperature. The difference is given by

$$\Delta E = \Delta H - p\Delta v.$$

ΔE shows the part of the heat of transformation which has been made potential inside the mass. The rest of this heat goes into performing mechanical work. The results are shown, with the steps of the calculation in Table XIV. and in Figure 21, in direct comparison with the changes of volume. The energy decreases from the start along the equilibrium curve in such a way that its ratio to Δv is nearly constant. There seems no particular reason why this is not as significant a quantity on the melting curve as the latent heat. It is true that it does not enter the equations so simply, but the behavior at the points of special interest on the melting curve is just as simple as that of ΔH. At the critical point, for example, if there is one, where Δv becomes zero, but

TABLE XIV.

THE CHANGE OF INTERNAL ENERGY OF MERCURY ON PASSING FROM THE LIQUID TO THE SOLID AT DIFFERENT TEMPERATURES.

Temp. °C.	Pressure, kgm./cm.²	Δv cm.³ per gm.	Work ($=p\Delta v$) gm. cal.	Latent Heat.	ΔE ($\Delta H - p\Delta v$) gm. cal. per gm.
−28.66	2000	0.002525	0.118	2.848	2.732
−18.48	4002	2512	.235	2.951	2.717
− 8.31	6005	2486	.350	3.041	2.693
+ 1.87	8018	2443	.459	3.114	2.657
+12.06	10034	2377	.557	3.154	2.599
+22.24	12064	2296	.649	3.165	2.516

$\frac{d\tau}{dp}$ remains finite, it is known that $\Delta H = 0$. In this case $\Delta E = 0$ also. If the melting curve shows a maximum, which is characterized by $\Delta v = 0$, and $\frac{dp}{d\tau} = 0$, then ΔH remains finite, as does also ΔE, which in this case equals ΔH. The trend of ΔE on the melting curve would seem therefore to be just as valuable an indication as the trend of ΔH as to the possible existence of a critical point or a maximum.

The present data offer no evidence, therefore, as to the existence of a critical point, except to show that if there is one it must be at pressures higher than can be reached directly. The positive result is obtained, however, that at high pressures there is a reversal in the behavior of the latent heat such as to invalidate Tammann's argument for a maximum.

As opposed to Tammann's argument for the impossibility of the existence of a critical point solid-liquid, it may be shown that it is possible to conceive of a molecular mechanism by which the continuous passage from the "molekular ungeordnet" assemblage of the liquid to the regular arrangement of the crystal on a space framework may be accomplished. We may think of the molecules as cubes, for example, with intense centers of force at the corners. As the molecules wander past each other in the liquid state, there will be a tendency for them to linger with the edges or the faces in contact. If the time during which they so linger in symmetrical positions with respect to each other becomes appreciable with respect to the total time, we will have some approach to the properties of a crystal, the approach becoming closer as the time of contact becomes relatively greater. And entirely aside from the question of molecular structure, there seems no difficulty in conceiving that if the temperature and pressure are varied properly on a cubical crystal, for example, that the three elastic constants might so change with respect to each other that two of them should eventually become equal, giving the two distinctive constants of an isotropic body, and so continuous passage from a crystalline to an amorphous body.

Summary of Results.

Data have been collected in this paper sufficient to give the p-v-t surface of the liquid, and the location and magnitude of the discontinuity in this surface when the liquid freezes to the solid over a pressure range of 12,000 kgm./cm.^{2}, and a temperature range from $-38°.85$ to $+20°$. Tables of various quantities of thermodynamic interest are given for the liquid; such as the compressibility, the dilatation, the change in the specific heats, the adiabatic compressibility, the heat of compression, and the change in internal energy both along an isothermal and along an adiabatic. For the change liquid-solid the co-ordinates of the melting curve are given, the change of volume, and the latent heat. The results do not show any behavior in the liquid at high pressures opposite in character from that which would be predicted from their trend at lower pressures. But for the change solid-liquid, it seems that at high pressures there is a change in the trend of the latent heat which invalidates Tammann's argument for the existence of a maximum. If either a maximum or a critical point exist for the transition solid-liquid it is at pressures higher than are at present open to direct realization. Such a point could not well exist at less than 50,000 kgm./cm.^{2}.

Apart from the main work of the paper a number of other quantities relating to mercury have been determined, some of them very roughly, either by a direct calculation from the new data given here, or else by

438   PROCEEDINGS OF THE AMERICAN ACADEMY.

a recomputation from the data of others, using in this recomputation some of the improved values found here. These new quantities are: a more accurate value for the density of the solid at $-38°.85$, a rough value for the thermal dilatation and the compressibility of the solid, an improved value for the specific heat of the solid, the electrical resistance of the liquid up to the freezing pressure, and the resistance of the solid under pressure, including the pressure and the temperature coefficients.

Finally, the computations made necessary the determination of another quantity of interest in itself, namely, the compressibility of steel. This was determined up to 10,000 kgm. and at two temperatures, from which the temperature coefficient of compressibility is found.

Acknowledgement is here made of several liberal appropriations from the Rumford Fund of the American Academy of Arts and Sciences with which the expenses of this investigation were partially defrayed.

JEFFERSON PHYSICAL LABORATORY,
HARVARD UNIVERSITY, CAMBRIDGE, MASS.
OCTOBER, 1911.

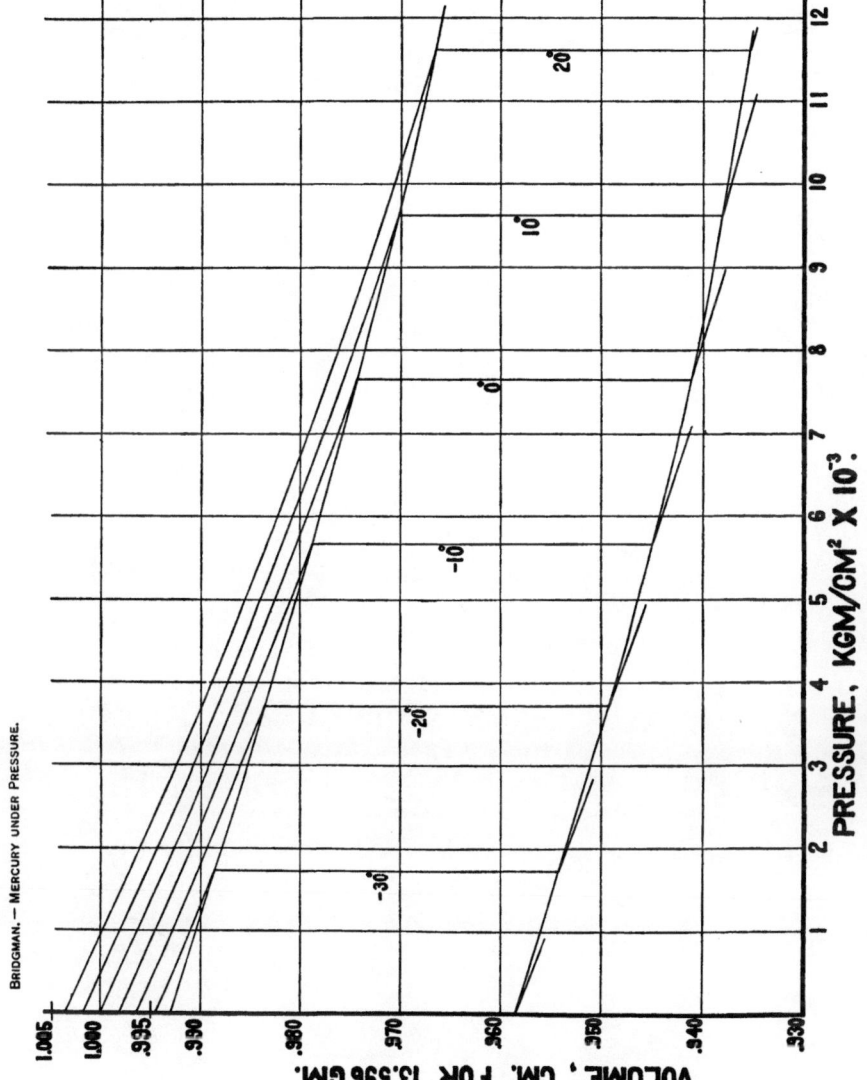

PLATE. Shows the isothermals at 10° intervals of liquid and solid mercury. The upper part of the diagram is for the liquid, the lower for the solid. The initial slope of the isothermals for the solid was not obtained by direct experiment, but was found approximately by a computation.

CONTRIBUTIONS FROM THE JEFFERSON PHYSICAL
LABORATORY, HARVARD UNIVERSITY.

WATER, IN THE LIQUID AND FIVE SOLID FORMS, UNDER PRESSURE.

BY P. W. BRIDGMAN.

Presented by G. W. Pierce, October, 11, 1911. Received October 6, 1911.

INTRODUCTION.

THE purpose of this paper is the same as that of the immediately preceding paper on mercury, — to indicate the nature of the fundamental facts which must be taken account of in any theory of liquids valid for high pressures, and to find something about the nature of the equilibrium between crystal and liquid. Both of these subjects are as yet practically untouched. All present theories of liquids are incompetent to explain known facts, and as for the theory of the equilibrium liquid-solid, even the fundamental facts are unknown. It seems, moreover, that these two subjects will find their best development side by side; one involves the other. Thus, any normal liquid may be made to crystallize by the application of high pressure. The internal forces producing crystallization must be present to some extent continuously in the liquid, modifying its behavior more and more as the crystallization point is approached. Yet no present theory of liquids takes account of these vector forces in the liquid which ultimately produce crystallization, but supposes instead that the force between molecules is uniform in every direction.

To be of significance the experimental study must be made over a comparatively wide pressure range, the order of the pressures being many fold greater than the pressures involved in the corresponding study of the liquid-vapor change. The results of this paper reach to 20,500 kgm./cm.^{2}, whereas the highest previous pressure under which water has been studied was only 3500 kgm.

As compared with mercury, the results for water are much more varied and richer in suggestion. It is well known that under ordinary conditions water is abnormal in many respects. The effect of high pressure is to wipe out this abnormality. The manner in which the abnormality disappears with increasing pressure is interesting. Beyond the disappearance of abnormality, the pressure has been pushed far enough to suggest in one or more particulars what the behavior of any liquid must be under very high pressures. The results with mercury did not suggest this so strongly, because the effect of pressure is much less on the properties of mercury than on those of water, so that to produce the effects peculiar to high pressure will require a much higher pressure for mercury than for water. Corresponding to the abnormal behavior of the liquid at low pressures, and probably connected with it, the solid also shows abnormal behavior at low pressures, appearing in no less than five allotropic forms. In addition to these five forms of ice with known regions of stability, there may possibly be two others with no domain of stability whatever. All of these forms, except ordinary ice, are more dense than water. With higher pressures, accompanying the return of the liquid to normality, the tendency of the solid to take new forms apparently disappears, the modification of ice stable at high pressures giving indications of being the last form, corresponding to the completely normal liquid. Here again the last form of ice has been studied over so wide a range as to suggest what may be the effects peculiar to high pressure for the equilibrium between any normal liquid and its solid.

The experimental study of the allotropic forms of ice involves the mapping of their regions of stability, by locating the transition curves, whether to the liquid or to some other solid form, and the measurement on these transition lines of the change of volume and the latent heat of transformation. Five of the six possible stable triple points have been found, and ten of the eleven possible stable transition lines of equilibrium have been followed. The sixth triple point and the eleventh equilibrium line lie at temperatures so low and at pressures so high that the slowness of the reaction makes them practically impossible to determine.

The methods used gave evidence on other points of interest, such as the variation in the reaction velocity with changes of temperature and pressure, the possibility of subcooling or superheating, and the compressibility and dilatation of the solid under pressure.

The data on these different curves are so numerous and bewildering in variety that there is considerable difficulty in choosing the order of presentation. This difficulty is increased by the fact that the paper itself was not planned from the beginning, but has been a growth. The first intention was to measure the isothermal compressibility of water at $0°$ and room temperature. The existence of a new modification of ice above 0° was not then suspected. When this was discovered in the course of the compressibility measurements, the plan of the paper was extended so as to take in the investigation of this new form of ice over the same temperature range as the compressibility determinations. The compressibility measurements were then completed and the investigation of this new form of ice taken up. This led into all the unsuspected complications of the equilibrium conditions between the five different forms of ice, which necessitated the expenditure of much more time than had been anticipated. It then became necessary, for the sake of completeness, to measure the compressibility of water in its region of existence as a liquid below 0°. During this work with the various modifications of ice, experimental familiarity with the possibilities of high pressure apparatus had been increasing, so that it became possible to extend the work on the ice first found to nearly twice the pressure range reached in the compressibility measurements. This is the state of the work as presented here. To be complete, the compressibility of water should now be redetermined over an increased temperature range corresponding to the increased pressure range, but this process of advancement might be continued indefinitely and a stop has to be made somewhere. For instance, it would probably be possible with the means now at hand to measure directly the compressibility and the dilatation of these various forms of ice, and also the adiabatic compressibility of the liquid. As it is, the pressures have been pushed far enough on both the liquid and the solid to indicate what the general nature of the effects for the highest pressures may be. The pressure required to indicate this is higher for the solid than for the liquid, so that the lack of complete parallelism between the results for the liquid and the solid is partly justified.

The order of presentation finally chosen arranges the subject matter in two parts. The first deals with the compressibility of the liquid at different temperatures. The second, by far the longer, gives the data for the different modifications of the solid, including the quantities involved in the transition solid-solid as well as in the transition solid-liquid. The order of presentation in the second part will be the natural order, proceeding systematically from the lower to the higher pressures.

This systematic order was not, as has been stated, the actual order of the experiment. The existence of another form of ice stable at high pressures and at temperatures above zero was first discovered by the anomalous results of the compressibility determinations. The existence of this form was then made certain by the measurement of the electrolytic conductivity of very dilute solutions, there being a change in the conductivity when the transition occurs. Not until then was the present method adopted by which the equilibrium pressure and change of volume were measured simultaneously, but even with this new method the curve was followed down, from high temperatures and pressures to lower temperatures and pressures. The domain of temperatures and pressures in which Tammann worked (up to 3500 kgm.) was approached with the distinct prejudice, therefore, that Tammann's work was incorrect, because there seemed no possible connection between the curves given by Tammann and the curves at the higher pressures. This prejudice makes more valuable, therefore, the essential verification of Tammann's work found at the low pressures. The expected discrepancy was avoided by the remarkable versatility of ice in appearing in different forms.

The experimental methods are in large part the same as those used in the preceding paper on mercury. Reference is made to that paper for detailed discussion. Where new methods have been necessary, discussion is given here in the appropriate place.

In presenting the data, the aim has been to give enough so that any one could, if he wished, check the computations for himself. Every one of the original observations, except those marred by obvious accidents, has been given. The table of contents should, however, enable one to omit the generally uninteresting discussion of details of methods and critical examination of data, and proceed to the discussion of the general results. In connection with the equilibrium of the different solid forms, it is suggested that some may find it clearer to read first the short history of the experiment, the description of the general nature of the experimental methods, and the description of the manner of appearance of the new forms of ice given first under the heading "The curve VI–L," and then under "The curve V–L."

Contents.

Introduction	441
Characteristic Surface of the Liquid	446
The Methods, Particularly that Used below 0°	447
The Data	451
Compressibility at 0°	451
Compressibility at 22°	453
Dilatation below 0°	458
Equilibrium between the Various Solids and the Liquid	462
Discussion of Tammann's Work	463
The Method of the Present Investigation	463
The Numerous Checks on the Accuracy of the Data	465
The I–L Curve	467
The Data, $p, t, \Delta v, \Delta H, \Delta E$	468
The Initial Compressibility of Ice I	472

The I–III Curve 473
    The Apparatus for Low Temperatures 474
    The Special Method for Measuring Change of Volume 476
    The Conditions under which III Appears 478
    The Data, p, t, Δv, ΔH, ΔE 481
The III–L Curve 485
    The Data, p, t, Δv, ΔH, ΔE 486
The I–II curve 487
    The Order of Experiment — Proof that II is Distinct from III 487
    Conditions under which II Appears 489
    The Data, p, t, Δv, ΔH, ΔE 490
The II–III Curve 493
    The Special Method for Points on this Curve, both p, t, and Δv 493
    The Data, p, t, Δv, ΔH, ΔE 496
The III–V Curve 497
    Conditions under which the Curve may be Realized 497
    The data, p, t, Δv, ΔH, ΔE 499
The II–V Curve 500
    General Characteristics of the Curve 500
    The Data, p, t, Δv, ΔH, ΔE 501
The V–L Curve 502
    Conditions under which V Appears 502
        The Actual Order of Experiment 503
    The Data, p, t, Δv, ΔH, ΔE 508
The V–VI Curve 509
    General Characteristics 509
    The Data, p, t, Δv, ΔH, ΔE 512
The VI–L Curve 513
    History of the Experiment 513
        The Compressibility Measurements 513
        Electrolytic Conductivity 514
    The Apparatus for the Higher Pressures 516
    The Data p, t, Δv, ΔH, ΔE 518
        Correction for Δv Points at High Pressures 520
    Digression on the Elastic Behavior of the Cylinder under High
        Pressures 521
    The Difference of Compressibility between Solid and Liquid . 523
The Triple Points 524
    Values of p, t, Δv, ΔH, ΔE, at these Points 524
Are the Various Forms really Solid? 526
Other Possible Forms of Ice 527
    Different Modifications of Ice I 527
    Another Possible Form at High Pressures 528
Subcooling and Superheating 530
    Superheating with Respect to the Liquid Impossible 530
    Realization of Unstable Forms, Persistent Nuclei in Solid and Liquid 531
    Metastable Limit 532
Reaction Velocity 533
    Enormous Variation with the Temperature 534
Compressibility of the Various Forms of Ice 535

Volume on the Equilibrium Curves 535
Approximate Compressibility 536
Discussion of the Results 538
   For the Liquid 538
      The Nature of the Abnormalities of Water already Known . . 540
      Compressibility and Dilatation of Water over the Range of this
         Paper 541
         Table of Volume for Equal Intervals of Pressure and Temperature 539
      Change from an Abnormal to a Normal Liquid 541
         The Manner of Transition, Particularly below 0° 543
      Comparison of Present Data with Values Extrapolated from
         Previous Formulas, Probable Behavior at High Pressures . . 548
      Difference of Specific Heats 550
      Change of Internal Energy on an Isothermal 551
      Adiabatic Compressibility and Temperature Effect of Compression 552
   For the Change of State Liquid-Solid 553
      The Present Theories of Liquid-Solid; the Points at Issue . . 553
      The General Shape of the Various Equilibrium Curves in their
         Bearing on this Question 554
      The New Evidence given by the VI-L Curves and the Probable
         Course of the Curve at Higher Pressures 555
      Bearing on van Laar's Recent Theory 556
   For the Change Solid-Solid 557

The Compressibility of Water.

A complete thermodynamic knowledge of any substance over any range of temperature and pressure is afforded by a knowledge of the characteristic equation of the substance over that range (that is, the relation connecting volume with temperature and pressure), and a knowledge of the specific heat along some curve of the pressure-temperature plane not an isothermal. For water, the specific heat is known in its dependence on temperature at atmospheric pressure, that is, along a line not an isothermal, so that theoretically all that is needed for a complete thermodynamic knowledge of water is the knowledge of the characteristic equation. This knowledge is evidently given by the change of volume with pressure along various isothermals, together with the change of volume along some line not an isothermal. Practically this means that the characteristic equation may be found by finding the isothermal compressibility at several different temperatures and combining with the known dilatation at atmospheric pressure.

It does not follow, however, that all the thermodynamic data are given with equal accuracy by the knowledge of these quantities. Thus the specific heat, which involves a second derivative of the p-v-t relation, can evidently be found much less accurately than the compressibility, for example. Therefore, while some hold is obtained by the data of this paper on all the thermodynamic quantities, some of these will be found more accurate than others.

Two methods were used in determining the compressibility. The first is the same as that used in determining the compressibility of mercury, and has already been fully described and critically discussed in the previous paper. Briefly, it consists in enclosing the water to be measured in a steel bottle, communicating at the upper end through a very fine channel with a mercury reservoir. The bottle and the reservoir are completely immersed in a fluid to which pressure is applied. The water contracts under pressure, mercury from the reservoir runs in to take its place, falls to the bottom of the bottle, and is weighed after pressure is removed. Compressibility was measured by this method at two temperatures, 0° and 22°, and the same method evidently might be used at any temperature above zero. Under pressure, however, water may exist as the liquid at temperatures considerably below zero, so that for a complete knowledge of the p-v-t surface of water, the volume must be measured throughout this extended region. The method just described is evidently not applicable here without a troublesome variation of both temperature and pressure together during a single measurement, and a second method was adopted.

The second method, which was used to obtain the changes of volume below zero, assumes as correct the compressibility at zero as found by the first method. The data given by this second method are the thermal dilatation from zero at constant pressure for a number of different pressures. Combined with the known volume at zero, this is evidently sufficient to give the volume at any temperature and pressure.

The apparatus used was the same as that already described in determining the data on the freezing curve of mercury, and consists of a lower cylinder placed in a thermostat, and an upper cylinder in which the pressure is produced by a moving piston. The water under experiment is placed in a cylindrical steel shell in the lower cylinder. The lower cylinder also contains the manganin resistance coil with which pressure is measured. The measurements of pressure by this method have been described in a previous paper. The remainder of the lower cylinder and the entire upper cylinder is filled with gasolene, by which pressure is transmitted to the water.

The experimental procedure was as follows. The temperature of the lower cylinder was maintained at some value below zero by the thermostat, and the pressure varied over the region of existence of water for this temperature. The piston displacement was plotted as a function of the pressure for this temperature. The temperature of the thermostat was then raised by a suitable amount, and the piston displacement found as a function of the pressure for this new temperature. This was done for four different temperatures, including zero. From these four curves, with the help of an interpolation for temperature, the piston displacement may be found for any temperature and pressure in the region of the liquid water below zero. This displacement was found at 5° intervals for several different pressures, increasing in steps of 800 kgm. At each pressure, therefore, the thermal dilatation may be found by taking the difference between the displacement at the desired temperature and zero.

If the experiment were performed in the simple manner described, evidently several very serious errors would be introduced. No account whatever has been taken of the thermal dilatation of the steel cylinder, which is not very large, or of the gasolene transmitting the pressure, which is comparatively more important. These two disturbing effects were corrected for by two auxiliary experiments. In the first, the lower cylinder was almost entirely filled with a cylinder of bessemer steel, and the displacement found for any temperature and pressure of the region in question, exactly as for water. In the second auxiliary experiment, the lower cylinder was entirely filled with gasolene, and the displacement found as before. In applying the correction, the entire interior of the apparatus may be thought of as consisting of two parts; one portion is that which would be occupied by the bessemer cylinder, and the second portion is all the rest, including the remainder of the lower cylinder and the upper cylinder. It is to be noticed that this "first part of the volume" is purely fictive: it need not be actually occupied by the bessemer. In the auxiliary experiment in which the lower cylinder is filled with gasolene, this "first part of the volume" is filled with gasolene; in the actual experiment with water, the "first part of the volume" is filled partly with the steel of the shell, partly with water, and partly with enough gasolene to make up the difference. The advantage in thus thinking of the volume as split up into two parts is that the second part remains the same in the three experiments. The displacement of the piston at constant pressure due to change of temperature of the thermostat will evidently include, for "the second part," the thermal dilatation of the cylinder, and what is important, will be independent of the position of the piston in the upper cylinder, since only the lower cylinder is involved in the change of temperature. By taking the difference of the apparent dilatations at constant pressure for the three experiments we shall obtain, therefore, two differences, from which the effect of the "second part of the volume" has been eliminated. These two differences involve the change of volume with temperature at constant pressure of the material occupying the same volume that the bessemer steel would have under the same conditions of temperature and pressure. The quantities entering into the two differences are, therefore, the thermal dilatation of bessemer steel, of gasolene, and of water at various pressures. Now the thermal dilatation of bessemer steel is very small comparatively, and is furthermore known to be little affected by changes of pressure. For the accuracy required in this work, therefore, the change of volume of the steel cylinder, both with temperature and pressure, may be assumed to be known. This leaves only two unknown quantities in the two differences mentioned above, so that either may be found. The thermal dilatation of the gasolene is found immediately from the difference of the displacements of the two auxiliary experiments. It should be noticed that the dilatation is used here in a sense slightly different from the usual one in thermodynamics. The quantity given here is the change in volume in cm.^{3} for 1° lowering of the temperature of the quantity of gasolene which at that pressure occupies 1 cm.^{3}. To find the change of volume of the quantity of gasolene which originally at 0° and atmospheric pressure occupied 1 cm.^{3} we should need to know in addition the compressibility of the gasolene. It is an advantage of the method that it is not necessary to know this compressibility. It is now obvious why a knowledge of the compressibility of water at 0° is necessary. The quantity of water is fixed. As the pressure is raised, the part of the "first part of the volume" occupied by the water decreases in a way known, because the compressibility is known, and the part occupied by the gasolene correspondingly increases. At any pressure, therefore, the number of cm.^{3} of gasolene concerned in the total dilatation of the "first part of the volume" is known, the dilatation per cm.^{3}, and so the total dilatation of the gasolene is also known, and therefore the remainder of the total dilatation due to the water is given immediately.

There are several minor corrections to be considered. The change of volume as given in cm.^{3} by the displacement of the piston involves directly the diameter of the piston. This changes with pressure. This correction has been already discussed in the mercury paper and found to be 1.35 per cent for 10,000 kgm. The magnitude is almost negligible for the present work, but the correction was nevertheless applied in making the computations. Furthermore, the volume swept out by the piston at constant pressure during a change of temperature of the lower cylinder does not represent accurately the change of volume of the material of the lower cylinder. This is due to the fact that the upper cylinder was kept at the constant temperature of 8° instead of 0°. For example, let us suppose that the contents of the lower cylinder expand 1 per cent when the temperature is raised from −20° to 0°. This 1 per cent passes from the lower cylinder at 0° to the upper cylinder at 8°, and in so doing experiences an additional increase of volume due to the extra 8°. The error introduced by assuming the piston displacement to give the dilatation will be, therefore, in this example, about 0.5 per cent on the dilatation. This error was not corrected for in the present work, since the accuracy of the largest dilatation found, judging only from the number of significant figures, was not over 2 per cent.

One experimental source of error did require correction. This is due to the wearing away of the packing material of the piston by the enormous friction during the advance of the piston. The effect is evidently the same in its results as a leak, although there was absolutely no leak of the liquid and only this wearing away of the packing during actual motion of the piston. The effect was corrected for by making two runs at every temperature with increasing and decreasing pressure, and taking the mean. The discrepancy between the displacements during increasing and decreasing pressure gave the amount of wearing of the packing. This was of the order of 0.01 inch on a total stroke 3 inches. The maximum correction was 0.013 inch. This correction was applied to the next run at higher temperature. The correction so determined incidentally covers the effect of hysteresis, elastic after-working, or of viscous yield in the steel, all of which must be very small at these comparatively low pressures.

The method used in making the calculations from the data was a combination of arithmetic computation with graphical representation. The graphical representation alone would not have been accurate enough. This is because the changes of volume due to temperature are very small in comparison with those due to changes of pressure in the region in question, so that it was necessary to retain all the significant figures given by the measurements. Piston displacement was measured to 0.0001 inch on a total of 3 inches, but since the pressure measurements were sensitive to only 1 part in 3000, the displacement readings were retained only to 0.001 inch. The method was to pass the best parabola through the experimental points at the various constant temperatures; calculate the displacement given by the parabola at the pressure of the experiment; plot on an enlarged scale the difference between the observed displacement and the calculated displacement; pass a smooth curve through these difference points; and finally by combining the results given by the smooth deviation curve with the computed values on the parabola, to find the displacements at equal pressure intervals. This has the advantage of giving a curve perfectly smooth to the last of the significant figures. Curves thus obtained for the different temperatures were then combined by plotting the displacement corresponding to the same constant pressure against the temperatures of the different curves, and smooth curves were drawn through these points. This last step could be made entirely graphically, since the change of displacement with temperature at constant pressure is comparatively slight. From these two sets of smooth curves, the displacement for uniform pressure intervals of 800 kgm. and uniform temperature intervals of 5° was found over the entire region. These points were found in this way for each of the three experiments, that with water and the two auxiliary ones, and the difference of displacement at corresponding pressures and temperatures combined in the way already described.

The actual data above and below 0° follow. All the details of the calculation of the compressibility at 0° and 22° by the first method were the same as that used for determining the compressibility of mercury. In fact, the two determinations, compressibility of water and compressibility of mercury, were made at the same time and each involves the other. It is to be noticed that the various disturbing factors due to irregularities in the behavior of the steel bottle are much less important in the case of water than they were in the case of mercury. This is because of the higher compressibility of water as compared with that of the mercury. Nevertheless, the effect of these irregularities was distinctly felt here also, particularly if the pressure had been allowed to reach so high a value as to freeze the water. The best results were obtained with piezometers in which the water had never been allowed to freeze.

At 0°, four different piezometers were used, one before and after annealing, so that there are really five independent sets of determinations. Three independent points with a sixth piezometer, No. 1, were also found, but were not used in the computations, because this was the very first trial ever made of the method, and the piezometer had been badly strained by a pressure considerably beyond the freezing pressure, the existence of which was not then suspected. The actual data are given in Table I. To afford a comparison of the regularity of these results as against the values obtained for mercury, the actual points for the second best piezometer, No. 6, are shown plotted in Figure 1. If comparison be made with the mercury points it will be seen that these of Figure 1 are distinctly better. The changes of volume at even pressure intervals, obtained from smooth curves from the data of each of the five piezometers, is shown in Table II., together with the weighted means. Finally, the deviation of the weighted mean from a parabolic formula is plotted against pressure, and a

TABLE I.

DATA FOR COMPRESSIBILITY OF WATER AT 0°.

Pressure, kgm. cm.^{2}	$\frac{\Delta v}{v_0}$.	Pressure, kgm. cm.^{2}	$\frac{\Delta v}{v_0}$.	Pressure, kgm. cm.^{2}	$\frac{\Delta v}{v_0}$.	Pressure, kgm. cm.^{2}	$\frac{\Delta v}{v_0}$.
Piez. No. 6.		Piez. No. 7.		Piez. No. 3.		Piez. No. 5'.	
2190	0.0795	2190	.0831	5270	0.1467	6810	0.1589
2970	0979	2970	0993	4520	1348	6090	1505
3740	1137	3740	1139	3750	1184	6810	1603
4500	1282	4500	1289	3000	1012	6800	1618
5270	1412	5270	1417	2220	0832	6070	1523
6020	1496	6020	1528	1450	0590	5340	1412
6760	1618	6760	1626	6050	1552	4540	1270
7110	1657	7110	1666	6770	1651	3770	1116
1420	0595	2960	0981	1000	0423	3020	1003
2180	0796	3740	1152	500	0221	2230	0806
2960	0995	6570	1636			1440	0564
3740	1152	6750	1605	Piez. No. 5.		1050	0421
6570	1624	6010	1524	1410	0.0568	1830	0667
6750	1629	5260	1416	2180	0776	2610	0919
6010	1547	4490	1290	3090	0989	3400	1052
5260	1429	3740	1139	3740	1265	4140	1208
4490	1301	2050	0078	4500	1265	4010	1366
3740	1151	2200	0786	5290	1391	5680	1457
2950	0985	1410	0547	5270	1365		
2200	0797	1000	0412	6030	1457	Piez. No. 1.	
1410	0561	500	0227	6760	1578	6750	0.1710
1000	0420			7500		6050	1501
500	0226					5260	1408

smooth deviation curve drawn. In only two instances does the smooth deviation curve fail to pass through the points of the weighted mean. The value given by this smooth deviation curve is to be taken as the final value, and is also given in Table II. The change of volume is to be found at every pressure by combining with the change of volume calculated by the formula

$$\Delta v = ap + bp^2,$$

where
$$\log a = 5.5942 - 10,$$
and
$$\log (-b) = 1.3512 - 10,$$

the small correction term given by the deviation curve of Figure 2.

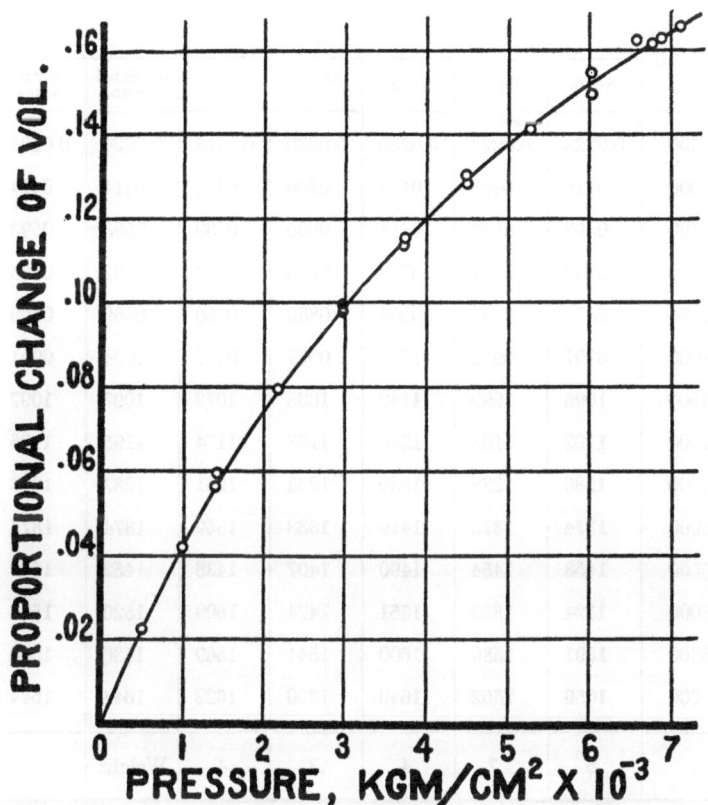

FIGURE 1. The change of volume of water at 0° C. as obtained with piezometer No. 6.

The data for 22° were obtained in exactly the same way. Only three piezometers were used here instead of five as at 0°. The results are slightly less irregular, owing to greater familiarity with the method and greater care exercised to keep the maximum pressure below the freezing pressure. The data are given in Table III.; the smoothed values for each piezometer, the weighted means, and the final values

TABLE II.

COMPRESSIBILITY OF WATER AT 0°. COMPARISON OF BEST RESULTS WITH DIFFERENT PIEZOMETERS, WEIGHTED MEAN, AND FINAL VALUES.

Pressure, kgm./cm.²	$\frac{\Delta v}{v_0}$.						
	No. 6.	No. 7.	No. 3.	No. 5.	No. 5'.	Weighted Mean.	Final Value.
500	0.0228	0.0222	0.0230	0.0230	0.0218	0.0224	0.0224
1000	0416	0408	0429	0409	0414	0414	0414
1500	0597	0586	0615	0605	0589	0593	0593
2000	0742	0721	0762	0736	0732	0732	0735
2500	0879	0859	0904	0860	0866	0869	0869
3000	0997	0992	1030	0967	0977	0994	0991
3500	1095	1099	1142	1068	1072	1097	1097
4000	1192	1193	1246	1168	1174	1195	1195
4500	1286	1288	1339	1253	1261	1287	1287
5000	1374	1375	1419	1333	1359	1374	1374
5500	1453	1454	1490	1407	1438	1452	1452
6000	1524	1523	1551	1474	1509	1520	1520
6500	1591	1589	1600	1541	1569	1586	1586
7000	1659	1652	1640	1600	1622	1644	1644
....	.6	1.7	.4	.3	.4	Weight	....

in Table IV. The change of volume at any pressure is to be found by combining with the value given by the formula

$$\Delta v = ap + bp^2,$$

where
$$\log a = 5.5139 - 10,$$

and $\log(-b) = 1.1239 - 10$,

the correction given by the deviation curve of Figure 3. It is to be noticed that both of these formulas, for 0° and 22°, take as the unit of volume the volume which the given quantity of water occupies under unit pressure at 0° or 22° respectively. To find the change of volume of 1 gm. of water, a correction must be applied for thermal dilatation between 0° and 22°.

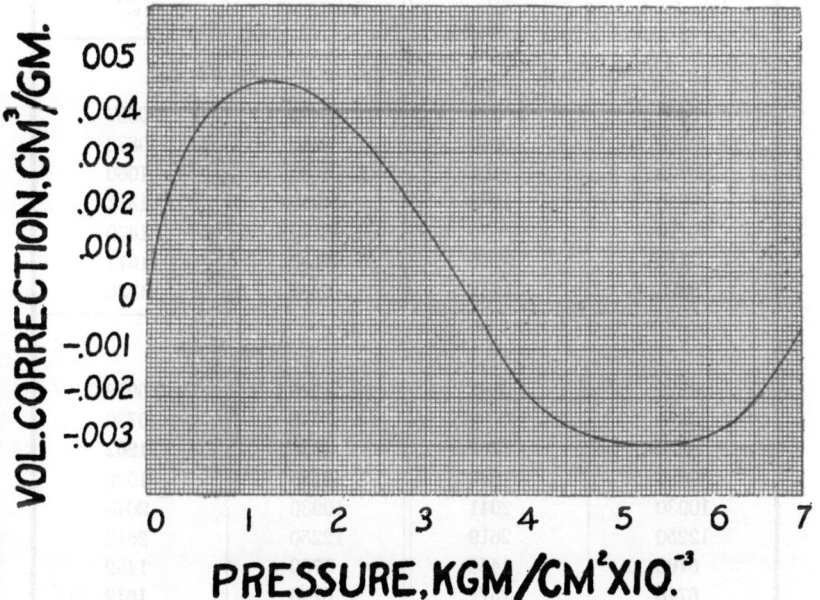

FIGURE 2. The deviation curve for water at 0°. This is to be used in combination with the formula on page 453 in determining the change of volume at 0° C.

These data may be compared with those of Amagat [1] (Table V.). In making this comparison, it must be borne in mind that this is a disproportionally severe test of these data. Values taken from the low end of a large range cannot be expected to show as small a percentage of error as will values obtained with apparatus designed only for this smaller range. In the present work very few actual experimental points were found below 1000 kgm.; the lower end of the curve is virtually an

---

[1] Amagat, Ann. de Chim. et Phys. (6), **29**, 68–136, 505–574 (1893).

extrapolation to a known zero. The values of the present work are, with only one exception, lower than those of Amagat. The absolute value of the discrepancy becomes less at the higher pressures, as it should,

TABLE III.

DATA FOR COMPRESSIBILITY OF H_2O AT 22°.

Pressure, kgm./cm.³	$\frac{\Delta v}{v_0}$	Pressure, kgm./cm.³	$\frac{\Delta v}{v_0}$
Piez. No. 6.		Piez. No. 7.	
1460	0.0522	1460	0.0537
2660	0853	2660	0836
3790	1101	3790	1060
4950	1288	4950	1311
6120	1470	6120	1470
7180	1615	7180	1614
8290	1732	8290	1730
		Piez. No. 9.	
3220	0981	3220	0.0978
2170	0732	2170	0730
4350	1206	4350	1202
9530	1889	9530	1999
10930	2041	10930	2046
12250	2619	12250	2612
5700	1458	5700	1469
6740	1571	6740	1012
7830	1706	7830	1754
495	0203	495	0204
995	0391	995	0403
6170	1493	6170	1488
8960	1839	8960	1832
10340	2394	10340	1983
11570	2634	11570	2147

since the comparison with the present values is made under more favorable conditions at the higher pressures. The fact that the absolute value of the discrepancy tends to become less, not greater, shows that the fundamental unit of pressure must be essentially the same for the two determinations. Furthermore, the dilatation between 0° and 22°, as found by taking the difference of the compressions at 0° and 22°, is evidently going to show much less absolute difference for the two sets of determinations than the compressibilities, suggest-

TABLE IV.

COMPRESSIBILITY OF H_2O AT 22°. COMPARISON OF BEST RESULTS WITH DIFFERENT PIEZOMETERS, WEIGHTED MEAN, AND FINAL VALUES.

Pressure, kgm./cm.²	$\frac{\Delta v}{v_0}$				
	No. 6.	No. 7.	No. 9.	Weighted Mean.	Final Value.
1000	0.0383	0.0379	0.0391	0.0383	0.0383
2000	0683	0667	0684	0679	0679
3000	0933	0914	0933	0928	0928
4000	1141	1123	1145	1137	1137
5000	1316	1304	1324	1314	1314
6000	1462	1460	1485	1465	1465
7000	1595	1594	1631	1600	1600
8000	1723	1702	1761	1728	1728
9000	1850		1885	1855	1855
10000	1963		2000	1967	1967
11000	2036		2078	2042	2042
	4	2	1	Weight.	

ing that the two sets are each consistent with themselves, since they show the same temperature effect.

Since this paper was written, Parsons and Cook [a] have published data for the compressibility of a few liquids up to 4500 atmos. Their results for water at 4° are shown in Table Va. compared with the results of this paper. The agreement is closer than 1 per cent at

[a] Parsons and Cook, Proc. Roy. Soc. A, **85**, 332–349 (1911).

the higher pressures. The results of Parsons and Cook differ from those of Amagat in the same direction as those found here and by about twice as much. The values given in the table as the result of the present work were calculated from the data of Table XXXI. on page 539.

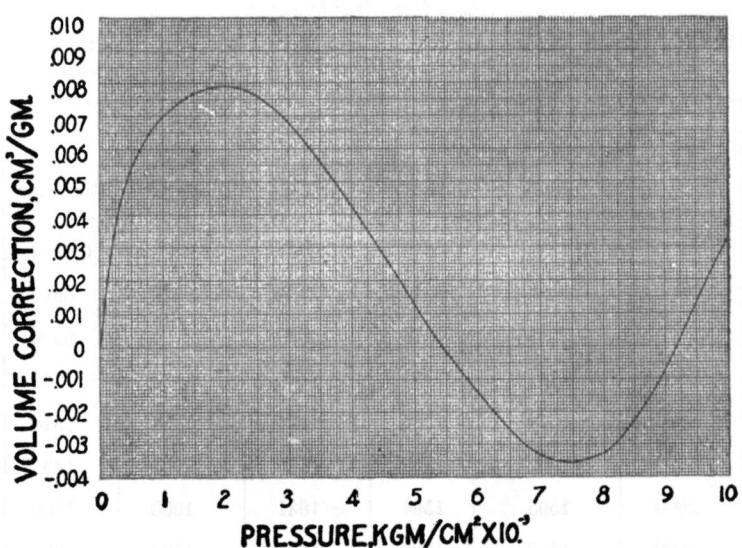

FIGURE 3. The deviation curve for water at 22°. This is to be used in combination with the formula on page 454 in determining the change of volume at 22°.

The results obtained by the second method below 0° next concern us. To give the original data would occupy so much space as to be out of the question. The methods and the details of the calculation have already been described. It is sufficient to give here some information about the actual data, and the probable accuracy. Four runs were made with water, at $-15°.8$, $-10°.2$, $-5°.0$, and $0°.0$, and displacement and pressure measured at respectively 8, 10, 14, and 17 points. Each of these pressure measurements was the mean of two readings, and the displacement measurements the mean of four. The auxiliary experiment with the lower cylinder full of bessemer was also made at four temperatures: $-17°.7$, $-12°.2$, $-6°.0$, and $0°.0$, and the number of points at these temperatures was 14, 15, 13, and 16 respectively. The auxiliary experiment with the lower cylinder full of gaso-

TABLE V.

Comparison of Results for Compressibility of Water with those of Amagat.

Pressure.		Δv.					
		At 0°.			At 22°.		
Atmos.	$\frac{kgm.}{cm.^2}$	Amagat.	This Work.	δ.	Amagat.	This Work.	δ.
500	516	0.0237	0.0231	− 6	0.0216	0.0220	+ 4
1000	1033	0440	0425	−15	0402	0394	− 8
1500	1549	0611	0599	−12	0565	0552	−13
2000	2066	0763	0753	−10	0710	0697	−13
2500	2583	0898	0891	− 7	0839	0829	−10
3000	3099	1017	1013	− 4	0957	0950	− 7

TABLE Va.

Pressure.		ΔV at 4°.	
Atmospheres.	kgm./cm.²	Parsons and Cook.	This Work.
0	0	0.000	0.0000
500	516	0.022	0.0226
1000	1033	0.042	0.0418
1500	1549	0.059	0.0591
2000	2066	0.073	0.0742
2500	2583	0.086	0.0879
3000	3099	0.098	0.1002
3500	3615	0.109	0.1103
4000	4132	0.120	0.1203
4500	4648	0.130	0.1306

TABLE VI.

DILATATION OF WATER BELOW 0° — DISPLACEMENT OF PISTON AT REGULAR INTERVALS OF PRESSURE AND TEMPERATURE.

Pressure		Piston Displacement, inches.				
Slider Displacement, cm.	$\frac{kgm.}{cm.^2}$	−20°.	−15°.	−10°.	−5°.	0°.
		Lower cylinder full of water.				
3	800	....	0.642	0.636	0.627	0.623
6	1600	....	1.137	1.133	1.125	1.112
9	2400	1.542	1.533	1.521	1.508	1.483
12	3200	1.861	1.848	1.834	1.818	1.797
15	4000	2.126	2.118	2.108	2.093	2.073
18	4800	....	2.342	2.339	2.332	2.307
21	5600	....	....	....	2.544	2.521
24	6400	....	....	....	....	2.718
		Lower cylinder full of bessemer.				
3	800	0.427	0.423	0.419	0.415	0.411
6	1600	0.721	0.717	0.712	0.706	0.697
9	2400	0.936	0.932	0.927	0.920	0.912
12	3200	1.112	1.110	1.105	1.098	1.089
15	4000	1.268	1.265	1.262	1.257	1.251
18	4800	....	1.402	1.398	1.393	1.387
21	5600	....	....	1.524	1.520	1.516
24	6400	....	....	....	....	1.629
		Lower cylinder full of gasolene.				
3	800	0.904	0.884	0.855	0.818	0.762
6	1600	1.494	1.470	1.442	1.411	1.378
9	2400	1.916	1.893	1.867	1.838	1.805
12	3200	2.276	2.256	2.228	2.203	2.176
15	4000	2.554	2.538	2.520	2.499	2.473
18	4800	2.815	2.797	2.778	2.756	2.733
21	5600	3.036	3.023	3.008	2.991	2.971
24	6400	....	....	....	....	3.177

lene was made at only three temperatures: −15°.2, −7°.1, and 0°.0, with 16, 16, 16 points respectively. All of these points, with very few exceptions, lay on smooth curves consistently to 0.001 inch.

The results of the graphical computations of displacement and pressure at equispaced points are shown in Table VI. The displacement at zero pressure is arbitrary in these tables. The data of these tables, together with the quantities of steel, water, and gasolene, are sufficient to enable anyone to check up the computations. The weight of the steel cylinder filling the lower cylinder was 286.345 gm. The weight of the steel shell containing the water was 51.818 gm. The weight of the water

TABLE VII.

DILATATION OF WATER BELOW 0° AT VARIOUS PRESSURES.

Pressure, kgm./cm.^{2}.	Change of Volume in cm.^{3} per gm. on changing from Tabulated Temperature to 0°.				
	−20°.	−15°.	−10°.	−5°.	0°.
			−0.00173	−0.00057	0
800		−0.00015	00031	00062	0
1600	0.0000	0000	0000	0000	0
2400	0045	0038	0028	0017	0
3200	0052	0043	0029	0017	0
4000	0045	0040	0031	0017	0
4800		0034	0033	0022	0
5600				0023	0

was 26.775 gm. The density of steel was 7.840 at 17°, and it was assumed that the cubical dilatation of steel was 0.00036 and the cubical compressibility 0.0_66 in kgm./cm.^{2}. The volume in cm.^{3} of the bore of the upper cylinder, 1 inch long, was 4.117 cm.^{3}. With increasing pressure this increases 1.35 per cent for 10,000 kgm.

The results of the computations from the data of these tables are given in Table VII. and Figure 4, showing the proportional change of the original volume at 0° and atmospheric pressure by which the water shrinks at various pressures on passing from 0° to the temperature listed.

From these results, both above and below 0°, the actual volume of the water for regular pressure and temperature intervals has been found and is given in Table XXXI. on page 539.

THE SOLID PHASES OF WATER — THEIR RELATION TO EACH OTHER AND TO THE LIQUID.

In order to present these data systematically and clearly, reference is made at once to the final equilibrium diagram for water and the solids given at the end of this paper. (Plate 1.) It may produce less confusion in referring to this diagram to state briefly the identical relations which must be satisfied in every such diagram. There are three of these relations. At every triple point the relative positions of the three equilibrium curves must be such that if any one is produced into the region of instability, it shall fall within the angle made by the other two. If one phase is carried into another by passing across an equilibrium line at constant pressure from a low to a high temperature, then the reaction runs with absorption of heat. And, if one phase is carried into another by passing across an equilibrium line at constant temperature from a low to a high pressure, the reaction runs with decrease of volume.

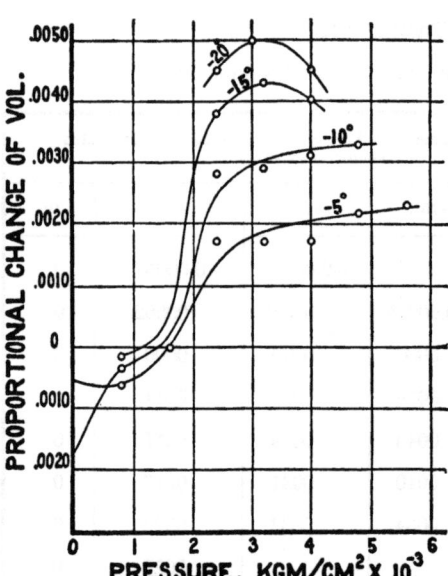

FIGURE 4. This shows as a function of the pressure the change of volume of water on passing at constant pressure from 0° to the temperature indicated on the curves.

The different branches of the curves will be taken up in order, beginning with I–L,[3] I–III, etc. The data to be given are the freezing pressure at any given temperature and the change of volume. From the relation between temperature and pressure on the freezing curve at different points, the slope may be found, and this, combined with the change of volume, gives the latent heat by Clapeyron's equation. The data are presented in this order: observed change of volume, calcu-

---

[3] Throughout, the abbreviation L will be used in referring to the liquid.

lated slope, latent heat, and then other quantities of thermodynamic interest, such as the work done in passing from one form to the other, and the change of internal energy.

In the early part of this work up to 3500 kgm. constant reference will be made to Tammann,[4] since his are practically the only previous experiments on ice under pressure. It will be well, therefore, to briefly outline his work and show where it needs verification or completion. Tammann's work consisted essentially in following out the equilibrium curves without systematic measurements of the changes of volume. He discovered the existence of two new forms of ice (II and III), and studied their relations to water and ordinary ice (ice I). He obtained points on the equilibrium curves I-L, III-L, I-III, and I-II. The change of volume I-III and I-II was measured at a few points without any very great accuracy. The greatest possibility of question as to the results is in regard to the equilibrium curves I-III and I-II. Tammann found that these curves cross at −37° and 2200 kgm. Now it is necessary thermodynamically that a third equilibrium curve should start from the point of intersection of two equilibrium curves with a common phase. In this case the predicted equilibrium curve would be II-III. Tammann found no such curve and even maintained that no such curve was necessary, claiming that the thermodynamic argument was valid only when the two phases were present simultaneously, and that he had never been able experimentally to produce II and III together. The argument seems inadequate, however, and in this work the missing curve had been actually found. The other points of difference between this work and Tammann's are more or less minor in character; there seems to be an error in Tammann's pressure measurements of 100 kgm. at 2000, and he found a curious point of inflection in the I-L curve, which does not seem actually to exist.

For convenience, the method used will be briefly outlined. It was essentially the same as that of Tammann, and consists in plotting the displacement of the piston by which pressure is produced against pressure. Change from one form to the other is accompanied by change of volume at constant pressure. This change of phase is shown, therefore, by a discontinuity in the curve piston-displacement vs. pressure. The pressure at the point of discontinuity gives the equilibrium pressure and the volume swept out by the piston, which is determined by the amount of the discontinuity, gives the change of volume on passing from one phase to the other. The water on which these experiments

[4] Tammann, Kristallisieren und Schmelzen, pp. 315–344 (Barth, Leipzig, 1903).

were made was placed in a steel shell, open at the top, and completely surrounded by kerosene or gasolene by which pressure was transmitted. Pressure was measured by observing the change of resistance of a calibrated manganin wire immersed directly in the same chamber with the water. That the fluid transmitting pressure suffers no change of phase which is accompanied by change of volume within the limits of measurements was shown by direct experiment. Furthermore, that there is no reaction under pressure between the water and the oil, which are directly in contact, was shown by the sharpness of the freezing. See Figure 31, page 515. The effect of an impurity is to make the discontinuity less abrupt. In order to give additional evidence on this point, in the course of the experiments the water was placed in shells of steel or copper, directly in contact with kerosene or gasolene, in glass bulbs directly in contact with kerosene, and in a glass bulb with a mercury seal, the water coming in contact only with mercury and glass. The results in all cases were identical.

Three different forms of apparatus were used according to the temperature range. The first, for the middle range from $-25°$ to $+20°$, was the same as that used in the work on mercury. It consisted essentially of two parts: an upper cylinder in which pressure was produced by an advancing plunger, communicating through a heavy tube with a lower cylinder containing the water under experiment and the pressure measuring coil. The lower cylinder was placed in a thermostat. The various sources of error and the corrections have been discussed in the paper on mercury. One correction had to be determined anew, since the values used in the mercury paper were not for a corresponding range of temperature and pressure. This is the correction for the thermal dilatation of the kerosene or gasolene which passes from the lower cylinder at the temperature of the experiment to the upper cylinder at the temperature of the room. The manner of determining this correction was the same as before. One correction applied to the results for mercury was avoided here, namely the correction for the variation in the room temperature. This was made unnecessary by a water jacket at constant temperature placed around the upper cylinder.

The second piece of apparatus, for temperatures from $-20°$ to $-80°$, was essentially the same as the first, consisting of two parts. But the pressures for this temperature range are comparatively low, not reaching over 4000 kgm., so that it was possible to make the lower cylinder so small (1 inch o. d. by 8 inches long), that it could be placed in a thermos bottle. The connecting tube was also made much smaller so as to avoid conduction of heat into the thermos bottle. The method used in determining the change of volume had to be modified with this apparatus, since the reaction was so slow at the low temperatures. It will be described in detail later.

The third piece of apparatus for higher pressures and temperatures, 0° to 76° and 6000 to 20,500 kgm., consisted of only a single cylinder containing the water, the measuring coil, and the advancing piston. For these high pressures it was necessary to use only one piece of apparatus because of the impossibility of getting tubing to stand these high pressures over any extended time. The water was placed in a steel shell surrounded by kerosene as before. The cylinder, together with the lower part of the hydraulic press by which the piston was advanced, was placed in a thermostat. In this third form, therefore, the correction for the change of volume of the transmitting fluid on passing from one temperature to another is avoided. But at the higher pressures another correction is introduced because of the slow yield of the steel. This will be described in its place.

The data given for the ten equilibrium curves were all obtained independently of each other, at different times, with different fillings of the apparatus, and in a number of cases with different pieces of apparatus, either new pieces to replace those destroyed by explosions or pieces of a different type. The consistency of the data obtained under these various conditions is shown by the very close approximation with which the identical relations are satisfied by the independent data of the three curves at each of the three triple points.

At each triple point there are three independent checks. In the first place, the absolute values of the equilibrium pressures and temperatures are checked by the necessity for the three curves running together to a point. In the second place, the values of the change of volume are checked by the obvious condition that at the triple point I–II–III, for example, the change of volume I–II plus the change II–III must be equal to the change I–III. The third check is given by the condition for the latent heats analogous to that for the changes of volume. This third check involves, besides the values already given, the values of the slope of the three curves meeting at the triple point. This third check is the most difficult to meet and the most sensitive, as it is well known that the derivative of an experimental curve is subject to much greater error than the curve itself.

Such a triple check is afforded for the data of every one of the ten equilibrium curves at at least one point: Five of the curves are rendered still more secure by being tied down at both ends by a triple point, and the other end of a sixth curve, I–L, is checked by the already known data for 0°.

In the curves given for ΔV and the latent heat, the identical relations at the triple point were kept in view. In some cases the curve given does not seem to be the best possible through the individual points. The departure from the best curve was occasioned by the necessity of satisfying the identical relations at the triple point, and

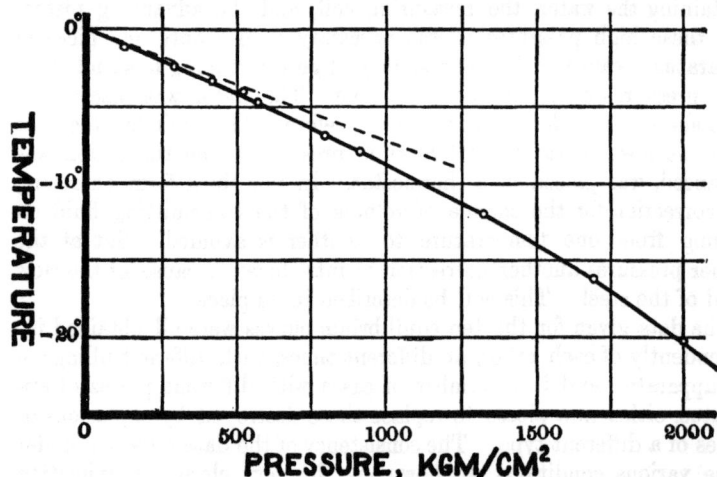

FIGURE 5. The freezing curve of ice I. The dotted line shows the slope at 0° computed by Clapeyron's equation.

the amount of departure gives a pretty good criterion of the self-consistency of the data for the different curves.

THE CURVE I-L.

The points on this curve were obtained with the same apparatus as that used for mercury. No special discussion is necessary. The data were obtained on three separate occasions: February 2, February 23–24, and March 15, 1911. The equilibrium points are shown in Figure 5 and ΔV in Figure 6; the actual data are given in Table VIII. The first set is not accurate. It was taken only for fun one evening after the supply of $CaCl_2$ for running the thermostat had given out, so that the temperatures are not sufficiently accurate. These points are not shown in the diagram.

The second and third sets of readings give the equilibrium points determined with sufficient accuracy. The second set suffices to give the general shape of the entire curve, while the third set was taken with more pains at the lower pressures in order to test more definitely the existence of a point of inflection at −4° as found by Tammann. This third set was made with the apparatus directly connected to a Cailletet pump of the Société Genevoise, with a capacity of 1000 kg./cm.² As a check, the equilibrium pressures as given by the Bourdon gauge of this pump are tabulated as well as the pressures given by the manganin resistance. The slide wire of the Cary Foster bridge was especially changed for this experiment, so as to give greater sensitiveness in the measurements of resistance. The readings of the manganin gauge are to be accepted as more trustworthy than those of the Bourdon. Within the limits of accuracy of the readings, there seems to be no point of inflection on the equilibrium curve as found by Tammann, but the curvature is perfectly regular, increasing more and more rapidly at the lower temperatures and higher pressures. The point of inflection

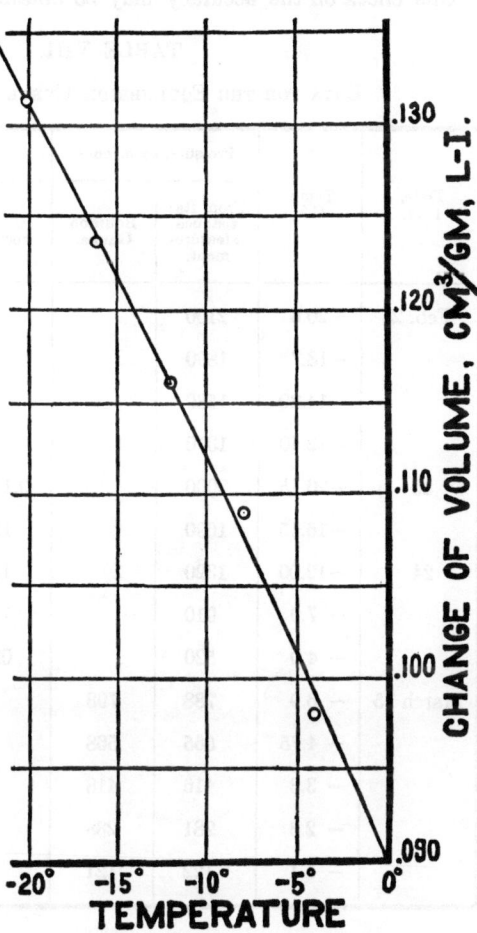

FIGURE 6. The change of volume when ice I melts to the liquid.

found by Tammann may very possibly be due to errors in the pressure measurements. His values for the pressures in the neighborhood of 2000 are nearly 100 kgm. higher than those found here. The effect of introducing such an error into the gauge readings at some point on the way up would be to give exactly such a point of inflection as he found.

One check on the accuracy may be obtained from the known slope

TABLE VIII.

DATA FOR THE EQUILIBRIUM CURVE ICE I–WATER.

Date, 1911.	Temp. C.°.	Pressure, kgm./cm.².		Δv, cm.³ per gm.		
		From Resistance Measurement.	Bourdon Gauge.	Uncorrected.	Corrected for Dilatation of Gasolene.	Corrected for Distortion of Cylinder. (Final value.)
Feb. 2	−20.6	2100				
	−18.7	1890				
	−14.80	1580				
	−12.40	1370				
23	−20.15	2000		0.1337	0.1312	0.1316
	−16.15	1690		1257	1236	1238
24	−12.00	1320		1178	1160	1162
	− 7.9	910		1103	1090	1091
	− 4.0	520		0991	0980	0980
March 15	− 6.9	788	798			
	− 4.75	565	568			
	− 3.3	416	418			
	− 2.3	281	288			
	− 1.1	122	121			

at 0°, calculated by Clapeyron's equation from the known latent heat and the change of volume. This slope is given by the equation

$$\frac{d\tau}{dp} = \frac{\Delta v \cdot \tau}{\Delta H}.$$

Using the following values: $\Delta V = 0.0900$, $\tau = 273.1$, $\Delta H = 79.82$ gm. cal., 1 gm. cal. = 42.66 kgm. cm., we find

$$\frac{d\tau}{dp} = -0.00722.$$

It is to be noticed that this value is lower than the calculated value usually given. This is because of the low value used for ΔV, which is taken from the most recent work of Leduc.[5] The effect of this change is to bring the calculated value into better agreement with experimental values. Thus Dewar[6] found recently for $\frac{d\tau}{dp}$, 0.0072. The slope at the origin as just calculated is shown by a dotted line in the diagram. The agreement is within the limits of error.

Even over this very low pressure range there are practically no other results to compare these with. Tammann is the only other observer who has gone to high enough pressures to find the direction of curvature of the curve, which agrees with that found here. Dewar has published results up to 250 kgm., and stated without publishing the data that the curve remains linear up to 700 kgm. The departure from linearity found above is about 6 per cent at 700. Several other accounts have been published of the effect of pressure on the freezing of ice, but the work has either been done only to a few atmospheres, or else the work to high pressures has been only qualitative, like that of Mousson.[7]

The second set of readings above gives the only determination of the change of volume made in the present work. The procedure in determining four of these five points was the same as that used for nearly all the other determinations with water as well as with mercury. This consists in plotting piston displacement against pressure during decreasing pressure. This has the advantage at higher pressures of almost entirely eliminating the effect of elastic after-effects in the containing vessel, but this procedure is less essential at lower pressures. For all ordinary substances this procedure means that the change of volume is measured during melting and has the advantage that it gives perfectly sharp values, it being impossible to overrun the melting curve. But here, due to the fact that ordinary ice is less dense than the liquid, the method gives the change of volume during freezing instead of melting, so that it is possible to considerably overshoot the mark before solidification sets in. The change of volume is so large during freezing, however, that this is no practical disadvantage, as it is with substances showing a smaller change of volume. The

---

[5] Leduc, C. R., **142**, 149–151 (1906).
[6] Dewar, Proc. Roy. Soc. Lon., **30**, 533–538 (1880).
[7] Mousson, Pogg. Ann., **105**, 161–174 (1858).

fifth point of the above series was obtained in the inverse way, during melting, proceeding from low to high pressures. The pressure corresponding to −4° is so low that the elastic after-effect must be small, but still it is noteworthy that this point shows the greatest deviation from the straight line. This inverse procedure at −4° was necessary in this case because of the low pressures, the friction of the packing being so great that pressure could not be reduced sufficiently below the equilibrium pressure to induce solidification from the subcooled liquid when the regular procedure of running from high to low pressures was adopted.

The correction applied to the change of volume for the thermal dilatation of the gasolene (the fluid transmitting pressure in this set of experiments), was not determined directly at all points of the curve, but only at the two end points. At 0° the correction is 0.9 per cent, and at −22° is 2 per cent. The correction was assumed linear for intermediate points. It makes no difference within the limits of error whether temperature or pressure is taken as the independent variable of this linear relation, and there seems little probability that the error so introduced can exceed this 0.1 per cent. The tables give the values corrected both for the thermal dilatation of the gasolene and for the elastic deformation of the cylinder. This latter correction is 0.2 per cent at the maximum.

The values of ΔV are shown in Figure 6. The open circles are the observed points. The solid circle at −22° was not an observed point, but is the value required at the triple point by the conditions of consistency with the other data. In the ΔV diagrams for the other equilibrium curves the values at the triple points are indicated in the same way.

As an additional check on the values found here there is the known change of volume at 0° as found by other observers. The recent work of Leduc seems the most accurate. He gives for the value of the density of ice at 0°, 0.9176, from which the change of volume is found to be 0.0900 cm.^{3} per gm. The value of ΔV at 0° as found by Leduc is taken as the origin in the diagram. It is seen that the other points are perfectly consistent with this value, thus confirming the accuracy of the method. The relation connecting change of volume with temperature is nearly linear, the curve being slightly concave toward the temperature axis. Four of the points, including Leduc's, lie sensibly on the curve; one is 1 per cent too high, and the other 1.5 per cent too low.

This value of Leduc is higher than that of most other observers, but the discrepancy seems to be due to occluded air, which Leduc took particular pains to exclude. In these experiments in which the water was frozen under pressure, any dissolved air, naturally, can have only very little effect. Leduc's value, therefore, seems best for this comparison.

It will be well here to explain the sign convention used uniformly throughout the paper. A change of volume is taken as positive if the

TABLE IX.

LATENT HEAT, ETC., ON EQUILIBRIUM CURVE ICE I-WATER.

Temp. °C.	Pressure, kgm./cm.²	Δv. cm.³/gm.	$\frac{dp}{dt}$.		$p\Delta V$. gm. cal. gm.	ΔH. (Latent Heat) gm. cal. gm.	ΔE. (Change of internal Energy) gm. cal. gm.
			Directly from Curves.	Adjusted.			
−20	1970	−.1313	74.6	74.0	6.06	57.7	63.6
−15	1590	−.1218	86.0	84.8	4.52	62.5	67.0
−10	1130	−.1122	99.0	98.4	2.96	68.0	71.0
− 5	610	−.1016	116.0	115.5	1.45	73.7	75.1
0	0	−.0900	138.5	138.5	0.00	79.8	79.8

volume increases when the reaction runs in the indicated direction. Thus ΔV L–I is positive (water expands when it freezes to ordinary ice) but ΔV I–L is negative. Similarly the latent heat, ΔH, is positive when heat is absorbed during the reaction, and ΔE, the change of internal energy, is positive when the internal energy increases. ΔH for the reaction L–I is negative, but positive for I–L. In general, as already explained, ΔV is positive when the reaction runs from high to low pressure and ΔH is positive when the reaction runs from low to high temperatures.

From the curves giving the equilibrium points and the changes of volume, the various data needed for computing the latent heat and the change of internal energy may be obtained. These are given in Table IX. for even temperature intervals. The values of ΔV were taken from the line drawn through the observed points. Two values of $\frac{dp}{dt}$ are tabulated. The first set of values was obtained by graphical construction from the equilibrium curve at different points, these values of $\frac{dp}{dt}$ plotted and a smooth curve drawn through them. The first tabulated values of $\frac{dp}{dt}$ are obtained from this smooth curve. With these values of $\frac{dp}{dt}$, ΔH was calculated. It was then necessary to adjust the values of ΔH slightly so as to satisfy the identical relations at the triple point. These new values of ΔH are tabulated. The change in the value of ΔH demands a corresponding change in the values of $\frac{dp}{dt}$ which are given in the second column. These latter values are to be taken as the most probable values. The difference between the two columns shows the amount of adjustment necessary, and gives an idea of the accuracy of the data. Of course, as already explained, the slope is very much more sensitive to slight errors than the observed values of p and t themselves. This same process of adjustment has been used on all the equilibrium curves. In some cases, where $\frac{dp}{d\tau}$ is very small, no change has been necessary. The data of Table IX. are shown graphically in Figure 7.

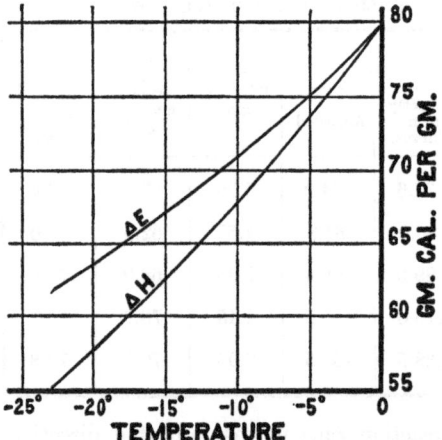

FIGURE 7. The latent heat and the change of internal energy on the I–L curve.

The data give some hold on the actual compressibility of the solid at different points on the equilibrium curve. At 0° this may be calculated without approximation from the observed data. The thermal dilatation and compressibility of water at 0°, and the thermal dilatation of ice at 0°, are already known. These present data give the change of ΔV with temperature and pressure on the equilibrium curve. These four quantities evidently suffice to determine the compressibility of the solid at 0°, for we have for small changes of p and t,

$$V_1(p,t) = V_1(0,0) + \left(\frac{\partial v_1}{\partial p}\right)_t dp + \left(\frac{\partial v_1}{\partial t}\right)_p dt,$$

$$V_2(p,t) = V_2(0,0) + \left(\frac{\partial v_2}{\partial p}\right)_t dp + \left(\frac{\partial v_2}{\partial t}\right)_p dt,$$

where the subscript (1) denotes the liquid and (2) the solid.

$$\Delta V(p,t) = V_2(p,t) - V_1(p,t) = \Delta V(0,0) + dp \left[\left(\frac{\partial v_2}{\partial p}\right)_t - \left(\frac{\partial v}{\partial p}\right)_\tau \right]$$
$$+ d\tau \left[\left(\frac{\partial v_2}{\partial t}\right)_p - \left(\frac{\partial v_1}{\partial \tau}\right)_p \right].$$

This holds for any values of dp and dt. On the equilibrium curve,

$$d\tau = \frac{d\tau}{dp} dp,$$

and we have

$$\frac{d\Delta V}{dp} = \frac{d\Delta V}{d\tau} \cdot \frac{d\tau}{dp} = -\left(\frac{\partial v_2}{\partial p}\right)_\tau + \left(\frac{\partial v_1}{\partial p}\right)_\tau + \frac{d\tau}{dp}\left[-\left(\frac{\partial v_2}{\partial \tau}\right)_p + \left(\frac{\partial v_1}{\partial \tau}\right)_p\right].$$

Everything in this equation is known except $\left(\frac{\partial v_2}{\partial p}\right)_\tau$.

We have $\left(\frac{\partial v_1}{\partial p}\right)_\tau = 0.0_452$ kgm./cm.^{2} (Amagat),

$$\left(\frac{\partial v_1}{\partial \tau}\right)_p = -0.0_457,$$

$$\left(\frac{\partial v_2}{\partial \tau}\right)_p = +0.0_3152 \text{ (Vincent)}.[8]$$

$$\frac{d\tau}{dp} = -0.00722 \text{ [already computed; see p. 468].}$$

$$\frac{d\Delta V}{dt} = -0.00200 \text{ [from these experiments].}$$

Substituting these values in the above expression, we find

$$\left(\frac{\partial v_2}{\partial p}\right)_\tau = -0.0_436.$$

The compressibility of ice, with the larger volume, is therefore about 1/3 less than the compressibility of the liquid. There seem to be no published experimental data with which to compare this value.

THE CURVE I-III.

The existence of this new variety of ice, III, was discovered by Tammann, who obtained it by increasing the pressure on ordinary ice to about 2500 kgm. at temperatures between $-22°$ and $-60°$. The nota-

---

[8] Vincent, Phil. Trans. A, **198**, 463–481 (1902).

tion used here is the same as that used by Tammann. In the present work this variety of ice was found in another way, coming from high to low pressures at about $-22°$, and from the domain of stability of a variety of ice stable only at higher pressures. It was obtained first with the same apparatus as that used on the I–L curve. Two points on equilibrium curve I–III and two unsatisfactory measurements of the change of volume were made with this apparatus, but all the subsequent determinations were made with the apparatus especially designed for low temperatures. A short description of this new apparatus will not be out of place.

This had three different pressure chambers instead of only two as formerly. The upper cylinder in which pressure was produced by the advance of the piston was the same as that used before. This was jacketed with running water from a large tank, so as to be maintained at a constant temperature. Communicating directly with this, and exposed freely to the air of the room, was a block of nickel steel with provisions for three connections. In one of these holes for connections was placed the manganin resistance plug with which pressure was measured. No temperature correction had to be applied to the readings of this, therefore. The third connection in the block was to the cylinder containing the water to be experimented on. This cylinder was of hardened nickel steel, 8 inches long, 1 inch o. d., and 1/2 inch i. d. The water to be experimented on was placed directly in the cylinder, nearly filling it. The remainder of the cylinder was filled with gasolene communicating through the connecting block with the pressure-producing cylinder. Some hesitation was felt at first about allowing the water to completely fill the interior of the cylinder, instead of being placed in a cylinder exposed to pressure on all sides. Fear was felt that the water in freezing might so expand itself against the sides of the cylinder as to be able to support an appreciable stress, so that a hydrostatic pressure applied by means of the gasolene to the upper part of the frozen ice cylinder would not be transmitted uniformly to all parts of the mass of ice. In the first few experiments an attempt was made to avoid the possibility of the effect by providing means for the gasolene to penetrate to the neighborhood of all parts of the ice. This was done with a tube of very thin sheet copper, closed at the bottom, but open at the top to the gasolene, placed axially in the cylinder, and extending throughout the mass of water. The precaution proved needless, however, since there were no discrepancies to be ascribed to this cause when the device was omitted.

It was necessary to use gasolene as the transmitting fluid because of the low temperatures. Kerosene becomes too stiff to transmit pressure, and the gasolene itself gave evidence of considerable viscosity at the lower temperatures. Attempt was made to use a still lighter oil to avoid altogether this viscosity. Pentane was tried, but after a number of failures had to be abandoned. There is evidently some action under pressure between the pentane and either the water or the ice. The action seems to be one of solution. In every case water found its way sooner or later through the long connecting tube to the manganin resistance coil, where it short-circuited the coil, making measurements impossible.

The nickel steel cylinder containing the water was supported from above by means of the connecting pipe in a cylindrical Dewar flask, 2 inches diameter and 10 inches long. The low temperature was obtained with solid CO_2 and ether, or usually CO_2 and gasolene, this being much cheaper and giving nearly as low a temperature as the ether. No thermostatic regulation was used with the apparatus. The thermal insulation provided by the bath proved sufficient so that the slight amount of regulation necessary could be performed from time to time by hand, by dropping in small masses of solid CO_2. The bath liquid was stirred by blowing bubbles of air into it through a tube reaching to the bottom of the flask. It was possible with very little effort to keep the temperature constant to $1/2°$, which was sufficient for most requirements, since in the region of these experiments the equilibrium pressure is affected only slightly by changes in temperature. Greater constancy of temperature was attained by the exercise of a little more care when this was necessary.

Temperature was measured with a nickel resistance thermometer. This was a coil of bare nickel wire wound in spiral grooves on a hard rubber cylinder, which was immersed in kerosene in a thin, tightly fitting tube of glass. The top of this tube was carefully stoppered to prevent the condensation of moisture around the leads. The glass tube, 3/8 inch o. d., was immersed directly in the temperature bath. The response to changes of temperature was always very prompt. The resistance of this wire was measured on the same bridge wire of the same Carey Foster bridge as that on which the measurements of the resistance of the manganin coil for pressure were made. Either the nickel or the manganin resistance, with the appropriate extension coils, could be connected with the bridge wire by two double-throw mercury switches in paraffin blocks. The temperature was ordinarily read both before and after the readings of pressure and piston displacement. The coil was calibrated by comparison at $0°$, $-50°$, and $-80°$ with a nickel resistance thermometer of Leeds and Northrup, which had been calibrated at the Bureau of Standards. The sensitiveness was sufficient to detect changes of 0°.03. The coil was tested for permanent changes, elastic after-effects, etc., by frequent calibrations in melting ice at 0°. That it was perfectly satisfactory is shown by the fact that the zero has not changed by 0°.03 since the coil was set up.

A special method for determining the change of volume at the low temperatures was made necessary by the extreme slowness of the reaction. At the higher temperatures, between −20° and −30°, the reaction runs rapidly enough so that the ordinary procedure is available, but at lower temperatures this is not possible. On one occasion, such an attempt was made at −70°. After four hours waiting at a pressure several hundred kgm. removed from the equilibrium pressure, the reaction had not yet shown any signs of drawing to completion, and the attempt was given up. The new method is made possible by the fact that the equilibrium pressure is very nearly constant. Figure 8 illustrates the method. The first part of the process consists in getting the change of volume by the ordinary method at some temperature where the reaction runs with convenient rapidity. This consists in measuring the piston displacement at constant temperature at several points on each side of the equilibrium line indicated at (1), (2), (3), (4), (5), (6). After reaching (6) the piston displacement is kept constant, but the temperature is reduced, the water now being in the form of ice III. The pressure is measured at several points with decreasing temperature until the point A is reached. The apparatus is so small that complete temperature equilibrium is attained in at most 5 minutes after changing the bath temperature. At the point A the pressure is reduced at constant temperature by changing the displacement, until the point B is reached, still to the right of the equilibrium line. From B, temperature

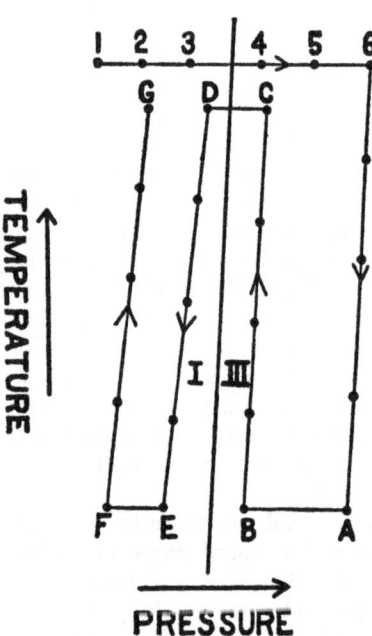

FIGURE 8. Shows the cycles described in finding the change of volume I-III.

is raised at constant displacement to the point C. Here the pressure is lowered to D, the reaction now running with sufficient speed because of the higher temperature. From D, two excursions to low and then back to high temperatures are made exactly as from (6), the water now being in the form of ice I. The result of all these manœuvres is that we have sufficient data to find the displacement corresponding to any temperature and pressure, and so the displacement on opposite sides of the equilibrium line corresponding to the two phases I and III. The difference of this displacement at any temperature is evidently the same as if the reaction had run at that temperature, but the slowness of the reaction is avoided. The change of volume is calculated directly from the difference of displacements as usual. It was in general necessary to measure the displacement at at least three points on each side of the equilibrium line, since the relation between displacement and pressure is not sufficiently linear to allow a linear extrapolation from two points to the equilibrium line. For greater security, four or five points were usually taken.

The data are such that they contain internal evidence of their self-consistency. After a complete cycle like the above, the starting point ought to be reached again, if there has been no change in the apparatus in the meantime. This condition was always approximately fulfilled, the piston returning within a few thousandths of an inch to the original position. Failure to return exactly to the starting point is probably due partly to elastic after-effects in the steel cylinders, and partly to wearing away of the rubber packing on the end of the piston. There was never any leak. The slight discrepancy was corrected for by assuming that it had grown uniformly with the time. It will be noticed that the cycle as described above is such that the effect of any such error is at a minimum, since the two paths nearest the equilibrium curve on either side were described in succession.

The method is applicable here because of the fact that the equilibrium lines run so nearly parallel to the lines of constant displacement that it is possible to approach very near to the equilibrium line over the entire range. The method evidently would not apply without modification to points on the I-L curve, for example.

The slowness of the reaction which made necessary a modified method of determining the change of volume is also evidently going to have its effect on the equilibrium determinations. At the low temperatures, the reaction runs so slowly that it requires a very long time for pressure to return to the equilibrium values, if it has once been changed. The behavior was such as to give the impression that the equilibrium pressure might never be reached, there being a domain of indifference within which the viscosity is sufficient to keep the reaction from running, even if the equilibrium conditions are not satisfied. This has been stated to be the fact by Tammann. The equilibrium points at the low temperatures were found by approaching as close as practicable to equilibrium from each side, and taking the mean. The tables of the actual data show how broad the band of indifference is and how rapidly it increases in breadth at the low temperatures.

For these determinations with this new piece of apparatus, ice III was usually obtained from ice I. The procedure was to freeze water to ice I at a low pressure, and then increase the pressure on ice I at a temperature in the neighborhood of $-30°$. It was always necessary to pass considerably beyond the transition curve before the reaction started. On several occasions at $-28°$ the pressure was carried so far over the I-III curve as to arrive at the prolongation of the I-L curve, at which melting took place immediately, it never being possible to carry a solid into the domain of the liquid. I melts to the liquid even if the liquid is relatively unstable with respect to another solid phase. At the temperature $-28°$ it was never possible to keep the unstable liquid long enough to make an accurate determination of the equilibrium pressure. The rough values fell within the limits of uncertainty on the smooth prolongation of the I-L curve. When the phase III did appear at temperatures above $-30°$, the reaction from I to III ran with explosive rapidity. Sometimes it was possible to hear a sharp click in the apparatus announcing the sudden change of volume.

Aside from the fact that it is possible to pass over the boundary line between two solid phases without the reaction running, the actual amount of possible trespassing has little significance. It is changed by alterations in the size and shape of the containing vessels, is affected by slight impurities, and most of all depends on the element of time. It may be mentioned, however, that under pressure the viscosity of the substances seems to be so great that mechanical shock has little if any part in starting the reaction, as it does under ordinary circumstances. It was never found possible to start a reaction from an unstable phase by smart blows on the outside of the apparatus with a hammer. This was tried both when the unstable phase was a solid and a liquid.

After the phase III had once been obtained, the other points on the equilibrium curve were obtained with decreasing temperature. At every new temperature, the piston was pushed in a little, the pressure then falling back to the equilibrium value, and then the piston was withdrawn slightly, pressure rising to the equilibrium value. The mean of these was taken as the equilibrium pressure. Sometimes at the higher temperatures, where the reaction velocity is high, equilib-

TABLE X.

Data for the Equilibrium Curve Ice I–Ice III.

Date, 1911.	Temp. C.°.	Pressure, $\frac{kgm.}{cm.^2}$	$\Delta v.$ $cm.^3/gm.$ Corrected Value.	Date, 1911.	Temp. C.°.	Pressure, $\frac{kgm.}{cm.^2}$	$\Delta v.$ $cm.^3/gm.$ Corrected Value.
Feb. 16	−22.80	2120	−0.1910	April 6	−52.5	2100−	
	−25.25	2120	−0.1861		−50.0	2110+	
Mar. 24	−32.0	2150−			−62.0	2000−	
	−31.7	2150+			−61.2	2100+	
	−24.7	2110±			−32.8	2150−	
Mar. 29	−33.8	2170−			−32.6	2150+	
	−28.7	2150+			−26.5	2120±	
	−26.7	2140ᵛ		April 18	−27.6	2110+	
	−23.5	2130ᵛ			−22.4		−.1831
April 5	−33.4	2170+			−29.1		.1903
	−33.0	2160−			−35.7		.1966
	−28.1	2150±			−41.7		.1948
	−23.1	2140±			−49.6		.2018
	−21.5	2140±			−56.6		.2042
April 6	−26.4	2120+			−63.8		.2059
	−25.9	2100−			−70.8		.2071
	−31.0	2110+		April 20	−23.3	2134±	
	−30.7	2110−			−24.5	2136±	
	−38.6	2120−			−27.8	2153±	
	−38.4	2130+			−31.5	2163±	
	−43.5	2130−			−37.2	2182±	
	−42.5	2150+			−46.0	2179±	

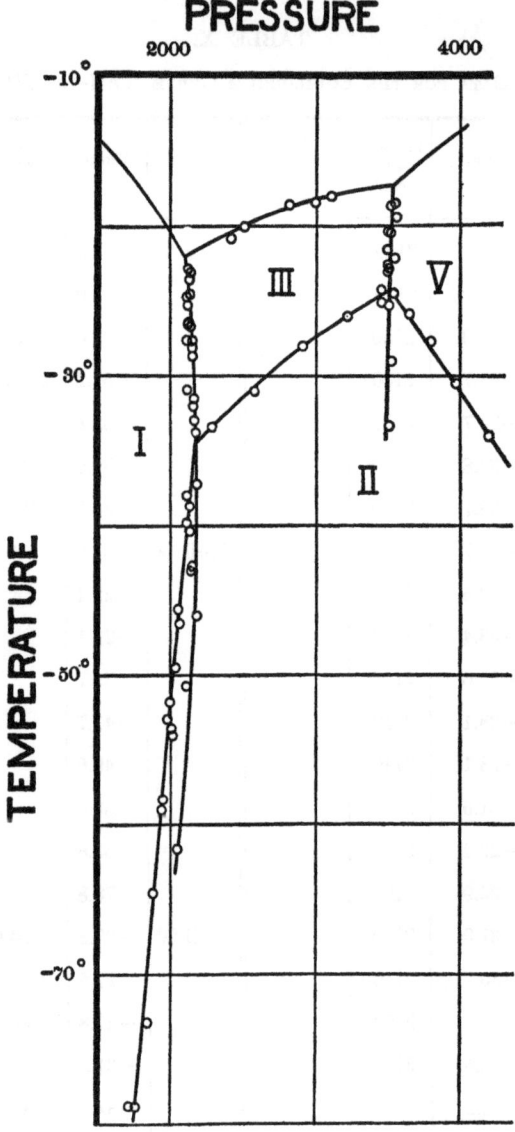

FIGURE 9. The various equilibrium curves between I, II, III, V, and the liquid.

rium was approached from only one side. By proceeding in this way from high to low temperatures, it was possible to prolong the curve considerably into the region of stability of II. A point was thus found at −62°.0, showing a subcooling of 28°. If the pressure had been increased on ice I at this temperature, ice II and not ice III would have separated out.

The actual data are given in Table X. Those of February 16 were obtained with the same apparatus as that used for the mercury; all the others with the smaller apparatus for lower temperatures. The plus or minus signs after the pressures show whether the value was approached from above or below. Three different manganin coils were used in making the measurements. The data of April 6 were obtained with a new coil which had not yet been sufficiently seasoned. For this reason the data of April 6 are slightly in error as to the absolute value of the pressure, but are competent to show the shape of the equilibrium curve. The data of April 20, taken with especial care to show the actual shape of the equilibrium curve, give more significant figures than one is entitled to for the absolute pressure, but the variation of pressure is accurately indicated by the figures given.

The equilibrium points are shown in Figure 9, in which are shown also the other equilibrium curves in the vicinity. The shape of the curve is given by the points of April 6 and 20, which agree, but the best value for the absolute pressure is to be given by the mean of the points at the upper end of the curve. On the same pressure and temperature scale as the complete diagram for ice and water this equilibrium curve appears nearly vertical, but on the scale shown here it is seen to have unmistakable curvature, being concave toward the temperature axis, and at one part parallel to that axis. At this point of parallelism, the latent heat of transformation is zero; above it is positive, below negative. This point is approximately at −40° and 2150 kgm.

The equilibrium curve I–III as found by Tammann shows the same general shape, concavity toward the temperature axis. But Tammann's vertical tangent is between −40° and −50°. The total change of pressure from the triple point at −22° to the maximum is practically the same, 50 kgm., but Tammann does not find so pronounced a curvature below the maximum as is found here. On the whole, however, the agreement in the shape of the curve as found in these two independent determinations is as good as could be expected, particularly when it is remembered that the discrepancies are at the low temperatures where the reaction runs very slowly. But the actual value of the pressure as found by Tammann at −22° is higher than that found here, 2200 against 2115. Tammann's pressure measurements were made with a Bourdon gauge, which is not accurate at high pressures. There are two pieces of evidence to show that Tammann's pressures may be wrong; the supposed inflection point on the I–L curve, which has been already mentioned, and the behavior of the I–III curve above the I–II–III triple point, which will be dealt with later.

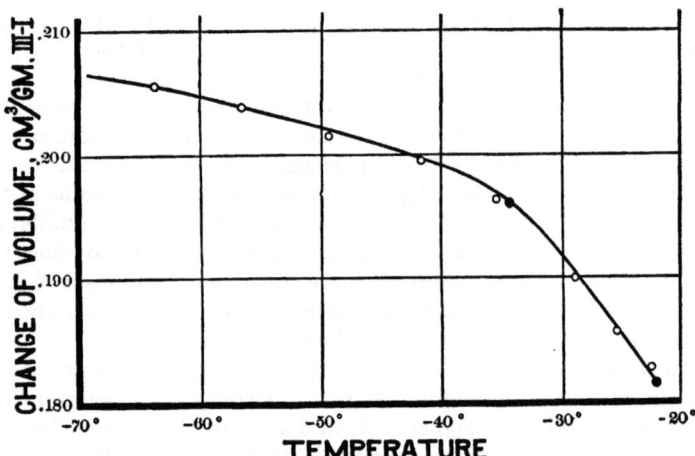

FIGURE 10. The change of volume III–I.

The values for the change of volume obtained on April 18 with the method described are given also in Table X., and shown in Figure 10. The most striking feature is the enormous value of the change, being about 20 per cent. The values extend into the unstable region over a temperature interval twice the interval of stability, which is from $-22°$ to $-35°$. The values given have been corrected in the usual way for the elastic distortion of the cylinder and the thermal dilatation of the gasolene. Owing to the wide temperature range, this latter correction is unusually large, rising to 6 per cent at the lowest temperatures. The curve drawn through the points is forced a little so as to give consistent results at the triple points, at $-22°$ passing about 1/4 per cent too low and at $-35°$ 1/4 per cent too high.

The actual procedure in getting these change of volume points was not exactly the ideal one described above. Below $-35°$, III is unstable, and the reaction to II might run at any time. The path 6 A (Figure 8) was described successfully and two points found on the lower end of BC before the reaction to II ran. On raising the temperature above the II–III curve, the reaction ran back to III, so that at C the phase present was III, as it should be. The line 6 A was perfectly straight; it was assumed, therefore, that the line CB would also have been perfectly straight if it had been possible to trace it out for its entire length.

TABLE XI.

LATENT HEAT, ETC., ON EQUILIBRIUM CURVE ICE I-ICE III.

Temp. C.°.	Pressure, kgm./cm.².	Δv. cm.³/gm.	$\frac{dp}{dt}$.	$p\Delta V$. gm. cal./gm.	ΔH. gm. cal./gm.	ΔE. gm. cal./gm.
−60	2117	.2049	5.4	10.13	−5.5	4.6
−50	2160	.2023	2.0	10.24	−2.1	8.1
−40	2178	.1992	−0.6	10.17	+0.7	10.9
−30	2156	.1919	−3.2	9.69	3.5	13.2
−20	2103	.1773	−5.3	8.74	5.6	14.3

The upper ends of the lines of equal volume for the phase I, that is, the lines ED and FG above the triple point at −35°, bear less and less to the right, and at their extreme upper ends are actually bearing to the left. This would mean a negative thermal dilatation for ice I at these temperatures. The effect deserves to be investigated for its own sake. The effect is so slight as to make very little difference for the purposes of this paper, however, and it was not investigated further. That there may be such an effect is borne out by the rather sharp change in the curvature of the ΔV curve at the triple point −35°, while the accuracy of the ΔV curve is made very probable by the closeness with which it checks at the two triple points. Pettersson[9] claims to have found that ordinary ice at atmospheric pressure shows a negative dilatation between −35° and −25°.

Table XI. gives the values taken from the equilibrium and the change of volume curves for the various quantities involved in the determination of the latent heat and the change of energy. The values of ΔE and ΔH are shown graphically in Figure 11. The internal energy of III is greater than that of I, and practically all this change of energy comes from the mechanical work absorbed when I

---

[9] Pettersson, Vega Expedition, vol. 2 (1883).

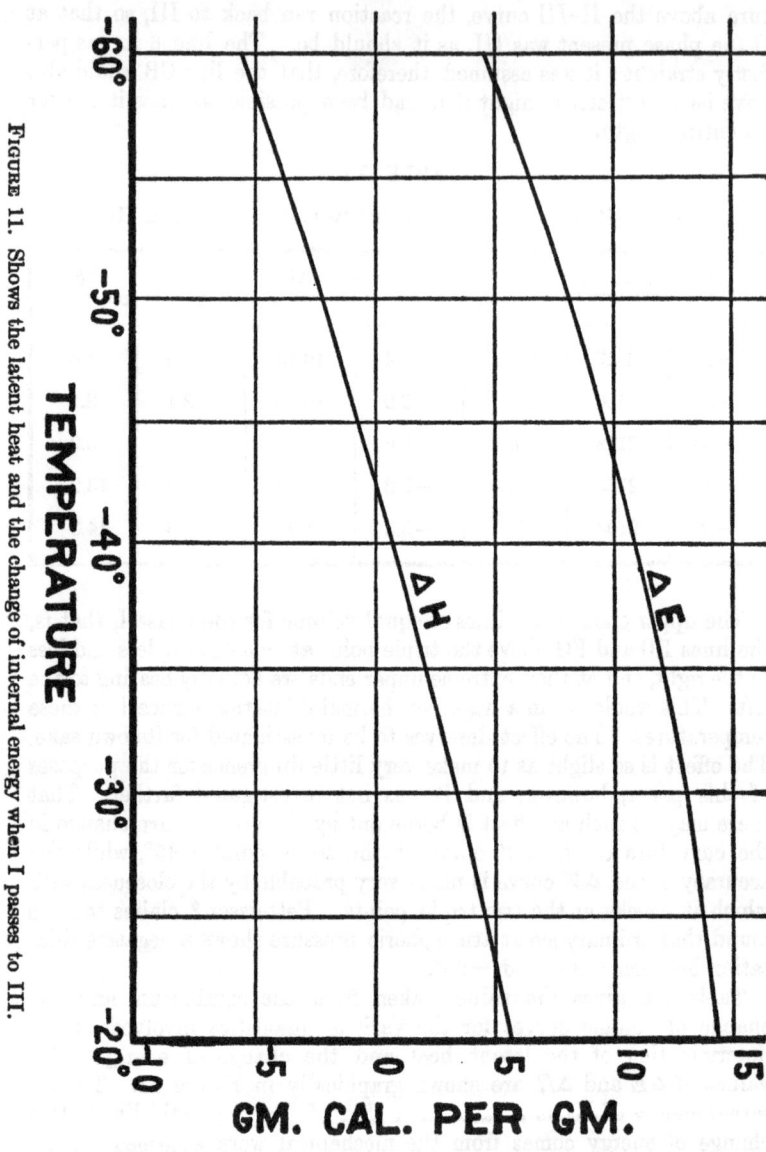

Figure 11. Shows the latent heat and the change of internal energy when I passes to III. The latent heat is small; above −40° I absorbs heat on passing to III, below −40° it gives out heat.

The Curve III–L.

The points on this curve were more difficult to obtain accurately than those on any of the other nine equilibrium curves. This is

TABLE XII.

Data for the Equilibrium Curve Ice III–Water.

Date, 1911.	Temperature, C.°.	Pressure, kgm./cm.²	ΔV cm.³/gm. Corrected Value.
Feb. 1	−18.55	2820	
	−20.00	2510	
	−17.95	3110	
Feb. 16	−18.40	3000	.0301
March 2	−20.80	2420	.0373

because of the extreme slowness of the reaction, accounted for in part by the comparatively small change of volume and the large heat of reaction. This does not prevent, however, a fairly accurate knowledge of the properties of the curve, since the curve is short, reaching over only 1500 kgm., and it is tied down by a triple point at either end.

The apparatus used for this curve was that of the mercury determinations, temperature being kept constant with a thermostat. This was necessary because of the great effect of temperature on the equilibrium pressure on this curve.

Five points on the equilibrium curve were obtained, and two on the change of volume curve. These are shown in Table XII. and Figures 9 and 12. The equilibrium points are satisfac-

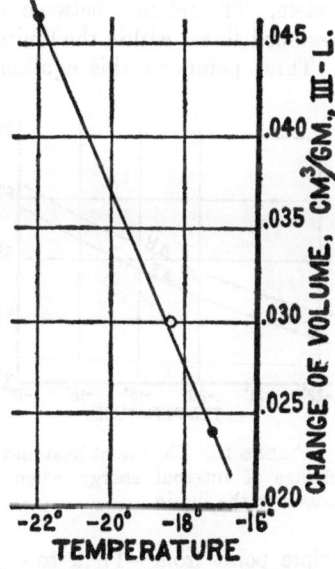

FIGURE 12. The change of volume when III passes to the liquid.

tory enough. The greatest difficulty was found with the ΔV points. These are evidently irregular, but with the two other known points on the curve at the triple points, the curve itself cannot be far from that

TABLE XIII.

LATENT HEAT, ETC., ON THE EQUILIBRIUM CURVE ICE III-WATER.

Temp. C.°.	Pressure, kgm./cm.²	Δv. cm.³/gm.	dp/dt.		$p\Delta V$. gm. cal./gm.	ΔH. gm. cal./gm.	ΔE. gm. cal/gm.
			Directly from Curves.	Adjusted.			
−22.0	2115	0.0466	180	186	2.31	50.9	48.6
−20.0	2510	.0371	240	246	2.18	54.1	51.9
−18.5	2910	.0301	326	320	2.05	57.4	55.4
−17.0	3530	.0231	450	443	1.90	61.4	59.5

drawn. The relation between temperature and change of volume was assumed linear within the limits of error.

Three points on this equilibrium curve were found by Tammann. The shape of his curve agrees with that found here within the limits of error, but the absolute values of the pressure found by Tammann are higher than those given, as already explained. Tammann gives no values for ΔV on this curve.

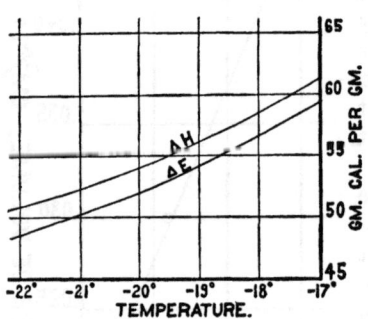

FIGURE 13. The latent heat and the change of internal energy when III passes to the liquid.

The values deduced from these curves for ΔH and ΔE are given in Table XIII. and in Figure 13. Very slight changes in the coordinates of points on the equilibrium curve make enormous changes in the value of the slope and ΔH. Thus, raising the upper triple point from −17°.2 to −17°.0 changed the calculated value of ΔH from 94 to 64. This latter value is taken as the more nearly correct, because it checks with the other values at the triple point, even though the V-L curve has to be slightly raised to pass through the point.

THE CURVE I–II.

We return now to the curve meeting the I–III curve at the lower temperatures, the I–II curve. The points on this curve were found with the low temperature piece of apparatus.

The data, shown in Table XIV., were obtained on five separate occasions with three different manganin resistance coils. The first three sets, on March 24, 29, and April 5, were obtained before the points on the I–III curve were obtained at the low temperatures.

The whole investigation at the low temperatures was approached with the prejudice that Tammann must have been wrong. In particular, the existence of two separate modifications of ice, II and III, with only slightly different properties, was thought to be explicable by some instrumental error. This prejudice was strengthened by Tammann's failure to find the triple point I–II–III, or any points on the melting curve of II, although he must have passed over this curve if there were such a one. His failure to find the II–L curve was difficult to explain because the reaction from solid to liquid always runs, and in this case the accompanying change of volume is fairly high. The work of the first three days still further strengthened the impression of error in Tammann's work. It showed conclusively that there was a modification of ice different from III in equilibrium with I at low temperatures, but seemed to show that this was not the variety supposed by Tammann. The points of the three days all lay on the same curve, which was headed exactly for the intersection of the I–L and the V–L curves both prolonged into the region of stability of III. It seemed probable, therefore, that Tammann's II was nothing but our V in the unstable region.

The determinations of these two days were made with two different coils, one a well seasoned one that had been maltreated in many explosions, and the other a brand new one. After several days' work suspicion was drawn to these coils and they were checked by measuring with them the equilibrium pressure on other already well established curves. Both on the I–III curve and the III–V curve they gave results consistent with each other but about 100 kgm. higher than the previous values. There seems no reason to doubt, therefore, that these coils were both in error and, by a stroke of luck, by the same amount. This disposed definitely of the idea that the new ice was V subcooled. The values tabulated are corrected by this quantity 100 kgm. There seems no reason why these readings should be entirely discarded; they

TABLE XIV.

Data for the Equilibrium Curve Ice I–Ice II.

Date, 1911.	Temp. C.°.	Pressure, kgm./cm.².	Date, 1911.	Temp. C.°.	Pressure, kgm./cm.².	ΔV. cm.³/gm. Corrected.
March 25	−64.7	1960	April 5	−63.7	1910+	
	−66.8	1950		−53.6	2000+	
	−65.2	1960		−52.2	1970−	
	−59.2	1900−	April 6	−69.5	2020+	
	−58.7	1980+		−76.5	1670−	
	−52.1	1980−		−58.7	1900−	
	−51.5	2020+		−58.0	2000+	
	−45.7	2040−		−50.0	2010−	
	−45.5	2060+		−49.0	2060+	
	−38.0	2110±		−40.1	2100−	
March 29	−78.8	1690−		−39.5	2120+	
		1840+		−53.4	2000−	
	−54.1	1990−		−54.5	2040+	
	−53.0	2030+	April 18	−36.9	2110±	
	−46.7	2050		−35.7		0.2187
	−46.4	2070		−41.7		.2172
	−40.8	2130−		−49.6		.2169
	−40.0	2130+		−56.6		.2159
April 5	−78.8	1780+		−63.8		.2154
		1640−		−70.8		.2145
	−65.6	1850−		−78.0		.2145

seem perfectly satisfactory in giving the shape of the equilibrium curve, on which the pressure varies only slightly, even if the absolute value of the pressure is somewhat high.

The last set of determinations on April 6 and 18 were with still another coil, a comparatively new one, but one which had been sufficiently seasoned. This coil gave readings consistent with the previous values on the I–III and II–V curves, and gave the same shape for the curve as that found with the two previous coils. There seems to be no question but that these are the correct values.

This variety of ice was obtained by increasing the pressure on I at low temperatures. The temperatures used here ranged from $-65°$ to $-80°$. At $-80°$ it is possible to run as much as 1000 kgm. beyond the transition curve before the reaction begins to run. Even so far removed from equilibrium as this, the reaction runs slowly and never seems to run to completion. The table shows the discrepancies in the stationary pressures when approached from above and below. The other points on the curve were obtained with regularly increasing temperature after the lowest. The temperature was raised as high as $-23°$ to get the last point. The succession of points found in this way lie on a curve running nearly linearly with increasing temperature and pressure up to about $-35°$, where it turns and proceeds upwards nearly vertically. This verifies the form found for the curve by Tammann. This curve as found by Tammann crossed his I–III curve. Tammann advances as an argument against there being a true triple point at the point of intersection the experimental fact that it was possible to start with II at low temperatures, to proceed along the equilibrium curve with increasing temperature across the I–III curve into the supposed region of stability of III, and then to retrace one's steps, following down the original curve into the region of II where the two curves I–III and II–III are unmistakably different. This possibility, surprising as it seems, was verified also by direct experiment. The points of April 6 at $-40°$ and $-53°$ on the I–II curve were separated by points at $-32°.5$ and $-26°.5$, which are in the supposed region of stability of III.

The whole mystery is cleared up by the discovery of the II–III equilibrium curve. It may be permitted to anticipate the results as they were found on this curve in so far as they are needed to explain the curious facts found on the I–II curve. In the first place, the existence of the II–III curve puts beyond question the fact that there are two varieties of ice, II and III. The point of intersection of the II–III curve with the I–II and the II–III curves is at $-35°$, the triple point. This point is also the point at which the change in direction of the supposed I–II curve was found. Above this point the modification II passes into the modification III. The absolute regularity with which this turning point was found, both here and by Tammann, is to be explained by the experimental fact that it is impossible to superheat II with respect to III. In the course of these experiments it has never been possible to carry II the slightest distance into the region of III, the reaction running as inevitably as the reaction from a solid to a liquid when the solid is heated to the melting point. Furthermore this reaction, when it does run, runs with great velocity, since the temperatures are sufficiently high. The change is furthermore accompanied by a comparatively small change of volume. This evidently accounts for the curve having been overlooked by Tammann. The supposed crossing of the two curves I-II and I-III found by Tammann is to be explained by slight errors in the pressure measurements. The greatest difference between the curves is 30 kgm. This is the second bit of evidence alluded to on p. 482, that Tammann's absolute pressure measurements may be in error. The turning point of the curve found by Tammann is at the same temperature as that found here, $-35°$. The impossibility of superheating II also explains Tammann's inability to find points on the II-L curve. Apparently it is impossible to reach this. II appears to be surrounded on all sides by other solid phases.

The behavior of III with respect to II is not the reverse of II with respect to III. It is possible to subcool III greatly with respect to II. The amount of subcooling possible depends on a variety of factors. One of the most important of these is the interval of time which has elapsed since the water had previously existed in the form II. It is a fact verified repeatedly on all the other curves as well as on this particular one, that it is very much easier to obtain a phase after it has once been present in the apparatus. Thus it is possible to subcool III at least as far as $-70°$ with respect to II, provided that II has never been formed. But if II has been formed previously, the reaction from III to II is very likely to run on passing from the III to the II region. This explains why it is possible to move up along the II-I curve, pass to the I-III curve, and then on decreasing temperature pass again to the I-III curve.

The points obtained experimentally have been classified with these facts in view. The points of March 25, 29, and April 5 obtained at temperatures above $-35°$ after coming from lower temperatures on the I-II curve have been listed on the I-III curve in the tables.

The points are plotted in Figure 9. These points are the mean of the points found with increasing and decreasing pressures at temperatures very close to each other. Within the limits of error the points lie on a straight line. The two lowest points are irregular because of the extreme slowness of the reaction. It is interesting to notice that the line as drawn is heading straight for the absolute zero, the equilibrium pressure becoming zero at this temperature. Because of the slowness of the reaction it would be well nigh impossible to verify this conjecture experimentally. Except for the constant pressure error already mentioned, the curve as found by Tammann is of approximately the same shape as that found here. These figures give a drop in the

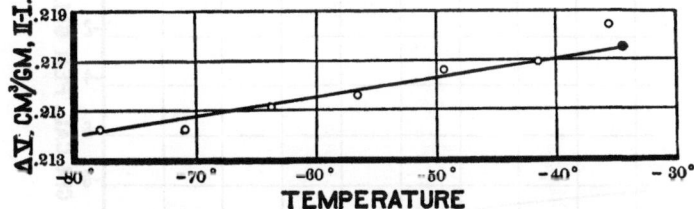

FIGURE 14. The change of volume when II passes to I.

equilibrium pressure of 360 kgm. between −40° and −80°, whereas Tammann gives 340 kgm. Tammann does not draw the line as straight, however, but prefers to make it slightly concave toward the temperature axis. This would bring the equilibrium temperature at atmospheric pressure up somewhat higher than the absolute zero.

The changes of volume, found by the method already described, are given in Table XIV., and shown graphically in Figure 14. Because of the fairly large difference in slope between the equilibrium curve and the lines of equal volume, the actual cycles described in measuring ΔV did not have the exact location shown in Figure 8. The lower parts of both DE and GF, which were fairly close together, projected over the equilibrium line into the region of stability of II. The upper end of the line CB was managed so as to just clear reaching into the region of I. This evidently introduces no essential error, since the compressibility and dilatation of one phase do not change on passing into the region of another. The difference merely means a somewhat wider extrapolation to determine the displacement on the equilibrium curve corresponding to the phase II, but this disadvantage is offset by the fact that what was an extrapolation for the phase I has now become an interpolation.

The change of volume is seen to be linear with temperature (or pressure) within a maximum error of 1/2 per cent. Only one point, that at −35.7°, shows a discrepancy as much as this, the average departure from linearity being in the neighborhood of 1/8 per cent.

Tammann gives three values for the change of volume which he lists as I-II, and two values listed as I-III. One of the supposed I-II values is above −35°, however, so that this must now be tabulated as I–III. He finds for I–II at −76°, 0.171 cm.³/gm., and at −55°, 0.180. These

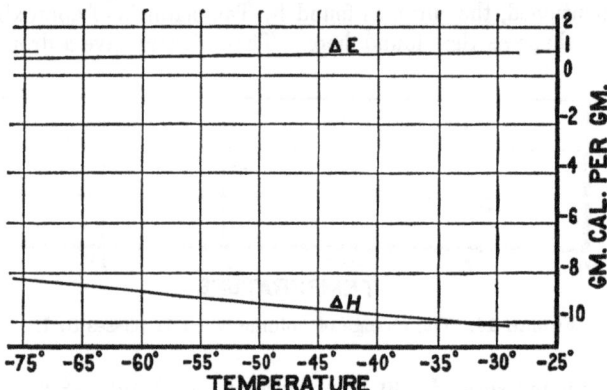

FIGURE 15. The latent heat and the change of internal energy when I passes to II.

values are considerably lower than the values found here. Tammann appears to have overlooked entirely the correction for the thermal dila-

TABLE XV.

LATENT HEAT, ETC., ON EQUILIBRIUM CURVE ICE I–ICE II.

Temp. C.°.	Pressure, kgm./cm.².	Δv. cm.³/gm.	$\frac{dp}{dt}$.	$p\Delta V$. gm. cal./gm.	ΔH. gm. cal./gm.	ΔE. gm. cal./gm.
−75	1794	0.2146	8.35	9.02	−8.31	0.71
−65	1886	.2154	8.35	9.52	−8.77	0.75
−55	1980	.2162	8.35	10.03	−9.22	0.81
−45	2072	.2170	8.35	10.54	−9.68	0.86
−35	2164	.2177	8.35	11.06	−10.15	0.91

tation of the fluid transmitting pressure; the effect of applying this correction would be to bring his values still lower. Tammann's results do agree with the present ones in showing an increase in the change of volume at higher temperatures, but Tammann finds a rate of increase four times that found here. Tammann states concerning his own data that his temperature coefficient is in all probability too high. His value at $-33.5°$ for the supposed change I-II, but really for I-III, is 0.193 cm./gm. This agrees exactly with the value found here at the same temperature for the known change I-III, giving another bit of evidence from Tammann's own data that above $-35°$ II changes spontaneously to III. Tammann's value for the change I-III at $-45°$ is 0.192, giving a temperature coefficient of the opposite sign from that found here. The actual values for the I-III change agree much better with the values found here than for the I-II change. The present values for I-III are 0.1904 at $-32.5°$ and 0.2005 at $-45°$.

The data taken from these two curves needed in the calculation of ΔH and ΔE are shown in Table XV. and Figure 15. I gives out heat on passing to II and absorbs work. The work is greater than the heat, so that the internal energy increases on passing from I to II.

The Curve II-III.

This has been already stated to have been the curve overlooked by Tammann, but demanded thermodynamically. The discovery of it places beyond question the essential difference between the two varieties of ice II and III. The change of volume on this curve is very slight, so that it is easy to overlook it altogether, as did Tammann. A special method had to be used here in determining these equilibrium points. It is the reverse of the usual method. The usual method consists in changing the piston displacement and so the pressure at constant temperature. Change of phase is indicated by change of displacement with constant pressure. In the modification of the method used here, the displacement is kept constant and the temperature changed. The resulting change of pressure is plotted as a function of the temperature. Change of phase, accompanied by a slight change of volume, is indicated by a discontinuity in the direction of the temperature-pressure line.

At any given temperature during this change of phase there is only one corresponding pressure, the equilibrium pressure, unlike the former method, where the piston displacement may have any value within a considerable range corresponding to the same pressure and temperature. The change of volume is so slight that the interval of discontinuity in the direction of the temperature-pressure line extended over only $1°$ and 100 kgm. The method was made practicable by two things, the smallness of the containing cylinder, resulting in a very rapid attainment of temperature equilibrium, and the fact already noted, that it is impossible to superheat II with respect to III. Evidently if it had been necessary to superheat II even a single degree in order to start the reaction, the change of volume would not have been sufficient to carry the pressure automatically back to the equilibrium value, and the reaction would have run to completion.

The procedure in getting the points was to start with II at some pressure and temperature below the equilibrium line, and to raise the temperature by small intervals. Arrival at the equilibrium line was shown by an abnormally large rise of pressure. This pressure and the corresponding temperature gave one point on the equilibrium line. The reaction was not allowed to run to completion, because of the difficulty of recovering II after it has once been changed to III, but the pressure was immediately raised from the equilibrium pressure, bringing the material back into the region of stability of II, and causing the reverse reaction III-II to run to completion. Another equilibrium point was then found starting with this higher pressure.

The job was a rather fussy one because of the narrowness of the critical temperature interval. The temperature had to be maintained by dropping solid CO_2 into the Dewar flask, and some little practice was necessary before satisfactory results were obtained. Determinations of the curve were made on three occasions. The first two sets confirm the last set, but sufficient adeptness had not yet been attained with the method, and the points were discarded.

The final points, four in number, are shown in Table XVI. and Figure 9. The curve rises with increasing pressure from lower to higher temperature and is convex upward; it is terminated at either end by triple points, at the lower end by the point I-II-III, and at the upper end by the point II-III-V. This curve is new; there are no previous results with which to compare it.

The change of volume determinations were made also with a slightly modified method. The usual method of measuring piston displacement at constant temperature would have been available if the measurements had been made with decreasing pressure from the region of II to that of III, but the method was undesirable practically for two reasons. There was the difficulty of maintaining the temperature in the Dewar flask constant for a sufficiently long interval of time by hand, and there was the necessity of running down to the very lowest temperatures in order to recover II for the next determination after the reaction to III had been allowed to run to completion. This would have required much time and would have been very wasteful of solid CO_2.

The method used was a modification of that used in finding the I-II and the I–III points, namely, measurement of pressure as a function of temperature at constant displacement. The changes II–III and II–V were measured at the same time, so that some anticipation is necessary in describing the method. The diagram (Figure 16) illustrates sufficiently the paths described. The start was made at the point A with the phase II. The approximately vertical lines show the paths at constant displacement on which the pressure, temperature, and displacement are all known. The dotted lines show the connecting paths on which it was not necessary to know these values. The result of describing all these paths is to give a knowledge of the displacement corresponding to any pressure and temperature in either the region of stability of II or III or V. The discontinuity of displacement on the three equilibrium curves can be found immediately, and so the change of volume. It is

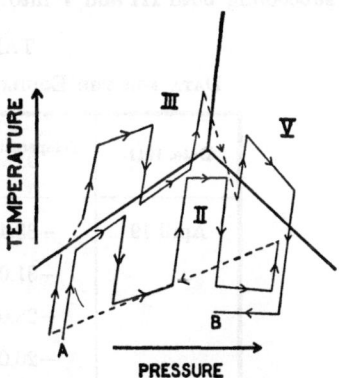

FIGURE 16. Shows the cycles described in finding the changes of volume II–III and II–V.

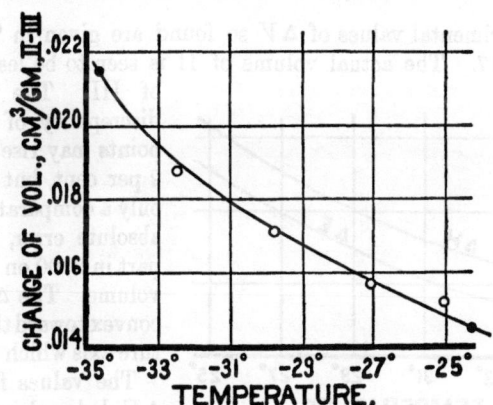

FIGURE 17. The change of volume when II passes to III.

to be noticed that the paths are described in such an order that one can easily apply the very small correction for wearing away of the packing or for viscous yield in the steel. The final point B is for the phase II. The diagram illustrates the impossibility of carrying II into the region of stability of either III or V, but the possibility of subcooling both III and V into the region of II.

TABLE XVI.

DATA FOR THE EQUILIBRIUM CURVE ICE II–ICE III.

Date, 1911.	Temperature, C.°.	Pressure, kgm./cm.²	ΔV. cm.³/gm. Corrected.
April 19	−33.4	2290	
	−31.0	2580	
	−28.0	2910	
	−26.0	3220	
April 20		2400	0.0188
		2710	.0171
		3050	.0157
		3390	.0152

The experimental values of ΔV so found are given in Table XVI. and Figure 17. The actual volume of II is seen to be less than that of III. The percentage discrepancy of two of the points may rise as high as 2 per cent, but this means only a comparatively slight absolute error, less than 1 part in 2000 on the original volume. The ΔV curve is convex toward the temperature axis which is unusual.

The values for ΔH and ΔE deduced in the usual way are given in Table XVII. and Figure 18. III gives out heat on passing into II and absorbs work. But the mechanical work is comparatively small, so that the internal energy of II is less than that of III.

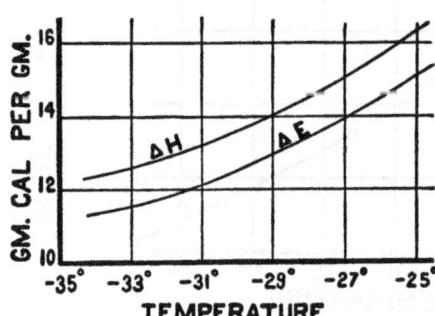

FIGURE 18. The latent heat and the change of internal energy when II passes to III.

This completes the description of the equilibrium curves of the varieties of ice already known. This last curve had not been found before, and the rest of the curves to be described involve either one or two of the new varieties of ice found at higher pressures.

TABLE XVII.

LATENT HEAT, ETC., ON THE EQUILIBRIUM CURVE ICE II–ICE III.

Temp. C.°.	Pressure, kgm./cm.².	ΔV. cm.³/gm.	dp/dt. From Curve.	dp/dt. Adjusted.	$p\Delta V$. gm. cal. / gm.	ΔH. gm. cal. / gm.	ΔE. gm. cal. / gm.
−34.0	2230	0.0206	100	107	−1.08	12.4	11.3
−31.0	2530	.0179	125	130	−1.06	13.2	12.1
−28.0	2910	.0164	152	154	−1.12	14.5	13.4
−25.0	3370	.0148	184	189	−1.17	16.3	15.1

THE CURVE III–V.[10]

The points on this curve were usually obtained from above, coming from the region of stability of V into that of III. The reaction does not usually run of itself at temperatures above −25°; between −25° and −18° it is usually possible to reduce the pressure on V as far as the V–L curve, where V melts completely to water without the appearance of III. Below −25°, however, III is pretty certain to separate from V on passing slightly beyond the equilibrium curve. The points were usually obtained first at low temperatures and then at higher. Proceeding in this way it is easy to get the changes of volume also, although this did necessitate the reaction running to completion, because the reactions on this curve proved particularly sensitive to the previous appearance of the phase desired. If III and V had both been present recently, the reaction would run almost immediately, in either direction, at any temperature, on passing over the transition curve.

It has been already noticed that it is possible to carry both III and V down into the region of stability of II. Corresponding to this possibility we have the realization of the prolongation of the unstable curve III–V into the region of II. A point on this curve has been found at

---

[10] For the notation V instead of IV for this new variety of ice, see p. 528.

—35°. The change in reaction velocity between III and V at different temperatures shows a greater and more striking change than that on any of the other curves. At the upper end, in the neighborhood of —20°, the velocity is explosive. It is possible to withdraw the piston

TABLE XVIII.

DATA FOR THE EQUILIBRIUM CURVE ICE III–ICE V.

Date, 1911.	Temperature, C.°.	Pressure, kgm./cm.².	ΔV. cm.³/gm. Corrected Value.
Feb. 1	—25.05	3460	
	—24.20	3460	
	—23.00	3500	
	—21.50	3500	
	—20.30	3500	
	—18.35	3500	
Feb. 16	—25.25	3510	0.0541
	—22.80	3510	547
	—20.40	3520	546
	—18.40	3550	547
April 20	—22.1	3549	
	—19.4	3559	
	—24.4	3542	
	—29.0	3527	
	—35.3	3514	
	—23.0	....	.0570

by an amount corresponding to nearly the entire volume change of the reaction, and not be able to detect any change at all in the pressure, so rapidly does the change occur. While only 15° lower, at —35° the reaction runs so slowly that the attaining of complete equilibrium is a matter of several hours. This is the reason that the attempt was not made to prolong the equilibrium curve further. It is evident that we have here something entirely different from the mechanism of ordinary chemical reactions which produces change of reaction velocity with temperature. For chemical reactions the temperature coefficient of velocity is almost universally in the neighborhood of 100 per cent for 10°, a value enormously lower than that found here.

The actual determination of the equilibrium points was made on three occasions, February 1 and 16, and April 20. The first two were with the apparatus for the middle temperature range, that used with the mercury.

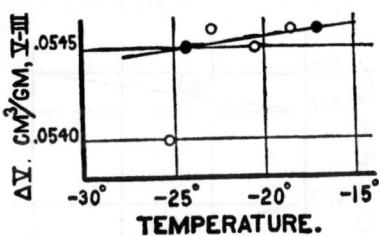

FIGURE 19. The change of volume when V passes to III.

These were made with the thermostat to keep the temperature constant. This apparatus was also used to give the changes of volume. The data of April 20 were taken with the low temperature apparatus. The particular object of this latter set was

TABLE XIX.

LATENT HEAT, ETC., ON EQUILIBRIUM CURVE, ICE III–ICE V.

Temperature, C.°.	Pressure, kgm./cm.².	ΔV. cm.³/gm.	$\frac{dp}{dt}$.	$p\Delta V$. gm. cal./gm.	ΔH. gm. cal./gm.	ΔE. gm. cal./gm.
−35.0	3470	0.05446	2.75	4.43	−0.83	3.60
−30.0	3495	5454	2.75	4.46	−0.85	3.61
−25.0	3508	5461	2.75	4.49	−0.87	3.62
−20.0	3522	5469	2.75	4.51	−0.89	3.62

to obtain as accurately as possible the slope of the transformation curve, the previous data having shown that it was nearly vertical.

The equilibrium points are shown in Table XVIII. and Figure 9. The slope of the equilibrium curve is the slope given by the data of April 20, but the absolute value of the pressure as given by the mean of the other points is more likely to be correct. The equilibrium line is straight within the limits of error. The slope is in the same direction as for the curve I–II, but is less. There is no possibility of this line heading for the absolute zero.

The changes of volume, four in number and taken with the mercury apparatus, are given in Table XVIII. and Figure 19. The change is constant so far as could be judged from the points themselves, but the best values at the triple points would demand a slight decrease of this volume with decreasing temperature. In addition to the four points listed above, there is a fifth determination on April 20 by the method described on p. 495. This is higher by 4 per cent (0.2 per cent of the original volume) than the previous point. In view of the fact that the first four points very approximately satisfy the requirements at the triple points, there seems little doubt that they are the correct ones.

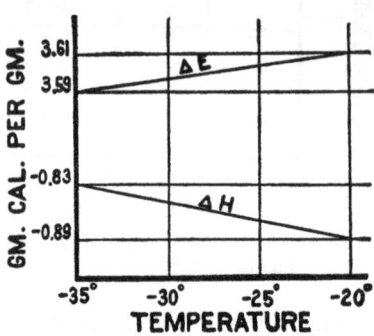

FIGURE 20. The latent heat and the change of internal energy when III passes to V.

The values of ΔH and ΔE calculated from these data are given in Table XIX. and Figure 20. III gives out heat on passing into V and absorbs work. The heat is very small, much less than the work, resulting in an increase of internal energy on passing from III to V.

THE CURVE II-V.

This curve has some of the properties of the II-III curve. It is characterized by small change of volume and by the fact that II passes into V immediately when the temperature is raised beyond the curve. The method of determining the points was the same as that of the II-III points. This was the last of all the equilibrium curves to be obtained. It was found after sufficient skill had been attained with the method, so that it was necessary to obtain the points only once. For this reason, the observation that it was impossible to heat II into the region of V was not tested by so many trials as the corresponding change II-III, and has therefore not quite so much probability of being absolutely true. The reaction velocity on this curve drops very much faster than it does on the II-III curve with decreasing temperature. This might, at lower temperatures, result in the apparent possibility of bringing II into the region of V, even if the reaction were actually running. This slowness of reaction makes very difficult the determination of the equilibrium points at the low temperatures, the determination being difficult enough anyway, because of the small change of volume. It would probably not be possible to extend this curve to very much lower temperatures than are given here. In partial explanation of the difference in the behavior of the reaction velocity with

TABLE XX.

DATA FOR THE EQUILIBRIUM CURVE ICE II–ICE V.

Date, 1911.	Temperature, C.°.	Pressure, kgm./cm.²	ΔV. cm.³/gm. Corrected.
April 19	−34.00	4200	
	−30.50	3970	
	−27.7	3800	
	−25.8	3650	
April 20		3700	0.0408
		4000	406
		4280	398

temperature on the two curves II–III and II–V, it is to be noticed that on the latter curve decreasing temperature is accompanied by increasing pressure, which of itself would tend to slow the reaction because of increasing viscosity, whereas on the former curve the retarding effect of decreasing temperature is offset by the accelerating effect of decreasing pressure.

The equilibrium points, four in number, are given in Table XX. and Figure 9. These points lie on a straight line, the temperature rising with falling pressure.

The change of volume points, obtained by the method already described for the II–III curve, are shown in Table XX. and Figure 21. The change is sensibly constant on this curve. This means that the relation between compressibility and dilatation is such that the changes of temperature and pressure on the equilibrium curve produce the same change of volume in both II and V.

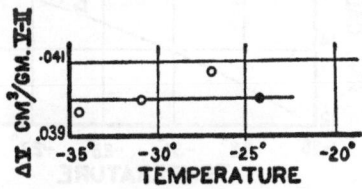

FIGURE 21. The change of volume when V passes to II.

The values of ΔH and ΔE, computed from these data, are given in Table XXI. and Figure 22. II absorbs heat on passing to V and absorbs work. The internal energy of V is greater, therefore, than that of II.

TABLE XXI.

LATENT HEAT, ETC., ON THE EQUILIBRIUM CURVE, ICE II–ICE V.

Temp. C.°	Pressure. $\frac{\text{kgm.}}{\text{cm.}^2}$	ΔV. $\frac{\text{cm.}^3}{\text{gm.}}$	dp/dt. From Curve.	dp/dt. Adjusted.	$p\Delta V$. $\frac{\text{gm. cal.}}{\text{gm.}}$	ΔH. $\frac{\text{gm. cal.}}{\text{gm.}}$	ΔE. $\frac{\text{gm. cal.}}{\text{gm.}}$
−34.0	4200	0.0401	75.6	68.4	3.95	12.4	16.4
−31.0	4010	401	69.8	68.4	3.77	13.9	17.8
−28.0	3800	401	62.4	68.4	3.57	15.1	18.7
−25.0	3570	401	55.0	68.4	3.36	15.9	19.2

THE CURVE V–L.

V was the second variety of ice found in this present work, the first being that stable at higher pressures, namely VI. It seems worth while to give here a short account of the actual order of the experiments, as it illustrates well the capricious character of these reactions. Ice V was first found at −8°. An experiment was being run on the change of volume of ice VI when passing to water. Pressure was being decreased; it had passed over the equilibrium pressure and a sufficient interval of time had elapsed to allow the pressure to automatically restore itself to the equilibrium value. The change to V occurred without warning. By good fortune, a reading at the bridge was being made at the particular instant, so that the entire

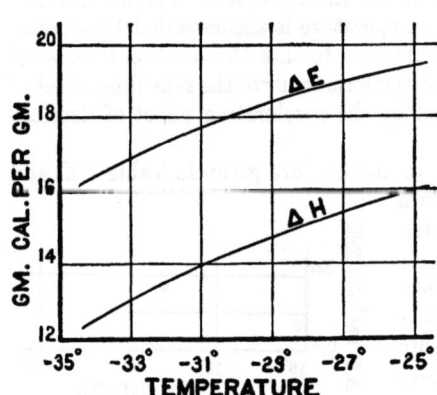

FIGURE 22. The latent heat and the change of internal energy when II passes to V.

process was observed. There was a sudden increase of pressure, followed by a slow fall past the equilibrium value VI–L to a final stationary value several hundred atmospheres lower down. This second stationary pressure had all the properties of a new equilibrium pressure, since the pressure was automatically restored to this value after withdrawal of the piston. After sufficient withdrawal of the piston, the pressure dropped in the normal way. The explanation, in view of the facts as they now appear with the complete diagram before one, is that ice VI at $-8°$ was unstable. It had partly melted when it flashed suddenly into V with increase of volume and increase of pressure. The form V now found itself in the presence of water beyond the equilibrium curve. The water froze to V with decrease of pressure back to the equilibrium curve V–L. At the time, however, this explanation was adopted with considerable hesitation. No experience had been had of the very high reaction velocity possible between solids. After this set of readings, pressure was left at 2000 kgm. over night, the temperature gradually rising.

The next morning another run was made at $-6°$. The liquid froze to VI at $-6°$ perfectly properly, with no trace of V. Pressure was increased considerably beyond the equilibrium value at $-6°$ and then decreased. On the way down a new transformation point was found at pressures higher than either the VI–L curve or the supposed V–L curve. The new reaction ran with unusual velocity, equilibrium being attained in the time usually required for temperature equilibrium. On decreasing pressure further, the VI curve was passed over without incident, but a new equilibrium pressure was found lower down on the prolongation of the probable V–L curve of the day before. The explanation suggested itself that the upper transformation point was on the V–VI curve, the lower on the V–L curve. Some slight irregularities in the data made it desirable to dismount the apparatus after this run. During these experiments, the water had been enclosed in a glass bulb, as this was one of the series of experiments to detect if possible any effect of the transmitting liquid on the water. The glass bulb was found crushed to fragments, evidently because of the reaction between two solid phases with sudden increase of volume.

In the apparatus as set up next time, the glass bulb was discarded, therefore, and the water placed directly in a shell of copper. With this apparatus no trace whatever could be found of the new phase V. Measurements were made on the equilibrium and change of volume VI–L to temperatures as low as $-20°$. Lower temperatures could not be easily attained because of the trouble with the $CaCl_2$ cooling arrangements. The subject was then dropped for a couple of months;

in the meantime the data required for the mercury paper were finished.

On taking the matter up again, the only difference that presented itself between the two runs, one giving V and the other not, was that in the former run water had been in the presence of glass. The apparatus as set up again was made to differ from the former set up, therefore, only in the fact that splinters of Jena chemical glass were placed in the water, which was again included in a copper shell. The very first attempt at $-10°$ was successful. Pressure was increased beyond the VI-L line, the reaction VI-L had started, and the equilibrium pressure had been reached, when the phase V put in an appearance, displacing VI. Briefly, the mere addition or subtraction of the fragments of glass appeared to be the determining factor. When it was desired to work on VI in the subcooled region, the glass was omitted; when V was desired, the glass was added. Even then the appearance of V was under surprising conditions. V was never formed directly out of the water, but always in the presence of the phase VI. Furthermore, V showed a preference for appearing only when VI, water, and glass were present together, usually during determinations of the equilibrium pressure VI-water, either during increasing or decreasing pressure. The facility with which V might replace VI was the source of some annoyance. Once a day's work was lost because V had successfully come in without being noticed. V might, however, form directly from the solid VI, as described above. All this holds only for the initial appearance of V, it being very much easier, as already explained, to produce it after it had appeared once. It was possible in this way to get V to separate directly from the liquid, if it had appeared recently before.

Figure 23, plotting piston displacement against pressure, shows well the way in which V might appear. With increasing pressure, the water froze regularly to VI, as shown at A. The point F shows the

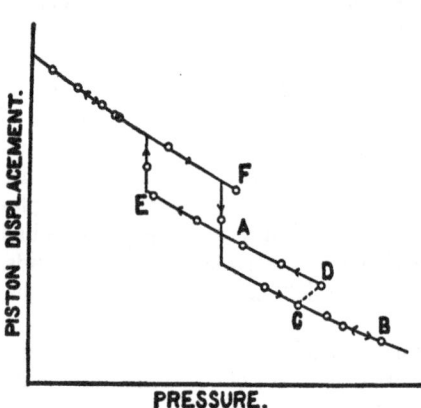

FIGURE 23. Shows the way in which V may appear. The dotted line C-D indicates the sudden transition from VI to V.

possibility of subcooling the water before the reaction starts. The pressure was then increased on the phase VI to the point B without incident.

TABLE XXII.

DATA FOR THE EQUILIBRIUM CURVE ICE V-WATER.

Date, 1910.	Temp. C.°.	Pressure, kgm./cm.²	ΔV. cm.³/gm. Corrected.
Nov. 11	−7.8	4860	0.0541
14	−6.0	5200	0512
19	−8.9	4140(?)	
1911			
Jan. 30	−10.1	4490	0702
	−18.75	3420	
	−16.65	3680	
	−11.00	4440	
Jan. 31	−14.5	3910	
	− 4.0	....	0578
Feb. 2	−20.6	3050	
3	− 2.0	5940	0552
4	−13.1	4030	0747
8	−15.45	3770	0765
	− 9.50	4600	0693
9	− 7.17	5010	0623
	− 4.3	5480	0586
	− 0.88	6120	
11	− 7.65	4890	0636
	− 4.60	5420	0587
	− 1.40	6020	0519
13	−10.8	4480	0693
	−10.5	4460	0663
	−14.2	3900	0724
14	−17.65	3460	0790
	−20.80	3040	0839
	−14.32	3780	0753
22	− 1.60	6040	0543

dent. On decreasing pressure beyond C, VI suddenly changed to V with increase of pressure to D. This change took place so far beyond the equilibrium line that the change of volume VI-V was not sufficient to raise the pressure to the equilibrium value VI-V. From D, pressure was regularly reduced to E, the melting point of V. Beyond E, the curve of increasing pressure was retraced.

These remarks about the manner of appearance of V hold only for temperatures above −25°. At lower temperatures, V may be made to

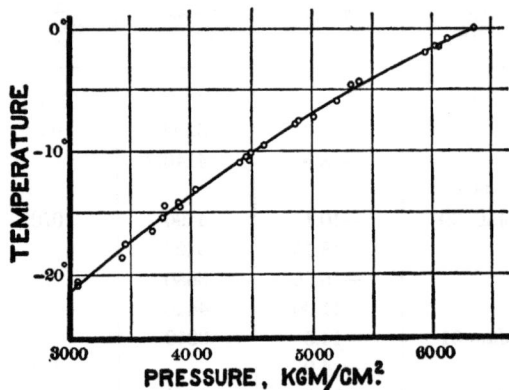

FIGURE 24. The freezing curve V-L.

separate from the solid phases III or II without the presence of glass. In any event, V was the hardest of any of the solid forms to obtain, the only certain way being via II, which demanded lowering the apparatus to below −60°.

The actual determination of the points on the V-L curve was made with the apparatus for the middle temperature range, temperature being kept constant as usual with a thermostat. The equilibrium points are shown in Table XXII. and plotted in Figure 24. As is seen, these points were obtained on a large number of occasions, with different pressure measuring coils, and with nearly all possible combinations of all the pieces of apparatus ever used with this general type of apparatus, part after part being replaced as it was destroyed by explosion. The lower part of the equilibrium line is a trifle high, still passing through a number of points, but not through the mean of all the points. This was demanded in order to obtain consistent values for the latent heat at the triple point L-III-V, the slope of the upper end of the III-L curve being extraordinarily sensitive to small changes, as already explained. The curve as given does not differ by over 0.2° from the best curve through the points for V-L alone. The curve as given extends to −21°, 4° into the region of stability of III. It is known, however, that it is possible to extend the curve as far as −25°. On one occasion a point was found here, but it is not sufficiently accurate to plot, owing to defective temperature control. The behavior at the upper end of the V–L curve is the exact opposite of that at the lower

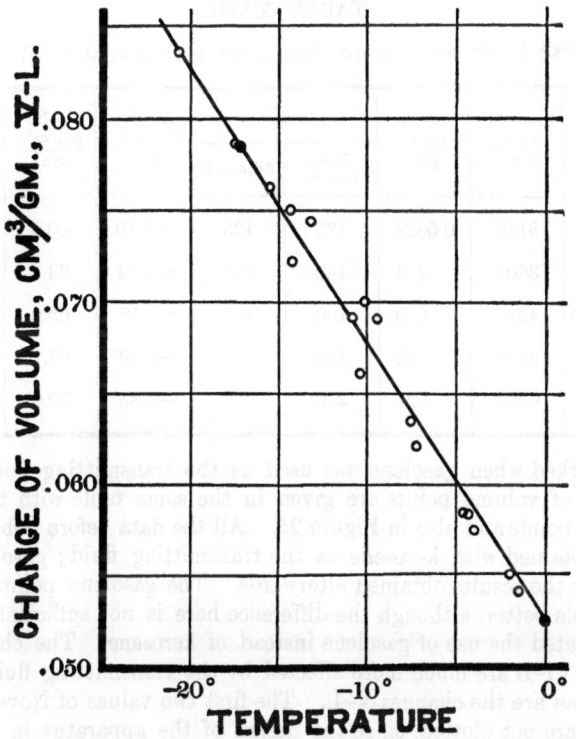

FIGURE 25. The change of volume when V passes to the liquid.

end. It was not found possible to prolong the equilibrium curve of V and L the slightest distance into the region of stability of VI, V always passing spontaneously into VI. This was tried several times; the reaction runs if the penetration into the VI region is even as slight as 1°.

More or less trouble was found in determining the change of volume. When kerosene was used as the transmitting fluid, there seems to be some action below 0°, either between the water or the ice and the kerosene. This results in a slight rounding off of the corners of the displacement-pressure curve, so that the actual discontinuity is difficult to determine. This would point to a solution of the water in the kerosene. The rounding off of the corners is not enough to affect appreciably the equilibrium pressure. The effect was on the whole

TABLE XXIII.

LATENT HEAT, ETC., ON THE EQUILIBRIUM CURVE, ICE V–WATER.

Temp. C.°.	Pressure, $\frac{kgm.}{cm.^2}$	ΔV. $\frac{cm.^3}{gm.}$	$\frac{dp}{dt}$.		$p\Delta V$. $\frac{gm.\ cal.}{gm.}$	ΔH. $\frac{gm.\ cal.}{gm.}$	ΔE. $\frac{gm.\ cal.}{gm.}$
			From Curve.	Adjusted.			
−20.0	3140	0.0828	123	123	−6.10	60.5	54.4
−15.0	3800	754	140	139	−6.72	63.4	56.7
−10.0	4510	679	160	157	−7.18	65.9	58.7
− 5.0	5440	603	184	180	−7.69	68.1	60.4
− 0.0	6360	527	210	208	−7.85	70.0	62.2

less marked when gasolene was used as the transmitting fluid. The change of volume points are given in the same table with the equilibrium points and also in Figure 25. All the data before February 10 were obtained with kerosene as the transmitting fluid; gasolene was used for the results obtained afterwards. The gasolene points are on the whole better, although the difference here is not sufficient to have necessitated the use of gasolene instead of kerosene. The changes of volume VI–L are much more affected by the transmitting fluid, below zero, than are the changes V–L. The first two values of November 11 and 14 are not plotted, since the design of the apparatus in the first few attempts was not good. Small portions of the water were likely to separate from the rest after a single freezing, so that unless the pressure was carried considerably beyond the freezing pressure, part of the water was likely to remain liquid, giving too low an apparent change of volume. The lowness of these first two points is to be explained in this way.

Within the limits of error, the relation between temperature and change of volume is linear. The lower end of the curve is about 1 per cent higher than what would be demanded by the points themselves. This was made necessary by the conditions at the critical point.

The values of ΔH and ΔE found from these data are given in Table XXIII. and Figure 26. Water gives out heat on passing into V, and absorbs work. The work is very much less than the heat, so that the internal energy of V is less than that of the liquid.

The Curve V–VI.

This is the last of the curves of equilibrium between two solid phases. The manner in which the curve was first found has been already described. The characteristics of this curve are the same as those of the other solid-solid curves; very high reaction velocity at the upper end, with enormously re-

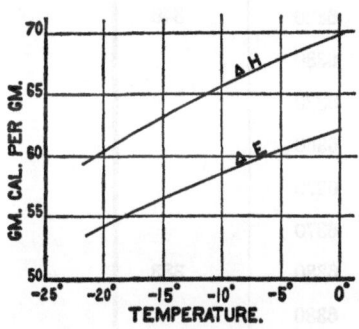

Figure 26. The latent heat and the change of internal energy when V passes to the liquid.

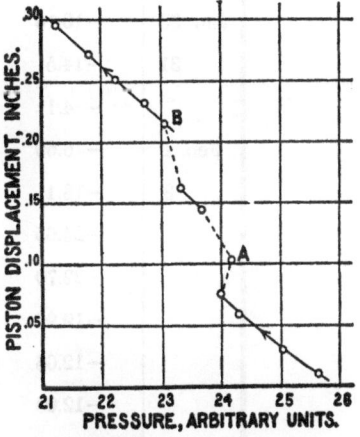

Figure 27. False equilibrium at low temperatures on the V–VI curve, due to the slowness of the reaction and the viscosity of the solids.

tarded reaction velocity at the lower temperatures. The slowing of the reaction was so great as to make impracticable the determination of points lower than $-25°$. Anomalous results obtained once or twice are to be explained by the slowness of the reaction velocity. The nature of the anomaly is shown in Figure 27, plotting piston displacement against pressure. It is as if there were two discontinuities, at A and B, and so two transition points. The explanation is probably as follows. The reaction from VI to V starts at the upper end of the steel shell. This reaction runs with increase of volume, so that the layer of the new ice V may form a protecting layer over the remainder of the old ice VI below. This protecting layer is capable of standing some appreciable stress at this high pressure and low temperature, so that changes of hydrostatic pressure do not reach the nest of ice VI below, until the pressure has been reduced by an amount corresponding to the strength of the ice V. An exactly similar effect was

TABLE XXIV.

Data for the Equilibrium Curve, Ice V–Ice VI.

Date, 1911.	Temp. C.°.	Pressure, kgm./cm.².	ΔV. cm.³/gm. Corrected.
Jan. 30	−10.10		0.0396
31	−14.5	6360	378
	− 4.1		400
Feb. 1	−16.60	6320	346
3	−18.1	6380	
	−24.65	6320	
	−22.70	6360	
	−19.90	6370	
	−12.05	6370	
	−12.08	6380	383
	− 2.00	6380	
4	−13.1	6360	396
20	−19.7	6360	386
	−16.8	6370	384
	−13.9	6370	384
21	−11.25	6370	386
22	− 8.15	6390	380
	− 5.00	6380	387
	− 1.60	6380	386

found once on the III-V curve at −25°. That the ice in passing from one solid to another is capable of exerting considerable stress is shown by the occasional rupture of the steel shells containing the water.

These were 9/16 inch o. d. and about 3/64 inch thickness of wall. In fact, it seems rather surprising that the effect did not prove troublesome on more than these two occasions.

The regular apparatus for the middle temperature range was used in finding these points. Some skill in manipulation was necessary to arrive on the curve at all. The procedure was to start with the water impregnated with glass splinters and raise the pressure at a temperature somewhere between $-10°$ and $-20°$. On sufficiently overpassing the unstable VI-L curve, VI would separate out. After some little while V would appear. With the first signs of its appearance, indicated by a rise of pressure, the pressure was increased as rapidly as possible, in a few seconds, to the V-VI curve, where the automatic change of volume was sufficient to ensure retaining of the equilibrium pressure after freezing of the remaining mass of liquid. This method of procedure was made necessary by the fact that at the temperatures at which it is easy to obtain V, it is possible to pass from either side of the V-VI curve into the region of instability on the other side, without the reaction running. At higher temperatures, nearer the triple point, it is not possible to carry V so far into the region of VI, corresponding to the fact already noted that at the triple point itself it is impossible to carry V any distance at all into the region of VI.

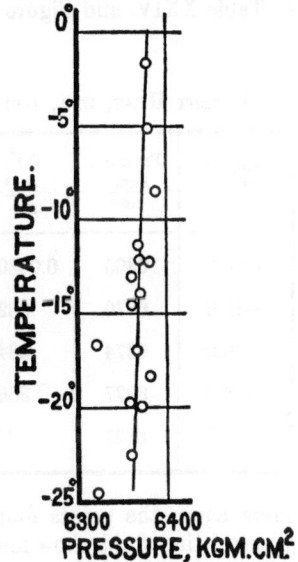

FIGURE 28. The equilibrium curve V-VI.

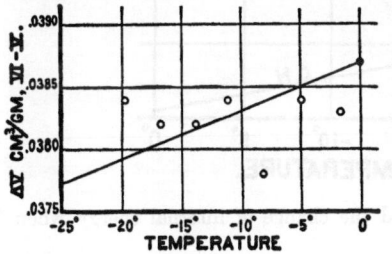

FIGURE 29. The change of volume when VI passes to V.

The equilibrium pressures and temperatures are shown in Table XXIV. and Figure 28. The equilibrium line is nearly at constant pressure, but there is an unmistakable though slight increase of pressure at the higher temperatures. Within the limits of error, the equilibrium curve is a straight line.

The change of volume points, obtained by the usual method of observing the piston displacement at constant temperature, are given in Table XXIV. and Figure 29, plotted on a very much enlarged scale.

TABLE XXV.

LATENT HEAT, ETC., FOR THE EQUILIBRIUM CURVE, ICE V–ICE VI.

Temperature, C.°.	Pressure, $\frac{\text{kgm.}}{\text{cm.}^2}$	ΔV. $\frac{\text{cm.}^3}{\text{gm.}}$	$\frac{dp}{dt}$.	$p\Delta V$. $\frac{\text{gm. cal.}}{\text{gm.}}$	ΔH. $\frac{\text{gm. cal.}}{\text{gm.}}$	ΔE. $\frac{\text{gm. cal.}}{\text{gm.}}$
−20.0	6365	0.03809	.80	5.682	−0.181	5.50
−15.0	6370	3828	.80	5.716	−0.185	5.53
−10.0	6374	3847	.80	5.748	−0.190	5.55
− 5.0	6377	3866	.80	5.779	−0.195	5.58
0.0	6381	3886	.80	5.812	−0.199	5.61

Here again the points found before February 10 with kerosene as the transmitting fluid show much less regularity than the points obtained afterward with gasolene. The kerosene points have been omitted from

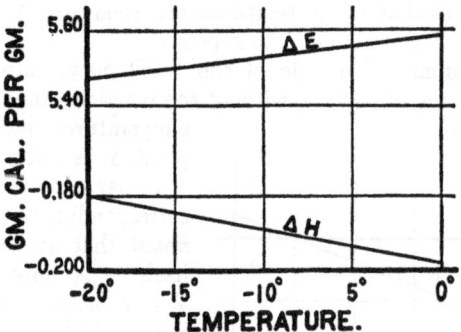

FIGURE 30. The latent heat and the change of internal energy when V passes to VI.

the diagram. The effect may be due in part to the increased viscosity of the kerosene under pressure. Within the limits of error, the relation between V and temperature is linear, the change becoming less at lower temperatures as one would expect.

The missing triple point, mentioned at the beginning, is evidently given by the intersection of this line, V–VI, with the line II–V. This point is apparently situated at about −65° and 6300 kgm. Evidently the reaction velocity will be so slow at these temperatures and pressures that it would be hopeless to try to reach it. The missing equilibrium curve starting from this point is evidently II–VI. There is even less chance of realizing this than of getting down to the triple point, always assuming of course that some new variety of ice does not appear.

The values of the latent heat and the change of energy calculated in the usual way from the data are given in Table XXV. and Figure 30. The latent heat, owing to the approximate perpendicularity of the equilibrium curve, is almost vanishingly small. V gives out heat on passing to VI, but absorbs work. The heat is almost negligible compared with the work, so that the internal energy of VI is greater than that of V by nearly the amount of this work.

The Curve VI–L.

This modification of ice was the first one discovered in the present work, and has engaged more attention than any of the other forms. It has been studied over a pressure range of more than 16,000 kgm., nearly three times the pressure range of all the other modifications put together. So far as can be judged, this is the final form of the solid. This ice shows characteristics in its behavior different from any of the others, characteristics which are particularly significant for the theory of the solid-liquid states. In fact, this is the only modification stable over a pressure range wide enough to indicate what may be expected at still higher pressures, as yet unreached.

The manner in which this variety of ice was first found in the course of the measurements of compressibility has been already described in the introduction. The appearance of this ice resulted simply in an irregular disturbance at the upper end of the compressibility curve. It was evidently impossible to make any accurate measurements by this method of the freezing pressure or of the change of volume. All that could be stated about the freezing pressure was that it was less than a certain value, the maximum pressure reached during a run when there was a disturbance, and all that was known about the change of volume was that it was higher than a certain value, the discontinuity in the compressibility curve. The lower and upper limits so set on these quantities were found to be consistent with the more accurate values found later.

At the time that these disturbances were found no great confidence was felt in the explanation that the effect was really due to the freezing of the water. All of the compressibility measurements were made in the presence of mercury. Now it had been already found that mercury freezes under pressure, and at pressures very close to those actually found. At 0°, for instance, the freezing pressures of water and mercury, according to the final accurate determinations, are about 6400 and 7600 kgm. respectively, and these two transitions seemed much closer together in the preliminary work because it was necessary to superpress the water considerably before freezing begins, while the superpressure required for mercury is very small. It is true that there were very great quantitative difficulties in the way of supposing the effect due to the freezing of the mercury, but still the neighborhood of the mercury freezing point produced a sensation of disquietude.

In order to definitely rule out the possibility of complications from the neighborhood of the freezing point of mercury, a form of experiment was devised in which there was no mercury present. This was the experiment also briefly alluded to in the introduction on the electrolytic conductivity of water. The water was placed in a shell of steel or glass, and the conductivity measured between two concentric brass cylinders attached to the insulating plug and suspended in the water. The pressure chamber was filled with kerosene by which pressure was transmitted directly to the water. There was no mercury anywhere about it. The water was not absolutely pure, ordinary tap water being used, which was sufficiently conducting for the purpose. In one case distilled water with 1/10 per cent HCl was used. The resistance was measured with a Kohlrausch bridge and a telephone in the usual way. At high pressures there was unmistakable evidence of some sort of change. The electrolytic resistance at first decreased with rising pressure, passed through a flat minimum, rose slowly, and then very rapidly to several times the former value, the rapidity being such as to be almost equivalent to a discontinuity. This pressure of sudden rise was taken as the freezing pressure, since it would be natural to suspect the solid incapable of conducting. The pressure of transition was measured in this way at several temperatures. In all, three sets of readings were made with different forms of apparatus, and the last one six months after the first. In the last determination, the existence of the effect independent of the mercury having been already established, the water was enclosed in glass with a mercury seal, the more effectively to prevent absorption of impurities from the kerosene. The three sets of determinations gave perfectly consistent results. A brief statement as to the existence of a new variety of ice as proved by these experiments was published in the Physical Review, vol. 31, 1910, p. 606. The values as found by these measurements lie on a straight line, and agree with the possible limits found with the compressibility measurements, except that at the highest temperature, 22°, freezing had once been found to take place during the compressibility measurements at a pressure lower than that found here. This was disconcerting.

The transition curve as found in this way did not give all the information needed for a complete statement of all the quantities involved in the change. A complete statement thermodynamically could be made if it were possible to measure also the change of volume during the change. This seemed at first sight of doubtful possibility. Tammann had made such measurements up to 2000 kgm., but had not been able to go much higher because of leak. As it was, even at 2000, a correction had to be applied for the leak, which was furthermore made as small as possible by the use of a heavy oil, such as castor oil. In the present work it was necessary to go to many times this pressure, the equilibrium pressure even at 0° being 6400 kgm. The difficulty was furthermore increased by the necessity for using some very light and mobile oil to transmit pressure, the castor oil freezing absolutely solid under such pressures as were to be used here. It was felt, therefore, to be a real step when it was found that the packing used in the previous high pressure work was entirely satisfactory when used for this purpose. It is possible with it to retain liquids as mobile as ether, CS_2, pentane, or mercury, absolutely without leak, up to the highest pressures attainable. Measurements of the change of volume have been made up to 15,500 kgm., and this limit was set by yield of the steel cylinders and not by failure of the packing. The readings of the piston obtained were consistent to a few ten thousandths of an inch. The only troublesome effect found with the packing was a slight wearing away by abrasion, owing to the enormous friction. This effect was troublesome only over a wide pressure range; for the limited motion of the piston during freezing or melting the effect was negligible.

It was also a part of the preliminary work to show that the freezing

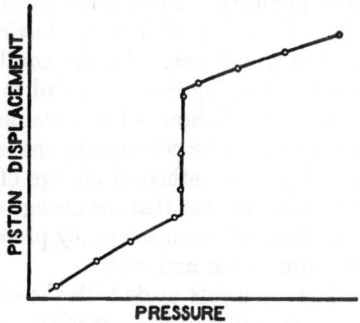

FIGURE 31. Shows the sharp change of volume during freezing.

is sufficiently sharp to give trustworthy measurements of the change of volume. The presence of impurity in the water will be shown by a rounding off of the corners of the melting curve and by a change in the value of the equilibrium pressure with the relative proportions of liquid and solid present. Figure 31, chosen at random from the earlier measurements, shows the sharpness of freezing and the consequent freedom from impurity. In the majority of subsequent measurements, the question of absence of impurity having been settled, it was customary to find only one value for the equilibrium pressure, corresponding to a mixture varying from one third to two thirds liquid.

The equilibrium points obtained by the change of volume method were found to be consistently lower than the supposed transition points found by the method of change of resistance. There can be no question whatever but that the change of volume points are the correct ones. The cause of the discrepancy probably is the formation of the eutectic mixture of ice and salt.

Measurements made with the electrolytic conductivity method after the completion of the rest of the experiments, with the new apparatus giving also the piston displacements, showed that the flat minimum in the conductivity curve previously referred to occurs at nearly the pressure of the beginning of freezing. Evidently at first pure ice separates out, leaving the remaining liquid richer in whatever salt produces the conductivity. The total quantity of this salt remains nearly constant, but its concentration and so its association are increasing, so that on the whole we have a decrease in the conducting power. When the pressure has been carried so far that the concentration reaches the eutectic value, the solid salt separates out with the pure ice and we have the sudden jump in the resistance. That this was the case was also suggested by a fact observed once or twice when using water in glass with a mercury seal. This was that the sudden discontinuity might be either a short circuit or an open circuit. Evidently the short circuit is due to the penetration of the mercury through the mass of ice along the core left by the freezing of the pure ice to the walls of the tube.

The points on the VI–L curve were obtained with two different pieces of apparatus. The points between −20° and 20° were found with the apparatus already described as used for the middle temperature range. For the higher pressures, up to 20,500 kgm., another form was used. The particular point of weakness in the middle temperature apparatus is the connecting tube. The new form for the highest pressures was made in one piece, therefore. This has been already described as the third piece of apparatus. It is to be noticed that with this there is no correction necessary for the thermal dilatation of the transmitting fluid. These measurements to the highest pressures were made after the completion of all the others. It was confidently expected to find a new variety of ice, but none was found.

The pressure measurements with the high pressure apparatus were

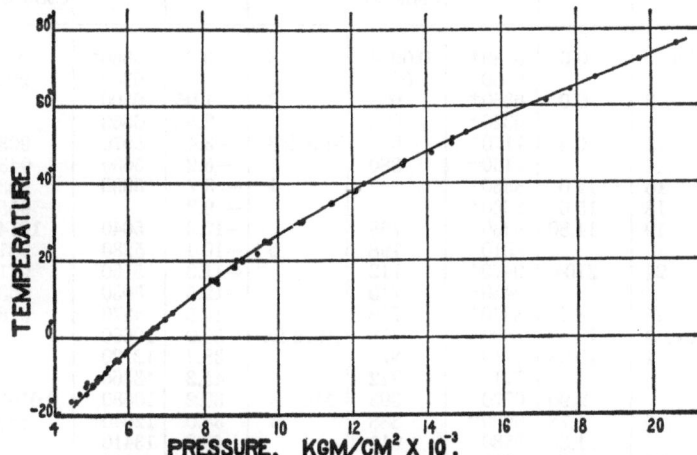

FIGURE 32. The freezing-curve VI-L.

made in the usual way by measuring the resistance of a coil of manganin wire. This had been calibrated by comparison with an absolute gauge up to 13,000 kgm., which was about the limit of the gauge. The pressures above this were obtained by a linear extrapolation, therefore. There does not seem much danger in doing this in view of the fact that the relation is linear within 1/10 per cent up to 13,000.

For the middle temperature range the high pressure apparatus offers the advantage of fewer connections over that actually used. There are practical difficulties in the way of its use below 0°, however. It is necessary to submerge the entire lower end of the hydraulic press with the cylinder into the constant temperature bath. There is considerable flow of heat out of the bath along the heavy connecting bars of the press. It would for this reason have been difficult and expensive to have maintained temperatures below zero with the $CaCl_2$ solution. Points have, however, been obtained with the high pressure apparatus over the entire range above zero.

The equilibrium points are given in Table XXVI. and Figure 32. A greater variety of groupings of apparatus have been used for the points

TABLE XXVI.

Data for the Equilibrium Curve Ice VI–Water.

Date, 1910.	Temp. C.°.	Pressure, kgm./cm.²	ΔV. cm.³/gm. Corrected.	Date, 1911.	Temp. C.°.	Pressure, kgm. cm.².	ΔV. cm.³/gm. Corrected.
Oct. 7	0.0	6380+	0.0914	8	0.0	6360+	916
		6360−	0915			6360−	906
8	0.0	6360+	908	9	1.05	6500	
		6340−	908		2.55	6660	
12	6.4	7170+	876	March 3	−3.4	5970	908
		7150−	880		−6.2	5630	943
15	14.0	8380+		4	− 3.2	5990	923
18	15.0	8360+			− 9.2		955
19	14.80	8330+	798		−12.2	5040	1014
		8310	798	6	−10.4	5280	981
24	2194	9470+	712		−11.3	5160	971
		9450−	710		−12.5	5030	963
25	18.4	8840+	776		−14.5	4870	978
Nov. 3	18.42	8800−	758	April 28	30.0	10660	
4	10.63	7710+	878		38.2	12120	
		7710−	842		45.2	13360	
	2.90	6770−	903	May 8	30.2	10680	0.0593
5	4.78	6960−	888	9	38.0	12040	556
	1.65	6580	911		45.8	13410	
11	0.0	6360	916		53.4	15050	
	− 7.8	5470		13	38.4	12120	618
14	− 4.0	5910			48.0	14140	541
	− 4.2	5910	938		55.5	15670	566
17	−11.7	4890	853		61.6	17200	
18	−16.5	4440	900		67.5	18500	
	−19.6	4100	936	25	35.0	11450	647
19	0.0	6370	914		30.0	10590	671
	− 8.9	5390	994		25.0	9790	713
	− 6.0	5630±	931+	26	25.0	9680	705
			959−		20.5	9050	758
21	−12.2	5080+	1035		15.0	8260	795
		5050−	1004		0.0	6350	928
22	−10.9	5160+	959	29	0.0	6360	902
		5120−	916		15.1	8180	797
Jan. 28 1911.	− 5.7	5660+	997		20.1	8890	754
		5690−	914	30	25.1	9650	707
30	−10.1	5200	1044		30.1	10450	676
	− 6.0	5750+		31	35.0	11470	629
		5740−			40.0	12330	590
31	−14.5	4680			45.0	13380	512
Feb. 1	−16.60	4590		June 1	45.0	13380	563
2	−12.75	4850		13	51.8	14700	516
7	0.0	6340+	932		64.30	17840	
		6350−	895		72.15	19670	
					76.35	20670	

on this curve than for any other. It was on this curve, for instance, that the test was made of the possibility of interaction between water and the substances in contact with it. The water was used in steel or copper or glass vessels, in contact with kerosene or gasolene or mercury. No difference in the equilibrium pressures could be found with any of

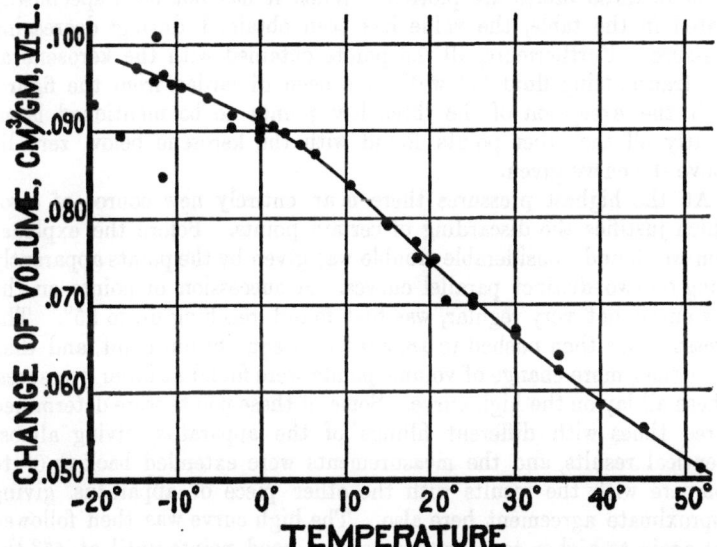

FIGURE 33. The change of volume when VI passes to the liquid.

these different forms of apparatus, but below zero it did seem desirable to use gasolene instead of kerosene. Figure 31 shows piston displacement against pressure for one of the points on this line. The corners of the curve were seldom more rounded than here. The pressure measurements on this curve were made with six different manganin coils.

Figure 32 shows that the equilibrium curve runs perfectly smoothly, without incident of any kind, from the lowest temperature of the subcooled region, −17°, to the highest temperature and pressure, 76° and 20,500 kgm. The curve is convex upward, like all the other curves of equilibrium between the liquid and the other solid forms of water. Attention is called to the five bad points lying off the equilibrium curve at the lower end.

The change of volume is shown in the same table with the equilibrium points, and graphically in Figure 33. All the values that were ever obtained have been given in the table, but not all these are plotted. In the earlier work, ΔV was found both during freezing and melting, that is, with increasing and decreasing pressure. In the paper on mercury a detailed discussion is given, showing why the values found during decreasing pressure are very likely to be in error. None of these values are plotted. When it has not been specifically stated in the table, the value has been obtained during decreasing pressure. Furthermore, all the points obtained with the kerosene as the transmitting fluid below 0° have been discarded from the figure, with the exception of the three low points to be mentioned later. Nearly all the other points found with the kerosene below zero lie above the curve given.

At the highest pressures there is an entirely new source of error which justifies the discarding of certain points. Before the explanation was found, considerable trouble was given by the points apparently lying on two distinct parallel curves. A succession of points on the low curve, not very regular, was first found, reaching up to 55°. The pressure was then pushed to 18,500 for an equilibrium point, and then after that, more change of volume points were found at lower pressures. These all lay on the high curve. Some of these points were determined three times with different fillings of the apparatus, giving almost identical results, and the measurements were extended back to 0° to compare with the results with the other piece of apparatus, giving approximate agreement here also. The high curve was then followed out again to higher temperatures, giving good points until at 45° the value jumped again to the low curve. The possibility of a new kind of ice was ruled out by the regularity of equilibrium points. Explanation was found in the viscous yield of the steel. This has the curious property of going on at a uniform rate during the time occupied by an experiment, not being asymptotic as one might expect. This yield had been carefully looked for at the lower pressures and not found. During the long course of the experiments so much practice had been obtained that during these last measurements the readings of pressure and piston displacement were made with almost clock-like regularity every six minutes. The yield was so slow as to be imperceptible during that interval of time, and the readings were made with such regularity that the effect was not discovered, as it otherwise would have been, by irregular points on the curve. Once the effect was discovered, it was found entirely competent to account for all observed discrepancies, since the melting occupies a fairly long time, between one and two hours. Another point at 45° was then found, applying a correction for the yield by noting its rate during the half hour before and after change of state. This corrected point lay consistently with the others on the high curve. Evidently the succession of high points found at the same pressures, as well as the previous low points, is to be explained by the seasoning effect of the maximum pressure of 18,500, which was applied between the two sets. The last low point is due to the passing to the viscous

TABLE XXVII.

Latent Heat, etc., on Equilibrium Curve Ice VI-Water.

Temp. C.°.	Pressure, kgm./cm.².	ΔV. cm.³/gm.	dp/dt.		$p\Delta V$. gm. cal. gm.	ΔH. gm. cal. gm.	ΔE. gm. cal. gm.
			From Curve.	Adjusted.			
−15.0	4790	0.0980	100.5	99.6	10.99	59.0	48.0
−10.0	5280	960	103.2	106.5	11.87	63.0	51.1
− 5.0	5810	938	112.4	113.8	12.77	67.1	54.3
0.0	6360	916	118.8	120.0	13.66	70.4	56.7
+ 5.0	7000	884	125.1	125.8	14.51	72.5	58.0
10.0	7640	844	132.0	132.7	15.13	74.4	59.3
15.0	8310	798	140.0	140.0	15.56	75.5	59.9
20.0	9000	751	148.3	148.5	15.84	76.6	60.8
30.0	10590	663	167.3	167.3	16.45	78.8	62.3
40.0	12390	590	189.0	188.7	17.14	81.7	64.6
50.0	14430	523	213.7	215.4	17.69	85.3	67.6
60.0	16690	477	243.0	242.9	18.68	90.5	71.8

state again, after the seasoning effect had disappeared in time or had been overcome by too high a pressure. After the point at 45° another point was found at 52°, applying the correction for yield, and found to lie on the high curve consistently with the others. It did not seem worth while to attempt points at still higher pressures, because the correction becomes more difficult to apply at the high pressures, both because the yield is becoming more rapid, and because the yield does not remain uniform in time at the highest pressures.

The behavior of this cylinder is interesting from the point of view of elasticity. It shows the impossibility of obtaining perfect elasticity very much beyond the original elastic limit. This cylinder had been previously seasoned for use in this work by subjecting it for some hours to 28,000 kgm. The hole was then bored out from the original 7/16 inch to 1/2 inch. The 28,000 kgm. had produced a stretching at the inside from 7/16 inch to perhaps 15/32 inch. Even this high pressure did not completely do away with viscous yield at pressures as low as 10,500. Another interesting feature is the approximate constancy of the rate of yield over a pressure range of 5000 kgm., from 10,500 to 15,500, as shown by the fact that the low points are nearly all the same distance below the corresponding high points. This, too, bears out an observation made before, that elastic after-effects, hysteresis, yield, etc., are greater in larger masses of metal. This cylinder was 4 1/2 inches o. d. and 1/2 inch i. d. No such viscous yield was found in the cylinder with which the change of volume of mercury was measured up to 11,000. This cylinder was 3 inches o. d. and 9/16 inch i. d., and had been previously subjected to 24,000 kgm.

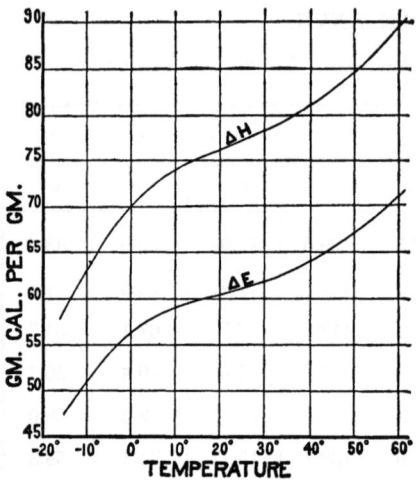

FIGURE 34. The latent heat and the change of internal energy when VI passes to the liquid.

In view of this yield, therefore, the values at the higher pressures which were evidently affected by this yield have not been included in the figure. The curve connecting ΔV with the temperature has a point of inflection, which is not shown by any of the other curves. This is of significance, as will be seen.

The values of ΔH and ΔE found in the ordinary way are shown in Table XXVII. and Figure 34. The point of inflection in both these curves is the particularly interesting feature. Water on passing to VI gives out heat and absorbs work. The work is less than the heat, so that the internal energy of VI is less than that of the liquid.

The data obtained with the high-pressure apparatus may be used to determine one other quantity which the other apparatus could not give, that is, the difference of compressibility between solid and liquid. This may evidently be found from the difference of slope above and below the melting point of the curves connecting piston displacement with pressure. If everything had been ideally perfect the other apparatus should have given it too, but the quantity to be determined is

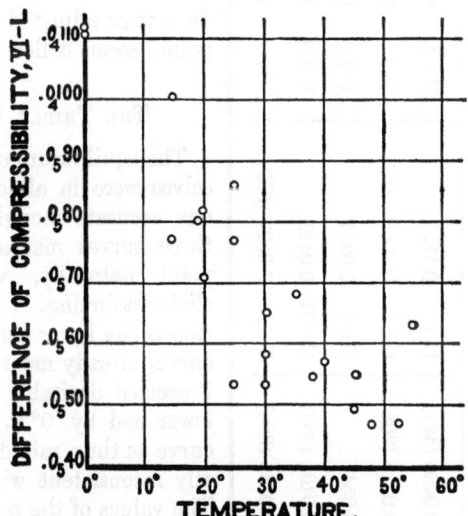

FIGURE 35. Showing the directly determined values for the difference of compressibility between VI and the liquid at various equilibrium temperatures.

very small, and the irregularities introduced by the transmitting liquid passing from one vessel to another at a different temperature, which might not be perfectly constant, were so great as not to admit of any results of value. With the high-pressure cylinder, everything is in one piece, and there are fewer complications. The values obtained even with this are very irregular, but they are given as having some value in pointing out the direction of the effect at high pressures. The points are shown with sufficient accuracy in Figure 35. This shows a very rapid decrease in the difference of compressibility at high temperatures and pressures.

This completes the presentation of the actual data. The points were determined separately, those on one equilibrium curve being independent of those on another. But the curves drawn through the points, particularly those for the latent heat and the change of volume, have not been in all cases the best curves through the points of the

TABLE XXVIII.

THE CO-ORDINATES AND CHANGE OF VOLUME AT THE TRIPLE POINTS.

Point.	Temp. C.°.	Pressure, kgm./cm.².	ΔV. cm.³/gm.			
III-L-I	−22.0	2115	III-L	0.0466 (461)	III-I	0.1818 (.1823)
II-III-I	−34.7	2170	II-III	0.0215 (208)	II-I	0.2178
V-III-L	−17.0	3530	V-III	0.0547	V-L	0.0788
V-II-III	−24.3	3510	V-II	0.0401 (408)	V-III	0.0546
VI-V-L	+ 0.16	6380	VI-V	0.0389 (388)	VI-L	0.0916
			V-L	0.0527 (522)		

single equilibrium curves in question, but have been influenced by the behavior of the two neighboring curves meeting at the triple point. Some discussion of the way in which these data were adjusted at the triple points seems called for.

THE TRIPLE POINTS.

The equilibrium curves themselves were in all cases, except one, accurate enough so that the three curves met at the triple point naturally, without the slightest forcing. The one exception is the lower end of the V-L curve, already mentioned, where it seemed desirable to raise the lower end by 0°.2. Even this curve as thus raised is not actually inconsistent with the data. The values of the co-ordinates of the triple points are shown in Table XXVIII.

In selecting the best values for the change of volume, the greatest weight was attached to the curve which was evidently most self-consistent. The values on the other two curves were then so adjusted as to give consistent results with the least violence possible. The full details of the process are shown in the table, which gives the final values, and in parentheses the values found from the individual curves before this process of adjustment. The signs are so chosen that the change of volume is an increase when the reaction runs in the direction in which the symbols read. At every point, the sum of the two smaller changes is equal to the larger.

An adjustment was necessary in seven out of thirteen possibilities. The average adjustment was about 5 parts in 10,000; the maximum, 7 parts in 10,000 of the original volume. The points on the III–L curve were so difficult to obtain, as already explained, that the entire course of this curve was in large measure determined by the behavior at the two triple points at either end.

With regard to the values for the latent heat and the change of internal energy, there was room for more difference of opinion as to the best way of making the adjustment. The latent heat, given by the formula $\Delta H = T \Delta v \frac{dp}{dt}$, is seen to depend on the change of volume, already fixed, and on the slope of the equilibrium curve. The slope was determined graphically from the equilibrium curves. There is chance for considerable error here, and the possibility of introducing comparatively large changes into ΔH by a very slight raising or lowering of the equilibrium curve, too slight in all cases except that of III–L mentioned to introduce a perceptible change in the co-ordinates of the triple point. On the equilibrium curves solid-solid, which run approximately vertically, the latent heat is usually so small as to have very little effect in the adjustment. It was possible to determine this latent heat with a fair degree of accuracy, however, and so this was left without adjustment, the other values being adjusted so as to be consistent with it.

Evidently when the values of the latent heat are properly adjusted, the values of the internal energy will also be consistent, because these differ from the latent heat only by a quantity involving the change of volume, which is already consistent at the triple point.

The tables already given, under the equilibrium curves separately, enable an estimate to be formed of the amount of adjustment necessary. Two values are listed both for $\frac{dp}{dt}$ and ΔH. The first value of $\frac{dp}{dt}$ was obtained directly by graphical construction from the equilibrium curves, and from this the first value of ΔH was calculated directly, point by point. The final value given for ΔH is the value taken from smooth curves so drawn as to check at the triple points. From this final value of ΔH the final value of $\frac{dp}{dt}$ was calculated, and listed in the column of final values. Table XXIX. gives the values of ΔH at the triple points. The III–L curve is the only one that it was necessary to juggle with in any way, and this has already been explained. The symbols are so chosen that heat is absorbed when the reaction runs in the direction in which the symbols read. Thus, the heat I–L is given as 79.8. This means that heat is absorbed when I passes to water.

The changes of internal energy at the triple points are also given in Table XXIX. The positive sign means that the reaction runs in the direction indicated with increase of internal energy.

TABLE XXIX.

LATENT HEAT AND INTERNAL ENERGY AT THE TRIPLE POINTS.

Point.	ΔH gm. cal./gm. (in the first lines). ΔE gm. cal./gm. (in the second lines).						
III–L–I	III–L	{ 50.9 48.6	L–I	{ −56.1 −62.8	III–I	{ − 5.2 −14.2
II–III–I	II–III	{ 12.3 11.2	III–I	{ − 2.2 −12.2	II–I	{ 10.1 − 1.0
V–III–L	V–III	{ 0.9 − 3.6	III–L	{ 61.4 59.5	V–L	{ 62.3 55.9
V–II–III	V–II	{ −16.0 −19.3	II–III	{ 16.9 15.7	V–III	{ 0.9 − 3.6
VI–V–L	VI–V	{ 0.2 − 5.6	V–L	{ 70.1 62.4	VI–L	{ 70.3 56.8

There are several considerations of interest in connection with the diagram that are not directly concerned with the data as given. These are the questions as to our knowledge that these forms are actually solid, as to the possibility of other forms of ice, the possibility of existence of any form out of its region of stability, and the question of reaction velocity.

ARE THE VARIOUS FORMS REALLY SOLID?

This is the first question that naturally presents itself as to this work. The experimental evidence so far has merely been that there is a discontinuous change of volume; the new modifications have not been seen; it is impossible to take the pressure off and examine them, because they are unstable; why then are we justified in assuming that they are solids and not liquids? No direct evidence on this point has been collected in this work; the improbability of their being anything except solids seems so great as to make needless the special arrangement of apparatus necessary to give a direct proof. This was doubly needless, since Tammann has already given direct proof that II and III are solid.

The most striking evidence of Tammann is for III. Tammann cooled a steel cylinder containing ice III to the temperature of liquid air, then released the pressure and took ice III out and examined it. The temperature was so low that the reaction from III to I did not run immediately. III was solid and gradually changed to ice I with large increase of volume. In another experiment Tammann placed an electric contact maker in the water. Freezing to a solid was shown by the refusal of the contact maker to work, both for III and II.

That V and VI are also solid is made probable by two bits of evidence. In the first place, the reactions between V and III and V and VI are exactly similar to those between I and III and I and II. There is no mistaking the difference between a reaction solid-solid and liquid-solid. In the second place, when the reaction VI–V runs with increase of volume, the containing vessel, whether a glass bulb or a fairly heavy metal cylinder, may easily be destroyed. In this reaction V has the larger volume. It, then, must be solid, and probably VI also.

The possible crystalline forms of the different modifications seem to be impossible of direct observation. It would be possible to insert windows in the apparatus and look at all the forms, but this would probably give no information whatever. On several occasions the apparatus was opened and the cylinder of ice I, which had frozen under pressure, was removed. This was always a perfectly structureless, translucent mass, capable of giving no information whatever as to any of its crystalline properties.

OTHER POSSIBLE FORMS OF ICE.

In a recent paper Tammann[11] has stated the probable existence of a fourth variety of ice, very much like ordinary ice in its properties. The evidence for this was very meagre. Slight discrepancies found on the equilibrium curve I–L could be explained by it; also a momentary rise of pressure on one occasion after the fall of pressure proper to melting had begun. Seven successive attempts of Tammann to get the same effect again failed. It is evident that the discrepancies on the I–L curve might be due to pressure errors; a new manometer used by Tammann gave pressures 50 kgm. lower than the former one, and we have already seen that the crossing of the I–III and the I–II curves found by Tammann must be set down to pressure errors. Finally, Tammann states that ice IV may separate out from water when cooled

---

[11] Tammann, ZS. Phys. Chem., 72, 609–631 (1910).

to −7°. The fact that there is a different modification is shown by placing the ice so formed in a dilatometer, and warming slowly. At −2° there is a sudden increase of volume of about 1/10 per cent, followed by regular increase of volume and melting at 0°. The ice IV appears, therefore, to have a slightly less volume than ice I. In this experiment, sufficient attention does not seem to have been paid to the possibility of there being internal strains in the ice. It seems plausible that the water, freezing suddenly as it must at −7°, might so freeze as to produce a slight volume compression. On increasing temperature, this strain is relieved by the softening of the ice in the neighborhood of the melting point. This weakening of the ice near the melting point has been established by experiment. At high pressures it is undoubtedly true that internal strains may produce anomalous effects, as was found on at least two occasions.

Entirely apart from Tammann, however, there is independent experimental evidence of the possibility of two forms of ice, differing in density by 1/10 per cent, the amount found by Tammann. Two modern experimenters, Nichols [12] and Vincent,[13] as well as several previous observers, have each found that the density of ice may have either one of two distinct values. The difference seems to be connected with the manner of formation of the ice, whether natural or artificial. The natural ice, when kept for some time, tends to assume the value for the artificial ice. Barnes,[14] however, failed to verify the work of Nichols, and the most recent work of Leduc suggests that the discrepancies may possibly be due to dissolved air.

The question is as yet unsettled, with the probability, however, that the ice does not exist. In order to leave open the possibility of the establishment of its existence, however, the numeral IV has been reserved for this according to Tammann, and the first of the two new varieties found in this paper has been called V.

Besides this, there have appeared from time to time spasmodic notices concerning ice existing in crystalline forms other than the hexagonal. This is usually natural ice, which has been subjected to intense cold in the far north. None of these statements seem to have been subsequently verified, however.

In the course of the present work evidence was obtained as to the possibility of the existence of another form at high pressures. It seems

---

[12] Nichols, Phys. Rev., 8 (Jan., 1899).
[13] Vincent, loc. cit.
[14] Barnes, Trans. Roy. Soc. Canada, 3, Sect. III, 3–27 (1909), and Phys. Rev. (July, 1901).

probable that the five discrepant points found on the lower end of the VI–L curve may be due to the presence of some other form of ice than VI. These points were obtained on two separate occasions more than two months apart. Every one of the equilibrium points which have ever been obtained has been plotted in the diagrams, except a few where the temperature control was defective, and which have been mentioned. There are nowhere any points lying so far off the curve as these five. It seems hardly probable that the five worst points should have all been bunched in the same place, all lying consistently on a new curve. The probability of there being a new kind of ice is strengthened by the change of volume measurements made at these same points. These points are shown on Figure 33 for the change of volume VI–L. They lie far below the smooth curve, further than even the wildest of the discarded points, and they also lie consistently on another curve. The probability seems very strong for another modification of ice. This modification, if it does exist, is unstable in the locality found, V being a more stable form. Whether this new form has any region of stability at all is open to question; there is no necessity for it.

Whether there are still other forms stable at the highest pressures is of course a matter of pure conjecture. No inconsistencies were ever found suggesting in any way the existence of another form in the region studied. The domain of stability of ice VI as found is already five times more extensive than that of any of the other modifications. Furthermore, the course of the freezing curve and of the change of volume curve is such that both of these curves could be extended without difficulty to infinite pressures and temperatures. VI seems at any rate suited to be a final modification.

In this connection, some comment seems called for as to the possibility of predicting new forms. Of course it is well known that there is no such possibility from the equations of thermodynamics alone; the domain of existence of any form may be extended indefinitely in either direction without running into thermodynamic inconsistencies. But every substance, besides satisfying the identical relations of thermodynamics, also satisfies its own particular characteristic equation. This characteristic equation is determined by the special internal mechanism of the substance in question. It seems *a priori* possible that the approach of a new form should be heralded by some change in the mechanism, which should have its effect on the characteristic equation. But no such effect as this has been noticed. One substance may be extended across the boundary line into the unstable region beyond, with no appreciable change in either the compressibility or the dilatation, the two quantities determining the characteristic equation. And on an equilibrium line, the neighborhood of a triple point does not cast its shadow before, either by a change in the direction of the equilibrium line, or of the change of volume line, or of the curves of latent heat or internal energy. So far as the data presented in this paper are concerned, there seems to be no way of making the prediction. Nevertheless the conviction remains that if one had a complete description of the internal mechanism, it would be possible to find some criterion as to the possible stability of other configurations. What additional data are needed to give a sufficient knowledge of the mechanism ? The question is an interesting one for investigation.

The Possibility of Subcooling or Superheating.

The facts regarding this subject have nearly all been mentioned incidentally in the course of the detailed description of the experiments. Although no observations of the possibility of subcooling or superheating were made separately for their own sake, yet the observations collected incidentally on this subject are nearly as numerous as all the other observations together. The effect was of course observed during every measurement of a change of volume, and frequently on other occasions. It was likely to be a very troublesome effect, preventing the appearance of the modification desired, so that at least sufficient familiarity had to be obtained with the slight regularities shown by the apparently confused mass of facts so that the modification of ice desired could be forced to appear.

First with regard to the solid phases. It has been found possible to go across nearly every one of the boundary lines into the region of instability on the other side. I has been found in the region of III and II; II in that of I (at low temperatures); III in that of I, II, and V; V in that of III, II, and VI; and VI in that of V. The only excursion of this kind which has not been found possible is that of II into the region of III or V. This latter reaction always ran immediately on passing the slightest amount into the neighboring country. For the other reactions no fixed limits could be set to the extent by which it was possible to overstep the boundary. The amount depends on the size and shape of the vessel, on the materials in contact with the ice, on the element of time, and on caprice. In general, however, the limits became narrower at high temperatures, as one would expect.

With regard to the passing from the solid to the liquid, the experience here is but a verification of the experience of everybody else; that it is impossible to superheat a crystalline phase with respect to the liquid. No good reason for this has ever been given, but no exception has ever been found, and it is coming to be regarded as a law of nature. One would think that if there were ever a chance to find an exception it would be here, where the materials are made viscous by the high pressure, and where the reaction must run with increase of volume against the pressure, doing considerable external work.

On the other hand, it is the easiest possible matter to subcool the liquid with respect to any one of the four solid phases bordering on the domain of the liquid. In fact, it is often a matter of some difficulty to start the reaction liquid-solid, superpressures of 1500 or 2000 kgm. being sometimes necessary. The amount of superpressure or subcooling necessary to start the reaction is again a matter of caprice. On the VI-L curve, however, where the greatest range was open to observation, there seemed to be in general a tendency for the subcooling to increase at high pressures.

In consequence of the possibility of carrying one phase into the region of another, it is possible to prolong certain of the equilibrium lines beyond the triple point into the region of instability beyond, thus realizing equilibrium points between two unstable phases. The equilibrium lines which have thus been extended are: I-L into the region of II, III-L into the region of I (by Tammann, not in the present work), III-L into the region of V, I-III into the region of II, III-V into the region of II, and VI-L into the region of V. It may also be possible to extend II-III into the region of I, although this was not tried. The only curves which it was found experimentally impossible to extend were II-III into V, II-V into III, and V-L into VI. There seems to be no obvious generalization about the possibility of extending these curves; everything seems to depend on the particular character of the substances in question. There was one plausible generalization from only two instances, which was missed by a very narrow margin, namely that it is impossible to extend a melting curve at the upper end into the region of another solid. The attempt always failed on the V-L curve, and until the very last day failed on the III-L curve. On this very last day, when the change of volume points II-III-V were being found, it was desired to pass from III to V. The greatest difficulty was experienced in doing this. Pressure was increased on III at about $-20°$ to 4500, and the temperature then raised until III began to melt, thus spoiling the generalization. It was finally necessary to lower the temperature to $-40°$ before V appeared. It is therefore conceivable that there are circumstances when V-L might be extended into VI, or II raised into the region of III or V.

It may be mentioned that at every triple point there is at least one equilibrium line which it has not been found possible to cross. At four of the five triple points there are two such lines, and at the fifth, I–II–III, the possibility of carrying II into the region of I was not tried. At three of the five points water is one of the substances, so that the statement is obviously true; at the other two points the solid II has the same relation to the solids above it that the solids above have to the liquid above them.

It was not found possible to extend any two of the curves so far into a region of instability as to give a triple point between three unstable phases, nor was the third unstable curve starting from this unstable triple point ever found. This third curve would have no region of stability whatever. The nearest approach to this was in the extensions of the III–L and VI–L curves. No especial attempt was made to realize such an unstable triple point, however, as it would have been a matter of considerable difficulty, but there seems no reason why such a point should not be found.

No such constancy was ever found in these subcooling experiments as to suggest the necessity of the existence of the metastable limit, as is claimed by many writers. With a particular piece of apparatus it may be possible to obtain fairly consistent results, but when the apparatus is being continually changed as it was here there seems to be no obvious regularity. There is no reason why there should be, if the formation of the nucleus of the new phase, which starts the reaction, is a matter of chance as seems likely. Owing, however, to the fact that there is at every triple point one equilibrium curve which it is not possible to cross, there are going to be lines reaching from every one of the triple points limiting the existence of one or more of the phases, which give the same impression as the metastable lines recently drawn by Tammann in the neighborhood of the triple point I–III–L for water, and also in the neighborhood of a triple point of phenol.[15]

In one respect the possible amount of subcooling showed great regularity and is of sufficient significance to deserve special mention. This is the fact, already noted, that a form is much more likely to appear again if it has appeared recently before. This fact was of great convenience in obtaining the change of volume points, because here one phase has to be allowed to be completely replaced by another, and the first phase is then wanted again in obtaining the second ΔV point. Very little difficulty was experienced in getting the desired form to appear under these circumstances. This was particularly true on the V–VI and the V–L curves, the modification V being the hardest to

---

[15] Tammann, ZS. Phys. Chem., **75**, 75–80 (1910).

obtain initially. On the V–L curve a succession of ΔV points was obtained without difficulty, V always putting in an appearance when desired, separating directly from the liquid with very little subcooling, although V was never obtained directly from the liquid the first time, but only by way of VI after considerable subcooling. This ability of V to separate directly from the water could be retained for several hours at points removed 1000 or 2000 kgm. from the equilibrium curve. The ability was once retained over night, pressure not being far removed from the equilibrium curve. The disposition to react depends both on the element of time and on the amount by which pressure has been changed from the equilibrium value. This predisposition to react is lost if a third modification has intervened. Thus, on the occasion mentioned, when III was melted at 4500 kgm. in the endeavor to obtain V, only a half hour previously the reaction II–V had been running in either direction with the greatest facility. The subsequent conversion of II into III resulted in the complete loss of disposition of V to appear.

In explanation, there must be some structure in both the liquid and the solid not ordinarily accounted for, some nucleation or aggregation of the molecules left as a heritage from the previous modification, which is particularly adapted to fall back again into the old position. The possibility of such nuclei in the solid must show that the molecules in a crystal are not arranged in the absolutely symmetrical way usually thought of. The fact that these nuclei may persist for some time in the solid does not seem so surprising as the fact that they exist at all. For the liquid, the reverse is the case. The existence of nuclei might be expected; they are known to exist even in a gas, but that these nuclei may persist for some hours in an assemblage of molecules supposed to be in constant interchange with each other might not be expected at first. The disappearance of these nuclei is a matter of extraordinary slowness, considering the usual times involved in the motion of the molecules as a liquid. Doubtless the formation of these nuclei is intimately connected with the freezing of a liquid for the first time. The freezing can start only from one of these nuclei. The formation of one of these nuclei in an unimpregnated liquid is a matter of chance, and until the molecules do happen to fall together into the right position the freezing cannot start.

Reaction Velocity.

Here again no accurate measurements were made. The velocity depends on too many things to allow quantitative results of value without the expenditure of a great deal of time. These disturbing factors are such as the size and shape and material of the containing vessel, the rapidity of heat conduction, and the distance from the equilibrium line. But just as for the question of subcooling, so here, every measurement ever made, whether of the equilibrium pressure alone or of the change of volume, involved this matter of reaction velocity. One had to be sure before making a reading that the reaction had stopped running, and in the endeavor to waste no more time than was necessary the progress of the reaction was constantly watched, so as to make the reading as soon as possible after sensible equilibrium had been reached.

There are two distinct types of behavior of the reaction velocity, according as the reaction is between a solid and a liquid or between two solids. In general the reaction between solid and liquid was slower than between solid and solid. The time for the completion of the reaction liquid-solid was about two hours on the I–L, V–L, and lower end of the VI–L curve. No particular variation in this time was noticed from one end to the other of the I and V curves, but at the upper end of the VI–L curve, the velocity had been very appreciably accelerated, the time for completion of the reaction at the upper end being about one hour. On the III–L curve, as already mentioned, the reaction was very much slower. It was not practicable here to wait for the reaction to run to completion, but the equilibrium points were taken as the mean of values approached from above and below. This shows that the velocity depends on the form into which the water is transformed as well as on the heat of reaction, for the heat III–L is of the same magnitude as that of the neighboring modifications I and V. On all the curves the velocity seemed the same for melting as for freezing.

The most striking behavior of the reaction velocity is shown by the reaction solid-solid. The velocity ranges from explosive rapidity at the end near the triple point to such sluggishness 20° further down as to make further prolongation of the equilibrium curve out of the question. Of course the very low heat of reaction would lead one to expect, other things being equal, a high rate of transformation, but that the heat of reaction has practically nothing to do with it is shown by the enormous temperature coefficient of reaction velocity, while the heat of reaction is practically independent of temperature. The slowness of the reaction does not depend so much on the actual temperature as on the nearness of the triple point. Thus I–II is explosive at its triple point, −35°, while at −35° III–V is almost impossible; III–V is explosive at its own triple point, −17°, but at this temperature V–VI may take a couple of hours to run to completion.

The explanation does not suggest itself. The mere fact of such rapid reactions between solids is itself sufficiently surprising. It does not seem as if the mechanism could be the same as that of an ordinary chemical reaction. It is as if the molecules changed from one crystalline arrangement to another by snapping round on their axes, like the supposed molecular magnets of a piece of iron in a magnetic field. The high temperature effect is difficult to account for. Certainly no known viscosity effect has so high a temperature coefficient.

Some connection seems likely between this high velocity at the triple point and the impossibility of superheating a solid. There is no doubt but that at this point the molecules of the solid have a perfectly astounding possibility of motion. The passage to the liquid may still depend on the chance formation of nuclei in the solid, but the freedom of motion of the molecules in the solid may be so great as to secure the practically instantaneous formation of the proper grouping. This recalls the question proposed a few pages back as to the possibility of predicting the presence of a new phase from the behavior of a single other phase. Here we have a hint as to the possibility of predicting a third phase by an enormous reaction velocity between two others.

The Compressibility of the Different Forms of Ice.

These compressibilities have not, with one exception, been directly measured, but it is possible, nevertheless, to obtain some idea as to this magnitude for those varieties of ice which are anywhere in equilibrium with the liquid. This includes all the varieties except II. The compressibility of VI has been directly measured above 0°, and the initial compressibility of ice I has been computed. It has been already stated that the data do not possess the requisite accuracy to permit a direct determination of the compressibility of the other forms of ice. The approximate determination comes by finding the actual volume of the different kinds of ice along the equilibrium lines. The volume of water is known along these lines, and also the change of volume when the liquid passes into the solid, so that the actual volume of the solid may be obtained. This volume will change with temperature and pressure along the equilibrium line. The approximation to the compressibility is made by assuming that the change of volume due to changes in temperature along an equilibrium line is negligible compared with the change due to pressure. The error so introduced may be found at 0° for ice I at atmospheric pressure, where it is 3 per cent. Furthermore, at any point except in the immediate neighborhood of the origin, the thermal expansion of the ice is very probably less than that of the liquid with which it is in contact. This enables us to put an upper limit of 10 per cent as the probable error introduced by assuming all the change of volume along an equilibrium curve to be brought about by changes of pressure.

The complete values of the volume are shown in Table XXX., which gives the approximate compressibility along these curves. For comparison, the approximate compressibility of the water at the corresponding pressure and temperature is also given. In consequence of the change of volume due to temperature, the quantity listed as compressibility is too big on the ice I curve, and too small on the other curves which rise to higher temperatures with higher pressures. The compressibility of the ice is uniformly less than that of the water, as one would expect, varying from 1/3 to 2/3. On all the curves except the III-L curve, the compressibility shows a very marked decrease with rise of pressure, the decrease being more rapid proportionally for the liquid. The same decrease would probably also have been shown on the III-L curve, if it had been possible to make the measurements of the change of volume with greater accuracy. On the I-L curve, the change for a range of pressure of only 2000 kgm. is abnormally high. This might be accounted for by a negative temperature coefficient of expansion of ice at the lower end of the curve. The change of volume measurements I-III have already suggested this as a possibility. The change of compressibility with pressure is greatest at the low pressures, being particularly great in the region of instability of VI. In general, when one variety of ice replaces another, one would expect the new form stable at the higher pressures to show the lower compressibility, corresponding to the smaller volume; but this is certainly not true in the case V-VI, and probably not true when III replaces I. The compressibility of each modification seems to be a property inherent in that modification, depending probably on the symmetrical arrangement of molecules in the crystal, and not depending so directly on the volume.

As verifying these values very roughly, the directly determined difference of compressibility between L and VI has been shown in Figure 35. The values are evidently not of any great regularity. The only two points found below 15°, at 0°, are much too high. These would give a value for the compressibility at 0° of $.0_530$ as against $.0_56$ from the data above, $.0_56$ being too small. But in the region where the points are thicker, between 15° and 50°, the agreement is better, discrepancies being of the order of 0.0_51. These points run to higher pressures than the actual determinations of the compressibility of water, and do show strikingly the rapid decrease in the difference between the com-

TABLE XXX.

Volume and Approximate Compressibility of Ice on Equilibrium Curves.

Pressure, kgm./cm.²	Temp. C.°	Vol. of Water cm.³/gm.	ΔV. cm.³/gm.	Vol. of Ice cm.³/gm.	Approx. Compressibility of Ice.	Compressibility of Water at Same Point.
Ice I–Water.						
0	0	1.0000	0.0900	1.0900	0.0_435	0.0_452
500	− 4.1	0.9777	0998	1.0775		
1000	− 8.7	9588	1096	1.0684	0_416	37
1500	−14.0	9414	1201	1.0615		
2000	−20.3	9253	1318	1.0571	0_58	29
Ice III–Water.						
2000	−22.5	0.9250	0.0476	0.8774	⎫	0.0_430
2500	−20.1	9099	373	8726	⎬ 0.0_591	26
3000	−18.3	9874	286	8688		24
3500	−17.0	8867	231	8636	⎭	21
Ice V–Water.						
3500	−17.0	0.8870	0.0785	0.8085		
4000	−13.6	8781	733	8048	0.0_572	0.0_4190
4500	−10.1	8694	681	8013		
5000	− 7.0	8610	634	7976	53	164
5500	− 4.2	8543	590	7953		
6000	− 1.6	8478	549	7929	47	140
6500	+ 0.6	8418	516	7902		
Ice VI–Water.						
4500	−18.0	0.8689	0.0985	0.7705		
5000	−12.8	8604	968	7636	0.0_4102	0.0_4164
5500	− 7.7	8536	940	7596		
6000	− 3.2	8472	928	7544	0_572	140
6500	+ 1.1	8418	905	7513		
7000	5.0	8370	882	7488	55	126
8000	12.6	8271	816	7455	48	120
9000	19.5	8156	755	7401	43	110
10000	26.0	8055	697	7358	37	104

pressibility of water and VI with rising pressure. This point has some bearing on the discussion of the general nature of the change of state solid-liquid at high pressures.

Discussion of the Results.

So far in this paper merely the numerical results have been presented, without much discussion of their significance. For the liquid, the change of volume at different temperatures and pressures has been found, without discussing the shape of the p-v-t surface which these combine to give. For the solid states, the data have been presented for each variety of ice separately, without any consideration of the general features which may be common to all. The object of this discussion is to give a comprehensive survey of all the results, especially for the transition liquid-solid, and to point out the theoretical significance of these data at very high pressures.

For the liquid, the results will have their chief interest in showing how water passes from an abnormal liquid at low pressures to a normal one at high pressures. The data do not have so much suggestiveness for a theory of the liquid state as would those for some normal liquid, and in any case it would be dangerous to generalize from the behavior of a single substance, but the results at the higher pressures do suggest at least the nature of the effects to be expected in general at high pressures.

The results for the liquid at even temperature and pressure intervals have been collected in Table XXXI. In this table the smoothed results obtained separately above 0° and below 0° have been given without any attempt to smooth the values of either set so as to make connections with those of the other. But the fact that the two independent sets of determinations do run smoothly into each other makes probable the accuracy of the work. Above 7000 kgm. the value of the thermal dilatation used in computing the table was obtained by an extrapolation. This is probably good up to 10,000 kgm. in giving the actual volume at any temperature and pressure within the narrow range of existence of liquid water, but the use of the tables in any theoretical considerations demanding knowledge of the derivatives would be dangerous at the highest pressures. Above 3000 kgm., and indeed above 2000 kgm., the dilatation has been assumed to be independent of the temperature over the range of 22°, as has been sufficiently shown by the work of Amagat. For values below 2000 the variation of the dilatation as found by Amagat has been used in computing the table. Of course below 0° the variation of dilatation with temperature was given by the data of this paper.

TABLE XXXI.

The Volume of Water at Regular Pressure and Temperature Intervals.

Pressure, kgm./cm.²	−20°.0.	−15°.0.	−10°.0.	−5°.0.	0°.0.	+5°.0.	+10°.0.	+15°.0.	+20°.0.	+25°.0.
0	....	....	1.0017	1.0006	1.0000	0.9999	1.0001	1.0007	1.0016	1.0028
500	....	....	9800	9783	9776	9782	9791	9800	9812	9825
1000	....	....	9606	9592	9586	9596	9609	9623	9638	9654
1500	9233	9401	9413	9404	9407	9420	9435	9451	9467	9483
2000	9083	9240	9248	9257	9265	9281	9298	9315	9332	9349
2500	8957	9092	9102	9115	9131	9148	9166	9185	9203	9222
3000	....	8966	8978	8991	9009	9026	9044	9063	9081	9100
3500	....	8860	8872	8884	8903	8923	8944	8964	8984	9005
4000	....	8764	8772	8784	8805	8823	8842	8860	8878	8897
4500	....	....	8680	8691	8713	8721	8749	8767	8785	8802
5000	....	....	8593	8604	8626	8643	8661	8678	8696	8714
5500	....	....	....	8531	8548	8565	8582	8599	8616	8633
6000	....	....	....	8464	8480	8496	8513	8529	8545	8561
6500	....	....	....	....	8414	8429	8444	8460	8475	8490
7000	....	....	....	....	8356	8370	8384	8398	8412	8426
7500	....	....	....	....	....	8309	8321	8334	8346	8358
8000	....	....	....	....	....	....	8262	8273	8284	8295
8500	....	....	....	....	....	....	....	8208	8218	8228
9000	....	....	....	....	....	....	....	8149	8157	8165
9500	....	....	....	....	....	....	....	....	8099	8106
10000	....	....	....	....	....	....	....	....	8046	8050

Before discussing the shape of the p-v-t surface as given by the determinations of compressibility, it may be well to review briefly the known behavior of water, especially with regard to the differences it shows when compared with ordinary liquids.

All ordinary liquids show a decreasing compressibility with rising pressure, the compressibility decreasing faster than the volume, and they also show an increasing compressibility with rising temperature. The mathematical equivalent of this last statement is that the thermal dilatation decreases with rising pressure. This normal behavior is exactly as we would expect if we regard the liquid as composed of nuclei of more or less invariable volume, separated by spaces which may be altered in size by pressure and temperature. It is not necessary for a qualitative understanding of the phenomena even to inquire whether these nuclei are subatomic or atomic; that is, whether the major part of the compression is given by the change of volume of the spaces between the atoms or by changes in volume of the atoms themselves.

For water the effects are anomalous. The compressibility decreases with rising pressure, as it does for everything else, but with rising temperature the compressibility at first becomes less, passes through a minimum, and then becomes greater again. This minimum is situated at about 50°. The position of the minimum is nearly independent of pressure but the minimum itself becomes less and less pronounced with rising pressure, and at 3000 kgm. has entirely disappeared. Corresponding to this anomalous behavior, the dilatation shows anomalous behavior with rising pressure, becoming greater with greater pressure at temperatures below 50°. It has been recognized by Amagat as possible, however, that at temperatures below 50° the dilatation would decrease with rising pressure at pressures sufficiently high. In the immediate neighborhood of 0° and atmospheric pressure, there are special anomalies connected with the maximum density point. In particular, the temperature of maximum density, which at atmospheric pressure is at about 4°, is depressed by rising pressure. This depression of the maximum density point is nearly linear with the pressure, and is so rapid that at 300 kgm. it has fallen below the freezing temperature at that pressure. So much has been shown by Amagat, [16] who worked up to 3000 kgm. The results may be briefly summed up in the statement that water, abnormal at low temperatures and pressures, tends to become normal at high temperatures and pressures.

Now consider the information given by the present data, examining

---

[16] Amagat, loc. cit.

first the compressibility data above zero. Of course the rough facts already known appear immediately from an inspection of the figures. The compressibility decreases strikingly with rising pressure and is less at 22° than at 0°. Figure 36 shows the compressibility $\left(=\left(\frac{\partial v}{\partial p}\right)_T\right)$ at 0° and 22°. At 10,000 kgm. it has decreased to 1/4 of its initial

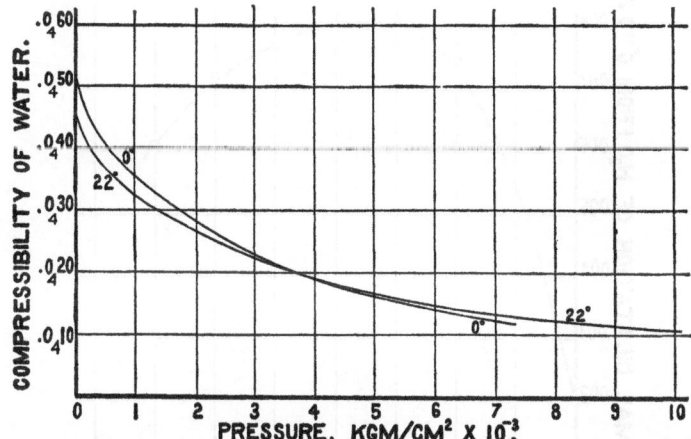

FIGURE 36. The compressibility $\left(\frac{\partial v}{\partial p}\right)_T$ of liquid water as a function of the pressure at 0° and 22°.

value. The figure shows also the crossing of the compressibility curves; at low pressures the compressibility at low temperatures is higher than it is at high temperatures, but with rising pressure the abnormality disappears, and beyond 4000 kgm. the compressibility is higher at the higher temperatures. The variation of thermal expansion between 0° and 22° is shown in Figure 37. This rises to a maximum at nearly 4000 kgm. and then falls again. The rise to the maximum is much more rapid than the fall away from it. This maximum verifies the surmise of Amagat that the dilatation at any temperature would ultimately decrease with rising pressure for pressures sufficiently high. The position of this maximum evidently corresponds to the crossing of the two compressibility curves at 0° and 22°, for we have

$$\frac{\partial}{\partial t}\left(\frac{\partial v}{\partial p}\right) = \frac{\partial}{\partial p}\left(\frac{\partial v}{\partial t}\right) = 0.$$

The usual explanation of the abnormalities shown by water at atmospheric pressure is on the basis of polymerization. The effect of decreasing temperature is to increase the polymerization. For the present purpose we may think of the molecule stable at higher temperatures as a single molecule and that at lower temperatures as a double molecule, although the latter is more probably triple and the former

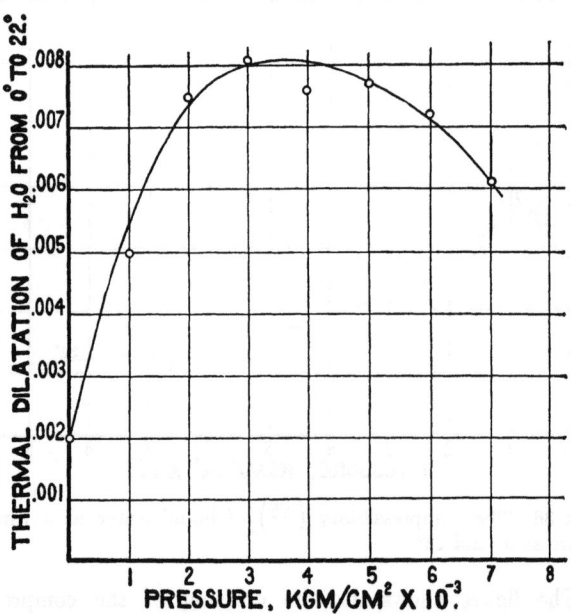

FIGURE 37. Shows the total change of volume of water between 0° and 22° as a function of the pressure.

double. The double molecule must furthermore be thought of as occupying more than twice the volume of each of the single molecules of which it is composed. The effect of a decrease of temperature on volume is twofold : a decrease such as takes place in any normal liquid, and an increase due to the clustering of single molecules into double ones. This increase becomes increasingly rapid at low temperatures because of the increasingly rapid polymerization. At 4° it has become sufficiently rapid to neutralize the natural decrease, and at lower temperatures volume increases with falling temperature. The way in which polymerization is affected by pressure can best be discussed after reviewing the data below 0°.

The data below 0° show still more strikingly than those above 0°

the abnormality at low pressures followed by normal behavior at high pressures. No previous measurements seem to have been made within this region, although the principal effects are within the reach of previously attainable pressures. At atmospheric pressure the study of water at low temperatures is prevented by the accident of freezing. It will pay us, however, to imagine what would be the relation between temperature and volume on the present theory of polymerization if it were possible to subcool the water indefinitely. At high temperatures we evidently expect the water to behave normally, for its molecules are all single molecules of the same kind, and similarly at very low temperatures, where the molecules are all double molecules, we should expect the behavior to become normal again. The curve connecting volume with temperature for a

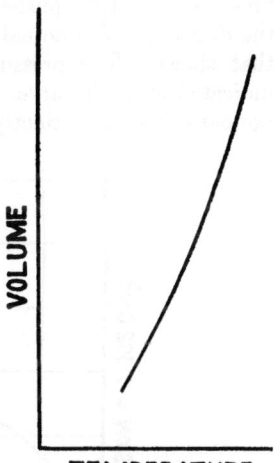

FIGURE 38. The relation between volume and temperature for a normal liquid.

normal liquid is of the form shown in Figure 38, the dilatation becoming more rapid at high temperatures. These considerations lead us to predict, therefore, a curve of the shape shown in Figure 39 for water. Experimentally it has been found possible to follow this only as far as $-10°$, not far enough to reach the first point of inflection. The effect of increasing pressure must be to change in some continuous way the curve of Figure 39 into that of Figure 38.

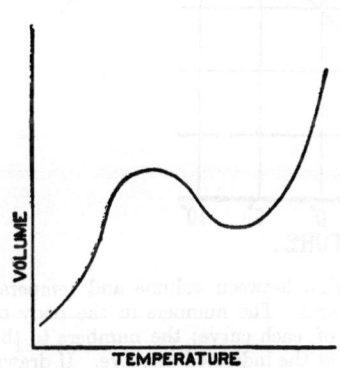

FIGURE 39. Hypothetical relation between volume and temperature for liquid water if it could be subcooled indefinitely without freezing.

The data indicate very strikingly the way in which the abnormality is effaced. Figure 40 shows this. In this figure the relation between volume and temperature for various constant pressures is plotted directly from the data of Table XXXI. Each separate curve is drawn to scale, but the curves for different pressures have been pushed together, so as to come within the limits of the drawing. The actual separation of the curves is about ten times that shown. The pressure and the value of the volume at 0° are indicated on each curve. The manner of transition from abnormal to normal is shown distinctly, and requires no comment. At 1500 kgm.

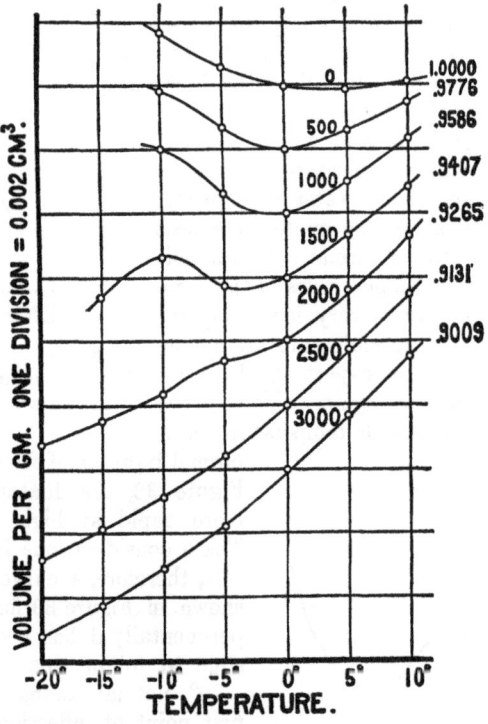

FIGURE 40. Curves showing the relation between volume and temperature of water for various constant pressures. The numbers in the body of the diagram show the constant pressure of each curve; the numbers to the right show the volume at 0° of the liquid at the indicated pressure. If drawn to scale the curves should be separated about ten times as much as shown.

we actually have realized a curve with both a minimum and maximum, like the conjectural curve at atmospheric pressure. The curves show one thing in connection with the previous experiments, namely, that the depression of the maximum density point with pressure cannot continue linear with pressure much beyond 300 kgm., the limit of the previous experiments, but the temperature of the maximum becomes nearly independent of the pressure until the maximum entirely disappears at higher pressures.

Figure 40 shows the results only to 3000 kgm. Up to this point the dilatation at 0° has been increasing with rising pressure. But the fact that the curves of Figure 4 pass through a maximum at about 3200 kgm. indicates a return to normal behavior. From here on the dilatation decreases with rising pressure. The position of this maximum evidently corresponds to the position of the maximum of the dilatation at 3700 kgm. found between 0° and 22°. One would expect some change in the position of the maximum toward lower pressures at lower temperatures, but even if there were none, the agreement is perhaps as good as could be expected when one considers how very small the experimental quantities are which are involved below 0°.

As far as 4000 kgm., the curves of Figure 4 have been in perfect accord with the manner of transition shown in Figure 40. Above 4000, however, if the curves are extrapolated in the direction in which they are heading, there will be new abnormalities. This extension is not actually possible physically, for the same reason that the hypothetical curve connecting volume and temperature at atmospheric pressure was not possible, namely because the water freezes. But the extension may be made, nevertheless, for the purposes of speculation. It is seen that this new abnormality at high pressures and low temperatures consists in a crossing of the lines again, so that at 5000 kgm., for instance, water would expand on passing from $-15°$ to $-20°$, just as for water at 0° and atmospheric pressure. The explanation suggests itself that this new abnormality is due to the new variety of ice which is about to separate out, either V or VI. At any rate, the idea seems perfectly plausible that each of the forms III, V, and VI is a solid form of water in a different state of polymerization, and that this polymerization should be shown by anomalous effects in the liquid. There certainly seems to be a strong presumption raised for this possibility by the present data. A more accurate experimental investigation would be well worth making, but would require new methods and apparatus.

We now have enough material in the behavior of the curves both above and below 0° to see what the role of pressure in wiping out the abnormalities must be. The one significant fact is that the pressure wipes out the abnormalities where they stand, without any perceptible shifting of their temperatures. Thus above 0°, the temperature of minimum compressibility is very nearly constant at 50°, independent of the pressure. Increasing pressure merely makes this minimum at 50° less and less pronounced until it has entirely disappeared. And below 0° the abnormalities remain confined to the 15° or 20° below 0°, the effect of pressure being merely to smooth out the variation of curvature. Now this is distinctly what one would not expect at first. The usual explanation of the normalizing effect of pressure is to suppose that the amount of polymerization, the cause of the irregularities, is decreased by rising pressure. This is the assumption made by Röntgen [17] and Sutherland.[18] The effect of this evidently would be to displace the region of abnormality from high to low temperatures, which is what we have seen does not happen. The explanation is rather to be found in supposing that the pressure merely reduces the effects of polymerization uniformly at every temperature without necessarily reducing the amount. This means that the difference of volume between the double molecules and the two single molecules becomes rapidly less at higher pressures; in other words, that the double molecules possess an abnormally high compressibility. This seems an entirely plausible hypothesis in view of the abnormally large volume of the double molecules. At high pressures, then, the polymerization, even if it occurs, is unable to produce volume effects, and might as far as we are concerned be entirely neglected. There may be effects on the specific heats, which cannot be detected from the present data. The explanation of the various pressure effects on this basis is simple. The anomalous decrease of compressibility with rising temperature is due to the fact that at the higher temperatures the double molecules with abnormally high compressibility are becoming fewer. When pressure has become so high that there is no longer distinction between the associated and the dissociated molecules, the behavior of the liquid becomes normal. This explanation leaves entirely open the question as to whether pressure actually increases or decreases the amount of polymerization. One would naturally expect an increase. The explanation also leaves open the possibility of the polymerization being to several different groups, more complicated than doublets or triplets. This possibility is suggested by the appearance of the several allotropic forms of ice.

It is to be noticed that in assuming that the molecules are compressible we have not by any means assumed that the actual change of size of the molecules under pressure is the chief factor in the change of volume of the substance as a whole under pressure. In fact it is almost certainly true that the greater part of the total change of volume is due to the closing up of the spaces between the molecules. The substance as a whole may have a greater or less compressibility

---

[17] Röntgen, Wied. Ann., **45**, 91–97 (1892).
[18] Sutherland, Phil. Mag., **50**, 460–489 (1900).

according as its structure is more or less compressible. Thus liquid water is more compressible than ice I, although it has the smaller volume and is composed of molecules which are themselves less compressible.

TABLE XXXII.

THE VOLUME OF WATER CALCULATED BY TUMLIRZ'S FORMULA COMPARED WITH EXPERIMENTS.

Pressure, $\frac{\text{kgm.}}{\text{cm.}^2}$	Volume. cm.³/gm.					
	0°			22°		
	Tumlirz.	Experiment.	Difference.	Tumlirz.	Experiment.	Difference.
0	0.9964	1.0000	−.0036	1.0030	1.0020	+.0010
1000	9560	0.9586	+.0026	0.9629	0.9636	−.0007
2000	9237	9265	+.0028	9318	9340	−.0022
3000	8983	9009	+.0026	9067	9090	−.0023
4000	8778	8805	+.0027	8861	8881	−.0020
5000	8607	8626	+.0019	8690	8703	−.0013
6000	8464	8480	+.0016	8544	8552	−.0008
7000	8341	8356	+.0015	8419	8417	+.0002
8000				8311	8288	+.0023
9000				8216	8161	+.0055
10000				8132	8049	+.0083

With regard to the bearing of these data on the theory of liquids, the theory itself does not seem to be at present far enough advanced so that these data can settle definitely any crucial questions. In fact, there do not seem to be any clear cut questions waiting for settlement. Nearly all the work done so far on liquids has been in modifying van der Waals' equation by making assumptions which seem more or less plausible about the way in which the forces between the molecules, the distances between the molecules, or the energy of the molecules vary with the temperature and pressure, and the chief aim of all this activity seems to have been the production of an equation with as few constants as possible which should accurately represent the behavior of as many liquids as possible under changes of temperature and pressure.

The bearing of these data for water on the theory of liquids as developed in this way may be best shown by testing how well the equations already proposed are applicable by extrapolation over this wider pressure range. For this purpose we may choose the equation given by Tumlirz [19] as perhaps the best. Tumlirz has applied his formula to the data of all the liquids studied by Amagat over a pressure range of 3000 kgm. and a temperature range of 40° or 50°, with really remarkable agreement. Tumlirz's formula has the form

$$(p + P)(v - a) = RT,$$

where a and R are constants for any given substance, and P is a function of the temperature only, to be determined by experiment. The significance of the assumption evidently is that the covolume (proportional to the total volume of the molecules) is independent of pressure and temperature, and that the internal pressure P is not affected by changes of volume at constant temperature. The results calculated by Tumlirz's formula, and the actual experimental results are shown in Table XXXII. The constants used at 0° in the calculation are those given by Tumlirz. At 22°, P was found by interpolation from Tumlirz's values above and below to be 7152. At 0°, the agreement is fairly satisfactory, and the discrepancies are of the same order throughout the entire pressure range. The discrepancies are greatest at the low pressures, corresponding to the abnormal behavior of water here. At 22°, for the lower pressures, the discrepancy is about the same as at 0°, but the most interesting thing about these values at 22° is the very evident failure of the formula at high pressures. The values given by the formula for the compressibility become small too rapidly at the high pressures.

This question as to the behavior of the compressibility at high pressures is the first one that would occur to one as of significance for the theory of liquids at high pressures. That is, does the volume shrink toward a limiting value in the way indicated at low pressures, or is there some other effect introduced by the high pressure? The physical picture of the mechanism of the liquid suggesting this question is that of an assemblage of molecules with intervening spaces. Is the change of volume of the molecules themselves under pressure sufficient to produce an appreciable effect after the intervening spaces

---

[19] Tumlirz, Sitzber. Wien, Bd. CXVIII, Abt. IIa (Feb., 1909), pp. 1–39.

have been shut up? The present conception of the atom as a planetary system of electrons would suggest that there are possibilities of enormous change of volume within the atom itself, and that the compression of the atom is going to continue uniform over a relatively enormous pressure and volume range, just as under ordinary circumstances a

TABLE XXXIII.

VARIOUS THERMODYNAMIC PROPERTIES OF WATER AT 22°.

Pressure, $\frac{\text{kgm.}}{\text{cm.}^2}$	$C_p - C_v$. $\frac{\text{gm. cal.}}{\text{gm.}}$	$\left(\frac{\partial E}{\partial p}\right)_t$ $\frac{\text{gm. cal.}}{\text{gm.}}$	$\left(\frac{\partial v}{\partial p}\right)_\phi - \left(\frac{\partial v}{\partial p}\right)_\tau$	$\left(\frac{\partial \tau}{\partial p}\right)_\phi$
0	0.0079	−0.00159	0.0_637	0.0016
1000	164	−0.00117	54	19
2000	279	−0.00104	75	23
3000	430	−0.00104	95	26
4000	498	−0.00071	95	26
5000	517	−0.00049	85	24
6000	492	−0.00019	71	22
7000	417	+0.00019	54	19
8000	279	+0.00068	33	15

gas obeys Boyle's law for a relatively very great range of volume and pressure. The fact is that at high pressures the compressibility does remain larger than the formula of Tumlirz demands, whether this is due to a compressible atom or not. The same thing is shown also by the curve of Figure 36 giving compressibility against pressure. The tendency of this curve is to become asymptotic to some value greater than zero, the compressibility changing very slowly at high pressures. At 5000 kgm. the compressibility has dropped to 1/3 of its value at atmospheric pressure, while at 10,000 it is 2/3 of its value at 5000.

The compressibility curves for 0° and 22° also indicate one other thing that would be expected, that at sufficiently high pressures the compressibility will become independent of the temperature, or in other words, that the dilatation will approach a value, probably zero, independent of the pressure.

The data obtained for the liquid are sufficient to enable us to calculate certain other quantities of thermodynamic interest. Two of these, the difference of the specific heats and the change of internal energy along an isothermal, may be calculated directly from the given data. Two others, the adiabatic compressibility and the rise of temperature

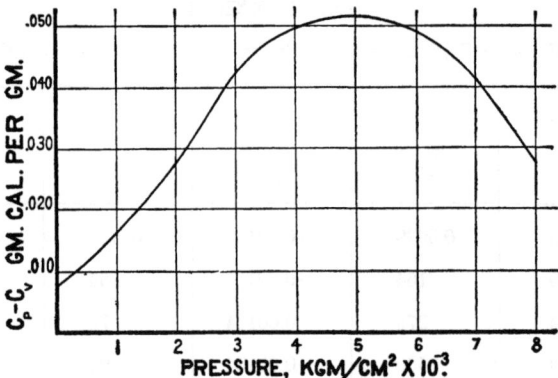

FIGURE 41. The mean difference of the specific heats of the liquid for the temperature range 0°–22°.

produced by compression, may be computed roughly, merely indicating the direction in which these quantities change under pressure.

The difference of the specific heats is given by the formula

$$C_p - C_v = \frac{-\tau \left(\dfrac{\partial v}{\partial \tau}\right)_p^2}{\left(\dfrac{\partial v}{\partial p}\right)_\tau}.$$

The quantities entering this equation have been determined directly. The computed values for 22° are shown in Table XXXIII., and graphically in Figure 41. The general behavior is a rise to a maximum at 5000 and then a decrease. For a normal liquid $C_p - C_v$ probably decreases continuously with rising pressure. This has been shown to be the case for mercury in the previous paper. The significance of the maximum is, then, merely a repetition of the old story that at high pressure water loses its abnormality and becomes normal. The measurements have not been made to high enough pressures to show the reversal of curvature on the descending branch of the curve, which would show complete attainment of normality. The values of $C_p - C_v$

found here directly differ from those given by the formula of Tumlirz. This formula, used by extrapolation, would demand a continuous increase in the value of $C_p - C_v$ to infinite pressures, suggesting again that there is an effect entering at high pressures not taken account of or foreshadowed by the behavior at lower pressures.

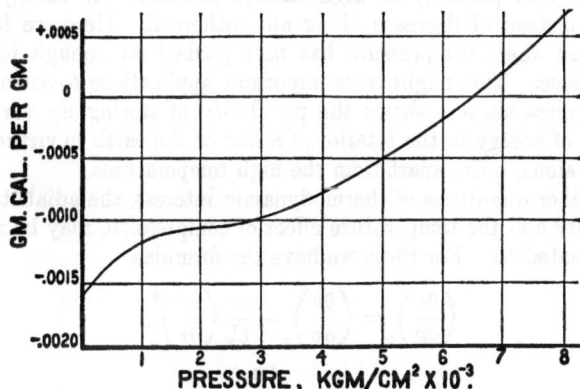

FIGURE 42. The change of internal energy of the liquid per kgm. rise of pressure along the isothermal at 22°.

The change of internal energy may also be found directly. For this we have the thermodynamic relation

$$\left(\frac{\partial E}{\partial p}\right)_\tau = -\left[\tau\left(\frac{\partial v}{\partial \tau}\right)_p + p\left(\frac{\partial v}{\partial p}\right)_\tau\right].$$

The computed values are shown in Table XXXIII. and Figure 42. Initially the internal energy decreases along an isothermal, but the rate of decrease becomes rapidly less with rising pressure, eventually changing sign, so that at the higher pressures the internal energy increases on an isothermal with increasing pressure. This means that initially the work done in compressing the water is more than lost by dissipation of the high heat of compression, but at higher pressures, the mechanical work per unit rise of pressure has increased so rapidly because of the high pressure that part of the mechanical work expended in compressing the water is retained as increased potential energy after temperature equilibrium has been restored. Except for a region of abnormal curvature, between 2000 and 5000, which is evidently due to the change from an abnormal to a normal liquid, the behavior of $\left(\dfrac{\partial E}{\partial p}\right)_\tau$ found here is probably typical for any liquid. For mercury, $\left(\dfrac{\partial E}{\partial p}\right)_\tau$ is initially negative, but it becomes greater algebraically with rising pressure, so that in the mercury paper it was suggested that probably at high enough pressures the energy would increase instead of decrease along an isothermal. Here we have an actual case where the pressure has been pushed far enough to secure this increase. This might have important applications to astrophysics or geophysics, since it shows the possibility of storing up very large amounts of energy in the interior of a star or the earth in virtue of the pressure alone, quite apart from the high temperatures.

Two other quantities of thermodynamic interest, the adiabatic compressibility and the temperature effect of compression, may be roughly approximated to. For these we have the formulae

$$\left(\frac{\partial v}{\partial p}\right)_\phi = \left(\frac{\partial v}{\partial p}\right)_\tau - \frac{\tau}{C_p}\left(\frac{\partial v}{\partial \tau}\right)_p^2,$$

and

$$\left(\frac{\partial \tau}{\partial p}\right)_\phi = \frac{\tau\left(\dfrac{\partial v}{\partial \tau}\right)_p}{C_p}.$$

Both of these involve the specific heat, which cannot be found from the data obtained, for the specific heat involves the temperature derivative of the dilatation by the well-known relation

$$\left(\frac{\partial C_p}{\partial p}\right)_\tau = -\tau\left(\frac{\partial^2 v}{\partial \tau^2}\right)_p.$$

Measurements of the compressibility at a number of temperatures would be necessary to obtain this. At high pressures, however, $\left(\dfrac{\partial^2 v}{\partial \tau^2}\right)_p$ becomes less very rapidly. Amagat's data for water show that already at 3000 kgm. and for a temperature range at least from 0° to 30° $\left(\dfrac{\partial^2 v}{\partial \tau^2}\right)_p$ has vanished within the limits of accuracy. We may assume, then, that at high pressure C_p shows a very slow change. For the rough approximation given here, C_p was taken as constant at 0.9. Tumlirz found C_p at 2000 to be 0.86, but, as already remarked, his value is probably too low. The merely suggestive values for $\left(\dfrac{\partial v}{\partial p}\right)_\phi - \left(\dfrac{\partial v}{\partial p}\right)_\tau$ and $\left(\dfrac{\partial \tau}{\partial p}\right)_\phi$ calculated in this way are shown in Table

XXXIII. The difference between adiabatic and isothermal compressibility increases to a maximum and then decreases. The rise of temperature produced by the application of 1 kgm. pressure also increases and then decreases again. The normal behavior of both these quantities, as shown by mercury, is a continuous decrease, so that here again, we have an effect of the transition from abnormal to normal.

The quantities involved in the change of state from one form to another are shown collectively in the folder at the end, where the equilibrium curves, the change of volume curves, and the latent heat curves are plotted on the same scale for all the modifications. The fundamental question as to the change of state liquid-solid may be stated much more definitely than any fundamental question for the theory of liquids. This fundamental question is as to the ultimate behavior of the liquid-solid curve. Does it end abruptly, indicating a critical point for the transition solid-liquid as many have maintained, or does it rise to a maximum and then descend, as Tammann has claimed in combating the idea of a critical point, or does it merely continue rising indefinitely to infinite pressures and temperatures? Evidently none of these things have happened within the domain of the present diagram for water, nor have they happened in the low range up to 3000 or 4000 used before for any other liquid. The only hold we get on this question is by an extrapolation. In this we are very greatly helped by the behavior of the latent heat and the change of volume, for evidently an extrapolation of the equilibrium curve alone is absolutely incompetent to decide whether it is going to stop abruptly or not. But if this curve has an end or a maximum, then the latent heat and the change of volume must behave in a definite manner with respect to each other. At a critical point, the latent heat and the change of volume must vanish together, while at a maximum, the change of volume becomes zero, the latent heat remaining finite.

Tammann's argument for the probable existence of a maximum comes from observing the general trend of the latent heat and the change of volume on the equilibrium curve. Tammann could not make any very accurate measurements of the change of volume, but they were accurate enough to show that for the substances tried up to 2000 or 3000 kgm. the change of volume becomes less at high pressure, but the latent heat remains nearly constant. The change of volume is approximately linear with temperature on the equilibrium line. Whence by an extrapolation, Tammann concluded that the change of volume would pass through zero before the latent heat, and that therefore the equilibrium curve has a maximum. He has calculated the probable position of this maximum for a number of substances, assuming the melting curve to be a parabola, but this extrapolation is open to very great question. He himself remarks that at high pressures the equilibrium curves tend to show less curvature than one would expect from their behavior at low pressures.

The idea of a maximum seems opposed to our common-sense feeling of what to expect. If there is a maximum, it is possible by taking the substance through an isothermal cycle from the domain of the liquid into that of the solid and back into that of the liquid again to find a necessary connection between the compressibility of liquid and solid over a wide pressure range. This is unexpected in view of our present experience that there is no necessary connection between the properties of liquid and of solid. It is to be noticed that the nearest approach to a maximum found here, on the II–L curve, was neatly avoided by the appearance of another form of ice.

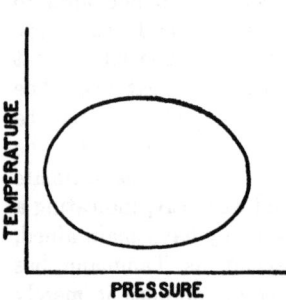

FIGURE 43. Tammann's complete equilibrium curve between liquid and crystal. The crystal is stable only within the closed region.

Proceeding from the probable existence of a maximum, Tammann has developed his well-known theory of the nature of the complete equilibrium curve between liquid and solid. The ideal curve (Figure 43) according to this theory is a closed curve, the crystalline solid having existence only in the interior of the curve. The complete curve may not be realizable for all substances, since part of the curve may fall at negative pressures or at temperatures below the absolute zero. As a matter of fact, only the two upper quadrants have been realized for known substances, and even then, no substance has been found in both the two upper quadrants. The upper left-hand quadrant is that for normal substances, while the upper right-hand quadrant shows the behavior of water and ice I.

We turn now to the evidence on these points afforded by the present work on water. First for the equilibrium curves alone. These all show curvature in the direction demanded by Tammann's complete diagram, on the I–L curve the fall of temperature becoming more rapid at higher pressures, and on the other curves the rise of temperature becoming less rapid with rising pressure. Except on the I–L curve, this behavior is just exactly what one would expect on nearly any conceivable theory, the effect of temperature becoming less at higher pressures. This is the behavior also on the liquid-vapor curve, which ends in a critical point. From the point of view of Tammann's theory, it is unfortunate that the form I gives place to III at higher pressures, for in the ideal diagram the behavior of I is the normal behavior at high pressure and that of III at low pressures, while here we have a form which should be adapted for the high pressures giving place at high pressure to one apparently appropriate to low pressures.

The change of volume curves next concern us. These also all show the general behavior demanded by Tammann's theory, the change of volume solid-liquid becoming algebraically less at high pressures. On the I-L curve this means that the change becomes numerically greater. The approximate reason of this has already appeared from the discussion of the compressibility of the solid to be merely that the solid is more incompressible than the liquid, whether it has the greater or smaller volume. The curvature of these change of volume curves is also everywhere, except for the curve VI-L, such as to suggest that the change of volume becomes zero at some finite temperature not very far removed from the temperatures actually reached.

The latent heat curves also bear out Tammann's point of view, for they all rise at the higher temperatures on the equilibrium curves. The direction of curvature of these latent heat curves appears to be governed by no such general rule as the change of volume curves, since the curve may be either concave or convex toward the temperature axis.

So far, for the forms of ice I, III, and V, which are stable at low pressures, everything seems as indicated by Tammann's theory. It should be remarked that the pressure range of existence of these forms is twice that reached before. It is on the VI-L curve, however, which reaches to much higher pressures, that we find the significant suggestion as to what to expect at still higher pressures. This suggestion comes from the change of volume curve, which shows a pronounced point of inflection in the neighborhood of 30°.[20] Below 30°, the curvature is like that for the other modifications at low pressures, indicating the vanishing of the change of volume at perhaps 50° or 60°, but beyond 30° the change of volume decreases less and less rapidly with rising temperature, with the possibility of becoming asymptotic.

---

[20] With regard to the effect of probable experimental error at the high pressure it is to be said that the effect of this would be to make the inflection shown in the diagram appear less pronounced than it really is. The change of volume at high pressures is too low if anything, because in making the correction for the change in the bore of the cylinder under pressure it was assumed that the increase of bore is linear with pressure, whereas, if anything, it increases more rapidly at high pressures.

That is, at the high pressures there is no indication that the change of volume will ever become zero. This inflection in the volume curve is also mirrored by a corresponding inflection in the latent heat curve, which rises more and more rapidly at the upper end. Also this change of direction of the volume curve occupies the same general locality as the region on the compressibility curves for the liquid where the compressibility ceases to decrease as fast as one would expect from the behavior at low pressures. This behavior of the volume curve, together with that of the latent heat curve, shows in the first place that the latent heat and the change of volume do not vanish together, so that there can be no critical point, and in the second place, that the change of volume apparently will not vanish at any finite temperature, so that we will not have a maximum as supposed by Tammann, but the curve will rise instead to infinite pressures and temperatures.

Recently J. J. van Laar [21] has been developing a theory of the solid state which is more far reaching than that of Tammann, in that it attempts to show the actual mechanism which makes a liquid pass to the solid. This theory explains the solid state by the association of the simple molecules to molecular complexes. For the sake of simplicity, the theory has been developed for the case where the complexes are double molecules, although this restriction is not necessary. Given, then, a liquid in which both single and double molecules may exist, van Laar has found, by writing down the thermodynamic potential of the two kinds of molecules, how the dissociation of the double molecules into single molecules varies with pressure, volume, and temperature. Accompanying the dissociation is a change of volume, for the volume of the double molecule is not in general twice that of the two single molecules from which it comes. This change of volume, due to dissociation, is found to so modify van der Waals' equation, which is still supposed to hold for either kind of molecule separately, that an isotherm now has two maxima and two minima, instead of the single maximum and minimum of van der Waals' original equation. This evidently means the existence of a new phase, the solid, the equilibrium conditions of which are determined in the same way as the equilibrium conditions liquid-gas of the ordinary equation, by applying the condition that the work done in a reversible isothermal cycle is zero.

By detailed numerical computation, van Laar has shown how on this theory the equilibrium pressure solid-liquid changes with increasing temperature. The results are similar to those of Tammann in that a

---

[21] van Laar, Proc. Amster., **11**, 765–780 (1909); **12**, 120–132, 133–141 (1909); **13**, 454–475, 636–649 (1910).

maximum melting temperature and a maximum melting pressure are both predicted. That is, in Figure 43, van Laar has the same maximum and the same right-hand vertical tangent as Tammann, but the results differ from Tammann's in that the minimum and the left-hand vertical tangent cannot exist, or at any rate if they do, they must always lie at temperatures below the absolute zero and at negative pressures.

These results of van Laar were obtained with the specific assumption that the actual volume of the molecules, and so the change of volume when a double molecule passes into two single molecules, is independent of temperature and pressure. This is almost certainly not the case. The value of the compressibility of water at high pressures, the way in which the abnormalities are smoothed out in the neighborhood of $0°$, and the point of inflection in the ΔV curve for VI above $0°$, all suggest most strongly that the assumption is not true, and furthermore that it is not approximately enough true to enable even the general character of the melting curve to be predicted at high pressures.

The conclusion of the whole matter seems to be that at high pressures, over 10,000 kgm. for water, we have a new effect appearing, probably connected with the compressibility of the atoms. This means that at high pressures the compressibility of the liquid and solid are going to become more and more nearly equal, which will have as a consequence that the equilibrium curve will continue rising indefinitely.

Besides the data just discussed for the liquid-solid curves, we have the corresponding data for the solid-solid curves. There is no theory at present of the equilibrium solid-solid, and the data here only bear out a remark of Roozeboom [22] that different allotropic solids would be expected to show every conceivable relation to each other. Two triple points between three solid phases had been found, I–II–III, and II–III–V. The first of these is of a type already known, but the second is of a type of which, according to Roozeboom, no examples have yet been discovered. This is Roozeboom's sixth type.[23] The equilibrium lines are for the most part straight, but this merely indicates that the compressibility and thermal dilatation of the solids are nearly constant over the range of temperature and pressure in question, as one might expect. The only curved equilibrium lines are I–III and II–III. In both of these III is involved. But it was to be expected *a priori* that III is a form of ice with more variable properties than the others, be-

---

[22] Roozeboom, Die Heterogenen Gleichwichte, vol. 1, p. 206. (Vieweg, Braunchweig, 1901).

[23] Roozeboom, loc. cit. p. 202.

cause of the close approach of its equilibrium curve with water to a maximum. In general, the internal energy increases on passing from a solid stable at low pressures to the solid stable at higher pressures, but III–II is an exception. The most interesting features found for the equilibrium solid-solid are the probable passing of the I–II curve through the absolute zero, the enormous increase in reaction velocity on approaching a triple point, and the existence in one crystalline form of nuclei about which crystallization to another form may begin.

Acknowledgment is here made of several liberal appropriations from the Rumford Fund of the American Academy of Arts and Sciences, with which the expenses of this investigation were partially defrayed.

JEFFERSON PHYSICAL LABORATORY,
HARVARD UNIVERSITY, CAMBRIDGE, MASS.,
    OCTOBER, 1912.

[The values of ΔH and ΔE on page 502 are in error.]

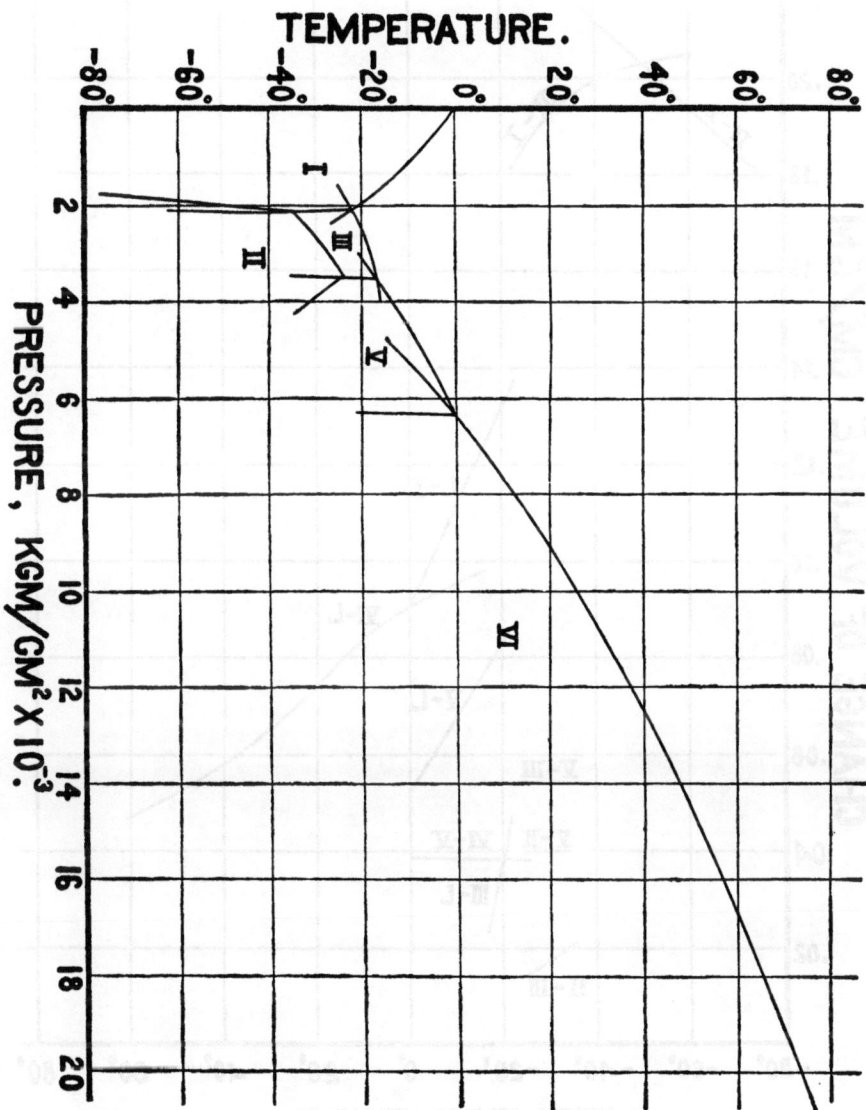

The equilibrium diagram between the liquid and the five solid modifications of water.

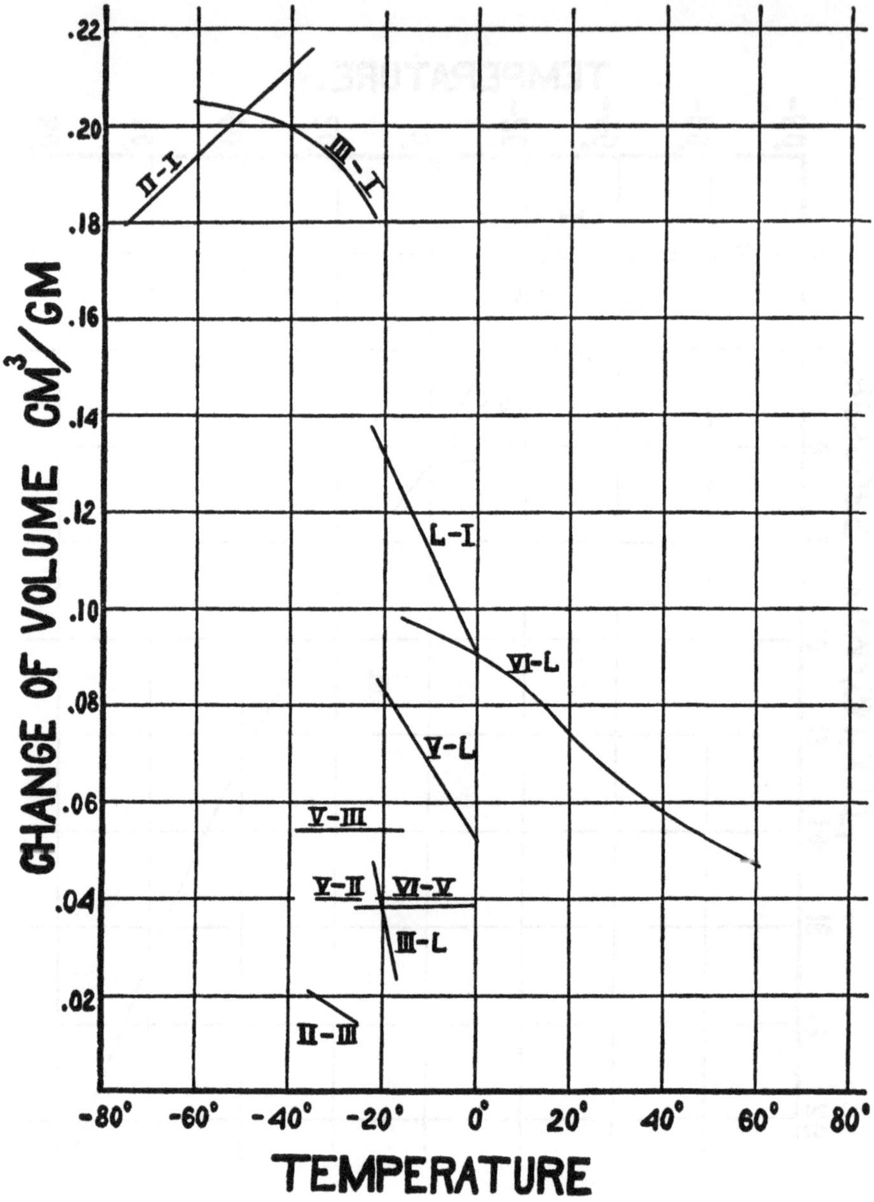

The change of volume when one modification passes to another under equilibrium conditions.

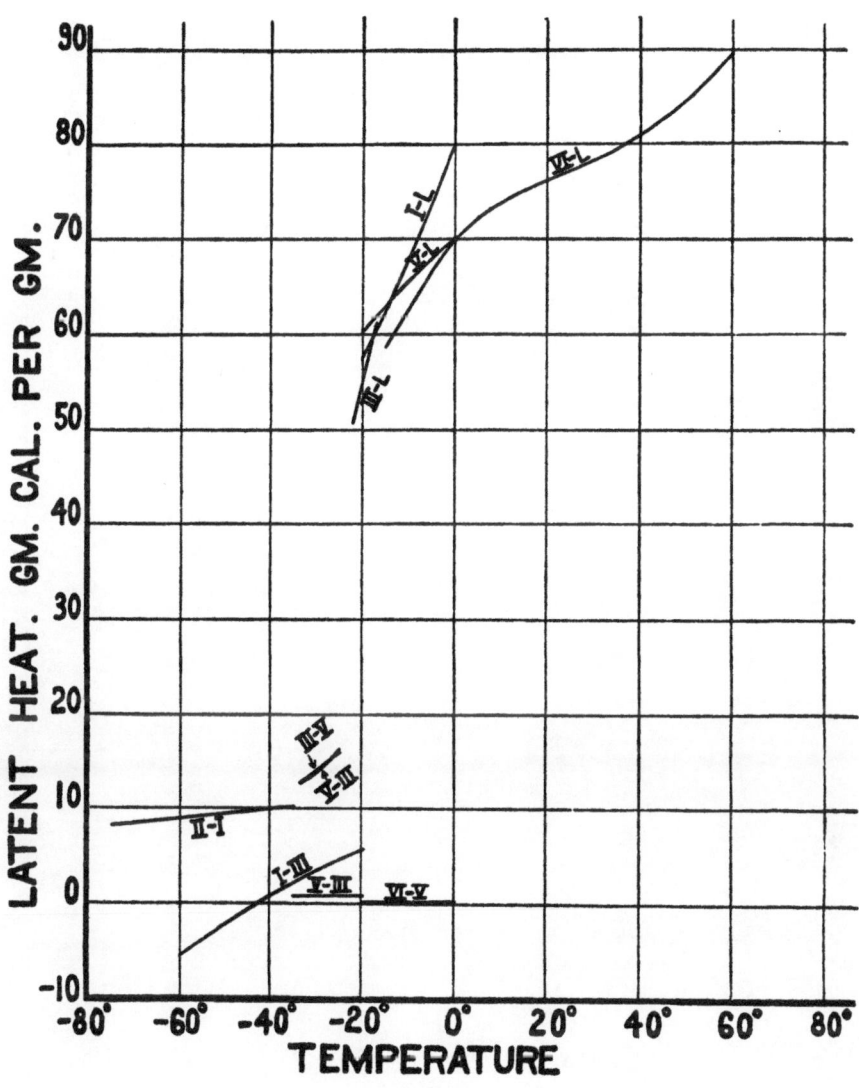

The latent heat when one modification passes to another under equilibrium conditions.

THE COLLAPSE OF THICK CYLINDERS UNDER HIGH HYDROSTATIC PRESSURE.

By P. W. Bridgman.

ALL metals exhibit, when exposed to stresses at all high, behavior other than the simple proportionality between stress and strain assumed in the mathematical treatment of elasticity. There are a number of effects invalidating this simple linear relation; such as elastic after effects, hysteresis, plastic yield, set, raising of the elastic limit by over strain, and one universal effect, rupture. Any complete description of the stress-resisting mechanism of a solid must provide explanations of all these effects. Conversely, from a more careful study and fuller knowledge of these obscure effects we may expect to be able to more completely characterize the internal mechanism of a solid. Our present knowledge of these effects is restricted, however, because of the narrow range of stress within which they have been studied. The reason for this is that all of these effects become of considerable magnitude only for high values of the stress, and all usual experiments with high stresses are brought to a speedy close by the rupture produced by the high stress. Thus, for example, a bar strained in tension shows, beyond a certain value of the tension, a yield point followed by an interval of stress within which the metal has the entirely new property of plasticity. But this new property of plasticity can be studied over only a comparatively narrow range, for rupture occurs very shortly after the yield point.

In this paper, experiments are described in which the applied stress is of such a nature that rupture will never occur, no matter how high the stress. In consequence, it has been possible to study these various effects over a range of stress very much higher than available under ordinary conditions of yield. Thus plastic yield has been observed over a range of stress twelve times as high as that required to produce the first beginning of flow. It is the purpose of this paper to describe these experiments somewhat in detail, and to comment on the interesting features displayed by the various effects. The paper is intended only to be suggestive; it would be demanding too much to expect a complete theory from experiments with a single type of stress.

The experiments are tests on thick hollow cylinders, closed at the ends, and subjected to hydrostatic pressure over the entire external surface. Similar tests, in which the walls of the cylinders are comparatively thin, are of familiar occurrence in engineering practice. Under such conditions, the tube fails by collapse, folding in toward the center in one or more creases. That failure takes place in this way is due to the fact that beyond a certain value of the stress the circular figure of the tube becomes unstable, so that very slight geometrical imperfections cause collapse. So slight is the requisite geometric imperfection after the pressure of instability has been reached, that it is possible to obtain very consistent results in collapsing tests of this type. Of course if the figure of the tube were absolutely perfect, collapse by an unsymmetrical folding could never occur. In the tests to be described here, on the other hand, the walls of the cylinders are so heavy that the figure does not become unstable, and yielding to pressure can not be asymmetric. The only conceivable method of yield under these conditions is by a uniform flow in toward the center, which is what actually does happen. It is to be noticed that flow is in such a direction as to reduce the possibility of still further flow under higher pressure. In the case of a rod under tension, the flow is in the direction of elongation, which is capable of indefinite extension. Too great flow of a bar under tension is followed by rupture, but with these cylinders, too great flow can be followed only by complete closing of the central cavity; rupture can never occur.

Two systematic sets of experiments were made on this subject. One, with copper and steel cylinders, was extended to 12,000 kgm./cm.^{2}. These experiments show the relation between plastic yield and the raising of the elastic limit over this range of stress. For the copper cylinders the range is sufficient to exhaust all the possibilities, since the hole was completely closed at less than the maximum pressure. The second set of experiments, to a lower stress maximum of only 7,000 kgm., was made on a series of steel cylinders. The arrangements of this set were such that the relation between pressure and internal volume could be measured continuously during application and release of pressure, whereas for the first set only the permanent set was measured after every application of pressure. The second set gives evidence, therefore, on such questions as hysteresis, elastic-after-effects, and the linear relation between stress and strain. The observations only of these two sets of tests will be presented at first, comment being reserved until the data are all in hand.

The first set of tests on both copper and steel was made on cylinders 2 inches long and 5/16 inch outside diameter. Seven of these cylinders of each metal were used for the tests, the internal diameter varying in steps of 1/64 inch from 1/16 inch to 5/32 inch. The holes were drilled first and then the exterior turned concentric with the hole. The copper cylinders were made from commercial rod, softened by heating to redness; the steel cylinders were made from a mild bessemer boiler plate which had been proved by special test to be particularly uniform in every direction. The cylinders were closed at the ends as shown in Fig. 1 by stoppers of hardened steel fitting into the ends, and leak was prevented by a thin rubber tube attached to the two stoppers and covering the entire cylinder. The advantage of this method of closing the ends is that after every application of pressure the arrangement may be easily taken apart, and both external and internal diameter measured. It is then possible to subject the cylinder to a higher pressure, so that a record of the relation between set and stress may be obtained after every advancing pressure step. The stoppers of course introduce an end effect, since the projections on the stoppers prevent colapse at the very end, but the effect of these was seldom sensible at a distance from the stopper greater than the original diameter of the hole. The cylinders were subjected to pressure, seven at a time, in a large pressure chamber. The fluid transmitting pressure to the cylinders was a mixture of glucose and glycerine. Pressure was measured by an absolute pressure gauge inserted directly into the same chamber with the cylinders. This is the same absolute gauge which has been described in detail elsewhere.[1]

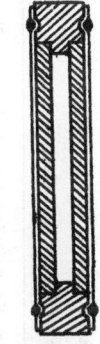

Fig. 1.

Shows the test cylinders and the manner of protecting the outside with a rubber tube so as to prevent leak.

Observations were made at intervals of approximately 1,000 kgm. Pressure was pushed to the maximum and kept there for a few hours, long enough for complete disappearance of the plastic yield, and then released. The cylinders were then measured outside and inside. The outside measurements were made with an ordinary micrometer reading to 0.0001 inch. The inside measurements were made by finding the diameter of a wire which would just slip through the hole. For this purpose a complete set of drill rods of all the different sizes was provided. Intermediate sizes were obtained by filing a rod in a lathe until it would

[1] Proc. Amer. Acad., 47, 1911, pp. 319–343.

Table I.
Collapsing Tests on Copper Cylinders.

Max. Pressure, Kgm/cm²	Cylinder No. 1		Cylinder No. 2		Cylinder No. 3		Cylinder No. 4		Cylinder No. 5		Cylinder No. 6		Cylinder No. 7		Results with Cylinders Fresh for Each Application.		
	O.D.	I.D.	O.D.	I.D.	O.D.	I.D.	O.D.	I.D.	O.D.	I.D.	O.D.	I.D.	O.D.	I.D.	Original inside diam. of cylinders in this column was .0625″.	Original inside diam. of cylinders in this column was .093″.	Original inside diam. of cylinders in this column was .140″.
0	.3130	.0625	.3132	.078	.313	.093	.314	.110	.313	.125	.313	.140	.312	.157			
2000	.311	.0515	.310 .0005	.063	.308 .0002	.076	.306 .0002	.089	.304 .0003	.098	.300 .0007	.1065	.292 .0087	.110	The final inside diameters are recorded.		
3000	.310	.045	.308	.055	.306 .0002	.063	.3025	.072	.299	.081	.296 .0005	.0855	.273 .0362	.054			
4000	.308 .0002	.038	.306 .0001	.0405	.303 .0003	.0465	.2985	.053	.293 .0005	.0565	.286 .0065	.058			.0275	leak	collapse
5000	.308 .0002	.0275	.305 .0002	.033	.301 .0002	.038	.296 .0002	.042	.291 .0012	.045	.283 .0070	.045			.020	.0285	.0265
6000	.307	.020	.303 .0002	.0275	.300 .0002	.029	.295 .0002	.0305	.289 .0006	.0305	.283 .0150	.0275			.022	.023	.022
7000	.307 .0002	.014	.303 .0002	.016	.300 .0002	.0195	.294 .0002	.020	.289 .0007	.019	.280 .0050	.0125			.010	.013	.008
8000	.306 .0002	.010	.303 .0002	.011	.299 .0002	.012	.293 .0002	.013	.288 .0003	.011	.279 .0050	.0075			.000	.000	.000
9000	.306 .0002	.005	.302 .0005	.005	.299 .0002	.0075	.293 .0005	.007	.287 .0005	.006	.303 .0045	—			.000	.000	.000
10000	Closed		Closed		Closed		Closed		Closed								

just fit. The smaller sizes were tested with wires drawn through a special draw plate. The internal diameter of the hole was uniform except at the very ends, and in general it remained perfectly round after flow, so that it was possible to measure the diameter of the hole to at least 0.0005 inch. After the measurements for one pressure, the cylinders were reassembled again with the stoppers, ready for a run to a pressure 1,000 kgm. higher. At every new higher pressure, two or three fresh cylinders were included with the original seven. The purpose of this was to determine whether the previous tests to lower pressures had affected the behavior at the higher pressures.

The results of the tests for the copper are shown in Table I. and are plotted in Fig. 2. The columns headed O.D. (outside diameter) contain

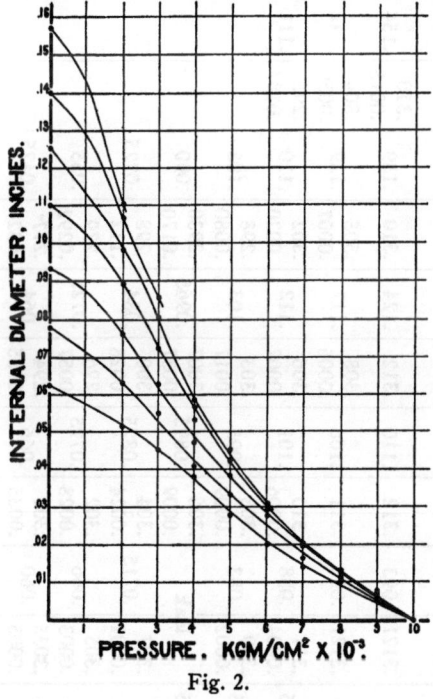

Fig. 2.

The results of the collapsing tests on the copper cylinders. The internal diameter is given as a function of the maximum pressure to which the cylinder has been subjected.

two figures corresponding to each pressure. The upper of these gives the mean external diameter of the cylinder after subjecting to pressure. The lower figure gives the eccentricity produced in the cylinder by the pressure, that is, the difference between the least and greatest external diameters. The results are somewhat irregular, as they always are for tests of this character, but there is sufficient regularity to show well the

TABLE II.

Collapsing Tests on Iron Cylinders.

Max. Pressure, Kgm/cm².	Cylinder No. 1		Cylinder No. 2		Cylinder No. 3		Cylinder No. 4		Cylinder No. 5		Cylinder No. 6		Cylinder No. 7		Results with Cylinders Fresh for Each Application.		
	O.D.	I.D.	O.D.	I.D.	O.D.	I.D.	O.D.	I.D.	O.D.	I.D.	O.D.	I.D.	O.D.	I.D.	Original internal diameter	Original internal diameter	Original internal diameter
															.0625″.	.093″.	.140″.
0	.3105	.0625	.3112	.078	.3121	.093	.312	.110	.312	.125	.311	.140	.311	.157		The final internal diameter is re-recorded.	
2000	.3105	.0625	.3112	.078	.3121	.093	.312	.110	.312	.124	.310 .0015	.139	.310 .0015	.154	.0612		
3000	.3105	.0625	.3112	.078	.3121	.093	.311	.106	.308 .0006	.117	.305 .0007	.129	.304 .0050	.142	.0582	.085	.099
4000	.3105	.060	.311 .0006	.0755	.311 .0004	.088	.310 .0006	.101	.306 .0007	.112	.302 .0020	.110	.295 .0187	.111	.055	.080	.089
5000	leak		.310 .0006	.0725	.309 .0002	.083	.308 .0009	.095	.303 .0017	.103	.298 .0080	.105			.053	.074	.062
6000	.3095	.0515	.308 .0002	.0675	leak	.0715	.306 .0009	.0895	.3007 .0037	.0965	.2930 .0170	.090			.051	.067	.034
7000	.308 .0002	.050	.307 .0002	.063	.307 .0001		.304 .0024	.0825	.300 .0028	.084	.298 .112	.0725			.046	leak	.0355
8000	.308 .0002	.046	.3065 .0005	.059	.305 .0003	.066	.302 .0028	.0725	.296 .0057	.0725	.279 .0292	.055			.041	.046	Collapse
9000	.3065 .0005	.0425	.3065 .0005	.056	.3035 .0005	.060	.301 .0035	.066	.294 .0055	.064	.277 .032	.0425			.035	.046	.023
10000	.3070 .0005	.040	.3055 .0015	.050	.303 .0002	.054	.2995 .0035	.0565	.292 .0070	.0565	.273 .0333	.032			.034	.0415	.023
11000	.3067 .0007	.037	.3047 .0017	.045	.3025 ?	.043	.2985 .0040	.053	.2905 .0075	.049							
12000	.3065 .0005	.033	.3045 .0015	.041	.3015 .0005	.035	.248 .0040	.037	.289 .0072	.0415							

8 — 340

general tendency of the results. The two cylinders with the thinnest walls showed the collapse characteristic of thin walled tubes, and the results with these two cylinders are not plotted in the figure. The observations were continued, however, on one of these cylinders in which the hole was initially 0.140 inch diameter, because the collapsing did not increase with rising pressure, but rather became less as the walls thickened under continued flow toward the center. The table shows how the eccentricity of this cylinder rose to a maximum and then decreased again. The results with the three cylinders which were freshly subjected to pressure each time were more irregular than the above. Within the limits of error, however, there were scarcely perceptible variations between the old and the fresh cylinders. If anything, the tendency was for the fresh cylinders to show the greater yield. The tests on the original six copper cylinders were not continued beyond 10,000 kgm., because at this pressure the hole had closed completely at the ends, and in order to find whether the hole was closed completely all the way through, it was necessary to cut the cylinders in two. The cylinders were found to be closed throughout the entire length except in a few isolated places where complete closing was prevented by a layer of scale which had cracked off from the inside of the hole. The tests on fresh pieces was continued up to 12,000, however. The results at 10,000, 11,000 and 12,000 were all the same, complete closing of the inner hole throughout the entire length of the cylinder.

The corresponding tests for the soft iron cylinders are given in Table II. and Fig. 3. The results with the iron were in general character the same as for the copper. The plastic flow was less, however, so that it was possible to push the pressure on more of the cylinders to the pressure of instability. Only the two heaviest cylinders have failed to begin the collapsing process at a pressure of 12,000 kgm. The results with the sets of three fresh cylinders put in for every new application of pressure were more regular than for the copper. In general, except for the irregularities attending collapse or growing eccentricity of figure, it made no difference whether the cylinder had been subjected to lower pressures previously or not.

Besides the measurements tabulated, measurements were also made of the length, both of the steel and of the copper cylinders. Only slight variations were found, not over 0.002 inch, and these were irregular. Evidently there was no change of length due to the collapsing of the cylinder, but the slight irregularities observed were due to the automatic seating of the hardened stoppers.

The second series of tests to lower pressures was made on seven steel cylinders simultaneously. These cylinders were of the form shown in Fig. 4. The diameter is 1/2 inch at B and 5/8 inch at A. The lower part was exposed to pressure from A to B, while at the upper end a stem 5/16

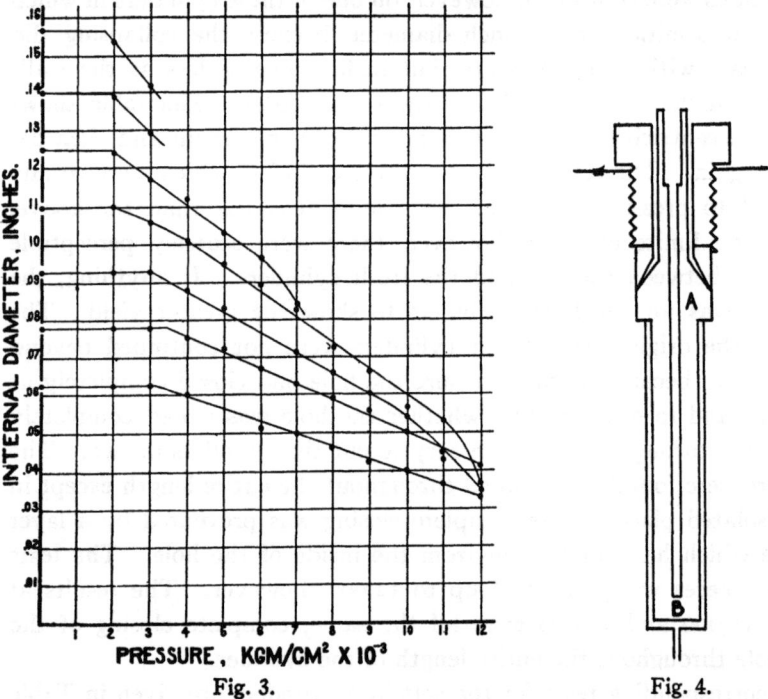

Fig. 3.

Fig. 4.

The results of the collapsing tests on the iron cylinders. Similar to the tests on the copper cylinders.

The form of steel cylinder with which the change of internal volume was measured as a continuous function of the pressure.

inch in diameter projected into the air through the conical steel packing as shown. This 5/16 inch stem was sealed directly to a calibrated and graduated glass capillary tube. The different cylinders differed in the material and the size of the inner hole. Two were of soft bessemer steel, with an interior hole of 1/4 inch and 1/8 inch respectively. The other five cylinders were of tool steel, 1.25 per cent. carbon, left soft, with holes ranging from 1/8 inch to 1/4 inch by steps of 1/32 inch. All of these seven cylinders were placed in a single large block of steel, the different cavities being in communication with each other, so that all seven were exposed to the same pressure, and the stress history of all seven was identical. This block of steel was a particularly homogeneous piece of open hearth steel presented to the laboratory for the purposes of this research by the Bethlehem Steel Co. Grateful acknowledgment is hereby made of this courtesy. In making the cylinders, considerable care was necessary to drill the hole accurately concentric the whole length. That sufficient accuracy was obtained was shown by the regularity of the results, and by the special examination of those cylinders which it was necessary to cut open after the tests. The size of the holes was found accurately by weighing the mercury which filled them to a known depth.

To make these readings, the entire interior of the cylinders, and the capillary up to a certain mark, were filled with mercury by exhausting the cylinder and working the mercury in through the capillary. The effect of an application of pressure to the outside of the cylinder is to decrease the internal volume, producing a rise of mercury in the capillary. The amount of this rise was recorded as a function of the pressure, and from the known dimensions of the several parts the equivalent change of internal volume or of internal radius was found. There was no change of length during the tests.

The procedure was to apply pressure in several steps to a maximum and then release pressure in steps to zero. After every change of pressure the position of the mercury in all seven capillaries was read, and recorded

TABLE III.

Pressure History of the Seven Steel Cylinders of the Second Set of Tests.

Time.		Successive Maxima and Minima of Pressure Kgm/cm².	Time.		Successive Maxima and Minima of Pressure Kgm/cm².	Time.		Successive Maxima and Minima of Pressure Kgm/cm².
Day.	Hour.		Day.	Hour.		Day.	Hour.	
April 24	10.05 A.M.	0	April 28	12.50 P.M.	} 6,490	Apr. 30	12.16 P.M.	1,760
	10.31	1,800		1.27			12.33	6,030
	11.00	0		3.00	0		12.41	3,680
	11.33	1,770		4.45	6,320		12.50	6,300
	12.03 P.M.	0		5.50	} 0		1.03	1,720
	12.55	3,380	April 29	10.13 A.M.			1.11	4,440
	1.21	} 0		11.47	} 6,400		1.20	} 1,530
	2.48			2.10 P.M.			1.58	
	3.18	3,050		2.44	0		2.09	0
	3.51	0		3.54	6,380			
	5.04	4,070	April 30	5.00	} 0			
				9.20 A.M.				
	5.37	} 0		9.50	6,380			
April 27	9.55 A.M.			10.18	0			
	12.43 P.M.	} 5,250		10.35	2,160			
	1.25			10.49	0			
	2.31	} 0		11.17	4,350			
	2.47			11.35	0			
	4.23	5,790		11.45	2,000			
	5.10	} 0		11.52	910			
April 28	9.44 A.M.			12.08 P.M.	4,060			

as a function of the pressure. It was necessary to wait after every change of pressure a sufficient length of time for elastic-after-effects or plastic yield to entirely subside. The pressure was measured with a mercury resistance gauge of the type described in Proc. Amer. Acad., No. 9, 1909. The pressure measurements could be made to 1/10 per cent., more than was necessary from the self consistency of the other readings. After describing a pressure cycle as above, another similar cycle was described reaching to a higher maximum pressure than the first, and then another, with a still higher maximum. In all, fourteen such cycles were described. The essential history of these cycles is shown in Table III., giving the maximum pressure of each cycle and the corresponding times. After this series of readings, the cylinders were removed from the block and the dimensions measured again. To show the complete record of the behavior of all these cylinders would take a great deal of space and is hardly necessary, because the characteristic features shown by all are the same. Two cylinders are taken as typical of the lot, and the various cycles of these shown in Figs. 5 and 6. In Fig. 5 the actual observed points are indicated, in order to show the general order of accuracy and regularity of the results. The observed points are omitted in the subsequent figures, however, as they tend to obscure the diagram. The succession of cycles described on April 30 was different from that on the other days; these cycles will be discussed in detail later.

Before discussing the results of these two sets of experiments, it will pay to examine the conditions of stress and strain produced in such a cylinder by the applied stress system. The solution is very easily obtained, and may be found in any book on elasticity. The precise formulation of the problem is as follows. Given a cylinder of external radius a, internal radius b, with a hydrostatic pressure P applied to the external surface, zero pressure to the internal surface, and a compressive stress across planes perpendicular to the axis uniform throughout the mass of the metal and of amount $P\dfrac{a^2}{a^2-b^2}$. This last statement amounts simply to neglecting the end effects and supposing the entire hydrostatic pressure exerted over the closed ends to be supported uniformly by the walls of the cylinder. The solution under these conditions is found to be as follows, using the ordinary notation.

For the displacements,
$$u_r = -\frac{Pa^2}{a^2-b^2}\left[\frac{b^2}{2\mu r}+\frac{r}{3\kappa}\right],$$
$$u_\theta = 0,$$
$$u_z = -\frac{a^2 z}{3\kappa(a^2-b^2)}P,$$

where μ is the shear modulus, and κ the compressibility modulus. That is, the radial displacement is negative, in toward the center, at all points

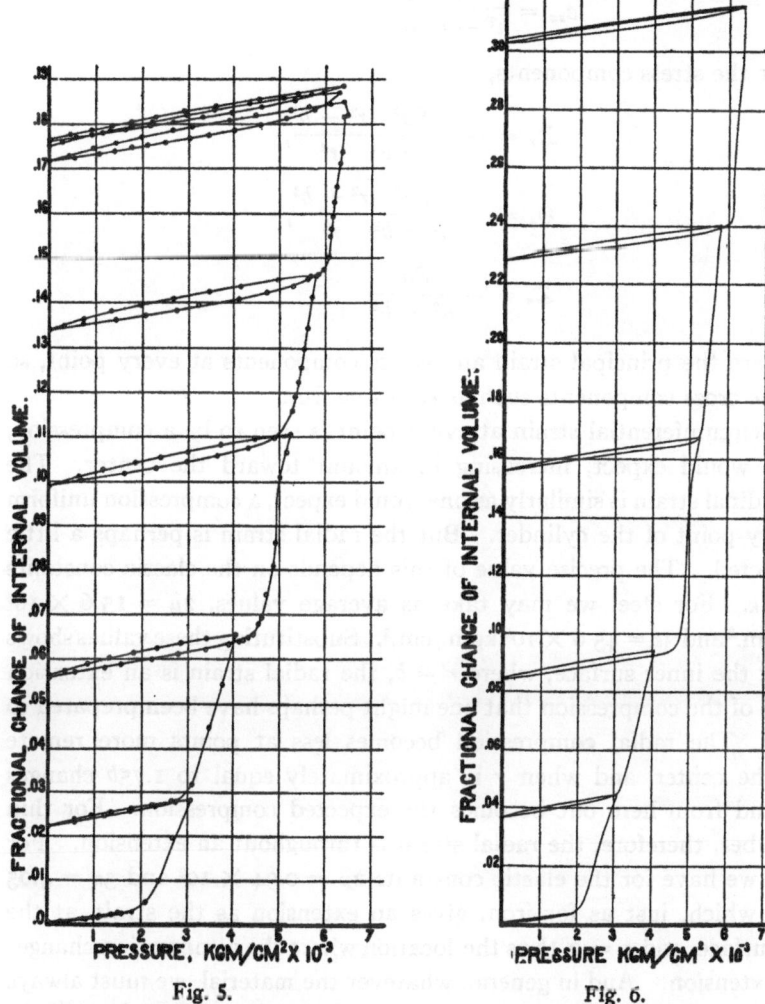

Fig. 5.

Shows the hysteresis loops obtained with a cylinder similar to that of Fig. 4. The internal diameter of this cylinder was originally 3/16 inch, the outside diameter 1/2 inch. This cylinder was of tool steel.

Fig. 6.

Another test like that of Fig. 5. The outside diameter of this cylinder was originally 1/2 inch, and the inside diameter 7/32 inch. This also was of tool steel.

of the cylinder, and the longitudinal displacement is a corresponding compression. The same solution gives for the strain components,

$$e_{rr} = \frac{a^2 P}{a^2 - b^2}\left[\frac{b^2}{2\mu r^2} - \frac{1}{3\kappa}\right],$$

$$e_{\theta\theta} = \frac{-a^2 P}{a^2 - b^2}\left[\frac{b^2}{2\mu r^2} + \frac{1}{3\kappa}\right],$$

$$e_{zz} = \frac{-a^2 P}{a^2 - b^2}\frac{1}{3\kappa},$$

and for the stress components,

$$R_r = -\frac{Pa^2}{a^2 - b^2}\frac{r^2 - b^2}{r^2},$$

$$\Theta_\theta = -\frac{Pa^2}{a^2 - b^2}\frac{r^2 + b^2}{r^2},$$

$$Z_z = -\frac{Pa^2}{a^2 - b^2}.$$

These are the principal strain and stress components at every point, so that the cross components such as $e_{r\theta}$ all vanish.

The circumferential strain at every point is seen to be a compression, as one would expect, increasing in amount toward the center. The longitudinal strain is similarly as one would expect, a compression uniform at every point of the cylinder. But the radial strain is perhaps a little unexpected. The precise value of this depends on the elastic constants μ and κ. For steel we may take as average values, $2\mu = 15.6 \times 10^5$ kgm./cm.^{2} and $3\kappa = 45.6 \times 10^5$ kgm./cm.^{2}. Substituting these values shows that at the inner surface, where $r = b$, the radial strain is an extension instead of the compression that one might perhaps have been prepared to expect. The radial compression becomes less at points more remote from the center, and when r is approximately equal to $1.75b$ changes sign, and from here out becomes the expected compression. For thin steel tubes, therefore, the radial strain is throughout an extension. For copper we have for the elastic constants $2\mu = 0.94 \times 10^6$ and $3\kappa = 4.05 \times 10^6$, which, just as for iron, gives an extension as the strain at the inner surface, and $r = 2.1b$ as the location where the compression changes to an extension. And in general, whatever the material, we must always have $3\kappa > 2\mu$, for this is merely the condition that Poisson's ratio be positive, and therefore always at the inner surface the radial strain will be an extension, whatever the material.

The circumferential strain is seen to be always a compression, and is greatest at the inner surface. The longitudinal strain is a compression, uniform everywhere, so that the sections remain plane. The greatest shearing strain comes at the inner surface and is equal to $Pa^2/\mu(a^2 - b^2)$, being greatest for thin-walled cylinders.

All the three principle stresses at any point are seen to be compressions, except the radial stress, which becomes zero at the inner surface only. The magnitude of the circumferential compression is greater at every point than the radial compression, and the magnitude of the difference is greatest at the inner surface.

These equations hold only while the metal is within the elastic limit. If this limit is exceeded, the relations break down. It is impossible to tell in our present state of knowledge what the new relation must be. It would be possible to make a variety of hypotheses and write down the consequent stress components. The only feature necessarily common to these various systems of possible stresses would be the satisfying of the conditions that at the interior surface $R_r = 0$, and at the exterior surface $R_r = -P$. We also have, by considerations of symmetry, that even after the plastic yield has begun, the directions of principle stress continue to be the same mutually perpendicular directions as before. The components of strain cease to have any definite meaning after plastic yield has continued for any time. The elastic components of stress and strain do enable us to see what to expect, however, when yield begins.

Let us inquire first what we can deduce from the simple fact that flow exists, and that plastic yield does come to an end, as shown by the first set of experiments. Yield begins at the inside surface as flow toward the center. This is evidently connected with the radial extension at the inner surface; when this radial extension becomes greater than the material can stand, there is viscous yield in the direction of the extension. It is to be noticed that at the inner surface where the yield is greatest there is no stress in the direction of yield. It is not necessary, then, that there should be stress in the direction of flow any more than that there should be stress in the direction of elastic yield below the elastic limit. The immediate cause of the flow is the existence of stresses at right angles to the direction of flow. The experiments also show the converse fact that flow which might be expected to take place in a given direction may be prevented from taking place by forces at right angles to that direction. This is shown by the fact that none of the cylinders showed any change of length under the longitudinal compressing force. This stress, in the case of the copper cylinders subjected to 12,000 kgm., is eight or ten times as much as would produce longitudinal set in a pure compression test. The longitudinal flow was evidently prevented from taking place by the force at right angles. What is a bit surprising is that one of these forces at right angles is always less than the longitudinal stress and may be zero. This is sufficient to show that the relations between the three principal stresses and direction of flow, to say nothing of the magnitude of the flow, are not perfectly simple. This relation must be established by experiment. It may be pointed out, however, that in this case the flow takes place entirely in the plane of the greatest and least principal stress, that is, in the plane of maximum shear stress.

After questions as to the fact of flow, the next natural consideration is as to how far the flow will proceed before it stops. This is a complicated matter and the question cannot be answered as yet. A complete solution must take account of the change of geometric shape with flow, of the hardening produced in the metal by flow, which varies in different metals, and of the unknown relation between the principal stresses for metals in the plastic state, which again probably varies for the different metals. In view of all these possibilities it would be impossible even to guess whether the flow would continue indefinitely or not; the experimental fact that the hole does close up is a new contribution which could not have been predicted. The mere fact that the hole does close up, moreover, is sufficient to enable us to rule out one relation between the stresses during plasticity which has been used tentatively a number of times in lack of any more probable hypothesis. This is the condition of maximum stress difference; namely, that during yield the difference between the greatest and the least principal stresses cannot exceed a certain value. This possible stress difference may not be perfectly constant, but may be increased slightly by the hardening of the metal under flow. In any event, however, the stress difference cannot rise above some definite maximum. This condition admits of precise mathematical formulation, so that we can completely solve the problem of stress distribution. We have the equations:

$$R_r = \Theta_\theta + K,$$

which is the hypothesis of maximum stress difference, where K is the value of this difference. We also have the equation of equilibrium

$$\int_b^r \Theta_\theta dr = rR_r.$$

The solution of these two simultaneous equations is readily found to be

$$R_r = -K \log \frac{r}{b},$$

which gives at the exterior surface:

$$P = K \log \frac{a}{b},$$

where P is the external collapsing pressure, a the external and b the internal radius. That is, under any given external pressure the dimensions of the cylinder will be so changed by the hydrostatic pressure as to satisfy the given relation. The equation shows that for every value of P there is a corresponding value of b. That is, the hole will never be closed up by any pressure no matter how large. The maximum stress difference criterion is not valid, therefore, for plastic flow under high stresses. The effect of hardening by flow would evidently be to decrease instead of increase the flow under any given pressure, so that we are not helped by taking account of the hardening. It must be, then, that flow beyond a certain point weakens the metal so that it is able to withstand considerably less stress difference than it could in the incipient stages of flow.

The problem cannot be attacked backwards. There are an infinite number of possible relations between Θ_θ and R_r which would give the experimental relation between pressure and internal radius. We can see roughly what the general nature of the stress distribution must be, however. The interior surface is to some extent a free surface, because yield can occur at this place. R_r at this surface is always zero, and from what we have just seen the value of Θ_θ at this surface cannot exceed a certain value, but must on the contrary tend to become less as the exterior pressure rises. But the average value of Θ_θ throughout the mass of the cylinder tends to become greater, for this must balance the external hydrostatic pressure. That is,

$$\int_b^a \Theta_\theta dr = -Pa.$$

Θ_θ must rapidly increase from the center out therefore. R_r similarly is zero at the inner surface but becomes equal to $-P$ at the outside. At the outer surface, therefore, the condition approaches one of uniform hydrostatic pressure. The possibilities of supporting stress when the metal is in such a state of hydrostatic pressure are indefinite. But at the inner surface the condition becomes one of greater and greater plasticity under a one-sided stress in a direction at right angles to that of yield.

Consideration of what happens when the external pressure is removed next concerns us. The pressure is supposed to have been applied so long that flow has ceased and the metal has entirely accommodated itself to the high stress. We may consider the strain, calculated from this new state of ease, to change roughly in the same way that the strain would in a condition of perfect elasticity when external pressure is removed. This is the reverse of the strain found above when pressure is applied to the outside. That is, when pressure is removed, the circumferential strain at the inner surface becomes an extension and the radial strain a compression; at the external surface the circumferential strain is also an extension, but the radial strain has changed from a compression to an extension. If the original strain had remained elastic under the applied pressure, the reverse strains when pressure was removed would have sufficed merely to wipe out the original strain. But here the strains are reckoned from a new state of ease at the maximum pressure, so that the metal is left with internal strains after the release of pressure. At the inner surface this strain will be a circumferential extension, which furthermore may be of considerable amount, because the pressure has been released through such a wide range. It is perfectly conceivable that this strain introduced by the release of pressure might be sufficient to exceed the elastic limit in extension, or might even be so great as to produce rupture at the internal surface.

That precisely this is the nature of the reverse strains has been verified by experiment. In an attempt to make tubing capable of withstanding high pressures, collapsing experiments like these were tried. Since it was not possible to obtain commercially drawn tubing with a hole small enough for some intended uses, it was thought that by collapsing the tube in this way it might be possible to make a tube of the requisite dimensions. None of the tubes so treated, however, were capable of standing as much pressure as the original tubes with the larger hole, and several of the treated tubes burst at practically no pressure at all. It was found on cutting these tubes open that the inside was cracked from the center practically out to the surface, evidently because of the internal strains set up by relieving pressure. The same thing was verified on several specimens of tubing which had been collapsed and then cut open without subjecting to internal pressure. These tubes were hard drawn of a steel with about 0.5 per cent. carbon. The effect was never found in the softer copper or steel tubes of the tests above.

In view of the internal strains evidently introduced after every release of pressure, one might well be prepared to expect the repeated application and release of the same pressure to be followed by progressive yield, the cylinder finally closing up under repeated applications of the same pressure. That this is not the case, however, is the first thing evident from an inspection of the diagrams of the second set of tests. With every successive application of pressure, yield is not resumed until the previous pressure maximum has been reached or exceeded. The same thing is shown in the first set of experiments by the approximate identity of the results obtained at any pressure with the fresh cylinders and with the cylinders which had been subjected a number of times previously to lower pressures.

The very fact, then, that the plastic limit can be so raised by pressure is unexpected and occasion for some thought. One might be prepared for a raising of the yield point to twice its original value. It has been shown, for example, that if the yield point of a bar be raised by overstrain in tension, the corresponding yield point in compression is lowered by the same amount, so that the total stress range, tension to compression, within which the bar behaves elastically remains constant. So here, one would be prepared for a shifting of the range from, in the first case, extension to equal compression, to, in the second case, zero extension to double the initial compression. But an increase in the range of ten or twelve times as in the case of the copper is unexpected. It points to some deep seated molecular rearrangement.

The second set of experiments shows instructive minor differences in the way in which the elastic limit may be raised. For instance, the correspondence is not perfect between the new yield point and the previous maximum; yield may occur either before or after the maximum. In Fig. 5 the old maximum was exceeded at 4,000 and 5,800 kgm., but was almost exactly reached at 3,000 and 5,000. On consulting the log of the experiment on page 9 it will be found that there is a distinct connection here with the time during which the metal has been resting. At 4,000 the cylinders had been resting under no stress for 64 hours, and at 5,800 for 16 hours, while at 3,000 and 5,000 the time of rest had been comparatively short, an hour or so. This is but a verification of common experience with tensile and similar tests, that it takes time for the hardening by over-strain to be affected. If the metal rests under no load, the elastic limit may be raised beyond the old value, whereas if stress is immediately reapplied without allowing any time for accommodation, the effect of the over-strain is to lower the elastic limit. The diagrams also show that the rapidity of recovery during rest may vary. In Fig. 6 the interval of an hour has been sufficient for complete recovery, but not in Fig. 5. The cylinder of Fig. 5 is of tool steel with a hole 3/16 inch in diameter, while Fig. 6 shows the bessemer cylinder with 1/4 inch hole. The yield of the tool steel cylinder to pressure is less than that of the bessemer, but the rate of recovery is slower. Corresponding to the slower yield are the heavier walls. This is only one case of the general observation that, beyond the initial stages, recovery is slower in the heavier masses of metal, irrespective of the total amount of yield, and, other things being equal, that the recovery is more rapid in the bessemer than in the tool steel. The internal structure of the bessemer is to be thought of as simpler than that of the tool steel.

The two diagrams Figs. 5 and 6 show one other characteristic of all the cylinders; that the initial rate of yield in the neighborhood of a previous maximum is slower than normal if the metal has not completely recovered by rest, but if the yield point has been raised by prolonged resting, the yield, when it does occur, is abnormally rapid. Both of these effects result in a resumption of the old yield curve a little way beyond the previous maximum. The release of pressure to zero and reapplication does not essentially alter the character of the yield curve, therefore, but is merely a temporary incident, the effect of which speedily disappears.

There is another effect shown by all the cylinders similar in many respects to that of plastic yield, namely elastic after effect. After an increase of pressure the result of this is that yield continues slowly for a while, gradually subsiding to zero. In this respect it is similar to plastic yield, but it may occur at lower values of the stress than is required to produce what is ordinarily recognized as plastic yield. After decreasing pressure, however, the elastic-after-effect produces a gradual creep in the contrary direction to that of plastic yield. This elastic-after-effect was observed in all the cylinders after the release of pressure to zero. Of course at the maximum pressure it could not be separated from the plastic yield. The effect was in general small, too small to show on the scale of the diagrams, and remained nearly constant whatever the range of the pressure. Here, then, is one of the unusual effects which does not become very much greater for a range of stress far beyond the elastic limit. The effect was greater in the heavier cylinders which showed the least set, and was noticeably less for the bessemer than for the tool steel.

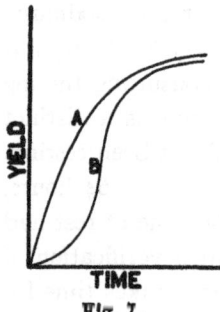

Fig. 7.
Shows two types of yield curve. *A* shows the normal type of curve; *B* the type after prolonged resting.

The time rate of yield was also observed a number of times. It is of course greater and extends over a longer interval of time at the higher pressures. The normal shape of the yield curve is shown at *A* in Fig. 7. But there is in addition a change in the character of the yield curves at the higher pressures. At the low pressures, the curve plotting yield against time is very steep at the origin, but as the yield pressure increases, the curves show a more uniform rise for a greater interval of time. Thus at low pressures, half the total yield might occupy only a tenth of the time required for nine tenths of the yield, but at higher pressures half the yield might consume one third of the time required for nine tenths. The same difference in character was also shown at the same pressure by different cylinders. The yield curves for bessemer steel were very markedly more steep than those for tool steel, and among the tool steel cylinders, those with the thinnest walls showed the steepest yield curves. This shows again that these various effects, such as elastic-after-effects, time rate of recovery, and viscous yield, take more time to run to completion in the larger mass of metal. All of them would seem due to some process of molecular readjustment, which must travel through the steel from places of greater distortion to those of less. The process takes more time when there is a greater mass of metal to be readjusted.

The yield curve in one case was found to be an exception in general shape to all the others. This is shown at B in Fig. 7. This was at 4,000 kgm., where yield was resumed after resting for 60 hours. The yield seemed particularly reluctant to get started in this case. It ran very slowly at first, then with gradually increasing velocity, and then slowed down again to an asymptote. The curve had the form shown. Evidently the effect is due to the gradual disappearance of the hardening under the previous load.

The change in the character of the yield curve with increasing pressure has an application to ordinary testing. Frequently two points are distinguished during a tensile test, for example: a point where the material first shows set and a point later on where it shows plastic yield. It seems probable that no sharp distinction between these two points can be maintained. The set point is merely a yield point at low stresses where the process of yield takes place so rapidly as to escape notice. With increasing stress the yield curve flattens out, until it has become slow enough to observe, when we have the ordinary yield point.

There is another feature shown by all the cylinders, the loops described when pressure is released and reapplied. These loops are imperfect in Figs. 5 and 6, being complicated by hardening effects and elastic-after-effects. The fact that the loops exist is perfectly evident, however, as also the fact that these loops become wider as the pressure range increases, the increase in width being more rapid than the increase in pressure. The loops are of a width much greater than could be produced by elastic-after-effects, although these effects do tend to produce loops when the metal is carried through rapid stress cycles. These are genuine hysteresis loops. To show the character of the loops more plainly after the metal had become perfectly seasoned so as to show no further hardening effects, pressure was increased and released a number of times over the range of 6,400 kgm., as shown in the upper part of Figs. 5 and 6, until the behavior had settled down to a reproducible cycle. A number of cycles were then described similar to the cycles in making tests for magnetic hysteresis. Three of these cycles are shown in Figs. 8, 9 and

10. They are different in appearance, but were selected because they all show certain traits in common. Fig. 8 is for the tool steel cylinder with 5/32 inch hole, Fig. 9 for the bessemer cylinder with 1/4 inch hole, and Fig. 10 for the other bessemer cylinder with 1/8 inch hole. The reversal in the normal direction of curvature of the cycle for the 1/4 inch bessemer is due to two causes. This cylinder collapsed under pressure, flattening

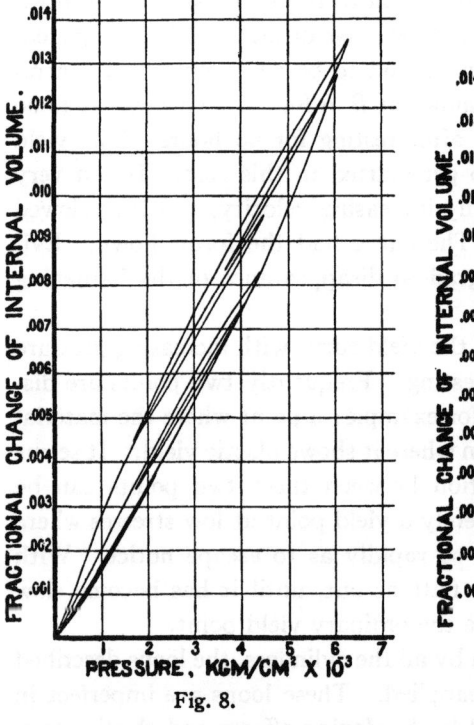

Fig. 8.

Hysteresis cycle after thorough seasoning. This cylinder was of tool steel with an internal diameter initially 5/32 inch.

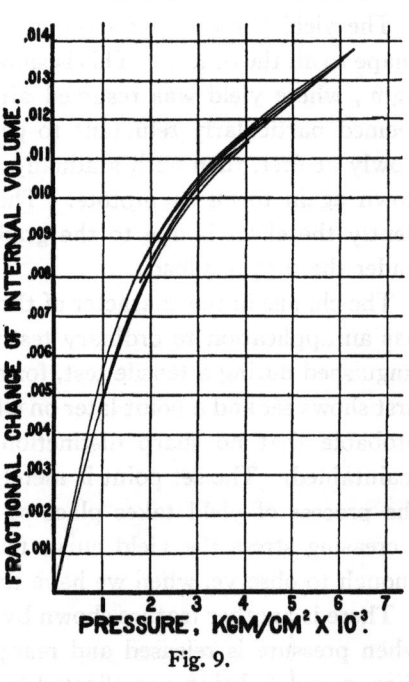

Fig. 9.

Hysteresis cycle of bessemer cylinder with hole initially 1/4 inch.

out until it received support from the walls of the steel block, and in addition there was placed inside this cylinder a piece of iron wire, so as to reduce the volume of the mercury and so the temperature coefficient. During collapse, the walls of the cylinder received support also from this piece of wire. Fig. 9 is not illustrative at all, then, of the normal behavior of a collapsed cylinder, but is included to show that the hysteresis of even so complicated a system as this shows the same general characteristics as the simpler cases.

The characteristics shown in common by all these hysteresis loops are also the same as those shown in magnetic hysteresis cycles. The branch of the loop described with increasing pressure lies at all points below the return branch with decreasing stress. If from any point a small cycle is described by reversing the direction in which stress has been applied before reaching that point, then on restoring pressure to the original value at the point in question, the strain will recover its initial value also. So much is common to all these cycles, in fact is common to all hysteresis cycles of whatever known cause. Beyond this, the loops show great differences of shape and size. The loops for the bessemer steel are narrower than for the tool steel, as indeed all these effects have been less in the bessemer. Among the tool steel cylinders, the widest loops are shown by the cylinders with the thickest walls, again confirming the

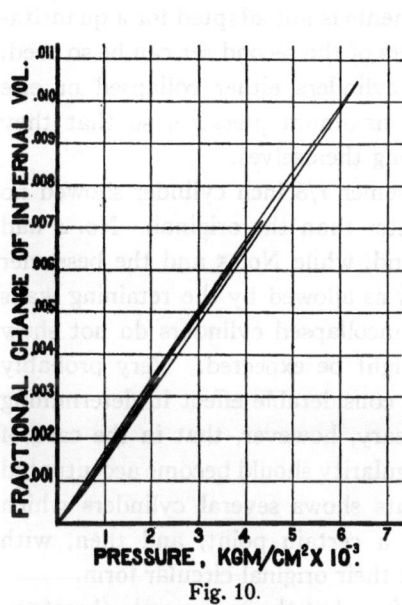

Fig. 10.
Hysteresis cycle for bessemer with hole initially 1/8 inch.

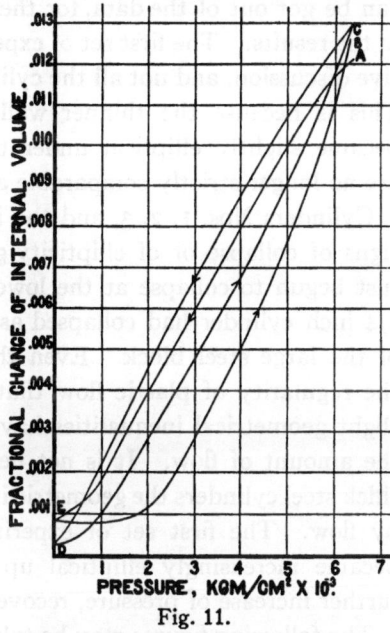

Fig. 11.
Abnormal hysteresis loops before complete seasoning for the tool steel cylinder with hole initially 1/8 inch.

experience with the other effects. The width of the hysteresis loops may be unexpectedly great. Thus in Fig. 8, the width of the loop is one seventh of the total strain under the maximum pressure, and is equal to 2/100 of the original volume, as much as the entire elastic deflection before yield began.

The behavior of one of these cylinders during the seasoning process preparatory to describing the hysteresis loops was so remarkable as to deserve special comment. This was cylinder No. 1, that of tool steel with the 1/8 inch hole. Fig. 11 shows the first cycle described during the seasoning process, and also the final shape to which the cycle settled down after several applications of pressure. The unusual features are shown at the lower end. The initial effect of the first increase of external pressure was to increase the internal volume, instead of giving the normal decrease. This abnormal behavior gradually disappeared with successive applications, but the memory of it survived in an abnormally low slope at the lower end. The precise mechanism of this anomalous effect is not at all clear. It is significant that the effect is shown in the cylinder with the heaviest walls. The next heavier cylinder, No. 2, shows also an abnormally low initial slope, although an actual increase of internal volume under external pressure was never observed here.

All this discussion so far has been qualitative. This is about all that can be got out of the data, for there is not much quantitative regularity in the results. The first set of experiments is not adapted for a quantitative discussion, and not all the cylinders of the second set can be so used. This is because the thinner walled cylinders either collapsed or else became slightly elliptical under the maximum pressure, so that they are no longer strictly comparable among themselves.

Cylinders Nos. 1, 2, 3, and the bessemer 1/8 inch cylinder showed no signs of collapse or of ellipticity greater than the original. No. 4 had just begun to collapse at the lower end, while No. 5 and the bessemer 1/4 inch cylinder had collapsed as far as allowed by the retaining walls of the large steel block. Even the uncollapsed cylinders do not show the regularity of plastic flow that might be expected. Very probably slight geometrical inequalities have a considerable effect in determining the amount of flow. It is not necessary, however, that in the case of thick steel cylinders the geometric irregularity should become accentuated by flow. The first set of experiments shows several cylinders which became increasingly elliptical up to a certain point, and then, with further increase of pressure, recovered their original circular form.

The following figures may be taken for what they are worth, therefore. Table IV. for the tool steel cylinders shows the ratio of the volume set of the different cylinders to that of the thickest cylinder, No. 1. The ratio for the different cylinders is seen to decrease with rising pressure. This is evidently due to the fact that cylinder No. 1 does not acquire the normal rate of flow until the pressure has been increased for a considerable way beyond the first yield point. This is shown by all the cylinders in the gentle curvature of the lower end of the yield curve. It is shown most strikingly of all by the bessemer 1/8 inch cylinder, where the pressure does not seem to have been pushed far enough to reach the steep part of the yield curve. The maximum yield for the 1/8 inch bessemer cylinder was 0.012 against 0.09 for the 1/8 inch tool steel. Evidently this first part of the curve is the location of the usual hardening effects of over-strain. It is only the initial gently rounded parts of this curve that have been available in ordinary tests, such as tension tests. As pressure is pushed beyond this initial part of the curve, the rate of yield seems to settle down to a steady value depending on the form of the cylinder. Beyond 5,000, the yield of cylinders 1, 2 and 3 progresses at nearly the same rate for each. The increase of the ratio beyond 5,000 for cylinders 4 and 5 is evidently connected with their collapse.

TABLE IV.
Plastic Yield of Steel Cylinders.

Max. Pressure.	Volume Set of No. 1.	Ratio to Vol. Set of No. 1.			
		No. 2.	No. 3.	No. 4.	No. 5.
3,100	0.005	1.80	4.40	7.20	9.80
4,050	0.022	1.48	2.54	3.78	4.77
5,250	0.051	1.25	1.94	3.06	4.61
5,800	0.070	1.23	1.91	3.24	4.93
6,400	0.090	1.23	1.94	3.38	5.83
		Ratio of Elastic Vol. Yield.			
0		1.21	1.14	1.34	0.99

The table shows no simple relation between the rate of plastic yield and the ratios of the original elastic yield before flow began. In fact the elastic yield does not show the simple dependence on the dimensions demanded by theory. This must be due to geometric imperfections of figure.

As far as the actual value of the elastic constants goes, the plastic yield seems to have made very little difference. This may be found by comparing the slope of the cycles described after accommodation to 6,400 with the initial slope before yield had begun, taking account also of the changed dimensions. Only cylinders 2 and 3 are available for this, because No. 1 shows an anomalous cycle and 4 and 5 have been collapsed. The elastic constants of 2 and 3 have not changed over 5 per cent., one being an increase and the other a decrease.

SUMMARY.

In this paper the behavior of hollow steel and copper cylinders subjected to external pressure has been examined over a range of stress many fold greater than the original elastic limit. It is found that:

1. The yield under any given pressure does not continue indefinitely, but stops after a while. With the next application of pressure, yield is not resumed until the old maximum is reached.

2. The cylinder cannot support an indefinite pressure, but closes up tight at some fixed pressure, the same no matter what the original dimensions of the cylinder. For copper, this is about 10,000 kgm. For soft steel, judging from an extrapolation, it must be at about 20,000

kgm. The fact that the cylinder closes up shows that the maximum stress difference criterion for flow is not valid. The maximum stress difference that the material can support decreases after prolonged flow.

3. There need be no stress in the direction of flow, and a stress ordinarily great enough to produce flow need not necessarily produce such flow under all conditions. In the cases considered here, flow can take place only in the plane of greatest and least principal stress.

4. The internal adjustment, or whatever it is that enables the metal to stand this greatly enhanced pressure without yield, does not result in a general raising of the resistance to all kinds of stress, but it is an accommodation only to the particular type of stress which produced the yield. A cylinder collapsed by external pressure bursts under less than the normal internal pressure.

5. The time rate of plastic yield undergoes modification as the pressure increases. At low pressure the greater part of the yield occurs in the initial stages, but at higher pressures the yield is more evenly distributed in time. The slower rate of yield is found in larger masses of metal, even although the total yield may not be so great, and the rate is slower in tool steel than in bessemer steel. The tool steel has a more complicated internal structure.

6. The usual hardening effect of over-strain is shown. A certain interval of rest after the over-strain is necessary for the hardening. This interval is greater for large masses of metal, and is greater in tool steel than in bessemer.

7. Elastic-after-effects are about normal. They do not increase markedly with increasing pressure.

8. Hysteresis is shown by all the cylinders. It becomes rapidly greater at the higher pressures, and is much greater than under ordinary conditions of test. The breadth of a hysteresis loop may amount to the entire elastic deflection before yield began. Hysteresis shows a tendency to be greater in the larger mass of metal, and is very much greater in tool steel than in bessemer.

9. The thickest tool steel cylinder showed anomalous results during accommodation. An increase of external pressure was followed by an increase of internal volume.

All of these effects are probably connected in some way with the break-up of unstable molecular complexes, and the formation of new complexes stable under the new conditions. This process takes a longer time in large masses of metal, and produces greater changes in the properties of a material with a complex structure than in a simpler material.

JEFFERSON PHYSICAL LABORATORY,
    HARVARD UNIVERSITY, CAMBRIDGE, MASS.

Breaking Tests under Hydrostatic Pressure and Conditions of Rupture. By P. W. BRIDGMAN.

[Plate II.]

WITHIN the last few years a number of papers[*] have appeared, either written by engineers or else of engineering interest, dealing with the conditions under which rupture is produced in the materials of ordinary practice. The objects of these papers has been to find if possible some criterion by which rupture may be predicted, whatever the type of applied stress. The present state of opinion seems to be that for ductile materials the maximum shearing stress plays the principal part, but that for brittle materials the maximum principal stress is the determining factor. At the same time it is pretty generally recognized that neither of these criteria is likely to be actually correct, but is at best only an approximation likely to give fairly good results for the materials of ordinary engineering practice under ordinary practical systems of load.

A general consideration of what may be the determining factors in producing rupture under so wide a variation in the nature of the applied stress that there is no immediate relation to the needs of engineering seems to have been neglected. Yet it is precisely such a consideration of the conditions of rupture under as wide a range of the conditions as possible that is likely to lead to a true theory of rupture, and so to a better formulation of the conditions for the range of ordinary practice.

In the course of a number of experiments on very high hydrostatic pressures, the author has observed many cases of rupture which have a bearing on the present question. The pressures which it has been found possible to reach are considerably in excess of any hitherto produced in fluids, pressures of 30,000 atmos. having been repeatedly attained. Under these pressures all ordinary theories as to the behaviour of elastic solids break down completely, and the entire subject had to be approached from the beginning. During the two or three years of preliminary work spent in acquiring familiarity with the methods by which these pressures might

[*] Scoble, Phil. Mag. xii. pp. 533–547 (1906); Hancock, Phil. Mag. xii. pp. 418–425, 426–430 (1906), and xvi. pp. 720–725 (1908); Gulliver, Proc. Roy. Soc. Edin. xxix. pp. 427–431 (1909); Smith, Eng. lxxxviii. pp. 238–243 (1909); Scoble, Phys. Soc. Lond., Nov. 26, 1909; xxii. pp. 130–146 (1910).

be handled, many cases of rupture have been observed. Several of these appear to be of types not observed before, and they have special bearing on the present question. In this paper three of these types of rupture will be described somewhat in detail, and a discussion given of the bearing of these tests on theories of rupture. No attempt will be made to develop a new theory. All that it is desired to do is to point out that in consequence of these tests the true criterion of rupture must be a much more complicated affair than is ordinarily supposed, and that considerations must be introduced which have been so far neglected, for these tests enable us to summarily dismiss all the conditions of rupture hitherto proposed.

The tests to be described in this paper fall naturally into three types, and the conditions of rupture to which they have application are also three in number, there being one other generally recognized possibility besides the two mentioned in the first paragraph. Each of the three types of tests gives some evidence on the validity of each of the three conditions of rupture, but it will be found that each of the three types of test has its most direct bearing on only one of the conditions. The argument of the paper is grouped about the three different types of test. The nature of the test is first described, then the bearing on the corresponding condition of rupture, and finally the incidental bearing on the other conditions of rupture. The argument is furthermore somewhat complicated by the necessity for keeping in mind the possibility of different criteria holding for brittle and ductile bodies.

The three conditions of rupture to be considered are the condition of maximum principal stress, of maximum stress difference, and of maximum strain. The first demands that rupture occur when either principal stress exceeds a certain value, whether this stress is compressive or tensile. The second demands that failure occur when the greatest shearing stress, or what is the same thing the difference between the greatest and the least principal stress, exceeds a certain value. The third criterion demands that rupture occur if the extension in any direction exceeds a critical value.

Another question of much practical interest, and under ordinary conditions closely related to the question of rupture, is the question as to the conditions under which a material will receive a permanent set. The three criteria just enumerated are very often used interchangeably as either criteria of rupture or as criteria of set.

The first criterion, as originally stated, evidently cannot

hold without modification, for it demands that a solid body when subjected to uniform hydrostatic pressure all over should break when the pressure rises too high. It is inconceivable in this case how the body can break, and the existence of the heavenly bodies with enormous internal pressures is conclusive proof to the contrary. It is still possible, however, that this criterion should be valid as determining set, and that the body might show volume set if the pressure exceeded a certain critical value. This is a question of some interest, and it might possibly seem reasonable at first sight to expect some such set. Kahlbaum* has published results on this subject showing the very surprising result that after the application of pressures up to 10,000 atmos. the density of most metals is increased, due to closing up of small pores, but that beyond 10,000 the density is decreased. Although no systematic investigation of this question has been made in the present work, several incidental determinations have been made up to 25,000 or 30,000 atmos., and no trace of a change of density has ever been found except in those cases where the metal was obviously porous. There seems to be no question but that Kahlbaum's results were due to the fact that he used castor oil to transmit pressure. This freezes under a few thousand atmos., a fact that Kahlbaum overlooked, so that beyond this, his pressure was no longer hydrostatic. This is shown most strikingly by Kahlbaum's own statement that the test cubes were often curiously distorted after application of pressure. Kahlbaum saw in this support of his theory that all metals tend to become plastic or even fluid under high hydrostatic pressures. But it can be stated unqualifiedly from the present work that, on the contrary, substances tend to become more rigid under high pressures. The distortion of the metal cubes found by Kahlbaum is to be explained by the fact that the effect of pressure in increasing rigidity is very different for different substances. It has been found in the present work that under 20,000 atmos. paraffin wax may become more rigid than Bessemer steel.

The first criterion is usually modified, therefore, so as to predict rupture when the tension alone, instead of either tension or compression, exceeds anywhere a critical value.

The first type of test has its most direct bearing on the maximum tensile stress criterion of rupture. In these tests cylinders were exposed to pressure over the curved surface

* Kahlbaum, Roth, und Seidler, Z.S. Anorg. Chem. pp. 29–30 254–294 (1902).

only, the ends being left unsupported. Fig. 1 illustrates the manner of applying pressure. The rod A, the subject

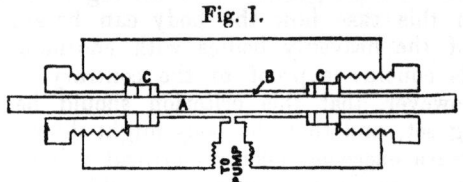

Fig. 1.

Apparatus for producing the "pinching-off" effect, that is, separation of the longitudinal fibres by the application of pressure to the curved surface of a cylinder. The specimen is shown at A, the fluid exerting the pressure by which rupture is produced is contained in the annular space at B.

of the test, passes completely through the cylinder B, projecting at either end through the packing rings C. The cylinder is connected to the pressure-pump through the indicated connexions and stress applied to the test specimen by the pressure of the fluid in the annular space between the specimen A and the interior wall of the cylinder. The specimen fails by separation of the particles across some plane perpendicular to the axis, the two disconnected ends of the specimen being expelled with violence through the packing rings. The fracture does not take place at the packing rings as might be expected, but, whether for brittle or ductile materials, takes place at some point well between the rings. The nature of the process of rupture is evidently merely that of squeezing the rod out sidewise. This type of rupture may therefore be referred to as the "pinching-off-effect." There is no longitudinal stress except that due to the friction of the packing, and this a stress of compression rather than one of tension, so that here we actually have the fibres separating against the direction of stress.

The nature of the fracture varies with the material. In the case of a rod of mild steel, the rupture looks very much like that of an ordinary tensile break, except that the necking down is likely to be a little more abrupt. Pl. II. fig. 2 shows a photograph of one such specimen. Other soft materials such as copper or brass show the same manner of rupture. Harder materials, such as hardened chrome-nickel steel or vanadium steel, show irregular fracture, a combination of necking down and of slip on shear planes at approximately 45° to the axis. Glass-hard tool-steel, on the other hand, shows a clean break at right angles to the axis without necking down. In the same way it is possible to break glass rod or heavy glass tubing. The fracture is beautifully clean, exactly at right angles to the axis.

In the cases of the ductile materials, the test is complicated somewhat by the fact that after the necking down has once begun there is a tensile stress tending to pull the bar apart. But this tensile stress cannot account for the beginning of the necking down, so that the tensile stress present during the actual rupture is only an incident due to the particular form of experiment, and is not at all the true cause of the rupture. Evidently there is no such complication in the case of the brittle materials which break with no necking down.

This first type of test disposes of the maximum stress criterion, therefore, as applied to either ductile or brittle materials. It shows *a fortiori*, therefore, that this criterion cannot be applied to brittle materials in contradistinction to ductile materials, as proposed by Scoble *.

Furthermore, the yield or set point and the rupture point practically coincide. No cases have ever been observed of a bar receiving set under this type of stress without rupture. The maximum stress criterion is applicable, then, neither to rupture nor to set.

Incidentally this test disposes also of the maximum stress difference hypothesis, although more direct evidence is afforded by another type of test. The principal stresses for this first type of test consist of a compression equal to the hydrostatic pressure on all planes including the axis, and a small compressive stress due to the friction on the plane normal to the axis. The maximum stress difference is equal to the hydrostatic pressure decreased slightly by the amount of the friction. In the similar case of a bar ruptured by tension, the maximum stress difference is equal to the tension. If the maximum stress difference theory holds, therefore, the "pinching-off-effect" should be produced by a hydrostatic pressure equal to the tensile strength in pounds per square inch. As a matter of fact, the stress to produce rupture always exceeded this by 25 or 50 per cent., except for the glass, when the condition was more nearly fulfilled.

Rupture of the first type has been encountered repeatedly in all this high-pressure work. It is the greatest hindrance to making connexions of any sort from the outside with the interior of a high-pressure cylinder, particularly when connecting one cylinder to another by tubing, or leading electrically insulated electrodes into the interior of the cylinder. Possible rupture of this sort has been the greatest

* Scoble, Phil. Mag. xix. pp. 908–916 (1910).

element of danger in this work. On one occasion a specimen 5/16 in. diameter and 3 in. long penetrated 5 inches of wood driven by a pressure of only 6000 atmos. This effect might very probably produce serious consequences for one designing apparatus for the highest pressures without previous experience at lower pressures. The effect is insidious because of the unexpected ways in which it may appear.

The second type of test is very similar to the first in the manner in which stress is applied, but diametrically opposite in its effects. The material for these tests is in the shape of a hollow cylinder, closed at the ends, and subjected to hydrostatic pressure over the entire external surface, on the ends as well as on the curved part of the surface. The tendency of the stress, as is well known, is to collapse the cylinder if the walls are comparatively thin. Such tests are familiar to engineers ; the tube folds in on itself in two, three, four, or more creases, depending on the dimensions of the tube originally and on the very slight departure from perfect geometrical symmetry. That there is collapse at all must evidently be due to some slight geometric imperfection. If the tube is made heavier, however, so that geometric irregularities have less effect, the tube does not show collapse by folding under pressure, but shows behaviour of a different sort, depending on the material. Tests of this sort do not seem to have been made hitherto, or at least are not well known, probably because the pressure required to produce the effect is fairly high.

If the material of the cylinder is a ductile metal like mild steel or copper, the effect of the pressure is to close up the hole uniformly, the cylinder retaining its geometric figure. Rupture is never produced in a test of this kind, the hole eventually closing up perfectly tight if the pressure is pushed far enough. This is perhaps as one would expect ; the interesting feature of this method of testing is the enormous raising of the elastic limit that it is possible to produce, and the unusual stress-strain relation below the yield point. At one time tests were made simultaneously on seven such hollow steel cylinders. The upper ends of these cylinders were led out of the pressure chamber and connected to graduated glass capillaries. The interior of the cylinders was filled with mercury, so that by observing the rise of mercury in the capillary it was possible to follow the change of internal volume with pressure. No change of length accompanies the closing of the hole, so that the rise of mercury in the capillary gives directly the change of internal volume with pressure. Fig. 3 shows a typical diagram of

Fig. 3.

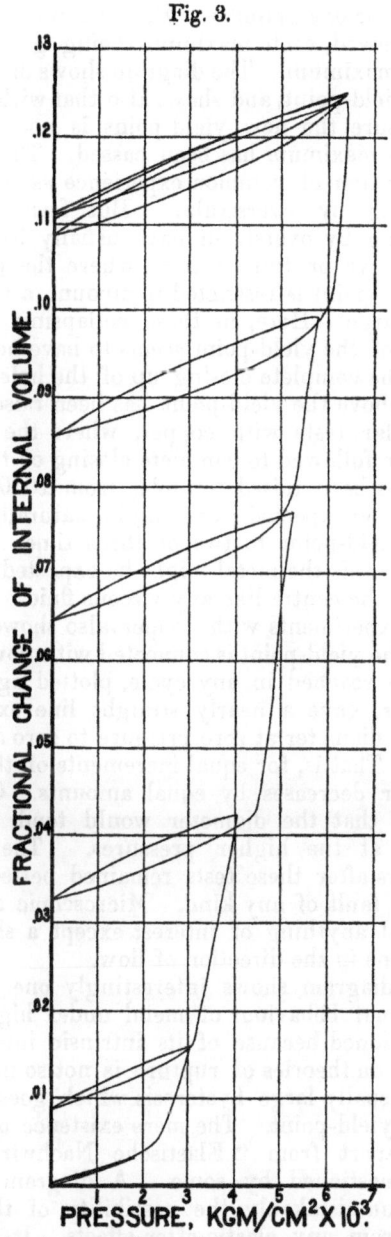

Shows the relation between internal volume and pressure of a cylinder strained beyond the elastic limit by the application of pressure over the entire external surface.

the test for one cylinder. Pressure was successively applied and removed, each maximum being greater than the preceding maximum. The diagram shows distinctly the location of the yield-point, and shows also that with every application of pressure the new yield-point is not reached until the previous maximum has been passed. This of course is only a verification of common experience as to the possibility of hardening by overstrain. But former experiments on hardening by overstrain have usually been with tensile or compressive or torsion tests, where the possible raising of the yield-point is restricted in amount, a rise of 50 per cent. being large. Here, in these collapsing tests, the possible raising of the yield-point seems to have no limit except that set by the complete closing up of the hole. In the diagram shown above the yield-point has been raised about six-fold; in similar tests with copper, where the complete process has been followed to complete closing of the hole, the yield-point has been raised ten-fold, from 1000 to 10,000 atmos. This was unexpected; one might naturally expect a raising of the yield-point to two or three times the original value, but after this the metal might be expected to flow uniformly towards the centre like any viscous fluid.

The experiments with copper also showed the manner in which the yield-point is connected with flow. The maximum pressure reached in any cycle, plotted against the internal diameter, gave a nearly straight line extending from the original diameter at zero pressure to zero diameter at 10,000 atmos. That is, for equal increments of the yield-point, the diameter decreases by equal amounts. One might expect perhaps that the diameter would tend to decrease more rapidly at the higher pressures. The material of the cylinders after these tests remained perfectly homogeneous, without fault of any kind. Microscopic analysis has failed to reveal anything of interest except a slight elongation of the grains in the direction of flow.

The diagram shows interestingly one other variation in the normal behaviour of metal under high pressure, which is mentioned because of its intrinsic interest, although the bearing on theories of rupture is not so immediate. This is the unusually large hysteresis which goes with the raising of the yield-point. The mere existence of hysteresis as an effect apart from "Elastische Nachwirkung" has even been questioned by some. A diagram like the present shows unmistakably the possibility of this effect entirely apart from any elastic-after-effects. It also suggests that the hysteresis may in some way be connected with an unstable configuration of the molecules set up in the metal by the overstrain. In the case of ordinary metals, where hysteresis appears before the yield-point of the metal as a whole has been reached, hysteresis is probably connected with local yield in the neighbourhood of the larger or more unstable crystalline complexes of the metal.

The application of all this to theories of rupture is immediate. It suggests, in the first place, that there is no necessary connexion between the yield-point and the rupture-point. Engineers, after some discussion, seem to have accepted the yield-point as the best criterion of rupture. The reason seems to be that the yield-point and the rupture-point are fairly close together, the yield-point is pretty definitely located, and it is possible to calculate the relation between stress and strain up to the yield-point and so to find a criterion at least for yield, if any exists; whereas it is well known that the usual relations between stress and strain break down in the region of viscous flow between the yield- and the rupture-points. Here, in these tests, there is a yield-point but no rupture-point at all, so that certainly, even if a criterion were found for yield, it could not be extended to rupture.

These tests further show that the maximum stress difference criterion of yield is no more valid than the maximum stress difference criterion of rupture. For if the distribution of stress in the cylinder be calculated on the maximum stress difference hypothesis, it will be found that the hole in the cylinder will never close up under any pressure, no matter how large. The mathematical solution of this problem is given in another paper in which this whole question of the collapsing of cylinders is taken up much more in detail*. The solution assumes that the maximum stress difference condition holds throughout the entire extended process of yield. To account for the observed complete closing of the hole it is necessary that the greatest stress difference which the metal can support become less in the last stages of yield. To inquire whether the initial yield always occurs at a definite stress difference would be of little avail, for it is well known, that even for the ordinary tests of engineering the material must first be put into a state of ease by subjecting it to considerable stress. If it is not so subjected to preliminary stress, it may show yield or departure from Hooke's law at values of the stress very much lower than normal. But in tests of this type there is no natural limit to the stress which shall be used to put the

* Phys. Review, xxxiv. pp. 1–24 (1912).

cylinder in a state of ease, and since the diagram shows plainly that the beginning of yield is not sharply defined, there is here no possibility of a natural initial yield-point to which a criterion might be applied.

So far this collapsing test has been discussed only in its application to ductile materials like copper or steel. The behaviour of a brittle material like glass under the same conditions is strikingly different. Cylinders of glass made of heavy capillary tubing sealed at both ends, and hollow spheres with thick walls, have been subjected to hydrostatic pressure up to 24,000 atmos. No permanent measurable change is produced, there being neither crushing nor alteration in the dimensions of measurable amount. It should be said, however, that there must be some slight amount of flow and of interior adjustment to the pressure, although too small to measure, because several of the glass cylinders have broken spontaneously several weeks or even months after the release of pressure. Others have been kept without fracture for a couple of years. One would expect that under the conditions of this test the glass would be crushed. This is the case if the cylinder has not been carefully annealed or if it is geometrically imperfect; the material may then be reduced to an almost impalpable powder. This complete destruction of the minutest fragments of the glass apparently is because the wave of expansion, travelling through the mass after the break has started, at any point is of such intensity that every minute portion is reduced to powder by its own inertia.

For a brittle substance like glass which shows no flow we may calculate the distribution of stress by the ordinary theory of elasticity. It is thus found that the maximum stress difference will occur at the inner surface of the cylinder, and for a thick cylinder will be equal in amount to the external hydrostatic pressure. We have already seen that if the maximum stress difference criterion were valid, the greatest stress difference that the material could support would be equal to its tensile strength. Therefore since the tensile strength of glass is seldom as high as 7000 lbs/in^2, we should have had crushing under the conditions of the test at a hydrostatic pressure no higher than 7000 lbs/in^2, if the maximum stress difference theory were valid for brittle substances. But fifty times this value has been reached without rupture.

The result of this experiment with the glass seems even more unexpected than that with the ductile materials. With the metals we have at least the rearrangement of the grains produced by the flow as a basis for the raising of the elastic limit, but in a perfectly amorphous substance like glass, which shows no alteration of shape, we have evidence of no such internal change to account for the increase in strength. The experiment suggests a difference between the mechanism of yield and the mechanism of rupture. One can perhaps see why rupture should not occur in the case of the glass cylinders, since there is no place for the fragments to go to in the case of rupture, no way for the rupture to get started. But the same considerations would seem to show also no possibility of yield towards the centre. Yet yield toward the centre does get started and rupture toward the centre does not get started; there must be some essential difference between the two.

These collapsing experiments also dispose of one other criterion of yield or rupture, better treated by the third type of test. This is the third criterion mentioned above, namely that rupture or yield will occur when the extension in any direction exceeds a critical value. If the strain is calculated in a cylinder exposed to hydrostatic pressure over the outside, the radial strain is found to be an extension at the inner surface. In the case of the ductile materials which flow toward the centre, this radial extension increases enormously with increasing flow, always without rupture or separation of the fibres in the direction of elongation. And in the case of glass under 24,000 atmos, where there is neither rupture nor flow, the elongation at the centre is greatly in excess of the elongation at rupture in an ordinary tensile test, and therefore in excess of the supposed critical elongation.

The third type of test is concerned with the rupture of heavy cylinders by internal pressure. The ordinary theory of the bursting or yield of cylinders under internal pressure is well known. At the inner surface the stress is a pressure on planes at right angles to the radius, and a tension on planes including the axis and radius; the corresponding strain is a circumferential elongation and a radial compression. At the inner surface the stress difference, the principal stress, the principal strain, and the strain difference all have their maximum values. On any theory of rupture, then, rupture would be expected to start at the inner surface. The precise value of the theoretical bursting pressure depends on the criterion accepted for rupture. If the principal stress criterion is accepted, rupture will occur when the hydrostatic pressure is equal to the tensile strength; if the maximum elongation is accepted, rupture will occur for a material like Bessemer steel at an internal pressure about 4/5 of the tensile strength.

The fact is, however, that for the ordinary materials of engineering practice rupture begins at the outside and runs in towards the centre. This rupture may take place either by tearing apart of the metal in an axial plane, or else by slip on a shear plane, the fracture in this latter event running in toward the centre approximately as an equiangular spiral.

Most of these bursting tests were made on drawn bars, which were pierced with the hole first and then turned off slightly on the outside so as to be concentric with the hole. Of course the most obvious suggestion with regard to this rupture from the outside is that there were flaws in the outer skin of the drawn bar penetrating more or less deeply into the interior, and that the rupture started from one of these flaws. In order to prove definitely that this is not the case, the following test was made. A bar of Bessemer steel, $4\frac{1}{2}$ in. diameter, was turned down to 4 in. so as to remove the outer layer. From this 4 in. bar a number of rings, 4 in. in diameter and 1/8 in. square in section, were turned at regular axial intervals of about 1 in. From the bar left after cutting off the rings, a cylinder for testing was made in the usual manner about $3\frac{1}{2}$ in. o. d. and $\frac{1}{2}$ in. i. d. and 8 in. long. From the other end of the same bar a similar set of rings and a similar cylinder were turned, only smaller, the cylinder being 2 in. o. d. and $\frac{1}{2}$ in. i. d. The rings were then tested to rupture on an expanding mandrel. No trace of flaw in the steel was found; the rings expanded between 10 and 15 per cent. before rupture, and the location of the fracture was haphazard, showing no longitudinal vein of special weakness in the steel. The cylinders were then tested to rupture in the usual manner. For this purpose, where the cylinders are made of very ductile material, it was found convenient to fill the cylinders with lead instead of with a true liquid, since it is easier to keep the lead from leaking after the cylinder has begun to stretch. The lead transmits these high pressures nearly hydrostatically. Pressure was produced by a hardened steel piston forced against the lead by the ram of an hydraulic press. A cup-shaped washer of Bessemer steel prevented the lead from leaking past the piston. Because of the very great stretching it was necessary to make several strokes of the piston before rupture was produced.

The results of these two tests with the two cylinders mentioned above are shown in the photographs. The larger cylinder, Pl. II. fig. 4, broke by separation of the fibres along an axial plane. Break occurred suddenly with explosive report, so that it was not possible to ascertain whether rupture really started from the outside or not. But with the smaller cylinder, figs. 5 and 6, break started more gradually, and it was possible to watch the whole proceeding. In fact, two strokes of the piston were necessary to enlarge the crack to the condition shown in the figure after the first beginnings of the fissure appeared at the outer surface. The fissure appeared first as a small longitudinal crack, which extended itself axially, the central portion gaping wider and wider as the metal on one side protruded by slipping out along a shear plane.

These two experiments showed that the rupture, which apparently begins at the outside, is not produced there by flaws in the steel. There might still be some question as to whether the rupture really did begin at the outside, since of course it is conceivable that the slip had begun at the inside and travelled to the outside, there first becoming noticeable. That the rupture actually does begin at the outside was shown by other experiments on nickel-steel and on copper cylinders. The nickel-steel cylinders do not stretch so much before rupture as the Bessemer cylinders, so that it was possible to produce rupture of these cylinders with a true liquid transmitting pressure. That the rupture really begins at the outside was shown by the appearance of the crack on the outside and its gradual growth, just as above, without any liquid leaking through from the inside, as of course it would have done under the high pressure, 30,000 atmos, if there had been the slightest crack reaching from the centre out. The copper cylinder showed the same thing. Here it was possible by a special arrangement of shrunk-on steel rings to make the necessary connexions and produce rupture with a fluid. Rupture again appeared at the outside first, and spread toward the centre along a shear plane; leak not finally occurring until the crack had opened to the dimensions shown in the photograph, fig. 7 (Pl. II.).

One feature common to all these tests, whether the rupture through the mass of the metal takes the form of a tearing of the fibres or a shear, is that at the inner surface the break is always along a shear plane for a short distance. If the rupture from the outside is by tear, at the inside there are almost invariably two shear planes, running down into the tear. The result is that a sliver of metal in the form of a triangular prism is expelled through the crack. This sliver has been caught on several occasions in a block of lead. It is difficult to see how this shear along two planes could get started if the crack originally started from the inside. Evidently the crack runs in from the outside until so close to the centre that the prism slips into the crack, driven by the high internal pressure. This universal manner of rupture affords additional evidence, in those cases where the rupture occurs so suddenly that it cannot be observed, that the crack starts from the outside.

The particular bearing of this third type of test on the theories of the conditions of rupture is in showing that the maximum extension criterion does not hold, although of course all the other criteria are also ruled out, if the stress-strain relation is calculated up to the rupture-point by the ordinary theory, because every one of the supposed critical quantities has its maximum value at the inner surface. But the striking feature of all the tests is the enormous stretching of the inner surface without rupture. The Bessemer-steel cylinders of Pl. II. figs. 4 and 5 show a circumferential elongation at the inner surface of 175 and 125 per cent. respectively. The tool steel cylinder of fig. 8 has an interior elongation of 120 per cent. The copper cylinder of fig. 7 shows an interior elongation of 300 per cent. and the lead cylinder of fig. 9 several thousand per cent. There does not seem to be any connexion between the value of the elongation at the exterior surface, where rupture actually does occur, and the elongation under pure tensile tests at rupture. We may have values either greater or less than the tensile elongation. The nickel steel cylinders, those of specially toughened steels, the copper cylinders, and the 3 in. Bessemer cylinder all showed a circumferential elongation at rupture very much less than the elongation at rupture under pure tension. The 2 in. Bessemer cylinder showed an elongation almost exactly equal to that showed by the rings cut from the same piece, while a drawn tube of annealed steel has shown an elongation of 100 per cent. at the outside, and the lead cylinder shows 300 per cent. The tensile value for the steel is about 20 per cent., and for the lead 25 per cent.

Tests on brittle materials like glass also show that the maximum elongation criterion is not fulfilled. Of course, in the case of a brittle material like glass, rupture comes so suddenly and is so complete when it does come that it is impossible to tell by any examination of the fragments whether rupture began at the inside or the outside. But the maximum value of the stress which the glass is capable of standing before rupture is considerably in excess of the theoretical limit. On no theory ought the glass to be able

Hydrostatic Pressure and Conditions of Rupture. 77

to stand more pressure on the inside than the tensile test limit, even if the walls of the cylinder are of infinite thickness compared with the bore. One can in cases like this, where there is no yield up to the rupture-point, safely apply the theory of elasticity in calculating the strain up to the rupture-point. This means that no glass capillary ought to be able to stand more than 500 atmos. Capillaries have been found, however, which have successfully withstood 1000 atmos. The question of annealing is of great importance here. Probably with more careful annealing it would be possible to exceed this limit. In any event, the circumferential extension, even when the pressure is only 1000 atmos., must be nearly twice the critical value under rupture in tension.

The manner of rupture described here, beginning at the outside first, is not universal for all materials, but probably holds only for those materials showing considerable plasticity. There can be little doubt but that the rupture of glass cylinders does begin at the inside, and one case in another substance has been actually observed in which the rupture did travel from the centre to the outside. A cylinder of transparent gelatine, about 1 in. o. d. and cast with a concentric hole $\frac{1}{8}$ in. diameter, was ruptured by blowing into it. The rupture took place exactly as the theory predicts ; it started at the inner surface as a tear across an axial plane, and travelled out slowly toward the outside as the interior was more and more distended by blowing.

It is worth while examining these bursting tests a little further, for in this case we can make plausible to ourselves why rupture does not occur at the inside, and so gain an inkling of a much more general criterion than any of those hitherto proposed which must always be satisfied when rupture occurs. If we consider the state of stress at the inside surface we shall find that initially as long as the equations of elasticity hold, the stress consists of a pressure across planes perpendicular to the radius, and a circumferential tension which is greater than the pressure. When the internal pressure becomes too great, however, the inner layers yield plastically, the hole taking a set. During this plastic yield the metal must be thought of as behaving like an imperfect viscous liquid, tending to transmit pressure hydrostatically in every direction. In consequence of the tendency to equality of stress in every direction, the initial circumferential tension becomes a circumferential compression. The mean stress, therefore, at any point of the interior surface becomes a compression after the plastic yield begins, whereas before yield the mean stress was a tension, because the circumferential tension was greater than the radial compression. The strain, then, after yield, consists of a volume compression at the inner surface, whereas before yield there was a volume extension. Consider now what would happen at the inner surface if rupture were to get started there. We are to think of the general action of rupture as one relieving the stresses at any point, for otherwise the rupture would have no excuse for existence. Before plastic yield, then, rupture at the inner surface would result in release of the volume distension, that is, the volume of the steel would become less. As a result, the compressing liquid would have room to become greater in volume, and the pressure would automatically fall. But if the rupture were to start after the plastic yield, that is after volume compression had taken place at the interior surface, release of the pressure would result only in increase of volume of the steel, the compressing liquid would lose volume, and the pressure would as a result be increased instead of decreased. In this case, rupture, instead of affording relief from the existing state of affairs, would only intensify them. That is, rupture would be an explosive affair, during which work was done against the applied forces instead of by them. This is opposed to all experience. This simply means, then, that even if rupture on a small scale should start at the interior surface during the state of plastic yield, it would be unable to spread. If rupture is to start at the centre and travel out it must start before the stage of plastic yield is reached.

The photograph of the 4 in. Bessemer-steel cylinder, fig. 4, shows this most strikingly. The interior surface is covered with numerous slip bands, some of them very prominent. where the rupture had started but was unable to spread, The pressure on the plastic metal at the inner surface has been so high that a process similar to cold-welding has gone on. The crystalline grains were probably torn apart so as to slip past each other, but the pressure was great enough so that the separation was never beyond the range of molecular attraction, and true rupture did not occur. The copper cylinder also shows the same slip bands at the interior surface.

Another interesting question in this connexion is that of the greatest possible raising of the elastic limit by permanent stretching. It is easy to see that the stress distribution in such a cylinder after the yield-point has been passed and the pressure released is exactly like that in a gun with shrunk-on

Hydrostatic Pressure and Conditions of Rupture.

hoops. One might expect, then, to be able to reach a pressure about twice that of the yield-point under pure tension, but as a matter of fact it is possible to exceed this, because of the hardening of the metal by over strain. Thus a copper cylinder, originally $\frac{1}{8}$ in. i. d. and 3 in. o. d. has been subjected to 10,000 atmos. without the yield extending throughout the entire mass. The inner hole has been stretched about 50 per cent. by this pressure. Pressures of 30,000 atmos. have been reached in cylinders of soft nickel steel or of tool steel, and of 40,000 in a cylinder of hardened nickel-steel. But the hardening process by over load is not nearly so complete or thorough as it is for cylinders collapsed by external pressure. The yield under high pressures continues for a very much longer period of time, and the raising of the elastic limit after complete disappearance of set is not permanent. After a period of rest, yield is likely to begin again at a pressure lower than the former maximum. Thus a cylinder of soft nickel-steel, originally 8 in. o. d. and $\frac{5}{8}$ in. i. d., had had the elastic limit apparently raised to 28,000 atmos. by repeated applications of pressure which stretched the inner hole from $\frac{5}{8}$ in. to $1\frac{1}{8}$ in. But now, after several years of use, it is unsafe to subject this same cylinder to more than 15,000 atmos. The initial yield-point of the cylinder was about 8000 atmos.

Summary.

In this paper the results of three types of tests with high hydrostatic pressures have been described in their bearing on theories of rupture. The tests of cylinders under pressure on the curved surface only ("the pinching-off-effect") show that the maximum principal stress criterion is not valid. The tests of hollow cylinders under external pressure show that neither the maximum shear stress nor the maximum shear strain criterion is valid, while the tests on the bursting of heavy cylinders under internal pressure show that the maximum elongation criterion is not valid. These conclusions apply to both brittle and to ductile materials. The tests do suggest that there may be some essential difference between a ductile and a brittle material (more generally between a crystalline and an amorphous one), but show that the distinctions proposed hitherto cannot hold. It is suggested that in every case there must be some more general condition satisfied than those usually considered. There is not enough material at hand to enable a formulation of the condition to be made, but the considerations with regard to the bursting of cylinders under internal pressure point out what the general nature of the conditions must be. Any entirely general criterion of rupture must demand, among other things, that there be a free space to contain the fragments when rupture occurs, and must also demand that the nature of the rupture be such as to relieve the applied stresses.

 The Jefferson Physical Laboratory,
  Harvard University,
   Cambridge, Mass., U.S.A.

FIG. 2.

FIG. 5.

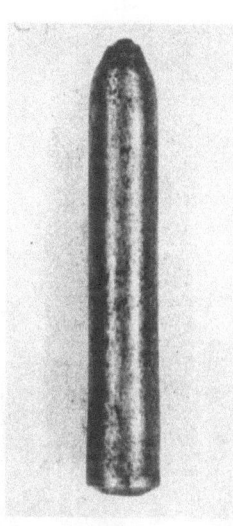

Cross section of a cylinder of Bessemer steel ruptured by the application of internal pressure. This cylinder was originally 2 in. outside and $\frac{1}{2}$ in. inside diameter. The inner hole has been stretched to $1\frac{3}{8}$ in.

Photograph of a "pinched-off" specimen, ruptured in the apparatus of Fig. 1.

FIG. 4.

FIG. 6.

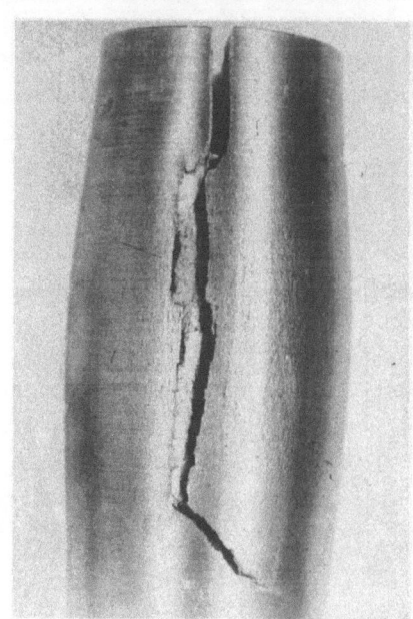

One of the halves of a cylinder of Bessemer steel (originally 4 in. in external diameter) ruptured by the application of internal pressure. The inner hole has stretched from $\frac{1}{2}$ in. to $1\frac{3}{8}$ in.

View of the outside of the cylinder of Fig. 5 taken before the section was made.

Fig. 7.

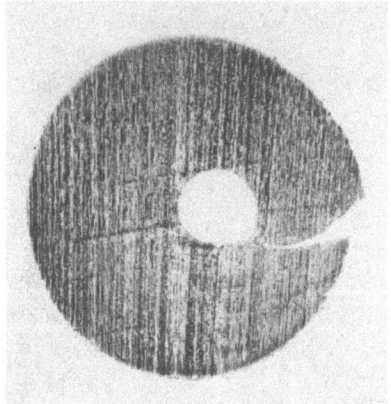

Cross section of a copper cylinder burst by the application of internal pressure. The inner hole has been stretched from $\frac{1}{8}$ in. to $\frac{3}{8}$ in.

Fig. 8.

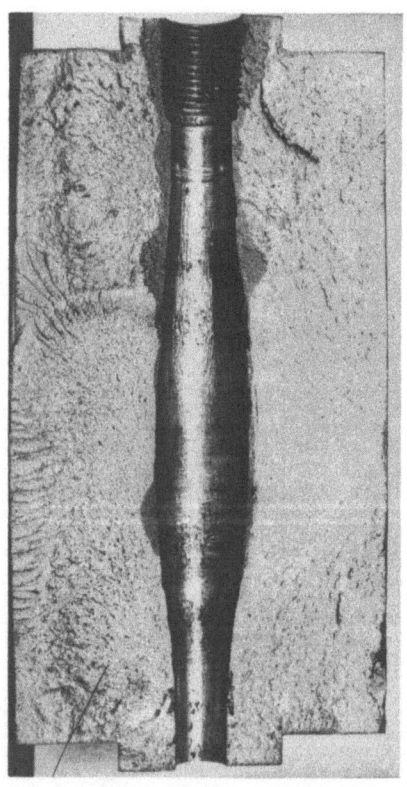

One of the halves of a cylinder of tool steel split by the application of internal pressure. The inner hole has stretched from $\frac{1}{2}$ in. to $1\frac{1}{4}$ in. The maximum pressure withstood by this cylinder was 30,000 atmos.

Fig. 9.

Longitudinal section of a cylinder of lead burst by the application of internal pressure. The elongation at the outside at the locality of rupture is over 300 per cent.

CONTRIBUTIONS FROM THE JEFFERSON PHYSICAL
LABORATORY, HARVARD UNIVERSITY.

THERMODYNAMIC PROPERTIES OF LIQUID WATER TO 80° AND 12000 KGM.

By P. W. Bridgman.

Received June 26, 1912.

Table of Contents.

	Page.
Introduction	310
Method	312
Previous Use of the Method	312
Description of the Apparatus	314
Correction for the Distortion of the Vessel	316
Experimental Procedure	319
In Determining Compressibility	319
Calibration of Manganin Coil	320
Formulas	325
In Determining Dilatation	326
The Data	328
Compressibility at Low Pressures	328
Dilatation at Low Pressures	330
Compressibility at High Pressures	331
Dilatation at High Pressures	334
Discussion of the Results	336
Table of Volumes	338
Method of Construction	336
Various Thermodynamic Quantites	337
Compressibility, $\left(\frac{\partial v}{\partial p}\right)_\tau$	340
Dilatation, $\left(\frac{\partial v}{\partial \tau}\right)_p$	344
Work of Compression, $W = -\int p \left(\frac{\partial v}{\partial p}\right)_\tau dp$	346
Heat of Compression, $Q = -\tau \int \left(\frac{\partial v}{\partial \tau}\right)_p dp$	347
Change of Internal Energy, $\Delta E = W + Q$	348
Pressure Coefficient, $\left(\frac{\partial p}{\partial \tau}\right)_v$	349
Specific Heat at Constant Pressure, C_p	351
Specific Heat at Constant Volume, C_v	352
Thermal Effect of Compression, $\left(\frac{\partial \tau}{\partial p}\right)_\varphi$	354
Adiabatic Compressibility, $\left(\frac{\partial v}{\partial p}\right)_\varphi$	355
Volume of Kerosene as a Function of Temperature and Pressure	356
Compressibility and Dilatation of Ice VI	359

Introduction.

This paper is in the nature of a supplement to a former paper on the properties of water in the liquid and the solid forms.[1] The solid forms were studied over a range of 20,000 kgm./cm.², and from $-80°$ to $+76°$, but the study of the liquid reached only from the lowest temperature of its existence to about $+20°$. Above $0°$, measurements were made on the liquid at only $20°$. The two measurements, at $0°$ and $20°$ were sufficient to give the mean dilatation between $0°$ and $20°$, but not the variation of dilatation with temperature. It was assumed in the earlier paper that the variation of dilatation with temperature became negligible at high pressures, since this seemed to be the most plausible assumption in view of all the data then available.

In this present paper the study of the liquid has been continued from $20°$ to $80°$, and to 12000 kgm. The pressure range is greater than that of the preceding paper by about 2,500 kgm. The range is not great enough to entirely cover the region of stability of the liquid, but it is as great as it was convenient to cover with the method used here, which is different from that of the former work. It has the advantage of very much greater rapidity of operation, but since it depends on the complete elastic integrity of the steel pressure cylinders it is not possible to reach so high pressures with it as with the former method. [The former limit of 9500 kgm. was set by the freezing of the liquid and was not due to any limitation of the method.] Nevertheless, it may be hoped that the present temperature and pressure ranges are both wide enough to give a fairly complete idea of the nature of the effects to be expected at high pressures with varying temperature.

Measurements of the dilatation have been made at four temperatures, so that it has been possible to find the variation of dilatation with temperature at any pressure. Perhaps the most unlooked for feature disclosed by the measurements is the fact, contrary to the assumption of the first paper, that the variation of dilatation with temperature does not become vanishingly small at high pressures, but reverses in sign. This means that while at low pressures the volume increases more and more rapidly with rising temperature, at high pressures the expansion becomes more slow at high temperatures.

The data of this paper are sufficient to completely map out the p-v-t surface over the domain in question: Both the first and second

---

[1] Bridgman, These Proceedings, **47**, 439–558 (1912).

derivatives are therefore completely determined, so that we now have all the data at hand for the determination of any one of the thermodynamic properties of the liquid. This means that we are in a position to find such quantities as the specific heats, change of internal energy; adiabatic temperature rise etc., as well as the more easily determined compressibility and thermal dilatation. The latter part of the paper, after the discussion of the method and the presentation of the data in the first part, is occupied with the computation of these various thermodynamic quantities. The accuracy of some of these is probably not very great, because the error in the second derivative of an experimental quantity may be considerable. It has, therefore, seemed best to give the general view of the nature of the quantities which is offered by a graphical representation, rather than to give tables, with the tacit assumption of greater accuracy which usually goes with a set of tables. In spite of the lower order of accuracy of some of these thermodynamic quantities, it has still seemed well worth while to give them, since even the general trend of some of the quantities, such as the specific heats, has not been hitherto known with relation to pressure.

The data presented here are only the beginning of a projected study of the characteristic surface under high pressures for a number of liquids. The measurements have already been carried through for twelve other liquids beside water. The purpose of this study is ultimately the development of a theory of liquids, since it would seem that a much more intimate grasp of the nature of the forces at work in a liquid would be afforded by a study over a wide pressure range, than over the comparatively low pressures hitherto used. It must be admitted, however, that this broader purpose is not particularly furthered by this work on water, because of the well known abnormalities of this substance. In the previous paper several abnormalities had been shown to exist at low pressures. In this paper, new abnormalities are found at higher pressures. Water gives the appearance of becoming completely normal only at the higher temperatures and pressures of the range used here, but of course whether this is really normal or not cannot be told until the behavior of normal liquids has been discovered. The full significance of the present data, in their bearing on such questions as the polymerization of the liquid, for example, cannot appear until after the discovery of the laws for entirely normal liquids. The investigation of water before that of normal liquids was undertaken for two reasons; firstly because of the desire to complete the work for water already begun, and secondly because in this and the following investigation a new method for determining the compressibility was to be used, which had not yet been proved to be reliable, but which could be tested by a comparison of the results obtained by this method with those already obtained by another method at lower temperatures for water.

In addition to the data for liquid water, two other quantities were determined incidentally in the course of the work, and are given at the end of the paper. One of these is the experimental measurement of the compressibility and thermal dilatation of ice VI between 0° and 20° and 6360 and 10,000 kgm. The other is the measurement of the volume of kerosene up to 12,000 kgm. and between 20° and 80°.

THE METHOD.

The method in its fundamental idea is as simple as it would well be possible to devise. The substance, whose compressibility or thermal dilatation is to be measured, is placed in a heavy steel cylinder in which pressure is produced by the advance of a piston of known cross section. The change of volume, given by the distance of advance of the piston, is measured as a function of the pressure. The method is simple, rapid, and above all, applicable to the highest pressures. But there are a number of corrections which must be made, often difficult to determine, which doubtless account for the slight use which has been made hitherto of the method. Apparently, with the exception of the present work, it has been used recently only by Tammann,[2] and by Parsons and Cook.[3] Tammann and Parsons and Cook applied it only to the measurement of compressibility, reaching pressures of about 4000 kgm. The author has previously applied it to the measurement of the thermal dilatation of water at temperatures below 0° C. over a pressure range of about 6500 kgm.

The most serious of the errors which readily occur to one is that of leak. It is almost essential to the success of the method to secure a piston absolutely free from leak, and this has hitherto been a matter of some difficulty at high pressures. Tammann did not entirely secure this freedom from leak, but avoided it in large measure by the use of a very heavy oil, such as castor oil, and still further lessened the error by correcting for the slight amount of leak by measuring the amount of liquid which escaped past the piston in a given time. This method would not be applicable to the highest pressures, however, because

---

[2] A. D. Cowper and G. Tammann, ZS. Phys. Chem., **68**, 281–288 (1909).
[3] Parsons and Cook, Proc. Roy. Soc. A, **85**, 332–349 (1911).

of the freezing of the oil. Parsons and Cook were able to secure entire freedom from leak up to 4000 kgm. by the employment of a cupped leather washer combined with a brass disc of special design. It has been the experience of all those who have worked with high pressures, however, that no leather washer is capable of standing pressures very much in excess of the limit of 4500 kgm., since the leather rapidly disintegrates under the pressure. In the present work the same form of packing was used which was used in the previous work on the freezing of water and mercury under pressure. This has been proved in the previous paper to be absolutely free from leak up to the highest pressures which can be sustained by the steel containing vessels. In the present work this same packing has proved itself to be reliable for the purposes of this method.

The question of the method of measuring pressure is also of considerable importance in using this method, since the usual measuring devices, such as a Bourdon gauge, cannot be applied, for reasons to be discussed later, and attempts to calculate the pressure directly from the force required to produce motion of the piston are likely to be in error because of the friction of the packing. Parsons and Cook did, however, adopt this latter method, and computed the pressure from the known force required to move the piston. The effect of the friction of the packings was allowed for in as large a degree as possible by taking the mean of the readings during increasing and decreasing pressure, assuming that the friction remained constant. The results obtained by Parsons and Cook in this way were surprisingly good. That the friction did remain fairly constant was indicated by the constancy of the results and the fact that the curve nearly always returned to the starting point; but it is doubtful if the method would work at very much higher pressures because of the increase of friction due to the flow of the softer parts of the piston. The brass washers used by Parsons and Cook would almost certainly have upset under two or three thousand more kgm., and it is the experience of the author that it is difficult to obtain even steel washers which will stand much more than 8000 kgm. without taking some set. In fact, at high pressure there must necessarily be some plastic yield, in order to follow the expansion of the cylinder. The result of this set in the washers is that the friction becomes very irregular, and cannot be assumed to be the same during increasing and decreasing pressure. Variations in the amount of friction due to this cause of as much as 200 or 300% have been found at the higher pressures of this work.

The only escape from the difficulty seems to be to measure the pressure directly inside the cylinder. This was done by Tammann by connecting a Bourdon gauge directly to the cylinder. But it is known that the errors of the Bourdon gauge become rapidly more serious at higher pressures,[4] due to the increase of hysteresis, so that this gauge could not be used for the pressures of this experiment. Furthermore, no Bourdon gauge has up to the present been made of sufficient sensitiveness which is capable of standing more than 6500 kgm. In the present work the pressure was measured inside the cylinder by inserting directly into it a coil of manganin wire, which had been already calibrated against an absolute gauge. This method of measuring pressure has been fully described in a previous paper.[5] It was necessary for the purposes of the present work, however, to make a somewhat more careful determination of the temperature coefficient than was done formerly, and this determination will be described in detail later. The method has shown itself perfectly satisfactory and reliable in every respect. One coil of wire has been used almost continuously for over six months, and occasional calibrations have shown no change. These calibrations were made by measuring with the coil certain fixed temperature-pressure points, such as the freezing pressure of mercury or of ice VI, at some fixed temperature.

The apparatus used in the present work is the same in most features as that used in the former work, a detailed account of which has already been given in the papers mentioned. Only the points in which this has been changed will be mentioned here. It was a disadvantage of the former method that the apparatus consisted of two parts; the lower part, a cylinder containing the liquid to be measured, was placed in a thermostat, and the upper part, a cylinder in which pressure was produced, was exposed to the temperature of the room. When temperature was changed in the thermostat below or pressure was changed in the cylinder above, liquid passed from the one cylinder to the other, experiencing in the transition a change of temperature, and so a change of volume also. This change of volume accompanying a known change of temperature varies in an unknown way with the pressure, and to apply the correction it was necessary to make an independent set of experiments. In the present form of apparatus the difficulty was avoided by including everything in one cylinder. This cylinder contained the liquid under investigation, the pressure measuring coil, and the piston by which pressure was produced. It

---

[4] Bridgman, These Proceedings, **44**, 201–217 (1909).
[5] Bridgman, These Proceedings, **47**, 319–343 (1911).

was placed in the lower part of the hydraulic press and, together with the lower part of the press, was placed in the thermostat. The dimensions were so small that this could be done without increasing to an unwieldly bulk the size of the apparatus, the four tie rods of the press being 1 1/8" in diameter and their centers 6" apart. It is the same form of apparatus which was used for the measurements on ice VI up to 20,500 kgm. The present experiments run to only 12,000 kgm., however, since it is evidently an absolute essential to the success of the method that there should be no permanent distortion of the cylinder. It would be easily possible to reach pressures much higher than those reached in this experiment, but it was felt that the risk and the extra time involved in the probable construction of new apparatus was not justified at present, when it seemed that the most important work was to map out the field, obtain data for as many liquids as possible, and determine the general nature of the significant problems. Later, if there are crucial points which need the use of much higher pressures, it will be a comparatively easy matter to obtain them.

The cylinder used in this experiment was not the same as that used in the previous work on water. This new cylinder is from a piece of chrome-vanadium steel made in the electric furnace by the Halcomb Steel Co., of Syracuse, N. Y. The steel itself is a wonderful product, and without it the present investigation would not have been so easily possible. It shows a tensile strength of 300,000 lbs. per sq. in. when hardened in oil, and an elastic limit of about 250,000 lbs. These figures are considerably in excess of those for the steel used in the previous investigation. The steel furthermore is remarkably homogeneous, because of its production in the electrical furnace. One of these pieces was pierced with a hole 1/8" diameter and 13" long, and the drill came through concentrically without any variation from the straight line. The dimensions of the cylinder used in the present work were 4 1/2" outside diameter, 13" long, inside diameter 17/32" for the greater part of its length, with an enlargement to 3/4" at the lower end for the reception of the manganin coil. The original inside diameter was 7/16". The cylinder was prepared for use by hardening in oil and then subjecting to a pressure much in excess of that contemplated for the actual experiment. The seasoning pressure was over 30,000 kgm. Even under this high seasoning pressure the cylinder showed very little permanent change of internal dimensions, not stretching as much as 1/32." This is less than the amount of stretch which has been found for any other grade of steel. The effectiveness of the treatment is shown furthermore in the fact that in over six months of continual use the inside has not stretched by so much as an additional 1/10000″. The hole was enlarged to a final size of 17/32″, instead of keeping it as small as possible, because of the difficulty of reaming out the hole so as to give a satisfactorily smooth surface after the seasoning process. The difficulty was occasioned by the hardness of the steel, and several attempts were necessary before the desired result was produced.

The pressure measuring coil was the same as that used in the last part of the work on ice VI. The construction of the insulating plug was also the same as that used there. During the course of the work it was necessary to take this plug apart several timess, because water had reached the mica washers, and once or twice the mica washers themselves have given way. These mica washers are the weakest part of the entire apparatus as at present used, since they gradually disintegrate and fail by shear after prolonged use, but it is a matter of only a few hours to replace them. Every time after the insulating plug has been freshly set up it has been tested for insulation resistance, both during application of pressure and after release. The resistance was in all cases as high as several hundred megohms, the limit of the measuring devise. The steel of the insulating plug has also failed once or twice by the "pinching-off effect"[6] after long use. This also is an easy matter to repair. Failure of this type is attended with some danger, however, because of the violence of the explosion with which the ruptured plug is expelled. The surest way of avoiding this danger is to so mount the apparatus that the plug points at the floor or other indestructible object.

The hydraulic press, the method of measuring the displacement of the piston, and the details of the packing of the moving piston, were the same as that used in the former paper.

In the use of the apparatus to determine compressibility there is one serious error which did not enter into its use in the determination of the change of volume during change of state, namely the correction for the distortion of the cylinder in which the piston moves. At low pressure the correction is relatively unimportant, and may be computed from the theory of elasticity, if one is willing to assume that the theory is sufficiently accurate for this type of stress. But at higher pressures the correction becomes more important, increasing in percentage value directly with the pressure, and is almost certainly

---

[6] Bridgman, Phil. Mag., **24**, 63–79 (1912).

not calculable by the theory of elasticity, because of the entrance of such effects as hysteresis. To determine the correction an auxiliary set of experiments is necessary. Evidently if the true value of the compressibility of some one substance were sufficiently well known, then the apparent compressibility as determined by this method would give the correction for the distortion of the cylinder. No such compressibilities are known with any high percentage accuracy, but this is not necessary, provided only that the uncertainty in the standard compressibility is small in comparison with the distortion of the vessel. The substance which most readily suggests itself because of its small compressibility is steel, but this is a solid, whereas the method is applicable directly only to liquids, so that some modification of the procedure is necessary. Such a modification readily suggests itself, and has been used by the author in the previous determinations of the thermal dilatation of water at temperatures below 0°, and has also been used by Parsons and Cook. The modification is to replace part of the liquid under investigation by a steel cylinder, and determine the compressibility of the liquid and the steel together. The difference of two determinations, the one for the liquid alone, the other for the liquid and the steel, gives a value for the difference of compressibility between the liquid and the steel from which the effect of the distortion of the vessel has been almost entirely eliminated. Furthermore, the compressibility of the steel is so small in comparison with that of the liquid that the slight uncertainty in the value for the steel is of no account, so that the compressibility of the liquid is given directly.

The application of this method would demand, then, that the interior of the cylinder be filled first with water and the apparent compressibility determined, and then part of the water replaced by steel and the apparent compressibility determined again. But this demands that the coil of manganin with which the pressure is to be measured come directly in contact with the water, which evidently cannot be allowed because of the short circuiting produced by the water. It seemed to be necessary, then, to devise some sort of protection for the coil, which should not occupy so much volume as to introduce a serious correction, and which should at the same time transmit the pressure readily to the innermost parts of the coil. Considerable time was spent in trying to devise such a protection. The scheme adopted was to surround the coil with a small mass of vaseline enclosed in a flexible sac, formed from the finger of a silk glove, and rendered impervious to water by painting it over with several coats of the collodion of surgeons. This sac was tied with silk thread directly over the end of the insulating plug. It was proved by trial that the vaseline did not become so viscous under pressure as to refuse to transmit the pressure with sufficient freedom, but the arrangement did not prove itself as trustworthy as was to be desired. The collodion might leak after several applications of pressure, which made it necessary to reassemble the insulating plug and redetermine the elastic constants of the apparatus, because the distortion included in the plug itself was sufficient to introduce appreciable error. The device probably could have been made usable with a little more effort, but it would always have been more or less unsatisfactory, and would have been applicable only to those liquids which do not attack the collodion, whereas most of the organic liquids which it was desired to use in the future do so attack the collodion. The attempt to protect the coil was abandoned after a month's work, therefore, and the method replaced by another, which at first sight introduced additional complications, but which is really just as simple as the first, and has the advantage of being applicable with only slight modifications to the investigation of other liquids.

The modified method used two liquids in every determination, one beside the one whose compressibility is to be measured. The water under investigation is placed in a thin shell of steel fitting the inside of the cylinder. This shell, when in position in the cylinder, is surrounded on all sides and above and below by kerosene, which below transmits pressure to the manganin coil, and above reaches to the moving piston with which pressure is produced. In the auxiliary experiment to eliminate the effect of the distortion of the cylinder, the shell with water is replaced by a solid cylinder of steel, and the quantity of kerosene remains the same as before. The motion of the piston due to the change of volume of the kerosene remains the same in the two experiments, therefore, and the difference of readings of the two sets gives directly the difference of compressibility between the water and the steel. The disadvantage of the method is that it is not possible to use so large quantities of water as in the former method, because the steel shell containing the water remains invariable in length under pressure, and enough kerosene must be introduced originally to take up the change of volume of the water in this shell as well as the distortion of the other parts of the apparatus. The reduction in the quantity of water under experiment is not greater than 30%, however, and the other advantages more than outweigh this comparatively small loss of accuracy.

The procedure in using the apparatus in this finally modified form is as follows. The manganin coil is first screwed into the lower part of the cylinder. The rubber washer used to make this plug tight is one cut with a standard set of cutters, so that all the washers used for this purpose are always the same in size. This insures that the distortion due to the compression of the washers shall always be the same. The steel shell with the water in it is next introduced from above. The quantity of water is previously determined by weighing. It is desirable not to fill the shell to closer than 1/4" of the top, experience having shown that otherwise water is likely to spill out and find its way to the manganin coil. The kerosene is next introduced into the cylinder from above. To ensure entire filling of all parts of the apparatus and the exclusion of air, only part of the kerosene is at first poured in, the air is then exhausted by attaching the mouth of the cylinder to an air pump, or simply by exhausting with the lungs, and then the remainder of the kerosene poured in. The amount of kerosene is determined by weighing the dish from which it is poured before and after filling. Because of the wetting of the dish by the kerosene it is not always possible to obtain exactly the amount of kerosene desired each time, but the variation is seldom over 0.02 gm., and the very slight effect of this discrepancy may be corrected for, as will be described later. Finally the movable plug is introduced into the top of the cylinder, taking particular pains not to allow any of the kerosene to escape in the process. Here again the rubber washer used has been cut with standard cutters, so that the amount of rubber used here is also the same in all the experiments. The cylinder is then placed in the thermostat, and the zero of the manganin coil read at the temperature of the room. The thermostat is then adjusted for the desired temperature and the cylinder seasoned for the run by the application of pressure.

A preliminary seasoning is necessary because of the hysteresis shown by the cylinder, and this hysteresis is shown with respect to both pressure and temperature. Many of the early results were somewhat in error because the necessity of this seasoning for temperature as well as for pressure was not clearly recognized. The method of seasoning to be adopted depends on the kind of data which it is desired to obtain from the run, whether the compressibility at constant temperature or the thermal dilatation at constant pressure. If it is desired to determine the isothermal compressibility, the seasoning consists simply in raising the pressure through the entire range and releasing several times. It was found by experiment that three such preliminary excursions were sufficient; after this the cylinder settles down into a state in which the normal hysteresis cycles are retraced with perfect regularity. Of course it is necessary to make the compressibility determinations immediately after this seasoning, as the effect gradually disappears with time. The time occupied in making the final readings to 12,000 kgm. and back with increasing and decreasing pressure, making in all 20 readings, might vary from two to three hours. After every change of pressure it was necessary to wait for the temperature effect of compression to disappear; this time was from 5 to 7 minutes.

If the thermal dilatation under constant mean pressure is to be determined, the seasoning consists simply in taking the cylinder once through the temperature range contemplated as well as through the pressure range. A word of description as to the general procedure in determining the thermal dilatation at constant mean pressure will not be out of place. The general plan is to change the temperature while the piston is kept invariable in position, and therefore while the volume is also approximately constant. The rise of temperature produces a rise of pressure, so that after the rise of temperature it is necessary to bring the pressure back to the former value by withdrawing the piston if the change of temperature has been an increase, or advancing it if the change of temperature has been a decrease. The amount by which the piston is withdrawn, as also the new final pressure, is noted. The temperature is then changed again, and the same set of readings made again. Thus every observation at any given temperature involves two readings of the position of the piston and the corresponding pressure. The slight change of pressure during the changes of temperature carries with it hysteresis effects, which it is necessary to avoid by previous seasoning, exactly as for pressure changes over a wider range. Two processes of seasoning are necessary for temperature, therefore, one a larger one for the entire temperature range, and another smaller one for the slight changes of pressure incident to the changes of temperature. This second seasoning is made after the first more extensive seasoning simply by running the pressure back and forth several times through the small range of pressure to be met with during the temperature changes. This small range was determined by preliminary experiment.

In the actual calculation of the results there are a number of corrections to be applied. These will now be discussed in detail separately. In the first place the temperature coefficient of the manganin coil has to be determined with particular care. This is because the pressure changes brought about by changes of temperature during the determinations of the thermal dilatation are comparatively slight, so that any change of the pressure coefficient of the coil brought about by the change of temperature appears in the result greatly magnified. Thus for the sake of example, we will suppose that a change of temperature of 20° produces a change of pressure of 400 kgm. at 10,000 kgm. total pressure. This figure is a fair average of the results to be met with in practice. If now the pressure coefficient of the coil is changed by 1% by this same rise of temperature, the pressure will thereby appear to have risen 500 kgm. instead of the actual 400, introducing an error of 25% for a change in the constant of the coil of only 1%. In addition to the effect of the temperature coefficient of the coil, there is an effect due to the change of the zero of the coil with temperature, but this change can be determined by observations of the temperature coefficient of the coil at atmospheric pressure and is easy to measure with the requisite accuracy.

The change in the pressure coefficient of the coil with temperature is more difficult to determine with the desired accuracy. It would not be possible to determine this by a direct calibration against the absolute gauge with which the mean value of the coefficient has been determined, for the reason that the absolute gauge itself is not accurate to better than $1/10\%$, and this would still leave a possible error in the thermal dilatation of 2.5%. To affect the desired calibration, some standard of pressure must be used which can be relied on to remain absolutely constant. Such a standard pressure is evidently afforded by the transition point of the liquid to the solid form of any convenient substance at some fixed temperature. In previous work the transition points of both water and mercury have been determined at various temperatures with an accuracy in the absolute pressure of $1/10\%$. To make the calibration it is only necessary to keep the pressure constant automatically at this known value by placing in communication with the chamber in which is the manganin coil to be calibrated another chamber in which are the liquid and solid forms of the substance whose transition temperature and pressure are known. This second chamber is to be kept at constant temperature accurately enough so that slight changes in this temperature will not produce changes of more than the allowed amount in the transition pressure. For this purpose the most convenient fixed temperature seems to be that of melting ice at atmospheric pressure, and the most convenient substance to use mercury, because of the sharpness of the freezing, and the ease with which it can be obtained pure.

The actual arrangements in making this calibration for the temperature coefficient of the pressure coefficient of the coil were as follows. The upper cylinder of the hydraulic press in which pressure was produced contained in addition to the moving plunger a steel shell in which was as large a quantity of mercury as convenient, about 150 gm. This upper cylinder as well as the entire lower part of the press was surrounded by a tank containing ice and water, by which the temperature of the mercury could be kept continuously and accurately at 0°. A heavy nickel steel tube led out of the lower end of the upper cylinder through the bottom of the tank, and connected with the lower cylinder in which was the manganin coil under examination. This lower cylinder was placed in an oil bath with thermostatic regulation, by which the temperature could be set at and retained at any desired value. The experimental procedure was as follows. The temperature of the lower bath was set at any desired value, and the pressure increased until the freezing point of mercury at 0° was slightly passed. The mercury then froze with decrease of volume, thus bringing the pressure back to the known equilibrium value at 0°. After equilibrium had been reached, the resistance of the manganin coil was read. The pressure was then lowered slightly by withdrawing the piston. This was followed by automatic restoration of the equilibrium pressure, brought about by melting of the frozen mercury with increase of volume. The transition point was always so sharp that no difference could be detected in the equilibrium pressure whether approached from above or below. The temperature in the lower cylinder containing the manganin was then changed to another desired value. This change of temperature, if it were an increase, would naturally carry with it a rise of pressure, but the pressure is then automatically lowered by the freezing of the mercury. After a steady state is reached, the new value of the manganin resistance is read, and then the pressure lowered again by slightly withdrawing the piston, and the value of the resistance noted again after the equilibrium conditions have been restored from below. In this way the coil can be calibrated over the entire temperature range contemplated for the experiments. Of course this calibration is good only for one fixed pressure, but in view of the proved linearity of the pressure-resistance relation within $1/10\%$ from 0° to 50°, it seemed safe to let the calibration go at this one determination, particularly since no effect could be found.

The calibration of the manganin was carried out at five temperatures; 25°, 45°, 65°, 85° and 110°. No appreciable change of the coefficient could be found for the four lower temperatures, but between 85° and 110° there is a very perceptible change of 1%. But since the range of temperature of the actual experiment did reach over 80°, no correction was applied to the observations for this effect. It is to be noticed that this result is valid only for this one coil, since previous work, both by Lisell [7] and by the author, have shown that different pieces from the same spool of wire may show slight variations in the temperature coefficient, which is sometimes positive and sometimes negative.

In addition to this special calibration for slight relative changes in the pressure coefficient with temperature, the absolute value of the pressure coefficient has been checked from time to time during the course of the experiments. This could be done conveniently with the apparatus as used for the compressibility determinations by determining the transition point of ice VI, or of mercury at known temperatures. These calibrations have shown no change whatever in the pressure constant of the coil.

It has already been stated that the actual measurements involve two sets of readings, one with the apparatus filled with water, kerosene and a small amount of bessemer steel, and a second set with additional steel replacing the water. By subtracting the piston displacement at any given pressure for these two sets of experiments a value is obtained which gives approximately the piston displacement for the water alone, and from which the effect of the distortion of the vessel has in large measure been eliminated. But a moment's consideration will show that the effect of distortion has not been entirely eliminated, and it is necessary to apply a correction for the slight residual effect. The correction comes because of the fact that the position of the piston at corresponding pressures is not the same in the two sets of experiments, so that the subtraction leaves still uncorrected the distortion due to the part of the cylinder exposed to pressure in the one set of experiments and not so exposed in the other. This correction cannot be determined directly, and the only way seems to be to calculate it by the ordinary theory of elasticity, taking for the constant of the steel the values under ordinary conditions, which are known not to vary much even for the most different varieties of steel. There is undoubtedly some error in the correction as so determined, but the total value of the correction is at best small, and any such error is relatively unimportant.

---

[7] Lisell, Om Tryckets Inflytande på det Elektriska Ledingsmotståndet hos Metaller samt en ny Metod att Mäta Höga Tryck (Diss. Upsala, 1903).

The compressibility of the steel replacing the water also evidently enters as a correction factor. This compressibility is relatively slight, and it has been previously determined over a range of 10,000 kgm. The value of the compressibility of the steel also changes with the temperature, but this change has also been shown by direct experiment to be slight, so slight that it can be neglected. In the present work the value was assumed to be constant, independent of temperature and pressure, having the value 58×10^{-8} per kgm. per sq. cm.

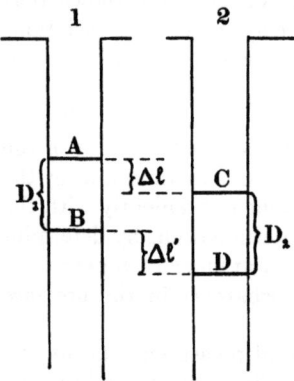

FIGURE 1. Diagram showing the position of the piston. To go with the computations for the corrections to be applied to the compressibility.

There is also a correction to be applied for the compressibility of the kerosene, if the amount does not happen to be the same in the two sets of experiments, and it was seldom that the amount was exactly the same. The variation was very small, however, and the correction is easy to apply if the compressibility of the kerosene itself is known. This was determined with sufficient accuracy for the purpose by an independent set of experiments, exactly the same in principle as those for determining the compressibility of water. The results of these independent experiments are given at the end of the paper.

The following formulas were used in making the corrections, and include all the corrections mentioned qualitatively above. Figure 1 shows the position of the piston at different times in the course of the experiment. The left hand part of the diagram (denoted by the suffix 1) is for the cylinder when it is filled with kerosene and bessemer steel only, and the right hand part (denoted by the suffix 2) is for the cylinder when it contains water, kerosene, and bessemer steel. A and C are the positions of the piston at the arbitrary zero of pressure in these two sets of experiments (this arbitrary zero was usually taken in the neighborhood of 2000 kgm. and will be denoted by p), and B and D indicate the position at some higher pressure, the same in the two sets, which will be denoted by p'. We now write down the expressions for the total volume of the cylinder beneath the piston.

At A, $V_1 = V_{1k} + V_{1s}$
At B, $V_1' = V_{1k}' + V_{1s}'$
At C, $V_2 = V_{2k} + V_{2\,H_2O} + V_{2s}$
At D, $V_2' = V_{2k}' + V_{2\,H_2O}' + V_{2s}'$

where the suffixes K, H$_2$O, or S indicate that the volume is for the kerosene, the water, or the steel respectively.

Subtracting the equations above from each other, we obtain

$$(V_1 - V_2) - (V_1' - V_2') = (V_{1k} - V_{1k}') - (V_{2k} - V_{2k}')$$
$$- (V_{2\,H_2O} - V_2'{}_{H_2O}) + (V_{1s} - V_{1s}') - (V_{2s} - V_{2s}').$$

We now denote by Δl the difference of displacements at the two positions A and C, and by $\Delta l'$ the corresponding difference at the positions B and D. We now assume that V_1 and V_2 differ only by the volume of the cylinder of length Δl, and similarly V_1' and V_2' differ only by the cylinder of length $\Delta l'$. This assumption is justified if only the positions of the pistons at A and C are so far removed from the end of the cylinder that the end effects in the distortion of the interior are the same in the two cases. This condition has been shown by the theory to be satisfied when the distance is two or three diameters, as it always was in these experiments. Hence we may write,

$$V_1 - V_2 = s_0 (1 + a\,p)\,\Delta l$$
$$V_1' - V_2' = s_0 (1 + ap')\,\Delta l',$$

where s_0 is the initial section of the cylinder at atmospheric pressure, and a is the factor of proportionality by which this is changed with pressure. Now if we call the displacement form A to B, D_1 and from C to D, D_2, then

$$D_2 + \Delta l = D_1 + \Delta l'$$

and the above equation may be thrown into the form

$$V_1 - V_2 - (V_1' - V_2') = - s_0(D_2 - D_1)(1 + a\,p') + s_0\,\Delta l\,a\,(p - p')$$

We now make use of the fact that the total change of volume of any substance under pressure is proportional to its mass. If Δv (positive for a decrease) is taken as the change of volume of 1 gm. between p and p', then,

$$V_{1k} - V_{1k}' - (V_{2k} - V_{2k}') = \Delta v_k\,(m_{1k} - m_{2k})$$
$$V_{2\,H_2O} - V_2'{}_{H_2O} = \Delta v_{H_2O}\,m_{H_2O}$$
$$(V_{1s} - V_{1s}') - (V_{2s} - V_{2s}') = \Delta v_s\,(m_{1s} - m_{2s})$$

This enables us to solve the equations for the compressibility of the water and the kerosene, giving,

$$\Delta v_{H_2O} = \frac{1}{m_{H_2O}}\{s_0(D_2 - D_1)(1 + ap') - s_0\Delta la(p - p') + \Delta v_k(m_{1k} - m_{2k}) + \Delta v_s(m_{1s} - m_{2s})\}$$

and for the kerosene, when the two runs are both made with kerosene, as in determining the data for kerosene given at the end of the paper,

$$\Delta v_k = \frac{1}{m_{2k} - m_{1k}}\{s_0(D_2 - D_1)(1 + ap') - s_0\Delta la(p - p') - \Delta v_s(m_{2s} - m_{1s})\}$$

The considerations so far apply only to the measurement of compressibility at constant temperature. The thermal dilatation is determined in the same way as the compressibility from the difference of the thermal dilatation as given by two sets of experiments, one with the water replaced by steel. The piston displacement is not the same at corresponding pressures here, either, and a correction is to be applied for the thermal dilatation of the part of the cylinder which is exposed to pressure in the one experiment and not so exposed in the other. But this portion of the cylinder to which the correction is to be applied was seldom more than 1" in length, and the correction for this amount of steel is negligible in comparison with the thermal dilatation of the total quantity of water. There is also a correction to be applied for the dilatation of the steel replacing the water, and this correction is small but not negligible. It was assumed that the dilatation of the steel remains independent of the pressure over the pressure range used, and the value for ordinary mild steels at atmospheric pressure was employed. This value is 0.000039 for the cubic expansion per degree Centigrade.

The corrections to the measurements of the thermal dilatation are not so serious or so important as those for the compressibility, since the total effect is much smaller and most of the corrections become negligible. The method of determining the thermal dilatation has already been explained to be that of observing the change of pressure brought about at constant volume by a known change of temperature. From this the change of volume with temperature at constant pressure can be immediately determined if the slope of the p-v curve at that point is known, for $\left(\frac{\partial v}{\partial \tau}\right)_p = -\left(\frac{\partial v}{\partial p}\right)_\tau \left(\frac{\partial p}{\partial \tau}\right)_v$; $\left(\frac{\partial v}{\partial p}\right)_\tau$ is evidently given directly from the curves for compressibility at constant temperature. The slope of this curve changes somewhat with the temperature, so that a correction should be applied for this, but the change is so slight at the higher pressures that for this purpose the compressibility can be assumed constant. At the lower pressures, below 2500 kgm., the change cannot be neglected, and another method of computation must be applied.

The thermal dilatation at low pressures was determined by taking directly the difference between the isothermals traced out at different temperatures. This method is not applicable at high pressures because the irregularities of isothermals traced at different times is sufficient to make their difference an inaccurate measure of the slight change of volume with temperature, but at the low pressures, the errors introduced by hysteresis and other irregular action of the steel cylinder are so slight that the method may be used directly to give the value of the compressibility, and by taking the differences, the value of the thermal dilatation. In fact it would seem that the method would be applicable with slight modifications to the determination of the compressibility of a great variety of substances at low pressures, and it is very much more rapid than the methods hitherto used.

A special setting up of the apparatus was necessary for the experiments at low pressures, because in order to be able to reach low pressure on release of pressure it is necessary that the friction in the movable plug be not too high, and if the pressure has once been run to so high a value as to upset the plug, the friction becomes so great as not to permit release of pressure to much below 1500 kgm. For these experiments, then, the plug was made initially a push fit for the hole, by making it about 0.0015" smaller than when used for the higher pressures, and in performing the experiment the pressure was never pushed beyond 2500 kgm. In other respects the experiments at low pressures were the same as those at higher pressures. It was not necessary to take quite so elaborate seasoning precautions at these low pressures, however.

With regard to the amount of hysteresis or elastic after-effects to be met in the experiments, the difference of the displacment with increasing or decreasing pressure usually amounted at the middle of the range to 0.03 in. This amount was very uniformly consistent, indicating that the cylinder had really settled down to a steady behavior. The piston always returned to the starting point to within the limits of accuracy of reading, indicating that there was no leak or permanent set, or wearing of the packing in appreciable amount.

Of course the experiments at low pressures showed very much less hysteresis, in fact it was so small as to be almost imperceptible. The effect of hysteresis was eliminated as far as possible by using for the displacement at any pressure the mean of the results with increasing and decreasing pressure. The hysteresis was so constant that it would probably have been sufficient to have used consistently the results either at increasing or decreasing pressure. The actual procedure has, therefore, the weight of two independent determinations. In the determinations of thermal dilatation, on the other hand, the hysteresis effects were so much smaller, that except for one run initially to show that there was no effect of this kind, the readings were always made either only with increase or only with decrease of temperature for any mean pressure, never with both increase and decrease.

The Data.

Three independent sets of experiments were performed to give the change of volume with temperature and pressure over the entire range; namely the isothermal compressibility at pressures over 2500 kgm., the isothermal compressibility and the thermal dilatation at pressures below 2500 kgm., and the thermal dilatation at pressures over 2500 kgm. This is the actual order of experiment, but for the purposes of presentation it will be better to use the natural order, proceeding from low to higher pressures.

Compressibility at Low Pressures.

The method with the present form of apparatus is not very sensitive at the low pressures, and not many measurements were made over this range. Two sets of determinations of compressibility were made, the first at 20°, 40°, 60°, and 80°, and the second at only 20° and 80°. Here, just as for the measurements at the higher pressures, there is always sufficient friction in the packing after the pressure has once been applied not to permit of close enough approach to the zero to make an extrapolation back to the zero justifiable. And if the extrapolation to the zero is to be made from the readings during first application of pressure, special effort has to be made to design the washers so as to avoid small initial distortions. For this reason only the second of the above sets could be used by extrapolation back to the zero of pressure. The readings of volume at 20° and 80° were corrected back to 40° from the thermal dilatation as determined by this same set of experiments, so that we have from the above two values for the compressibility at 40° up to 2200 kgm. The first set of readings at five temperatures is consistent with this latter set above 1000 kgm., but at the lower pressures gives values for the compressibility which are doubtless too high. To find the best value for the change of volume at low pressures we now have three sets of data, those of the

TABLE I.

VOLUME OF WATER AT 40° AND LOW PRESSURES BY DIFFERENT METHODS.

Pressure, kgm. cm.²	Change of Volume, cm.³/gm.			
	Steel Piez.	Piston.	Amagat.	Final Mean.
0	.0000	.0000	.0000	.0000
500	.0200	.0202	.0204	.0203
1000	.0368	.0377	.0378	.0376
1500	.0527	.0527	.0534	.0532
000	.0669	.0667	.0676	.0673

present determination, those of the previous work by the method of the steel piezometers, and the results of Amagat. The most probable value for the change of volume has been found by comparing these three sets of values. These values are given in Table I, as also the mean selected from them as the most probable value from the data at present in hand. In taking this mean, the greater weight has been given to the values of Amagat at the lower pressures, since his method of measurement was doubtless more accurate for the low pressures than the present method, which was intended only for high pressures, but at the upper end of the range in the neighborhood of 2000 kgm., more weight has been given to the present determinations. It is to be noticed that the mean value taken as final is lower than that found by Amagat. This divergence is in the same direction as that found by Parsons and Cook, who worked with a method like the present one. The deviation found by them from the results of Amagat is greater than that adopted here.

DILATATION AT LOW PRESSURES.

For the thermal dilatation at low pressures, two sets of determinations were made; one was the series of isotherms at four different temperatures already mentioned, and the second was by the method adopted for the higher pressures, namely variation of temperature at constant mean pressure. The method of calculation for this lower

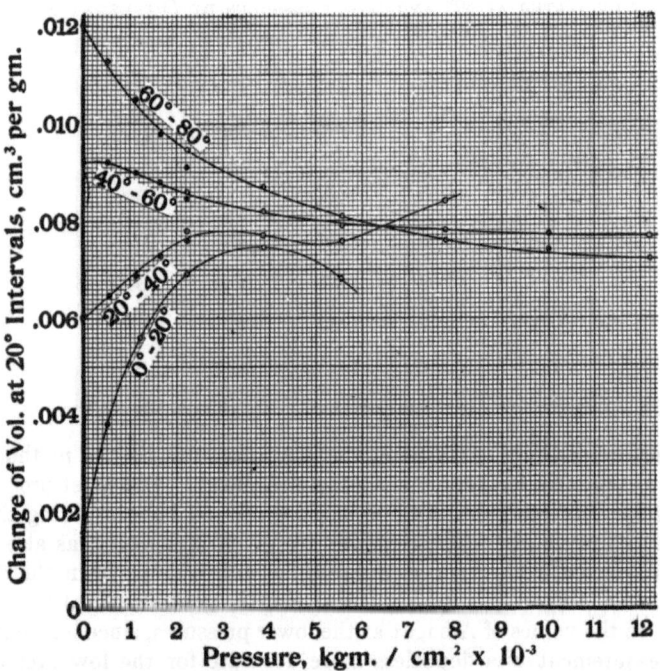

FIGURE 2. The change of volume of water for intervals of 20° plotted against pressure.

range was not the same as that employed for the higher pressures, as already explained, due to the fact that the slope of the isothermals is not sufficiently independent of temperature at the lower pressures. The method of computation adopted here was a graphical one, by plotting the observed volume and pressure points for the different temperatures and taking the difference between adjacent curves graphically. The temperatures at which the different determinations were made were not exactly the even temperatures desired, namely 20°, 40°, and 60°, and 80°, but they were in all cases within a few tenths of a degree of these temperatures. The results were corrected to these even temperatures by assuming the mean variation with temperature over the whole temperature range to hold for the few tenths of a degree on either side. The final result given by the data is the total change of volume for an interval of 20°; from 20° to 40°, from 40° to 60°, and from 60° to 80°. The mean of the results of the two sets of experiments is shown with satisfactory accuracy in Figure 2, on which are plotted all the values obtained by the different methods. The results for the low pressures are shown in the full black circles. These values are seen to extrapolate, without forcing, to the values already found by other observers for atmospheric pressure, and they also make fairly good connections with the values found by the other method for the higher pressures. In view of this agreement it did not seem to be necessary to make further determinations of this quantity.

Compressibility at High Pressure.

The determinations of the isothermal compressibility at higher pressures extended over a considerable interval of time and are more numerous than any of the other determinations. In all, twelve determinations of this quantity were made, at five different temperatures. These determinations include those made during the early course of the experiment, when the attempt was being made to find the thermal dilatation directly from the difference of compressibility at different temperatures. A little work with the method showed that it was not sufficiently accurate for the purpose, but the results obtained then can be used to give the compressibility at the standard temperature, 40°, by applying the temperature correction found from the later more accurate results. The temperature of 40° was chosen as the standard because this is the lowest of the 20° intervals at which the water is liquid up to 12000 kgm.

The results of these twelve determinations, extending over a period of three months, are shown in Table II. The results as given are reduced to 40°, but the temperature at which the original measurements were made is given also in the table. Two of these sets of determinations differ considerably from the others, and were discarded in taking the mean, although as it happens one of these discarded sets is too high and the other too low, so that it makes very little difference

TABLE II.

RESULTS OF DIFFERENT DETERMINATIONS OF CHANGE OF VOLUME OF WATER AT 40°.

Pressure, kgm./cm.²	Change of Volume, cm.³/gm.													
	Jan 22–27.	Dec. 19–29.	Dec. 4–6.	Nov. 29–Dec. 1.	Nov. 27–28.	Nov. 24–25.	Nov. 9–10.	Nov. 8–9.	Nov. 6–8.	Nov. 3–4.	Nov. 2–3.	Nov. 1–2.	Final Mean.
2220	.0000	.0000	.0000	.0000*	.0000	.0000	.0000	.0000*	.0000	.0000	.0000	.0000	.0000
3330	.0255	.0256	.0266	.0268	.0255	.0262	.0257	.0243	.0260	.0253	.0256	.0250	.0257
4440	.0471	.0476	.0487	.0494	.0473	.0481	.0468	.0452	.0471	.0465	.0472	.0466	.0473
5550	.0656	.0666	.0675	.0687	.0660	.0670	.0652	.0635	.0650	.0648	.0658	.0652	.0659
6670	.0816	.0830	.0836	.0852	.0820	.0830	.0813	.0794	.0806	.0809	.0818	.0810	.0819
7780	.0956	.0974	.0979	.0999	.0962	.0970	.0956	.0937	.0947	.0953	.0961	.0952	.0961
8890	.1079	.1100	.1100	.1126	.1087	.1094	.1086	.1064	.1072	.1083	.1088	.1077	.1087
10000	.1190	.1211	.1213	.1238			.1201	.1178	.1186	.1198	.1201	.1189	.1199
11110	.1291	.1310	.1313	.1337			.1305	.1282	.1291	.1302	.1304	.1289	.1301
12220	.1386	.1398					.1412	.1377			.1413	.1408	.1403

* Discarded in taking mean.

in the final result whether they are included in the mean or not. For convenience in making the computations the pressure was taken in units given conveniently by the changes of the manganin resistance, the intervals of pressure corresponding to a displacement of the slider of the bridge wire of 5 cm.

TABLE III.

COMPARISON OF RESULTS BY TWO METHODS FOR CHANGE OF VOLUME OF WATER AT 20°.

Pressure, kgm. cm.²	ΔV.		Pressure, kgm. cm.²	ΔV.	
	Piston.	Piezometer.		Piston.	Piezometer.
2220	.0000	.0000	6670	.0814	.0821
3330	.0257	.0259	7780	.0954	.0964
4440	.0472	.0479	8890	.1078	.1105
5550	.0659	.0666	10000	.1190	.1229

These results, reduced to 20° are shown compared with the results of the previous determination in Table III. It is seen that the newer results are lower than the former ones, the difference being about 1%, except at the higher pressures, where the difference is greater. The agreement is perhaps not as close as could be desired, but at present there seems to be no way of choosing between the results. There is no consistent discrepancy, which would indicate a fundamental error in the present method, such as in the correction applied for the distortion of the steel cylinder, for example. If there were any such error it could be eliminated by so choosing the correction as to make the present results agree with the former ones. In the absence of any means of deciding between the two methods therefore, and since the results by the present method reach over a wider temperature and pressure range, and since also the method has been used much more extensively than the former one and with no greater discrepancy in the individual results, these present results have been accepted as the best ones. But it must be remembered that the absolute compressibility given here may be in error by as much as 1% at the higher pressures. This error, however, will not be found to invalidate any of the conclusions drawn from the data.

Dilatation at High Pressures.

The determinations of the thermal dilatation at the higher pressures were made on four occasions. The first two of these were preliminary, during which was discovered the necessity of seasoning for temperature as well as for pressure, and also the necessity for the secondary pressure seasoning over the small range of pressure accompanying the changes of temperature. These first two determinations, while confirming the results of the two later ones, were not given much weight in selecting the final value. The method of computation adopted in finding the thermal expansion from the data requires mention. At first an attempt was made to apply the same graphical method which has been already explained in its application to the determinations at the lower pressures. This method involves the drawing of a curve of the same general slope as the compressibility curve through the two points giving piston displacement against pressure at each temperature. But it was found that even after the seasoning for the small pressure range involved here, the points were too irregular to give good results by this method. The irregularities may be due to residual hysteresis, but are more probably due to slight irregularities brought about by the motion of the piston itself. These irregularities are too minute to have any effect on the compressibility determinations. The best way to avoid them is to utilize in the computations only those readings during which the piston remains stationary. This means that only the change of pressure accompanying a change of temperature is used in making the computations, the second reading at any temperature by which the pressure is brought back to the mean value being ignored. The change of volume at constant pressure for the given change of temperature is then computed from the known change of pressure at constant volume and the previously determined change of volume with pressure at constant temperature. In making this computation it is generally necessary to make two corrections; one to bring the temperature interval to the exact 20° desired for the final results, and the second to correct for the very slight change of measured piston displacement accompanying the change of temperature. This change of displacement is seldom over 0.003″. It is probably not due entirely to actual motion of the piston, but partly to temperature changes in the bars of the press which dip into the thermostat. That this method of computing the results is preferable to the graphical one previously mentioned is shown by the fact that this method gives very much more uniform and consistent results when applied to the same set of data than the graphical method.

The method of computation adopted was first to calculate independently from the individual observations of each set of readings the thermal dilatation at six mean pressures between 2200 and 12,000 kgm. Then smooth curves were drawn through these points for each set of readings, the curves being spaced in the best way so as to give regular variations with both pressure and temperature. The values given by the smooth curves of each set of readings were then combined into the grand mean. In taking this grand mean, as already explained, almost the entire weight was given to the last two sets of readings. The agreement between the different sets was best at the higher temperatures, 60° to 80°, and about equally good between 20° and 40° and 40° and 60°. All four sets of curves, while not agreeing very well as to the numerical value of the coefficient, do agree as to the general character of the results, which are, perhaps, not quite what would be expected. The unexpected feature is the change in the sign of the temperature derivative of the dilatation at the higher pressures. At the low pressures the dilatation is greater at the higher temperatures, but at the higher pressures the thermal dilatation becomes less at the higher temperatures. This essential feature is verified on all four sets of curves. There are indications that it may be an essential characteristic of the behavior of any normal liquid at high pressures, and that it is not peculiar to water alone. This is shown by the work on kerosene, and is also indicated by the work at present being done on still other liquids. This will be taken up in greater detail later. The other feature not to be expected is the increase in the value of the thermal expansion between 20° and 40° at the higher pressures. It is to be distinctly expected that the thermal dilatation will decrease with rise of pressure, as indeed it does for all the other intervals of temperature, but this rise between 20° and 40° is shown by all the sets of determinations and seems to be an undoubted fact. It is probably connected with some new abnormality in the behavior of water at the higher pressures, which may be connected in some way with the appearance of the new variety of ice.

The values finally taken as the best values for the thermal dilatation are the mean of the results of the four determinations, much the greater weight being given, as already explained, to the two latter determinations. Figure 2 gives these results, as also those of the other methods at the lower pressures. The agreement of the two best determinations at the higher pressures is about 5% for the lower temperature interval from 20° to 40°, 3% for the interval 40° to 60°, and 2% from 60° to 80°. The order of accuracy to be expected in these thermal measurements is not so great as that in the compressibility determinations, therefore, but perhaps the accuracy is as great as could be expected when one considers the smallness of the quantities involved, and the difficulty of making such measurements at high pressures. At any rate the absolute value of the coefficient cannot be very much in error. This is made probable by the agreement with the known values at atmospheric pressure. The accuracy is at least high enough to enable us to expect a fairly good quantitative description of the various thermodynamic quantities under high pressure, even those most sensitive to error. The calculation seems to be worth while carrying through in some detail, because such calculations seem never to have been undertaken for any substance, even for the low pressure range up to 3000 kgm., which is the range over which compressibility determinations have been previously made.

Discussion of Results.

The first necessity for a calculation of the various thermodynamic quantities is as accurate as possible a knowledge of the relation between pressure, temperature and volume over the entire pressure-temperature plane. It may be shown that this is sufficient to completely determine the thermodynamic behavior of the substance if in addition the behavior of the specific heat at constant pressure, for example, is known in its dependence on temperature at atmospheric pressure. This may be assumed to be known well enough for the present purpose. The first and the most important outcome of the present data is, therefore, the construction of a table giving pressure, volume, and temperature at sufficiently close intervals. In constructing this table the basis of computation was the compressibility as determined at 40°. This, together with the known value of the volume at 40° and atmospheric pressure, gave the volume as a function of the pressure down a line through the middle of the table at 40°. The values of the volume were tabulated for intervals of the pressure of 500 kgm., the values found graphically from smooth curves through the experimental points being so smoothed as to give smooth second differences. The values of the change of volume for intervals of 20° now were combined directly with these values to give the volume as a function of the pressure at 0°, 20°, 60°, and 80°. To find the intermediate values of the volume, smooth curves were drawn through these five points at every constant pressure, and the intermediate values so chosen as to given smooth values for the second differences over the entire temperature range. The values for the points below zero, which are also given in the table, were taken directly from the previous work, the values for the dilatation found there being kept without modification, but the present value for the compressibility at 0° being used. The differences so introduced may be seen by comparison of the two tables to be only slight.

The table gives the volume to only four significant figures, since this is as many as the variations in the values of the compressibility would entitle one to, but in making the calculations of the thermal expansion it was necessary to keep three significant figures for the expansion, which would mean five figures in the table.

The thermal dilatation per degree rise of temperature was determined from the values used in the construction of the table for the differences of volume at 5° intervals by dividing by 5, and using the result as the thermal expansion at the mean temperature. The values of the total change of volume for five degree intervals had been smoothed so as to give smooth second differences, so that the dilatation as found in this way was smooth also with respect to the second differences, and could be used directly to give the second temperature derivative of the volume at constant pressure.

The difference of thermal dilatation at different temperatures can evidently be combined with the known compressibility at 40° to give the compressibility as a function of the temperature.

These several quantities so determined; the compressibility, the thermal expansion, and the second temperature derivative of the volume, in their dependence on temperature and pressure, are the basis of most of the calculations of the quantities of thermodynamic interest to be given presently. The accuracy of most of these quantites is not so high but that they can be shown as well in figures as in tables, and this manner of presenting them has been chosen as giving the most ready general survey of the facts.

The tables and figures follow. The results are given simply for themselves, without much comment, except to call attention to the unexpected features, or those properties which seem to be peculiarly characteristic of high pressures. It would not be safe to generalize from the behavior of this one liquid, abnormal at low pressures, to the general behavior to be expected for any liquid for high pressures and the bearing on a possible theory of liquids. Such a general treatment must be reserved for another paper, when the data for more liquids are in hand.

TABLE IV.

Volume of Water as a Function of Pressure and Temperature.

Pressure kgm./cm²	−20°	−15°	−10°	−5°	0°	5°	10°	15°	20°	25°	30°	35°	40°	45°	50°	55°	60°	65°	70°	75°	80°
0			1.0017	1.0006	1.0000	0.9999	1.0001	1.0007	1.0016	1.0028	1.0041	1.0057	1.0076	1.0096	1.0118	1.0143	1.0168	1.0195	1.0224	1.0255	1.0287
500			.9795	.9778	.9771	.9778	.9786	.9796	.9808	.9821	.9837	.9854	.9873	.9894	.9916	.9940	.9965	.9992	1.0020	1.0049	1.0075
1000		.9598	.9584	.9573	.9578	.9589	.9602	.9616	.9630	.9646	.9663	.9681	.9700	.9721	.9743	.9766	.9791	.9816	.9842	.9869	.9896
1500		.9404	.9416	.9407	.9410	.9424	.9439	.9454	.9471	.9488	.9506	.9525	.9544	.9564	.9586	.9609	.9632	.9657	.9682	.9707	.9732
2000	.9228	.9235	.9243	.9252	.9260	.9276	.9293	.9310	.9327	.9345	.9364	.9383	.9403	.9423	.9445	.9467	.9489	.9513	.9537	.9561	.9585
2500	.9085	.9094	.9104	.9117	.9133	.9150	.9167	.9185	.9203	.9221	.9239	.9259	.9279	.9299	.9320	.9341	.9363	.9386	.9409	.9433	.9457
3000	.8963	.8972	.8984	.8997	.9015	.9032	.9050	.9068	.9087	.9105	.9124	.9144	.9164	.9184	.9205	.9226	.9247	.9269	.9292	.9314	.9337
3500		.8864	.8876	.8888	.8907	.8924	.8943	.8961	.8979	.8998	.9018	.9036	.9056	.9076	.9096	.9117	.9138	.9160	.9182	.9204	.9226
4000		.8766	.8774	.8786	.8807	.8825	.8843	.8861	.8880	.8897	.8916	.8936	.8956	.8976	.8996	.9016	.9037	.9058	.9080	.9101	.9123
4500			.8684	.8695	.8717	.8734	.8751	.8770	.8788	.8807	.8825	.8844	.8864	.8884	.8904	.8924	.8945	.8965	.8986	.9008	.9028
5000			.8599	.8610	.8632	.8651	.8666	.8684	.8702	.8721	.8740	.8758	.8778	.8798	.8818	.8838	.8858	.8879	.8899	.8920	.8940
5500				.8537	.8554	.8566	.8585	.8603	.8621	.8639	.8662	.8678	.8698	.8718	.8737	.8757	.8777	.8798	.8818	.8838	.8858
6000				.8464	.8480	.8494	.8509	.8527	.8545	.8564	.8583	.8603	.8623	.8643	.8662	.8682	.8702	.8722	.8742	.8762	.8781
6500					.8409	.8425	.8438	.8454	.8473	.8492	.8511	.8532	.8552	.8572	.8591	.8611	.8631	.8650	.8670	.8689	.8709
7000						.8370	.8386	.8404	.8424	.8443	.8464	.8485	.8505	.8524	.8544	.8564	.8583	.8602	.8621	.8640	.8640
7500						.8305	.8321	.8338	.8360	.8380	.8401	.8421	.8441	.8460	.8480	.8499	.8519	.8538	.8557	.8575	.8575
8000								.8259	.8275	.8298	.8318	.8340	.8360	.8380	.8399	.8419	.8438	.8457	.8477	.8495	.8513
8500									.8216	.8240	.8263	.8283	.8303	.8323	.8342	.8361	.8381	.8400	.8419	.8437	.8455
9000									.8160	.8185	.8208	.8229	.8249	.8269	.8288	.8308	.8327	.8346	.8364	.8383	.8401
9500										.8133	.8156	.8178	.8198	.8218	.8237	.8256	.8275	.8294	.8313	.8331	.8349
10000										.8083	.8107	.8129	.8149	.8169	.8188	.8207	.8226	.8245	.8264	.8282	.8300
10500											.8060	.8082	.8102	.8122	.8141	.8160	.8179	.8198	.8216	.8235	.8252
11000												.8036	.8056	.8076	.8095	.8114	.8133	.8152	.8170	.8188	.8206
11500												.7991	.8011	.8031	.8050	.8069	.8088	.8107	.8125	.8143	.8160
12000													.7966	.7986	.8005	.8024	.8043	.8062	.8080	.8098	.8115
12500													.7922	.7942	.7961	.7980	.7999	.8017	.8036	.8054	.8071

In presenting the results, the quantities have been arranged in order of simplicity of the thermodynamic formulae, which is also the order of directness with which they are derived from the experimental data.

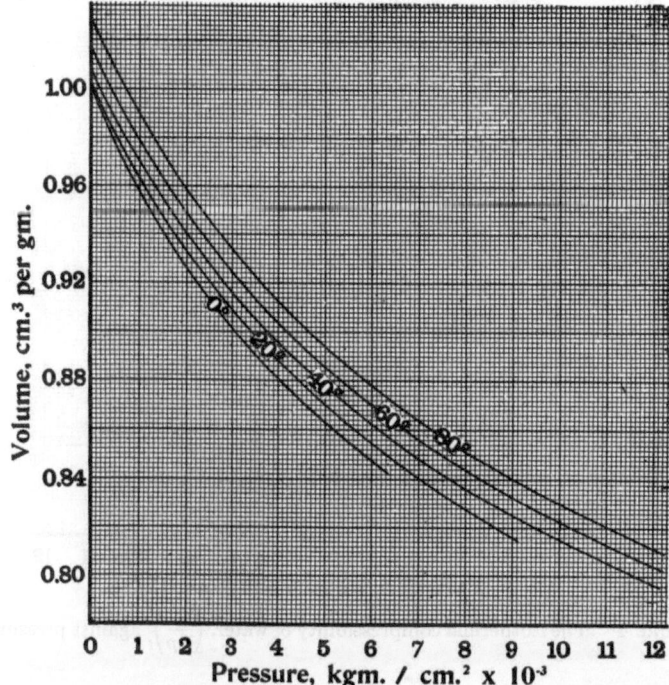

FIGURE 3. Isothermal lines for water, showing volume against pressure.

In Table IV are given the values of the volume for intervals of pressure of 500 kgm., and intervals of temperature of 5°. The table does not require comment. It was computed in the way already described. The values of the volume at intervals of temperature of 20° are shown as a function of the pressure in Fig. 3. The figure does not show the results as accurately as the table, but enables one to form a clearer mental picture of the nature of the results. The curves, on the scale of the figure, do not show any abnormalities to the eye, except in the neighborhood of the origin, where the well known negative expansion at 0° results in the curves drawing together.

There are various abnormalities besides those in the neighborhood of 0°, however, as will be shown by the other figures.

With regard to the compressibility there seems to be some variance of usage, so that it will be well to call attention to the fact that the quantity used throughout this paper in the sense of compressibility is

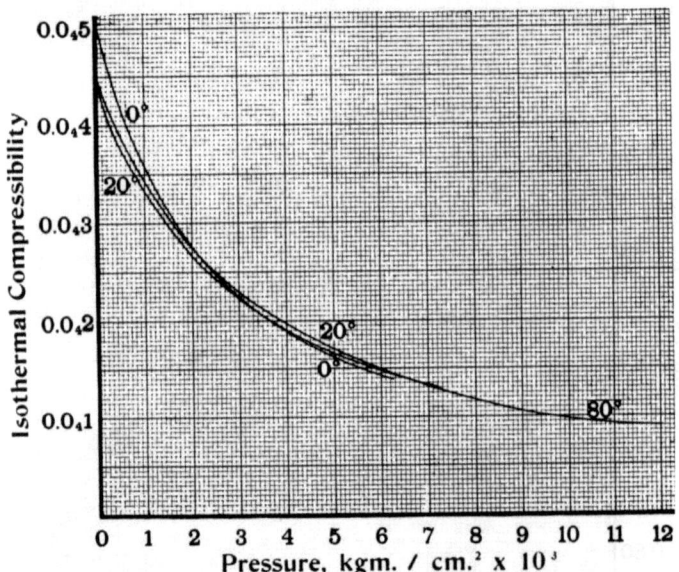

FIGURE 4   The isothermal compressibility of water, $\left(\dfrac{\partial v}{\partial p}\right)_t$, against pressure.

the derivative $\left(\dfrac{\partial v}{\partial p}\right)_t$. Sometimes the expression $\dfrac{1}{v}\left(\dfrac{\partial v}{\partial p}\right)_t$ is used in the same sense. Figure 4 shows the compressibility, that is, the analytic expression $\left(\dfrac{\partial v}{\partial p}\right)_t$, as a function of the pressure at 0°, 20°, and 80°. It would have made the figure too crowded to have tried to show the values for 40° and 60° also. The complete values for the five standard temperatures are shown in Table V separately, however. The figure shows the well known abnormality in the compressibility at the low pressures, namely a higher compressibility at the lower than at the higher temperatures. This abnormality disappears above 50°, and from here on the compressibility increases with rising temperature. The figure shows that at 80° the initial compressibility is higher than at 20°, although it has not yet risen to the value at 0°. In addition to the abnormality at low pressures, the curve shows also a slight

TABLE V.

COMPRESSIBILITY OF H_2O.

Pressure, kgm./cm.²	$\left(\frac{\partial v}{\partial p}\right)_t$, cm.³/gm.				
	0°.	20°.	40°.	60°.	80°.
	.0000	.0000	.0000	.0000	.0000
0	504	453	440	438	450
500	417	381	371	373	388
1000	360	336	328	332	346
1500	313	298	292	296	306
2000	276	267	263	267	274
3000	225	223	222	224	229
4000	189	191	191	193	198
5000	162	166	165	166	170
6000	143	148	146	147	149
7000		133	130	130	132
8000		121	117	117	119
9000			104	104	105
10000			096	096	097
11000			091	091	092
12000			089	089	090

abnormality at the higher pressures in the neighborhood of 6500 kgm. Here the compressibility at 20° rises and at the melting point of ice VI, it has become higher than the compressibility at 80°. The thermal dilatation shows abnormality in the same locality; it would seem to be connected in some way with the appearance of the new variety of ice, but the exact connection cannot at present be stated.

The large change in the value of the compressibility brought about by pressure should be noticed, amounting at 12,000 kgm. to a decrease of five fold. Furthermore the rapid flattening of the curve at the higher pressures also should be commented on. The curve gives the appearance, for the pressure ranges used here, of becoming asymptotic to some value greater than zero. Of course this cannot really be the case for infinite pressures, for otherwise we should have the volume completely disappearing for some finite value of the pressure, but it may indicate the entrance of another effect at the higher pressures, which may persist in comparative constancy for a greater range of pressure than will ever be open to direct experiment, such an effect as the compressibility of the atom, for example. This possibility has been already mentioned and made plausible from the data of the preceding paper.

If instead of the compressibility as defined above, the quantity $\frac{1}{v}\left(\frac{\partial v}{\partial p}\right)_t$, which in this paper will be called the relative compressibility, is plotted, a curve of the same general character as that shown will be obtained.

The compressibility may also be plotted against a different argument than the pressure. For many purposes the pressure is perhaps not the most significant independent variable that might be chosen. This is because the external pressure is not a measure of what is happening inside of the liquid. We conceive a liquid as composed of molecules in a state of constant motion and of collision with each other, acted on also by attractive forces between each other. The effect of these attractive forces is to produce at the interior points a pressure which may be much higher than the external pressure. The external pressure is equal to the interior pressure diminished by the amount of the attractive pressure drawing the molecules to the interior at the exterior surface, where the attraction is an unbalanced action in one direction. The amount of the unbalanced pressure at the outside depends in a complicated way on the law of attraction between the molecules, on their mean distance apart in this surface layer, and on the distribution of velocities in this layer. The external pressure required to hold the liquid in equilibrium is, therefore, largely a surface phenomonon, and is connected in a complicated way with the state of affairs at inside points. A more significant independent variable, therefore, would be one involving only the condition of the molecules on the average throughout the mass, and not one dependent on the surface layer. There are only a few such quantities depending on the state of the liquid at interior points. Any quantities involving in any way the constancy of pressure or of entropy, for example, do depend on the complicated action of the surface layer. One of the quantities which is independent of this surface layer, however, is the volume. In many theoretical considerations the use of the volume as an independent variable is known to produce simplifications.

If the volume, instead of the pressure is taken as the independent variable for the compressibility, curves are obtained of the same general appearance as when the pressure is used for the variable. The compressibility falls with decreasing volume, and the curvature is in the same direction as when the pressure is the independent variable. The same general characteristics are also shown if the relative compressibility instead of the compressibility is plotted against the volume. The two sets of curves, for the compressibility and the relative compressibility, do show one feature in common, however, different from the curves when the pressure is used as the variable. This is the fact that the compressibility is always lower for the same volume at the higher temperature. This is true throughout the entire range of volume used; there is no crossing of the curves indicating abnormalities, such as is the case when the pressure is used as the variable. This is what one would expect on the kinetic theory. A liquid, at two different temperatures but at invariable volume, differs only in the violence of the motion of its molecules. At the higher temperature, the kinetic pressure due to the motion is greater, and so the resistance offered to change of volume under a given increase of external pressure is greater when the temperature is higher.

Fig. 5 shows the thermal dilatation as a function of pressure at various temperatures. The thermal dilatation plotted in the figure is the expression $\left(\frac{\partial v}{\partial t}\right)_p$ instead of the expression $\frac{1}{v}\left(\frac{\partial v}{\partial t}\right)_p$, which is sometimes used as the dilatation. The usage adopted here for the dilatation is analagous to that explained above for the compressibility. The values listed in the figure were obtained from the table of volumes in the manner already described. The curve at 0° was obtained from the data of the previous paper for the low temperatures, but in that paper the mean value of the thermal expansion for the range 0°–20° was given, whereas here the instantaneous value at 0° is given instead. The substitution of the instantaneous for the mean dilatation produces no change in the general character of the curves, however.

The points at the higher temperatures were obtained from the data of this paper alone. There are two striking features that call for special comment. The first of these is the abnormal behavior of the curve for 20°. In the initial stages, the dilatation rises with increasing pressure, unlike normal liquids, but this merely indicates the return of water to the normal behavior to be expected at high pressures. At about 3500 kgm. the curve at 20° has reached a maximum and begins to descend with increasing pressure, as it does for the curve at 0°. But the descent continues for only a little way, and at 5500 kgm. the curve begins to rise again, indicating the entrance of a new abnor-

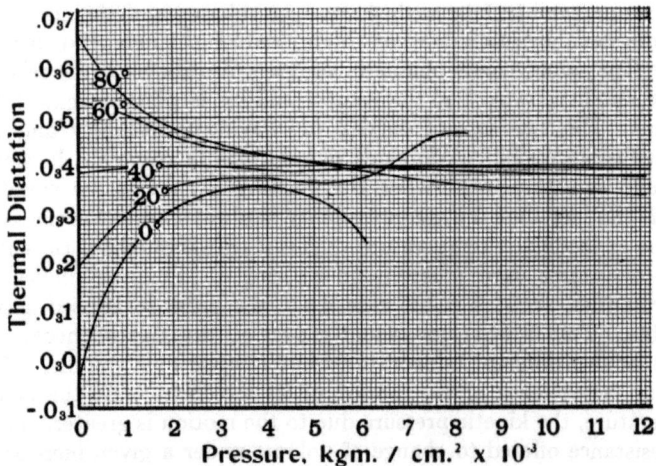

FIGURE 5. The dilatation of water, $\left(\frac{\partial v}{\partial t}\right)_p$, against pressure.

mality. The abnormality is not so striking or so great in amount as that in the neighborhood of 0° and atmospheric pressure. The abnormality at 20° continues for about 2500 kgm., up to 8000, where the curve is terminated by the entrance of the solid phase, but the direction of the curve has already begun to change, indicating that if it could be continued, this abnormality also would probably disappear at higher pressures. As to the question of experimental error here, there would seem to be no room for doubt as to the actual existence of this new abnormality, for it was shown by all four of the dilatation curves, even those taken before the method was got to running satisfactorily, and in which the accuracy was not very high. The curves at the higher temperatures behave as one would be prepared to expect in the region of low pressures. The curve for 40° shows vestiges of the abnormal behavior at the low pressures, namely slight initial rise of dilatation with rising pressure, followed by a fall, but the curves at the higher temperatures, 60° and 80°, show the regular initial decrease with rising pressure shown by all normal liquids. But at higher pressures, the behavior of all three of these curves, for 40°, 60° and 80° is different from what might be expected. The unexpected feature consists in the crossing of the curves, all in the vicinity of the same pressure, 5500 kgm., so that at higher pressures the thermal dilatation at the higher temperatures is lower than it is at the lower temperatures. It has been already remarked that there are indications, both from the present work and from that of Amagat, that this may be the behavior for any normal liquid at sufficiently high pressures. The comparative constancy of the thermal dilatation at the higher pressures, is also a matter perhaps not to be expected. Thus the expansion at 40° remains nearly constant over the entire range of pressure, while the compressibility has in the same range dropped from 44 to 9. It was distinctly expected, before these measurements were taken, that the dilatation would show the greater variation with pressure, so that the effect of temperature on the volume would tend to disappear at the higher pressures, but such is not the case.

The relative thermal dilatation may be plotted against pressure, as was the relative compressibility. The curve shows no striking features. The curve plotting relative dilatation against volume has also been plotted, and this is the same in general character as the others. The slight differences consist in an accentuation of the abnormalities in the neighborhood of 5500 kgm., and the fact that at the lower volumes, that is at the higher pressures, the dilatation against volume increases with decreasing volume for 40° and 60°, but decreases for 80°.

These figures for the thermal dilatation and the compressibility complete those which are obtainable directly from the table. Other quantities of thermodynamic interest may be obtained by combining these, however. Perhaps the simplest of these quantities are those connected with the absorption of energy when the pressure is changed at constant temperature. The first of these is the actual mechanical work done by the external pressure in compressing the liquid at constant temperature. This of course is simply the expression

$W = \int p \left(\frac{\partial v}{\partial p}\right)_t dp$. It was obtained by a mechanical integration from curves similar to the volume curves of Figure 3, drawn on a larger scale. For this purpose the integrating machine owned by the mathematical Department of Harvard University was used. The

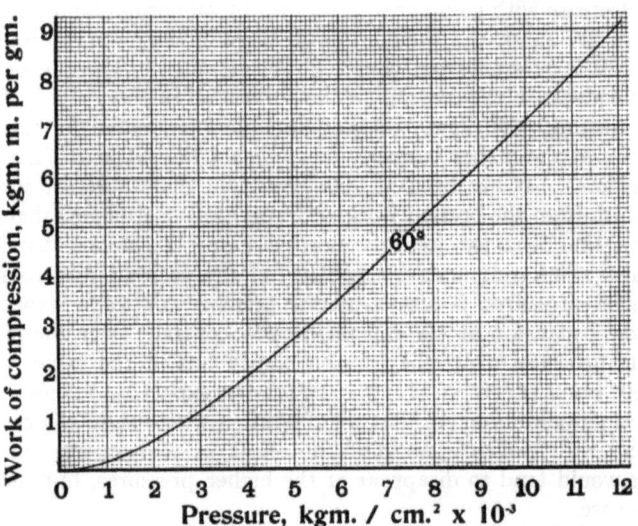

FIGURE 6. The mechanical work of compression at 60°.

actual value of the mechanical work at any pressure is of course dependent on the temperature, but since the variation is so slight that it would have been impossible to show it in the figure (see Figure 6), the work of compression is plotted for only the one temperature, 60°. Although the change of external work with temperature was too slight to show in the diagram, the change with temperature was nevertheless taken account of in making the calculations of the quantities depending on it to be described immediately. After the first 4000 kgm. it is seen that the curve becomes very approximately linear. The curve for a substance which retains the same compressibility unchanged over a wide pressure range, as steel for example, is a parabola, the work increasing directly as the square of the pressure. That this curve for water becomes linear, means that the compressibility decreases so fast with increasing pressure that the decrease in the yield of the liquid for a given increment of pressure decreases almost at the same rate that the pressure itself increases.

The total heat given out during an isothermal compression may be derived from the formula $\left(\frac{\partial Q}{\partial p}\right)_\tau = -\tau\left(\frac{\partial v}{\partial \tau}\right)_p$. This quantity is shown in Figure 7. The figure does not call for especial comment. The

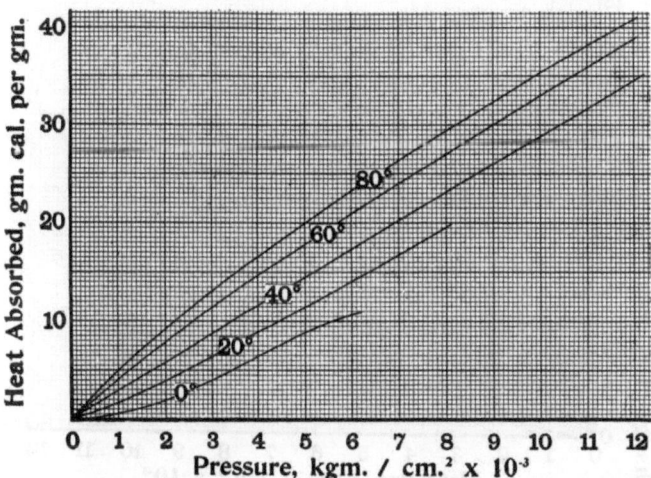

FIGURE 7. The heat given out by water during an isothermal compression.

rapid change in the direction of the isothermal lines in the vicinity of the origin due to the abnormal behavior at low temperatures and pressures is manifest from the figure, as also the slight abnormalities at the upper ends of the 0° and the 20° curves, already commented upon in other connections. Beyond 5000 kgm. the curves for all temperatures tend to become linear and parallel to each other.

These two quantities, the heat liberated in compression and the mechanical work, combine to give the change of internal energy along an isothermal, this change of energy being equal to the difference of the heat and the mechanical work. The change of energy so calculated is shown in Figure 8. The change is a decrease, which continues at all temperatures up to the highest pressures. In the previous paper a value of this quantity was given, confessedly inaccurate, since in the computation the mean thermal dilatation between 0° and 20° had been used instead of the actual dilatation at 0° or 20°. The curve so obtained had the characteristics of the curve now given for 0°, but the maximum at the top was much more strongly accentuated than in the present figure. It was surmised in the previous paper that at high enough pressures the internal energy of all liquids would probably increase instead of decrease along an isothermal. This surmise seemed

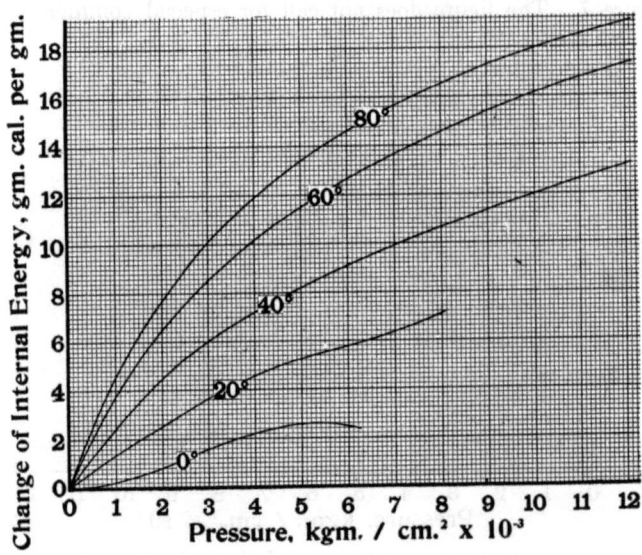

FIGURE 8. The decrease of internal energy of water during an isothermal compression.

plausible because one would expect that at high enough pressures the energy stored up as strain in the interior of the molecules in virtue of the extremely high pressures would more than counterbalance the work done by the attractive forces of the molecules themselves as they were brought closer together by the action of the pressure. This present figure shows that this is not the case, however, for the range of pressure reached here. The lower temperature, 0°, is the only one at which this reversal of the direction of the change of internal energy manifests itself, and this change, in comparison with the other curves, is now seen probably to be an effect of the other abnormalities shown at low pressures and temperatures. Nevertheless it would still seem as if at very high pressures the energy must increase instead of decrease along an isothermal, but the only indication of it from the present curves is in the direction of curvature, which is in the direction to indicate the possibility of a maximum and a reversal of direction at higher pressures. The pressure for a maximum, however, if there is one, is much beyond the reach of any at present attainable. Within the pressure range of these measurements, the attraction between the molecules still remains the dominant feature, so that the work done by the attractive forces and liberated as heat much more than suffices to overbalance the mechanical work of compression.

The internal energy of a substance is one of those quantities which depend only on the properties of the mass of the substance at interior points and do not involve the action of the surface layer. Change of energy plotted against volume shows in the first place that the change of internal energy is much more nearly a linear function of the volume than it is of the pressure. The average slope of the isothermal lines of energy increases rapidly with rising temperature for the lower temperatures, but the two curves for 60° and 80° run nearly parallel to each other for their length. Abnormalities are shown at the upper ends of the 0°, 20° and the 40° curves, and the 0° curve shows the same maximum as it does when plotted against pressure. The origin, of course, for the curves at different temperatures does not coincide as it does for the same quantities when plotted against pressure.

One other quantity may be simply determined in terms of the compressibility and the thermal dilatation alone, the so-called pressure coefficient, that is, the change of pressure following a rise of temperature when the temperature is raised by 1° at constant volume. This quantity is given immediately in terms of the compressibility and the thermal dilatation by the well known formula,

$$\left(\frac{\partial p}{\partial \tau}\right)_v = -\left(\frac{\partial v}{\partial \tau}\right)_p \bigg/ \left(\frac{\partial v}{\partial p}\right)_\tau.$$

It is shown plotted in Figure 9. The curves for 0° and 20° show anomalies, as is indicated by the unexpected direction of curvature. The other curves for the higher temperatures seem to be regular enough, though of course it cannot be told whether the course of these curves is the same as that which would be shown by a normal liquid or not. At the upper ends of the high temperature curves, the curvature is in such a direction that if they were continued far enough the pressure coefficient would decrease instead of increasing with rising pressure.

This quantity, the pressure coefficient, has been made the basis of theoretical speculation. It has been enunciated as a law, approximately true, by Ramsay and Shields, that the pressure coefficient is a function of the volume only. This means that if the coefficient were plotted against volume instead of pressure the curves for all five temperatures would fall together. That this is not the case for water at high pressures is shown very distinctly in Figure 10. At the lower pressures and the larger volumes, the curves for the different tempera-

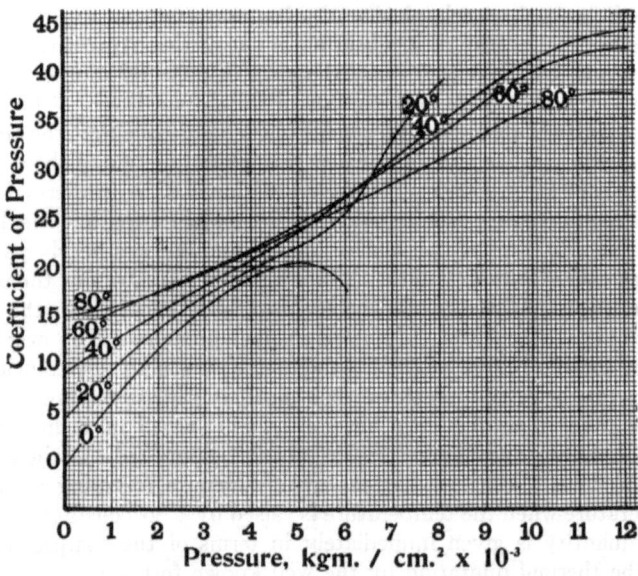

FIGURE 9. The pressure coefficient, that is the change of pressure accompanying a rise of temperature of one degree, as a function of the pressure.

tures are very widely separated. The abnormality on the curve at 0° in the neighborhood of the locality where the new variety of ice makes its appearance is very striking. At the higher pressures the curves do draw together, but they are not approaching coincidence, for they cross in the neighborhood of a volume of about 0.85. It does not seem likely that the entire failure of coincidence throughout the whole range of pressure can be due to abnormalities, since even at low pressures water is nearly normal at the higher temperatures, and certainly at the higher pressures and temperatures we have every reason to expect that its behavior is quite like that of other liquids.

This completes the list of quantities which can be deduced directly from the compressibility and the thermal dilatation. Other quantities of thermodynamic interest involve the specific heats, and these in turn involve the second temperature derivative of the volume.

The first of these quantities is the specific heat at constant pressure. This is given by the thermodynamic equation $\left(\frac{\partial C_p}{\partial p}\right)_\tau = -\tau \left(\frac{\partial^2 v}{\partial t^2}\right)_p$. It will be seen that only the derivative of the specific heat is given by the data as directly determined. In order to obtain the specific heat itself, the derivative, obtained from the tables in a manner already described, must be integrated. This integration was performed mechanically, in the same manner as the integration for the mechanical work of compression. The results are shown in Figure 11. The values for the specific heat as a function of temperature at atmospheric pressure were taken from the steam tables of Marks and Davis.[8]

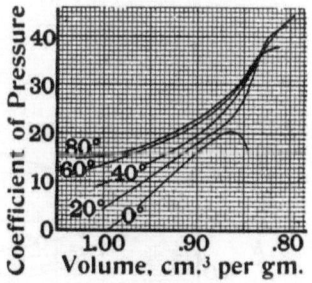

FIGURE 10. The pressure coefficient of water as a function of the volume.

These values seem to be open to some slight question at the present time due to experimental work done by Bousfield[9] since the publication of the tables, but in any event the possible error is slight, too slight to be visible on the scale of the figure. The curves show the now expected abnormalities at 0° and 20°. The striking feature about the curves for the higher temperatures is the very rapid increase of the specific heat with rising temperature at the higher pressures. The specific heat at first decreases on all the curves except at 0°, but passes through a minimum, and then increases. The pressure of the minimum rapidly becomes less with rising temperature, and is situated at 6500 kgm. for 40°, 5500 kgm. for 60°, and at 1100 kgm. for 80°. At 80° the specific heat rises rapidly beyond the minimum, reaching the value 1.17 at 12000 kgm.

Any valid characteristic equation should predict the behavior of the specific heat at high pressures as well as giving the volume in terms of pressure and temperature, since from the equation the second temperature derivative of the volume may be found. The equation of Tumlirz[10] has been mentioned in the preceding paper as giving perhaps as good agreement as any with the previously known facts over

---

[8] Marks and Davis, Steam Tables. (Longmans, Green, and Co.)
[9] W. R. and W. E. Bousfield, Trans. Roy. Soc. (A), **211**, 199–251 (1911).
[10] Tumlirz, Sitzber. Wien, Bd. **68**, Abt. IIa (Feb., 1909), pp. 39.

a pressure range of 3000 kgm. This equation would predict a continuous diminution in the specific heat up to infinite pressures, the limiting value being very approximately 0.5. It was shown in the preceding paper that there is some new effect introduced at the high pressures which does not make itself felt at the low pressures, with the

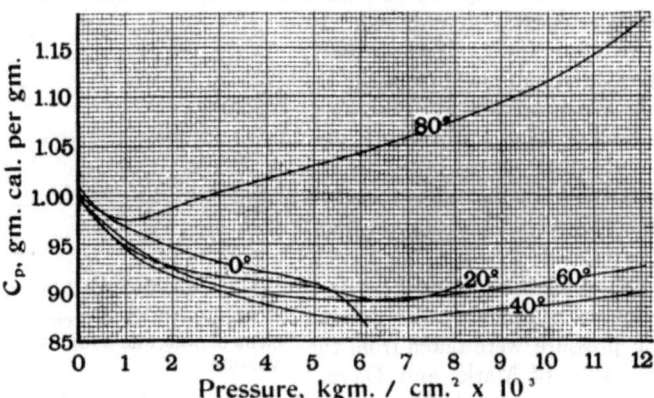

FIGURE 11. The specific heat at constant pressure of water as a function of the pressure.

result that an extrapolation to infinite pressures from the behavior for the first 3000 kgm. is not safe. This was shown in that paper by the behavior of the volume, which tended to decrease more rapidly at the high pressures than was predicted by the formula. The present data also show that there is a new effect at the high pressures, and indicate that the effect, whatever it is, is such as to have a much greater influence on the specific heats than on the volume itself.

The specific heat at constant volume may be found from the specific heat at constant pressure by means of the formula, $C_p - C_v = -\tau \dfrac{\left(\dfrac{\partial v}{\partial \tau}\right)_p^2}{\left(\dfrac{\partial v}{\partial p}\right)_\tau}$.

This quantity, so calculated, is shown in Figure 12. The same abnormalities are shown at 0° and 20° as were shown in the curves for C_p. The curves for 40° and 60° decrease for nearly their entire lengths, although they are just beginning to rise at the very highest pressures, but the curve for 80° shows the same sharp turning point and the same rise through the greater part of its length as the curve for C_p. This quantity, the specific heat at constant volume, has more theoretical significance than the other specific heat, since this represents the heat going into the rise of internal energy of the liquid when the temperature rises, and does not involve the work done against external pressure in expanding the liquid. The external work in-

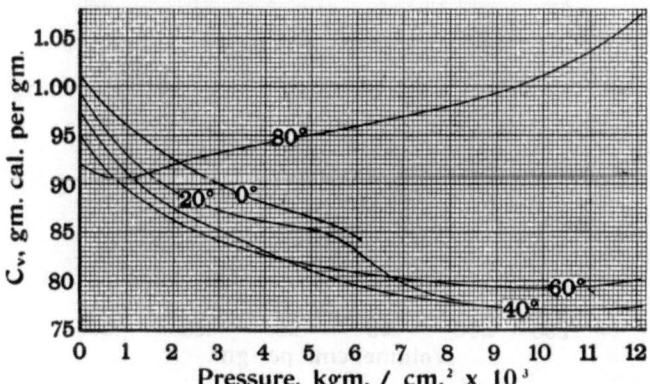

FIGURE 12. The specific heat at constant volume of water as a function of the pressure.

volves in a complicated and at present unknown way the action of the surface layer, while the specific heat at constant volume does not contain this surface effect. This specific heat is therefore one of the quantities mentioned in the beginning as having significance because it does not involve the unknown attractive forces between the molecules as displayed in the surface layer. In order to show this independence of the surface layer, of course C_v should be plotted against a variable not itself involving the action of the layer. It is evidently not adequate, therefore, to plot C_v against the pressure as as been done in Figure 12. C_v plotted against volume may be expected to show this independence of the action of the surface layer. It is shown so plotted in Figure 13. The figure is of the same general character as that in which it is plotted against pressure, but the separation of the curves for the different temperatures is greater, partly because the curves do not start from a common origin. The minimum on the curves for 40° and 60° comes at a lower pressure than it does in the former figure, and the upper end of the 80° curve is perhaps a trifle steeper at the upper end, but there are no essential differences. The entire behavior of the curves is not what one would expect from the ordinary theoretical considerations, however. It is usually considered that when the volume of a substance is kept invariable all, or else a fixed fraction, of the heat put in during a rise of temperature goes toward increasing the kinetic energy of the molecules. This is because the temperature is supposed to be proportional

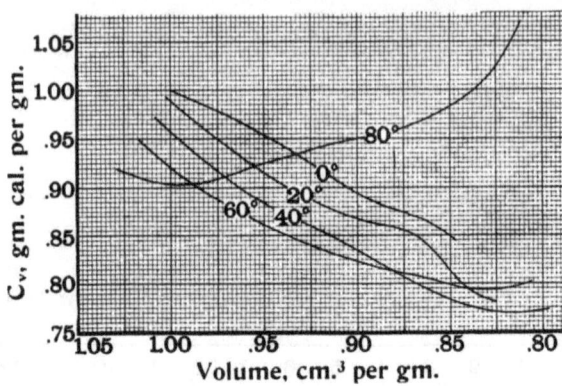

FIGURE 13. The specific heat at constant volume of water as a function of the volume.

to the energy of translation of the molecules, and therefore, because of the law of the equipartition of energy, to the total energy of the molecules. We should expect, therefore, that the input of energy required to raise the temperature by a specified amount would involve only the interval of temperature, and would be independent of the absolute value of the temperature and of the volume. The curves show most convincingly that this is not the case. This suggests that in formulating a theory of liquids it would be well to scrutinize pretty carefully several assumptions that underlie the above considerations, namely that the temperature is proportional to the kinetic energy, that a fixed fraction of the total energy of the molecules is kinetic, and that the law of the distribution of velocities is independent of temperature.

Another quantity of thermodynamic interest which may be found in terms of the specific heats is the thermal effect of compression, that is the rise of temperature in degrees accompanying a change of pressure adiabatically of one kgm. per sq. cm. This may be computed by the thermodynamic formula $\left(\frac{\partial \tau}{\partial p}\right)_\varphi = \frac{\tau \left(\frac{\partial v}{\partial \tau}\right)_p}{C_p}$. The results so calculated are shown in Figure 14 for the five standard temperature intervals. The character of the curves is the same as that shown so many times before, namely a rise to a maximum and then a fall at 0°, the abnormal behavior at the upper end of the 20° curve, and the more or less regular behavior of the three curves for the higher

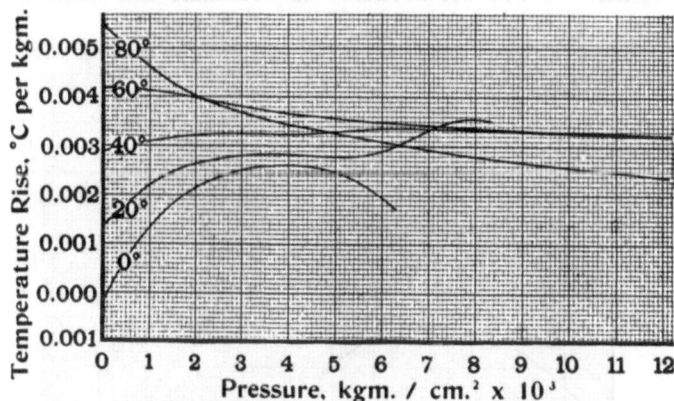

FIGURE 14. The adiabatic rise of temperature of water against pressure.

temperatures, with the crossing of the high temperature curves below the low temperature curves at the higher pressures. In the preceding paper only the approximate values for the very lowest temperature interval could be found. The calculation was based on the mean value of the dilatation between 0° and 20°. The general character of the curve was the same as that found here for 0°, namely a rise to a maximum and then a fall.

Finally it is possible to compute from the quantities in hand the difference between the isothermal and the adiabatic compressibilities. This is found from the formula $\left(\frac{\partial v}{\partial p}\right)_\varphi - \left(\frac{\partial v}{\partial p}\right)_\tau = \frac{\tau}{C_p}\left(\frac{\partial v}{\partial \tau}\right)_p^2$. The results are shown in Figure 15. The general character of the results is exactly the same as those previously given for the temperature effect of compression. Here again, the results at the lowest temperature agree with those of the previous paper which were based on a mean value for the dilatation.

PROPERTIES OF KEROSENE UNDER PRESSURE.

In the course of the experiment other data were gathered incidentally which are of interest for themselves, and which will now be given. First of these is the compressibility and the thermal dilatation of kerosene. It was not necessary to determine this quantity in

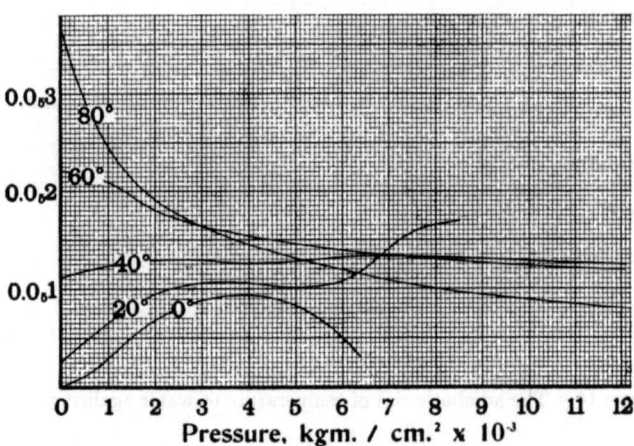

FIGURE 15. The difference between the adiabatic and the isothermal compressibilities of water.

order to find the corrections to be made for the distortion of the vessel, but since half the work was already done in determining the effect with the cylinder partly filled with kerosene and the other part filled with bessemer steel, it seemed worth while to make the additional run necessary to determine the pressure and temperature effects on the kerosene. Not so many determinations were made of these quantities for kerosene as were made for the water. The results are given in Table VI. The curves showing the total thermal change of volume for 20° intervals are shown in Figure 16. This figure is the analog of Figure 2 for water. The results are very different. At the lower pressures the dilatation is greater at the higher temperatures, as it is for all normal substances, but with rising pressure the effect is reversed, the dilatation becoming greater for the lower temperatures. This is the same behavior which takes place for water at higher temperatures after it has regained normality. But above 5000 kgm. the kerosene shows other abnormalities quite different in their character from those of water. This is shown plainly in the figure as a separation and then a drawing together again of the curves. The curve for 20°–40° between 6000 and 8000 and the curve for 60°–80° beyond 9000 accomplish this separation and drawing together again

TABLE VI.

VOLUME OF KEROSENE AS A FUNCTION OF TEMPERATURE AND PRESSURE.
(The volume at 0° and atmos. pressure is taken as unity.)

Pressure, kgm. cm.²	Volume.			
	20°.	40°.	60°.	80°.
0	1.0221	1.0412	1.0611	1.0819
1000	.9643	.9763	.9885	1.0010
2000	.9274	.9376	.9468	.9553
3000	.8995	.9086	.9165	.9239
4000	.8781	.8861	.8931	.8997
5000	.8606	.8681	.8747	.8807
6000	.8456	.8529	.8592	.8647
7000	.8323	.8396	.8456	.8508
8000	.8201	.8275	.8334	.8384
9000	.8090	.8161	.8220	.8269
10000	.7989	.8057	.8115	.8164
11000	.7897	.7960	.8017	.8069
12000	.7815	.7872	.7928	7982

by rising with rising pressure, exactly as do some of the curves for water. The abnormality is doubtless due to an entirely different cause, however. In this case the effect is to be explained by the delayed freezing of the kerosene. Kerosene is not a simple pure substance, but is a mixture of several components with different melting points. Freezing under these conditions is not sharp, but is spread out over a considerable interval of temperature or pressure as the case may be. Neither is there any necessity that the freezing should ever be perfectly complete, as indeed it is probably not. This may be shown at atmospheric pressure by plunging the kerosene into solid CO_2. The effect is to change the kerosene to a white pasty mass, like white vaseline. The pressure at which this transition occurs will rise with increasing pressure. The existence of a transi-

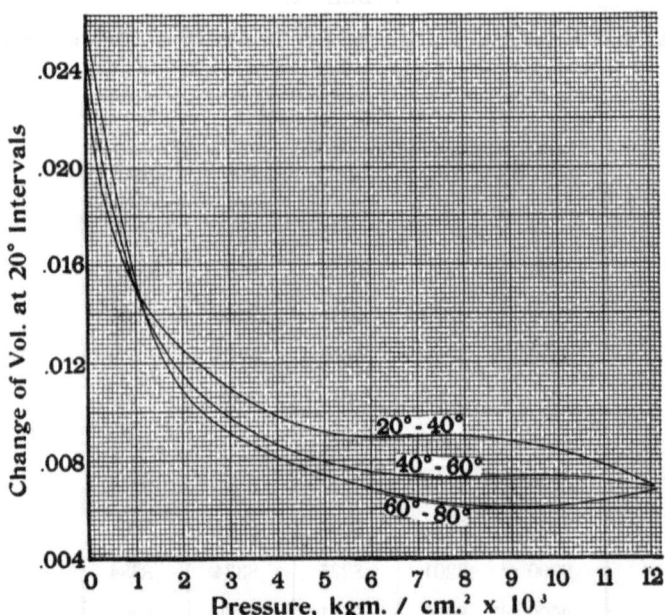

FIGURE 16. The change of volume of kerosene at constant pressure for a rise of temperature of 20°.

tion point, if there were one perfectly sharp, would be shown by an abrupt rise of the curve by an amount corresponding to the change of volume on freezing. But with the delayed freezing which takes place here due to the separation out of the separate components from a solution of varying strength, this abrupt rise becomes converted into a gradual rise extending over a fairly wide pressure range. Furthermore, the mean pressure at which this rise takes place increases with rising pressure, just as the ordinary freezing point is raised by increasing pressure. These features are all clearly shown in the diagram. At the extreme right of the diagram, at pressures over 12,000 kgm., there is evident the beginning of the reversal of the effect, that is, the curves are going to cross again, and the thermal dilation become greater at the higher temperatures. This may possibly indicate a reversal of the reversal of the effect mentioned above for liquids, but more probably the meaning is simply that at pressure above 12,000 the substance is practically a solid, and that for solids the reversal of the effect found in liquids at high pressures does not occur.

There is one bearing which these observations have on the previous data which should perhaps be mentioned. This is in connection with the delayed freezing. Whenever freezing takes place there is usually the possibility of subcooling before separation to the solid form takes place. The amount of subcooling usually taking place depends on the nature of the liquid. In some it is very considerable, while in others it is negligible. If such subcooling took place here, it would produce irregular results, because the change of volume in the kerosene transmitting pressure to the water would not always be the same under the same pressure. The only answer to be made to this objection is that in this experiment the subcooling was not great enough to produce sensible irregularity. No discrepancies were found in the data suggesting that they were due to this effect. It was feared in the beginning of the work that the effect might be very troublesome, but such did not turn out to be the case.

Also with respect to the solidification of the kerosene, the experiments showed that the solidification could not be complete, but the kerosene, even at the highest pressures, must remain a pasty mass like vaseline in nature, always capable of transmitting pressure nearly hydrostatically. But that on the other hand the kerosene does undoubtedly become pretty stiff under pressure has been already shown in the course of some measurements on the linear compressibility of steel rods.

The second bit of data collected incidentally in the course of the work was a measurement of the expansion and the thermal dilatation of the high temperature variety of ice.

THE COMPRESSIBILITY AND THERMAL DILATATION OF ICE VI.

Although these data are not directly concerned with the properties of liquid water, which forms the subject matter of this paper, still it was so easy to obtain them without any modification in the arrangement of the apparatus, that it was thought worth while to measure them. In the previous paper on the properties of water and the several varieties of ice, a very rough experimental value for the compressibility was given, as also a computation of the approximate compressibility, neglecting the thermal dilatation of the ice, for which no experimental value was found at that time. These measurements here include a direct measurement of the thermal dilatation, and two different determinations of the compressibility by two different methods. The value for the dilatation may be combined with the already determined values for the volume of the liquid and the change of volume when ice VI separates out, to give a third independent value for the compressibility.

The determinations of the dilatation will first be described. This was found in the same manner as the dilatation of the liquid, by changing the temperature at constant mean pressure, and measuring the change of pressure brought about thereby. Three determinations of this were made for the combination of ice and kerosene, and two for the combination of kerosene and bessemer. The agreement of the different determinations was within 2% of the mean. The dilatation was found between 0° and 20° at a mean pressure of 10,000 kgm. The correction introduced by the thermal dilatation of the bessemer cylinder in the control experiment is fairly large here, being about 25% of the entire effect. The value assumed for the cubic dilatation was 0.000036, which is the value for atmospheric pressure. The effect of pressure is to decrease this number slightly, which would result in a larger value for dilatation of the ice. The effect of pressure on this quantity is, however, very small, and the error so introduced is probably negligible. The mean dilatation found in this way for the 20° above 0° at 10,000 kgm. was 0.00241 cm.^{3}/ gm., or 0.000120 cm.^{3}/ gm. per degree. This is considerably less than the dilatation of the liquid in this neighborhood, for which the value 0.00040 has been found previously.

This value for the dilatation may now be combined with the other data for the liquid and the solid to give the compressibility of the solid along the equilibrium curve. For this we have the following data: vol. of 1 gm. of water at 0° and 6360 kgm., 0.8428 cm.^{3}, and at 20° and 9000 kgm. (these are the equilibrium pressures at these temperatures) 0.8160 cm.^{3}. For the change of volume when the liquid freezes to the solid we have at 0°, 0.0916, and at 20°, 0.0751. This gives for the volume of ice at the equilibrium pressures at 0° and 20° the values 0.7512 and 0.7409 respectively. The decrease of volume of the ice along the equilibrium curve is 0.0103. Part of this is an increase due to rise of temperature, which according to the above data is 0.0024. This leaves a decrease of 0.0127 to be accounted for by the increase of pressure of 2640 kgm. which gives a mean compressibility over this range of 0.0000048, a little more than one third of the compressibility of the liquid over the same range.

The direct determination of the compressibility of the ice was made by two different methods. One of these was the same as that used roughly in the preceding paper, that is by finding the difference of the slope of the curves plotting piston displacement against pressure above and below the transition point to the solid. The values obtained in the preceding paper for this were very rough. In these determinations the cylinder was very much more carefully seasoned, and the readings were made with all the precautions which had been suggested by all the experience of this paper. Two determinations of this quantity were made at 0° and also two determinations at 20°. The two values for the difference of compressibility differed by 2.5% at 0° and by 0.7% at 20°. The value found for the difference was 0.0000087 at 0° and 0.0000067 at 20°. Combining with the values given already for the compressibility of the liquid, this gives for the compressibility of ice VI 0.0_549 at 0° and 6360 kgm., and 0.0_543 at 20° and 9000 kgm. Mean 0.0_546.

The second method for determining the compressibility was exactly the same as that for finding the same quantity for the liquid, comparing the displacements when the apparatus was filled with ice and kerosene with those when the ice was replaced by bessemer steel. This determination was made over a wider pressure range, to find if possible the variation of compressibility with pressure. No variation with pressure could be found over a range of 4500 kgm. at 0° and 3300 kgm. at 20°. The absolute values do not agree with those found by the two other methods, however, the figures being 0.0_531 at 0° and 0.0_535 at 20°. The cause of the discrepancy is not clear, but is probably connected in some way with the hysteresis of the cylinder. The hysteresis was not regular for these small pressure ranges, being at times almost negligible, and again being as large as for almost the entire pressure range from atmospheric pressure to the maximum. There seems little question but that the greater weight is to be attached to the values found by the first two methods. This third determination does show, however, that the variation of the compressibility with pressure and temperature over this range is so small as to be beyond the accuracy of these measurements. In selecting the best probable value for the compressibility the only weight that will be assigned to this third determination is in slightly lowering the mean of the other two.

The final most probable values for Ice VI are as follows: for the compressibility 0.0_545, and for the thermal dilatation 0.000120 cm.^{3}/ gm. over the range 6360–10,000 kgm. and 0° to 20°.

The cost of much of the apparatus used in this investigation was defrayed by an appropriation from the Rumford Fund of the American Academy.

Jefferson Physical Laboratory,
Harvard University, Cambridge, Mass.

[In Table IV, page 338, the volumes in the 30° column from 2500 to 8000 kg/cm² inclusive should be increased by 0.0020.]

THERMODYNAMIC PROPERTIES OF TWELVE LIQUIDS BETWEEN 20° AND 80° AND UP TO 12000 KGM. PER SQ. CM.

By P. W. Bridgman.

Presented November 13, 1912. Received December 30, 1912.

Table of Contents.

		Page.
I	Introduction	4
II	Experimental Method	7
III	Methods of Computation	18
	The Table of Volumes	18
	Thermal Dilatation	27
	Isothermal Compressibility	29
	Work of Compression	30
	Heat of Compression	32
	Change of Internal Energy	32
	Specific Heat at Constant Pressure	33
	Specific Heat at Constant Volume	36
IV	Numerical Details of Experiment and Computation	37
	Methyl Alcohol	38
	Ethyl Alcohol	42
	Propyl Alcohol	46
	Isobutyl Alcohol	48
	Amyl Alcohol	52
	Ether	55
	Acetone	58
	Carbon Bisulphide	61
	Phosphorus Trichloride	64
	Ethyl Chloride	66
	Ethyl Bromide	68
	Ethyl Iodide	72
V	Discussion of the Thermodynamic Properties	74
	Volume	74
	Thermal Expansion	76
	Isothermal Compressibility	83
	Pressure Coefficient	89
	Work of Compression	90
	Heat of Compression	93
	Change of Internal Energy	95
	Specific Heat at Constant Pressure	100
	Specific Heat at Constant Volume	101
VI	General Discussion of the Bearing of the Results on a Theory of Liquids for High Pressures	104
VII	Summary	113

I. Introduction.

The experimental material of this paper consists of direct measurements of the volume of twelve liquids at different temperatures and pressures. The pressure range is from atmospheric pressure to 12000 kgm. per sq. cm., and the temperature range from 20° to 80°. The measurements were made at enough points to determine the volume at any pressure and temperature. These data are presented in the first part of the paper. The second part of the paper contains a discussion of a number of quantities of thermodynamic significance which have been computed from the data of the first part. The discussion is concerned only with the more important thermodynamic properties, namely the isothermal compressibility, thermal expansion, work of compression, heat of compression, change of internal energy, and the specific heats at constant pressure and constant volume.

Apparently the only other work of similar character at even comparatively high pressures is that of Amagat,[1] published in 1893. Amagat measured the volume of twelve liquids up to 3000 kgm. and between 0° and 40° or 50°. Beside the volume he tabulated the compressibility, dilatation, and pressure coefficient for some of the liquids, but the tabulation was by no means systematic or complete.

It is hoped that the material in this paper will afford the means for a renewed attack on the problem of the nature of the mechanism of a liquid. Theoretical speculation has been concerned hitherto chiefly with phenomena of liquids at low pressures, such as the latent heat of evaporization or the surface tension. At high pressures there are effects of another order, just as significantly descriptive of the internal mechanism, but as yet hardly touched by speculation. The data of this paper cover four times any previous pressure range, and should be sufficient to show the general nature of high pressure effects. Furthermore, the systematic presentation of different thermodynamic properties should afford different points of view for the attack.

The actual state of affairs at high pressures was found to be exceedingly complicated, contrary to what we might expect. We might suppose that the molecules would become so closely packed at high pressure as to allow less variety in their response to external changes, so that a liquid would approximate to a solid, in which compressibility and expansion change only slightly with pressure and temperature. A hypothesis of this character, backed up it is true by some experi-

---

[1] Amagat, Ann. Chim. et Phys., **29**, 68–208 (1893).

mental evidence, has been made the basis of a recent empirical theory of liquids by Tammann,[2] who assumes that at high pressures variations in thermal expansion due to changes of temperature ought to become vanishingly small. Such, however, is by no means the case, but, on the contrary, at high pressures the thermal expansion varies with temperature in a more complicated way than at atmospheric pressure.

The irregularity of the effects at high pressures makes it evident that a complete theory of liquids must be very complicated. The first step, therefore, toward a theory would be to explain only the general features. With this in mind, the average of the various thermodynamic properties over the entire temperature range has been computed for each liquid. To facilitate comparison, the average of any one property, compressibility for example, is shown on the same diagram for all twelve liquids. One would expect to turn at first to these collected diagrams in seeking light on a theory of liquids.

In this paper no attempt has been made at the very considerable task of developing a quantitative theory to represent the data. The results suggest very strongly in some cases, however, that conceptions of the mechanism of a liquid which may be adequate at low pressures can no longer be adequate at high pressures. For instance, we shall probably have to modify our ideas of the mechanism accounting for pressure and temperature. Some discussion is given of modifications that may be necessary, and in several cases it is shown that the proposed modifications are competent to explain, at least qualitatively, the complicated effects found at high pressures.

A preceding paper on water[3] is very similar to the present one in the scope of its measurements and computations. The hope was expressed in the introduction to that paper that the projected study of twelve liquids (that is, this paper) would show what we might expect of a normal liquid, and that the results for water might then be compared with these results, and yield information about the abnormalities of water. This hope now appears to have been unfounded because no liquid is really normal at high pressures; all show individual peculiarities. It is true, nevertheless, that at high pressures these twelve liquids do become more nearly alike in a general way, and that water becomes increasingly like them. In many cases it will be found instructive, therefore, to compare the properties of water at high pressures with those of the liquids of this paper.

---

[2] Tammann, Ann. Phys., **37**, 975–1013 (1912).
[3] Bridgman, These Proceedings, **48**, 307–362 (1912).

The initial abnormalities of water have such a far-reaching effect, however, up to 6000 kgm., that it seemed desirable not to include water directly in the same diagrams with the other liquids.

The apparatus and the experimental method are in large measure the same as were used in the preceding work on water, and will not be described again in any detail. The methods of computation, however, are somewhat different, and are discussed at some length. It is impossible to give the original data because of lack of space. However, a few sample curves of the original data are given, and the average error of the compressibility and the expansion measurements is stated for each liquid.

The liquids used are with two exceptions the same as those used by Amagat. This choice of liquids has two advantages. In the first place, the present method does not give accurate results at the very lowest pressures, so that it is desirable to supplement the measurements with others at low pressures. For this purpose the values of Amagat for the change of volume between atmospheric pressure and 500 kgm. have been used. And in the second place, it was not desired to complicate this study, which is concerned with the liquid only, by the possibility of freezing under pressure. The freezing points of all the liquids used here are very low. It was hoped, therefore, that none of them would freeze under pressure at the temperatures of this investigation. The anticipation of no freezing was justified except in the case of acetone, which, however, froze at such high pressures that it was not necessary to discard it.

Comment should perhaps be made on the very extensive use of diagrams instead of tables. A table is capable of greater accuracy than a diagram, but a diagram has the advantage of presenting a large collection of results in a form immediately grasped. By the use of diagrams with fine rulings, the attempt has been made to combine the general grasp afforded by a diagram with the accuracy of a table. In most cases the numerical values may be read directly from the diagrams within the limits of experimental accuracy. The only exception is for the fundamental data, volume against pressure and temperature, for which tables have been given to four significant figures. The diagrams accompanying these tables are quite secondary in importance, giving merely a general survey of the trend of the results.

A brief indication of the plan of presentation may be helpful to the reader. The apparatus and the experimental method are first briefly described, indicating the points of departure from the previous work on water (pages 7 to 18). Then the methods of computation are taken up in detail, describing in succession the method used for each one of the thermodynamic properties (pages 18 to 37). A somewhat detailed analysis of the data for each liquid is next given, including any special features of the experiment for that liquid, the experimental error, and the source and probable accuracy of data by other observers which have been used in the computations (pages 37 to 74). Here are included the fundamental data for each liquid, that is, the tables and diagrams of volume against pressure and temperature. The various thermodynamic properties are then discussed, with comments on the peculiarities of the individual liquids, and on the general features common to them all (pages 74 to 104). The diagrams for this general discussion are given on folders at the end of the paper, all the diagrams for any one property being collected on one sheet. It is hoped that this arrangement will produce less confusion in the text, and at the same time, permit a more ready general survey of the facts. Finally, there is given on pages 104 to 113, a discussion of the bearing of the results on previous theories, and of the possible effects on our ideas of what the actual mechanism of a liquid may be.

II. Experimental Method.

The method is essentially similar to that used in the preceding work on water. As in the earlier work, so here, the liquid under investigation is placed in a cylinder closed by a piston which moves absolutely without leak. The pressure on the liquid may be varied by changing the position of the piston, and the temperature may be varied by altering the temperature of the surrounding bath. The volume is found directly by measuring the position of the piston in the cylinder. The fundamental data to be obtained with the apparatus are the values of the volume as a function of pressure and temperature at a sufficient number of points to allow the calculation of the volume at any temperature and pressure within the range.

In actual use the simple procedure suggested above is complicated by the necessity of introducing the instrument for measuring the pressure into the same cylinder with the liquid under investigation. The pressure in these experiments was measured with a coil of manganin wire, the changes in the resistance of which give the pressure by a previous calibration.[4] Although this manganin gauge is compact

[4] Bridgman, These Proceedings, 47, 319–438 (1911).

and simple, its use introduces complications, because it must come in contact only with a liquid that is an insulator. The water of the preceding paper and nearly all the twelve liquids of this are not insulators. Some auxiliary liquid must be used therefore, to transmit pressure to the manganin coil. Kerosene has shown itself perfectly adapted to the purpose, and has been used in all this work. In the former work on water, the kerosene and water could be allowed to come directly in contact with each other, but in this investigation, nearly all the liquids are more or less miscible with kerosene, so that means had to be provided to prevent the liquid under investigation from coming into contact with the kerosene. This necessitated slight changes in the design of the apparatus, and the use of still a third liquid, mercury, to keep the liquid and the kerosene from direct contact. The modified receptacle for holding the liquid is shown in Figure 1. The liquid is enclosed in a steel bulb A, the stem of which dips into the cup B containing mercury. The cup B is bored out at the lower end so as to protect the manganin pressure gauge, which is shown at C. The bulb, cup, and gauge together occupy the lower part of the pressure cylinder. In order to obtain as large a quantity as possible of the liquid to experiment on the diameter of the bulb etc. was made larger than that of the moving piston ($\frac{11}{16}$ inch against $\frac{1}{2}$ inch). To allow this the cylinder had to be bored out at the lower end to the larger diameter. In other respects the cylinder is similar to that used for water, in fact it is the same cylinder, the only change being the enlargement of the lower part to meet the new requirements.

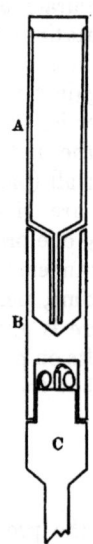

FIGURE. 1. The compressibility bulb with attachments. A is the bulb containing the liquid under investigation, B the mercury cup, and C the insulating plug and the manganin coil with which pressure is measured.

The final form of receptacle shown in Figure 1 was arrived at only after a considerable number of failures. Early attempts to use glass cylinders, which from the point of view of purity would be more desirable, had to be abandoned because of the invariable fracture of the glass. This is doubtless because kerosene under pressure becomes so viscous as not to transmit small changes of pressure hydrostatically immediately after changes in the position of the piston. The enormously increased viscosity both of the kerosene and of the liquid under investigation within the bulb afforded the only satisfactory explanation of many of the capricious misfortunes of the preliminary work. For instance, in the early work the bulb was closed at the upper end by a cap put on with soft solder. Several times after the application of pressure this soft solder was found eaten away by the mercury. The only apparent way in which this could happen would be by the liquid (ethyl alcohol in the preliminary investigations) becoming so viscous as to crack as the volume decreases with rising pressure, thus allowing the mercury to rise through the cracks from below. Another very troublesome effect in the preliminary work was the frequent short circuiting of the manganin coil by small drops of mercury. These were probably forced out of the cup by the viscous motion of the kerosene during changing pressure. The difficulty was avoided by making the lower part of the mercury cup in the form of a protecting cap for the manganin coil, and by enlarging the channels of communication between the cup and the bulb from the kerosene to the mercury. It was found also that it was necessary to give the bulb fairly thick walls. Otherwise, when pressure is relieved, the bulb expands under the viscous motion of the fluid inside and tightly fills the cylinder. Furthermore, the bulb must not fit the cylinder too closely, for otherwise the pressure is not transmitted rapidly enough to the interior of the bulb, which may thereby become collapsed at the upper end. Still again, the upper end of the bulb must be provided with radial grooves to give access from the lower to the upper part of the cylinder, or else when pressure is released, the entire bulb rises with the viscous kerosene against the ledge on the cylinder, acting effectively as a valve which permits only comparatively slow release of pressure, with the result that eventually, when pressure is released, the top is blown off the bulb and forced into the smaller bore of the cylinder above.

It must be a matter of experiment to find the dimensions for any particular piece of apparatus which will avoid all these difficulties. Complete success was attained, however; the last twenty out of a total of twenty-four runs being completed without accident of any kind.

Although the method outlined above is exceedingly simple, yet there are evidently a number of corrections for the distortion of the steel containing vessel and for the auxiliary liquid. These corrections have been discussed in great detail in the preceding paper on water. The corrections to be used in this paper are exactly similar to the former ones, with the exception of the correction for the effect of the mercury. It will be recalled that two sets of measurements are necessary to obtain all the data needed in making the corrections; the first is with the steel bulb filled with the liquid under investigation, and the second with the liquid and bulb replaced by a cylinder of Bessemer steel. The quantities of kerosene are approximately the same in the two sets of measurements. If the quantity of mercury used in the two sets were exactly the same, it would not be necessary to apply any correction for it, but since it was not easy to use exactly the same quantity, a correction for the difference had to be applied. This correction was exceedingly small, being merely the change with pressure and temperature of the small difference, and could be obtained with sufficient accuracy for this purpose from measurements already made on mercury up to 12000 kgm. at 0° and 20°.[4] The variation of compressibility and dilatation with temperature is so small at high pressures that it is perfectly safe to extrapolate to 80°. The quantities of mercury in the two determinations seldom differed by as much as 0.3 gm., in which case the correction can be entirely neglected.

The altered design of the containing vessel made necessary a modification of the procedure in filling the cylinder ready for a run. The bulb was filled by boiling the liquid into it at reduced pressure at room temperature. During the filling the liquid came into contact with a gum stopper for a few seconds. The bulb was then disconnected from the air pump, and gently warmed by holding it in the hands, thus forcing a slight quantity of liquid out of the stem. After the bulb had come to the temperature of the hand, the outside was carefully wiped dry, and the stem immediately inverted into the cup of mercury. In this way the complete exclusion of all air bubbles was ensured. After the bulb had once more come to the temperature of the room, the combined weight of the filled bulb, mercury, and steel cup was found. The weight of the liquid in the bulb can then be found immediately by subtracting the known weight of the mercury, cup, and bulb. The quantity of the liquid used in the majority of the experiments was about 12 c. c. The mercury cup with the bulb still in place was now slipped over the manganin resistance coil,

---

[4] Bridgman, These Proceedings, **47**, 319–438 (1911).

and the coil, mercury cup, and the liquid under investigation were inserted together into the cylinder. This naturally had to be done from below, and for this purpose the cylinder was held vertical during this operation in a special vise, projecting at right angles from the wall. The packing and the retaining screw were also inserted while the cylinder was in this position. The cylinder was now filled with a weighed quantity of kerosene through the open upper end. This filling with kerosene was accomplished in two operations; first, half of the kerosene was poured in, and all air extracted from the lower part of the cylinder by gently exhausting it, and then the remainder of the kerosene was poured in. The moveable plug was then inserted, with special care not to spill any kerosene, and finally the cylinder, kept always upright, was put in position in the lower part of the press. After a run the bulb was cleaned for the next run by heating it nearly to redness. That the cleaning was thorough was shown by the constancy of weight of the bulb, which seldom varied more than one milligram from experiment to experiment.

The essential idea of the experimental method remains the same as in the work on water, but there are slight modifications. The purpose is to obtain by direct measurements the change of volume with pressure over the entire pressure range at some one constant temperature, and then by another independent set of measurements to obtain the change of volume with temperature over the entire temperature range at a sufficient number of pressures to completely cover the field. This method has a distinct advantage over that employed by Amagat, for example. Amagat obtained the volume as a function of pressure and temperature by measuring the change of volume with pressure isothermally at a number of temperatures. This has the disadvantage that the thermal dilatation can be found only by taking the difference of compressibility determinations at different temperatures. Obviously this makes it necessary to measure the compressibility with a much greater degree of accuracy than the desired final accuracy of the dilatation, and this is a difficult matter to accomplish. But the present method gives both compressibility and dilatation with the accuracy to be expected of direct measurements.

The procedure in measuring the change of volume with pressure at constant temperature was the same as in the work on water. The temperature chosen for this determination was 40°, because it is the same as was used for water. Reference is made to the former paper for the details of manipulation. In order to cover completely the pressure range, it was necessary to make the readings in two series, one at lower pressures up to 2000 kgm., and the second between 2000 kgm. and the maximum, 12000 kgm. A similar method was found necessary in the previous work on water. The reason for this is that at high pressures the moveable piston takes a permanent set, and moves with extreme friction against the sides of the cylinder. The result is that if the pressure has once been pushed to the maximum the piston will not return far enough to reach the low pressures again. The lowest pressure that it was possible to reach after the maximum varied somewhat, but was usually in the neighborhood of 1000 kgm. In order completely to cover the field, therefore, two sets of experiments were necessary, one for the high and the other for the low pressures. To make the results comparable, the low pressure measurements were carried some distance into the high pressure domain, the upper limit of the low pressure measurements being about 2000 kgm.

The chief variation from the method of the earlier work is in finding the thermal expansion. The method of the earlier work involved two steps. The first was to change the temperature at constant volume. This change of temperature, if it were an increase, produced as a secondary effect a rise of pressure. The second step was to lower the pressure to the original value preparatory to the next change of temperature. The withdrawal of the piston necessary to effect this lowering of pressure was always attended by slight irregularities; apparently the piston does not begin its regular march until after some slight motion has taken place. So that although perfect regularity was found in the larger motions of the piston of the compressibility determinations, yet the initial irregularity was sufficient to be disturbing with the smaller motions accompanying the temperature changes. The difficulty was avoided by keeping the piston constant in position during a series of temperature changes, and allowing the pressure to vary accordingly. The variation of pressure for a change of temperature of 60°, from 20° to 80°, was about 1000 kgm. at the highest pressures, and proportionally less at the lower pressures. Temperature measurements were made at six values of the average pressure with the apparatus set up for the higher pressures, above 2000 kgm., and at two mean pressures with the apparatus set up for the lower pressures, below 2000 kgm. Readings were made at 20° intervals, as in the work on water. At 2000 kgm., for example, the temperature was raised from 20° to 40° and the increase of pressure noted, then to 60° and the increase of pressure noted again, and finally from 60° to 80° and once again the increase of pressure noted.

The pressure was then increased at 80° to the next higher value, about 3500 kgm., and the corresponding set of measurements made at 20° intervals, this time with decreasing temperature. The other four sets of readings were made in a similar way. The highest pressure reached, on the last reading at 80°, was about 12500 kgm. From here the temperature was decreased, so that the final reading at 20° was made at a pressure of about 11500 kgm. Here is an incidental advantage of the modified method of making the expansion measurements, because mercury solidifies at 20° at less than 12000 kgm., so that if the former method had been used a smaller value of the mean maximum pressure would have been necessary for the last reading at 20°. Of course the modified method necessitates a slight change in the method of making the computations, which will be described later.

All the measurements on each of the liquids were repeated to ensure greater accuracy. The measurements were made in two series; all twelve liquids were first measured, and, except for the preliminary experiments, no repetitions made until at least one set of readings had been made for each liquid. This has the advantage of separating the measurements on any one liquid by a considerable interval of time, and thereby eliminating any possible effect of temporary variations in the apparatus. During the first extended series of readings the two sets of readings for each liquid at high and low pressures were made with two separate fillings of the apparatus. At the lower pressures the viscosity effects do not play any part, so that it was not necessary to leave so large a space between the bulb and the walls of the cylinder as was necessary for the higher pressures. In this way a larger bulb could be used at the lower than at the higher pressures, so that somewhat greater sensitiveness could be obtained at the lower pressures because of the greater quantity of liquid. But this made it necessary to take the apparatus apart after the set of readings at low pressures, and replace the larger bulb by the smaller one for the run at high pressures. This made considerable trouble, which did not seem adequately compensated by the somewhat greater accuracy, so that in the repetition of the experiment the readings at the high and low pressures were all made with the same filling of the apparatus. The readings at low pressures were made before the readings at high pressures, and so before the piston had been upset.

The final detailed procedure was as follows. The readings at low pressures were made first. It is not necessary to describe them in detail as they were similar to the readings at the higher pressures, the only difference being that no seasoning precautions were taken. At high pressures the thermostat was first adjusted to 40°, and then the cylinder seasoned over the entire pressure range by advancing the pressure to the maximum and releasing it three times. The details of the seasoning process are described in the former paper. The compressibility measurements at 40° were then made. Readings were taken both at increasing and decreasing pressures, so as to correct for hysteresis. The range of these measurements was from 2000 to 12000 kgm., returning after the complete set of readings to the initial pressure of 2000 kgm. Next, the cylinder was seasoned for the thermal expansion readings by advancing the pressure to 11500 kgm. raising the temperature to 80° at constant volume, releasing the pressure to 2000 kgm. at 80°, and finally reducing the temperature to 20°. The thermal expansion measurements were then made by the method already described.

In the later form of procedure, in which the effects at high and low pressures were measured with the same set-up, all the measurements at the low pressures and the isothermal compressibility at 40° were made on the same day; on the next day the thermal dilation at six mean pressures was measured, and the apparatus taken down and set up again with a new liquid ready for another run on the next day.

By the use of large reservoirs of hot water and by nearly emptying the thermostat at every change of temperature, it was possible to make the readings with changing temperature fairly rapidly, the elapsed time between readings at successive temperatures being about fifteen minutes. A much longer time than this would have been necessary to secure temperature equilibrium throughout the mass of the cylinder if a special device had not been adopted. This consisted in running the temperature past the final value, and then returning to it. Thus, let us suppose that the temperature was to be changed from 20° to 40°. Water was drawn from the thermostat and enough hot water poured in to raise the temperature to 45°, and not until the lapse of several minutes, the exact time to be determined by experiment, was the temperature reduced to 40° and the regulator set at this final value. After some practise it was possible to reach temperature equilibrium in little more time than was necessary for the manipulations of drawing water and putting in fresh. Readings were never made, however, until at least three minutes had elapsed without change of pressure.

The actual experiments, after the preliminary work, occupied four months, from February through May, 1912. Many of the early measurements are due to the assistance of Mr. S. L. Gokhale. During the entire time there was no change in the internal diameter of the cylinder of so much as 0.0001 inch. To ensure further that there was no progressive change in the cylinder during the measurements, the comparison measurements with the liquid replaced by Bessemer steel were made at five intervals during this time. The maximum divergence of any of these readings from the mean was only 0.5%, better perhaps than would at first be expected from the method.

The original data are so numerous that it seemed undesirable to give them here in full. Every point recorded involves six readings, two of pressure and four of piston displacement. On the average each liquid involved 140 points, 75 for compressibility and 65 for thermal expansion. This makes a grand total of 12500 readings in the original data. It was thought to be sufficient to give a few sample curves of complete data, and to state for each liquid the average departure from the mean of the two series of compressibility and dilation measurements. The average departure from the mean of the two series of compressibility measurements for the twelve liquids was 0.15% of the maximum, and the departure from the mean of the thermal expansion measurements was 2%. The changes of volume due to changes of pressure are much greater than those due to temperature, so that the compressibility measurements determine the final accuracy of the volume.

In regard to the purity of the liquids, it was not thought necessary to take special precautions, because the properties studied here are not much influenced by the presence of impurities. The compressibility, for example, of a mixture of two liquids of small concentrations of the one in the other is an additive function of the compressibility of the two components. An example of this fact occurred in the preliminary work on ether. The first measurements were made with the ether enclosed in a glass bulb. This bulb broke on the application of pressure because of the great viscosity of the kerosene. The compressibility measurements of this preliminary run were measurements, therefore, on a mixture of ether and kerosene. In spite of the fact that this ether was very much contaminated the result showed that the compressibility of this mixture was only different by 4% from what it was when the measurements were made on the same quantities of ether and kerosene prevented from mixing.

The liquids used were obtained from Eimer and Amend. They were either the purest manufactured by them, or else Kahlbaum's purest. It was to be expected therefore, that only slight impurities were present, and that the errors due to them are beyond the limits of observational error.

Nine months after the completion of the experiments, the remaining samples were subjected to a rough analysis by determining the boiling points. If the liquid is pure, the boiling point should remain constant until the liquid is completely boiled away. The amount by which the boiling point changes during evaporation can be expected to give at least some clue as to the amount of impurity present. [A substance is considered good enough for most *chemical* purposes if it all boils within 1°.] Of course this analysis does not give the nature of the impurity present, nor does it give the impurity at the time the experiment was performed. During the nine months the liquids were left tightly corked, but there can be no question that some deterioration had gone on in this time, so that the liquids were all actually better than is indicated by the analyses. The deterioration with time is much greater for some liquids than for others.

The fractionations were performed by Mr. R. H. Patch at the Chemical Laboratory of Harvard University. The results of his examination are given below. It should be noticed that the temperature readings have not been corrected, so that all they can show is the constancy of the boiling point, not its absolute value.

Methyl Alcohol (Kahlbaum).
    93% boils between 64.5° and 65° C.
    7%   "   "   65° and 65.8° C.
    Free from ethyl alcohol and acetone, the most likely impurities, and was neutral to litmus. Excellent sample. (It would seem that the impurity was probably water, most of which had probably been absorbed while standing, as the stopper was not perfectly tight).

Ethyl Alcohol (Kahlbaum).
    3.8% boils between 77.3° and 77.8° C.
    96.2%   "   "   77.8° and 78.0° C.
    The sample showed some suspended inorganic matter, probably iron from the container. (The sample had been standing for the nine months tightly corked with a cork stopper in the tin vessel in which it came from Kahlbaum). With this exception, and admitting the presence of a small quantity of water the sample was a good one.

Propyl Alcohol (Kahlbaum).
    4.% boils below 96.° C.
    96.% boils between 96.0° and 96.8° C.
    A good sample.

Isobutyl Alcohol (Eimer and Amend).
    7% boils below 105.8° C.
    17% boils between 105.8° and 107° C.
    76% "   "   107.0° and 107.2° C.
    Thus 93% boils within 1.4°, a good sample.
Amyl Alcohol (*Kahlbaum, Free from Pyridin*).
    3% boils below 128.9° C.
    89.5% boils between 128.9° and 129.9° C.
    7.5 boils between 129.9° and 130.0° C.
    Sample free from foreign organic matter and neutral to litmus. An excellent example.
Ethyl Ether (*Kahlbaum, "Sp. gr. 0.720"*).
    The whole sample boiled between 34.5° and 35.0° C.
    Neutral to litmus, free from aldehydes, and sulphur compounds.
    Contained some water, and without doubt some alcohol.
Acetone (*Eimer and Amend. Marked "Pure"*).
    70.6% boils between 56° and 57° C.
    19.4% boils between 57° and 58° C.
    10% boils between 58° and 59° C.
    The sample left a dark brown residue in the distillation flask, probably aldehydic in nature. Free from water, and neutral to litmus. A fair sample. (The brown color developed on standing; at the time the experiment was performed, the liquid was perfectly colorless.)
Carbon Bisulphide (*From the store room of the Chemical Laboratory*).
    5% boils between 45.8° and 46.° C.
    95 boils at 46.0° C.
    The sample was free from hydrogen sulphide, sulphuric and sulphurous acids, but contains some foreign organic sulphur compounds and left a yellowish residue. The latter always results upon allowing the liquid to stand, especially upon exposure to light. (Here again the color had developed upon standing during the nine months. At the time of the experiment the liquid was perfectly colorless.)
Phosphorus Trichloride (*Eimer and Amend*).
    True boiling point is 78.3° C.
    Sample showed no constant boiling point.
    6.5° boiled below 77° C.
    11.2% boiled between 77° and 79° C.
    17.9% boiled between 79° and 80° C.

15.9% boiled between 80° and 81° C.
17.9% boiled between 81° and 87° C.
31.4% boiled between 87° and 102° C.

Chief impurity probably phosphorus oxychloride formed by combination of the trichloride with the moisture in the air, also some pentachloride is almost always unavoidably present, due to excess chlorine.

Ethyl Bromide (*Eimer and Amend*).

The whole sample boiled from 38.0° to 38.4° C. and with the exception of not over 2% at 38.4° exactly. Neutral to litmus, an excellent sample.

Analyses could not be made of the ethyl chloride and the ethyl iodide, because these samples were completely used up in the experiment. Both of these were obtained in sealed glass bulbs from Kahlbaum, were used immediately after opening, and were both colorless when used.

These analyses show that probably none of the liquids contained enough impurity to have any perceptible effect, except the phosphorus trichloride, and to a less degree the acetone. The phosphorus trichloride was very noticeably simpler in its behavior at high pressures than the other liquids; probably due to the fact that it is a mixture of different substances, and so can not be expected to show the well marked behavior of a single pure substance.

A few measurements were made on one liquid which are not tabulated here. These were measurements on chloroform, which was found to freeze at fairly low pressures, about 6800 kgm. at 40°, and 10000 kgm. at 80°. Further investigation of this liquid has been postponed until a systematic investigation of freezing curves can be taken up.

III. Methods of Computation.

The methods of computation were slightly different from those used in the work on water, in part because of the somewhat different experimental method. They are, in consequence, here described in detail. As in the computations for water, the first step was to prepare a table giving the volume at regular intervals of pressure and temperature. From this table the other thermodynamic quantities were then computed.

The Table of Volumes. In preparing the table of volumes, the first computation was of the volume as a function of pressure at 40°. The first step was to plot the piston displacement against the displacement of the slider of the bridge wire on which the resistance of the manganin coil was measured. Because of hysteresis effects it was necessary to make two sets of readings, one with increasing and the other with decreasing pressure. Care was taken that any two corresponding readings should be at as nearly as possible the same pressure, so that it should be allowable to take the average of two corresponding displacements as the best value of the displacement at the average of the two corresponding pressures. The next step was to draw a smooth curve through the average points, and from the curve to tabulate the displacement at regular intervals of pressures (5 cm. on the bridge wire, or about 1100 kgm.). The second series of measurements on the same liquid was then treated in the same way. These two series of measurements differed somewhat as to the quantities of materials used, that is, the liquid under investigation, the kerosene, the mercury, and the steel. But the amounts were so nearly the same that it was permissible to take the average displacement of the two series as the displacement that would have been found for the mean between the quantities of material used in the two series. In only a few cases did the amounts of material in the two series differ so much that it was not permissible to do this. In these cases the displacement for the mean quantity had to be determined in a way which need not be described in detail. The average of the displacements obtained in this way were now corrected for the effect of the kerosene, the mercury, the steel, and the distortion of the steel containing vessel. In applying the corrections, the mean of the five auxiliary experiments in which the liquid was replaced by Bessemer steel was used. The correction was applied in essentially the same way as for water. The corrected result gave the motion of the piston due to the compression of only the liquid under investigation. This, with the known cross section of the piston and the weight of the liquid, determined the change of volume at any pressure of 1 gm. of the liquid. Finally, by using the values for the density at 0° deduced from the recent Tables of Kaye and Laby, the results were reduced to the change of volume in c. c. of a quantity of liquid which at 0° C. and atmospheric pressure occupies 1 c. c. This is the unit quantity which is here adopted throughout, and seems to have been most usually used in work of this kind. In particular it is the unit quantity of Amagat.

The computation just described applies only to the measurements at the higher pressures. The results are tabulated as changes of volume from 2000 kgm. as the zero of pressure. If the pressure during the high pressure measurements went lower than this, as it usually did, the corresponding changes of volume were taken as negative. To obtain the volume at lower pressures, the measurements at the lower pressures were reduced in much the same way as has been described for the higher pressures. One difference is that the measurements at the lower pressures were at 20° instead of 40°, because several of the liquids boil at less than 40° at atmospheric pressure. It was not possible to reach entirely to the zero of pressure with the low-pressure measurements, because of the slight friction of the piston even when the pressure had not been pushed so high as to permanently upset the piston. It was not possible to get much nearer to zero than 100 or 200 kgm. after the pressure had once been pushed to 2000 kgm. In order to come still closer to zero, several measurements were made during the very first application of pressure, before the cylinder has been seasoned for hysteresis. During these first readings the pressure was increased to about 1000 kgm. and then seasoned for hysteresis as before. As a result the second set of readings between 200 and 2000 kgm. made after this seasoning process did not make close connection with the first set. The discrepancy was due in part to hysteresis, but also in part to better adaptation of the packing to all the crevices in the steel washers. The direction of the two curves was usually the same, however, within the limits of error. This allowed an extrapolation to zero by combining the results of the first set of readings with those of the second. In this way the change of the unit volume from atmospheric pressure up to 2000 kgm. was determined. It must be remembered, however, that the readings at the very lowest pressures are almost certainly in error, because the instrument is not designed for low pressures. But with increasing pressure the readings become more and more trustworthy, until above 500 kgm. they seem to merit entire confidence, if we can judge from self consistency. To get the total change of volume from atmospheric pressure it is desirable, therefore, to supplement these readings with others made with apparatus especially designed for low pressures. Such measurements are afforded by the data of Amagat, and in some cases by others.

At the time that the computations of this paper were made, the data of Amagat were the best that we had for the compressibility at low pressures. Between the time of computing these results and writing this account, however, there has appeared a paper by Richards,[5] in which the compressibility up to 500 kgm. of several of the liquids

---

[5] Richards, Jour. Amer. Chem. Soc., **34**, 971–993 (1912).

used here is determined. It is unfortunate that these values did not appear in time for use in the present computations. They are quoted, however, so as to afford a comparison with the results of Amagat, which are here used as the standard for low pressures. It makes no difference with the essential conclusions of the paper, however, which set of data is accepted as most probably correct. The only effect would be to change the initial values of some of the thermodynamic properties; their magnitudes at high pressures will not be altered.

The precise steps in combining into a final result the changes of volume computed from the two sets of readings at high and low pressures were as follows. From the high pressure readings, the changes of volume at 40° (ΔV, cm.^{3}/cm.^{3}) were plotted against the displacement of the slider of the resistance bridge, the zero of pressure being at 10 cm. displacement. The scale of the diagram was large; 2 cm. for 0.1 inch piston displacement, and 55 cm. slider displacement for the maximum pressure. From the low pressure readings ΔV in cm.^{3}/cm.^{3} was found at 20°, and was plotted against slider displacement, the zero being at 5 cm. A smooth curve was drawn through these points. From this curve, knowing the constants of the manganin coil, ΔV was found at 500, 1000, 1500, and 2000 kgm. and 500, 1000, 1500, and 2000 atmos. ΔV in terms of atmos. was now corrected so as to be reckoned from 1000 atmos. as zero. The values of Amagat for ΔV at 20° were next computed with 1000 atmos. as the zero. These two sets of data were compared, and the new values for the lowest interval adjusted so as to be in agreement with Amagat between 1 and 500 kgm. It will be noticed that this preliminary comparison with Amagat does not enter the final result; it was an orienting comparison for obtaining some idea of the accuracy at low pressures. Using Amagat's value for the lower interval, the changes of volume at pressures corresponding to 2.5, 5.0, 7.5, and 10.0 cm. were next determined, reckoned from zero pressure. These values were now corrected to 40° with the data obtained for the thermal dilatation and were recomputed with 10 cm. as the zero of pressure. The changes of volume at 40° obtained from both the low and high pressure sets of measurements were now plotted together on one diagram, and a smooth curve drawn through the points. From this curve the changes of volume with the kilogram as the unit of pressure were read off at 500 kgm. intervals, starting from an origin at 500 kgm. The changes of volume so found were now smoothed to give regular differences. The smoothing was performed in a manner somewhat different from the corresponding operation for water, and in a manner which has the advantage of preserving the irregularities which are shown by experiment actually to exist, but which would have been effaced if the attempt were made to give smooth second differences, as was done for water. In performing this smoothing, the fact was used that the change of volume of nearly all the liquids can be represented by a curve of nearly the same shape. Ethyl chloride is an exception, and a special computation had to be applied to it. As a first approximation the change of volume of the other eleven liquids was found to be reproducible by a formula of the type;

$$\Delta V = \alpha \left(\frac{p-500}{1000}\right)^{.8} + \beta \left(\frac{p-500}{1000}\right)^{.6} + \gamma \left(\frac{p-500}{1000}\right)^{.4} + \delta \left(\frac{p-500}{1000}\right)^{.2}.$$

In order to apply this formula to any one liquid it is necessary simply to multiply all four constants by the same factor. The meaning of this is that to this degree of approximation the chief difference of the liquids with regard to changes of volume is in respect to the absolute, not the relative magnitudes of the change. The constants of the above formula were computed, therefore, so as to apply to the average of the eleven liquids. This was done by finding the average change of volume for the eleven liquids at 1500, 3000, 7000, and 12000 kgm. and determining the four constants of the formula so that the curve should pass through these four points. It will be noticed that in this formula the zero of pressure is at 500 kgm., so that it applies directly to the changes of volume as found above. The formula was now applied to the eleven liquids in succession by multiplying the four constants by the appropriate multiplier for each liquid. The multiplier was so determined that the formula should give the observed change of volume at 7000 kgm. The changes of volume were now calculated with this formula at intervals of 500 kgm. up to 5000 kgm., and beyond 5000 kgm. at intervals of 1000 kgm., and compared with the observed values. The differences between the observed and the computed values were plotted on a large scale and a smooth curve drawn through the points. The values obtained from these smooth curves and the formula were combined to give the final values for the volume at 40°. The advantage of the method is that the final results lie on perfectly smooth curves, and that the curves show the various slight irregularities which correspond to the experimental facts but which would be smoothed out if the second differences were made uniform. The values for ethyl chloride were treated in the same way, except that it was necessary to start from a slightly different formula, to be given on page 67.

It would have been possible to dispense with this finer adjustment by means of a difference curve with very little change in the final result, because the changes of volume obtained from the original

TABLE I.

METHYL ALCOHOL.

DIFFERENCE BETWEEN VOLUME OBTAINED FROM ORIGINAL SMOOTH CURVE AND FINAL COMPUTED VALUE.

Pressure. $\frac{\text{kgm.}}{\text{cm.}^2}$	Difference.	Pressure. $\frac{\text{kgm.}}{\text{cm.}^2}$	Difference.
500	.0000	5000	−3
1000	−.0000	6000	+2
1500	−1	7000	−3
2000	0	8000	−1
2500	−1	9000	0
3000	+1	10000	0
3500	−2	11000	−1
4000	+1	12000	0
4500	0		

smooth curves were almost the same as given by the final computation. Table I shows for one liquid (methyl alcohol, chosen at random) the very slight changes made by this readjustment.

The values obtained by this computation start from 500 kgm. as the zero. The change of volume between atmospheric pressure and 500 kgm. has been taken directly from the results of Amagat in those cases where his data are sufficient. In some cases where Amagat does not give the data, it has been necessary to use the more inaccurate values of the present method. In the detailed presentation of data for each liquid, the values taken from Amagat are given. If it should happen at any future time that more probable values than these of

Amagat are found for the changes of volume at low pressures (which indeed is already the case for those liquids measured by Richards), then the results given here may be corrected by adding a constant to the volumes throughout the tables, except of course at atmospheric pressure. The addition of such a constant to the volumes will not alter the behavior of any of the thermodynamic properties at high pressures; it can at most affect those which involve integrations by a very small constant corrective term.

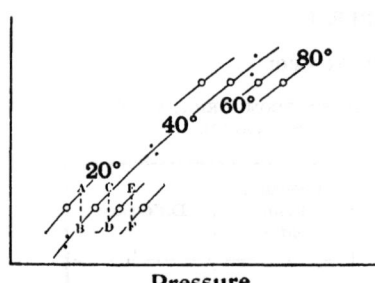

FIGURE 2. Shows a portion of the diagram for determining the compressibility and thermal expansion from the piston displacements. The piston displacements are plotted against pressure. The heavy points are the readings with increasing and decreasing pressure at 40° for the isothermal compressibility. The discrepancy between these points is due to hysteresis. The open circles show the readings at constant volume with changing temperature. The dotted lines, AB for example, show the piston displacement which would have been found if the temperature had been changed at constant pressure.

That it was possible to represent the approximate behavior of these twelve liquids by similar formulas is itself a somewhat surprising and significant fact. It seems to suggest that at extremely high pressures all liquids become alike. The greatest differences between different liquids are at the low pressures. The use of a separate formula in the case of ethyl chloride, which might appear to be an exception, was necessitated in fact only by its abnormal compressibility at low pressures.

It was necessary to compute the thermal dilatation also by a method slightly different from that used for water, because only one piston displacement was read at each temperature instead of two. The piston displacements were plotted against pressure on the same diagram as the isothermal compressibility at 40°. (See Figure 2.) Through each of the points a curve was drawn of the same general slope as the curve of pressure and volume at 40°. The slight changes necessary in the slope of this curve at different temperatures could be made graphically with sufficient accuracy. The difference of the piston displacement for every interval of 20° at the mean of the two pressures involved was now read from these curves. Thus in Figure 2, the line AB represents the piston displacement at constant pressure corresponding to a rise of temperature from 20° to 40°, CD from 40° to 60°, and EF from 60° to 80°. These displacements correspond to slightly different values of the mean pressure. The piston displacements at constant pressure obtained from a diagram like Figure 2 were then plotted against pressure on another diagram, the points for the two independent series of measurements on the same liquid being plotted on the same sheet. The quantities of liquid used in the two series were usually so nearly the same that the mean of the piston displacements could be assumed without error to be the piston displacement of the mean quantity of liquid. The mean of the two independent series of readings was found graphically from the diagram by drawing a smooth curve through the two sets of points. From this value of the piston displacement, after corrections had been applied for the kerosene, mercury, and steel, the change of volume per unit quantity for intervals of 20° was computed in a way analogous to the similar computations for the compressibility. Here again the values for the low pressures are most likely to be in error. The change of volume at 20° intervals at atmospheric pressure was taken directly from the tables of Landolt and Börnstein. Finally, the experimental points and the points at atmospheric pressure were plotted together on a single diagram, smooth curves drawn through them, and from these curves the changes of volume for intervals of 20° were obtained which were used in the construction of the tables of volume.

In plotting as above on a single diagram ΔV for 20° intervals, two independent series of measurements, namely those of this paper and those on which the formulas of Landolt and Börnstein are based were therefore brought together. The two sets of data should of course, if consistent, lie on a smooth curve, so that the amount of discrepancy might be expected to afford an indication of the order of accuracy at low pressures. The change in the dilatation is, however, so rapid at the low pressures that it was possible in nearly every instance to make smooth connection between the two sets of points, without departing from either of them. Furthermore, it would be possible in most cases to make just as smooth connection if somewhat different values were used at atmospheric pressure. Slight discrepancies between the smooth curves and the individual points do not therefore, give a reliable indication of the accuracy.

The thermal dilatation is probably not so accurate as are the changes of volume with pressure, because the dilatation is much smaller. The dilatation can be measured with no greater accuracy than the changes of pressure accompanying the changes of temperature at constant volume. These changes were of the order of 200 kgm. for 20° intervals and could be read on the bridge with an accuracy of about ⅜%. The agreement between the two independent sets of readings was not in general as good as this, averaging about 2%. This is better than was expected at first could be obtained with the method, and is certainly

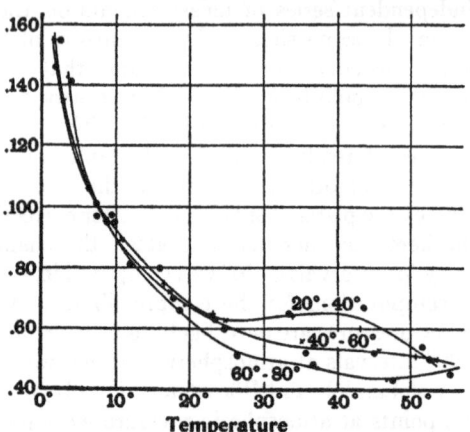

FIGURE 3. A sample set of observations on the change of volume with temperature at constant pressure. The ordinates are piston displacements in inches. Two independent sets of readings are shown on the diagram, those with the circles are the repeated set. The liquid shown here is carbon bisulphide. The accuracy for this is almost exactly the average accuracy for all twelve liquids.

much better than could have been obtained with the alternative method of determining the difference between isothermal lines for different temperatures. Figure 3 for carbon bisulphide shows a fair average of the order of agreement. A more detailed account of the order of accuracy will be given under the description of the individual liquids.

The material was now at hand for the construction of the table of volumes. Up to 5000 kgm. the volume was to be tabulated at intervals of 500 kgm., and above 5000 kgm. at intervals of 1000 kgm. The difference in the length of the pressure steps is desirable because at low pressures the volume changes more rapidly than at high pressures. The tabulation at each pressure was to be made at temperature intervals of 10°. The volume as a function of temperature at atmospheric pressure was first tabulated. This was taken from the formulas of Landolt and Börnstein, or from other sources to be described in detail under the separate liquids. The agreement between different observers, even for atmospheric pressure, is not always as close as could be desired. The next step was to tabulate the volume as a function of pressure over the entire pressure range at 40°, starting from the volume at atmospheric pressure and 40°. These values were then combined with the change of volume for 20° intervals, thus giving the volume for each pressure at 20°, 60° and 80°. To obtain the volume at the intermediate intervals of 10° a device was adopted which at the same time gave the material for determining the thermal dilatation. The change of volume to be expected at 40° and 60° if the change had been linear with temperature over the entire range was calculated from the total change of volume between 20° and 80°. The differences between the actual and the calculated changes of volume were plotted against temperature for each pressure of the table, and smooth curves drawn through these points. From these curves the departure from linearity at the intermediate intervals of 10° was found, and combined with the values computed by the linear relation to give the total change of volume at the desired 10° intervals. This method, as the method for computing the isothermal compressibility, has the advantage of giving smooth curves without smoothing off the second differences.

Thermal Dilatation.—From the table of volumes the next problem was to compute the more significant quantities of thermodynamic interest. The first of these was the thermal dilatation, or $\left(\frac{\partial v}{\partial \tau}\right)_p$. "Dilatation" is perhaps not generally used in this sense, $\frac{1}{v}\left(\frac{\partial v}{\partial \tau}\right)_p$ being more common, but it has the advantage of being the quantity which enters directly into the thermodynamic formulas. The quantity of material to which the former differentiation refers is the unit used throughout this paper, namely the quantity which at 0°C. and atmospheric pressure occupies 1 c.c.

Evidently if the dilatation were uniform over the entire temperature range it could be found from the change of volume between 20° and 80° by dividing by 60. The dilatation is not linear, however, but departs from linearity in a way which can be found from the curves used to determine the change of volume at 10° intervals. The correction to the linear value is obviously to be found from the slope of the difference curve, which can be found graphically from a large drawing with sufficient accuracy. The dilatation was determined in this way at 20° intervals, and was plotted as a function of the pressure for each of the twelve liquids.

The details of the computations for the volume at 10° intervals and of the dilatation are shown more clearly perhaps in Figures 4 and 5. The experimental values of the changes of volume at 20° intervals were first plotted. A represents the change from 20° to 40°, B the change from 20° to 60° (obtained by adding the change 40°–60° to the change 20°–40°) and C the change 20° to 80°. The origin was now connected to C by a straight line (this was done actually by a compu-

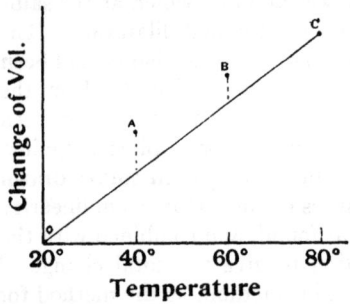

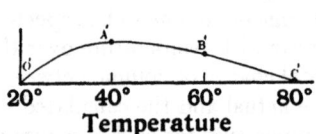

FIGURE 4. Shows the first step in finding the change of volume at intervals of 10° and the thermal expansion from the readings of the volumes at 20° intervals. The heavy line shows what the volume would be if the relation between volume and temperature were linear.

FIGURE 5. Second step in finding change of volume at 10° intervals and the thermal expansion. The point A' is the difference between the point A of Figure 4 and the straight line. The ordinates of this curve at intermediate points, when added to the ordinates given by the straight line of Figure 4 at corresponding points, give the volume at intermediate points. The slope at A' when added to the slope of the straight line of Figure 4, gives the thermal expansion at 40°, for example.

tation, not graphically) and the differences between the points 0, A, B, and C and this straight line were plotted on another diagram, Figure 5, on a larger scale. A smooth curve was drawn through these four points; from this curve the ordinates were read at the intermediate intervals of 10°, and combined with the straight line values of Figure 4 to give the volume at the temperature in question. The thermal dilatation at any temperature, 40° for example, was found by adding to the slope of the straight line OC the slope determined graphically at the point A' of Figure 5.

This method was also applied in determining the dilatation at atmospheric pressure. An alternative method would have been by differentiating the power series of Landolt and Börnstein for volume as a function of temperature. The graphical method was thought preferable, however, because a power series may often reproduce the experimental points with greater fidelity than the slope of the experimental curve.

The dilatation, computed in this way, was transferred directly to tables, and from the tables the curves were drawn which are given later for the dilatation. In order to save space, these tables are not given here. The thermal dilatation enters into many of the other thermodynamic quantities listed in this paper. In computing these quantities the values of the dilatation given in the tables have been used, not the values obtained from the curves given later. The same is true for all the other thermodynamic properties listed in the paper; tables were first computed for all of them before curves had been drawn for any. In this way any progressive error was avoided which might have been introduced by the use of diagrams. Although each diagram shows the property in question with as great an accuracy as is justified experimentally, it might be that if a computation involved the transference of points from one diagram to another several times, the error so introduced might finally mount up to more than the experimental error.

Isothermal Compressibility. — The compressibility, or the quantity $\left(\frac{\partial v}{\partial p}\right)_\tau$, was the next to be determined. This was found by a method somewhat analogous to that for the dilatation. Evidently the compressibility does not vary greatly with temperature. If the compressibility can be found as a function of pressure for one constant temperature, then the compressibility at other temperatures can be found by applying a small correction. The temperature chosen for the direct determination of the compressibility was 40°, since this was the temperature at which the change of volume with pressure had been found. The compressibility was determined graphically from a large scale drawing of the change of volume against pressure. An alternative method would have been to calculate mathematically the slope from the approximate formula for the change of volume, and then to correct this by the graphically determined slope of the difference curve. But this method would fail at the lowest pressure, 500 kgm., and at the higher pressures it did not prove necessary, because the simpler direct graphical method was sufficiently accurate.

It was now possible to correct the compressibility at 40° to the other temperatures of the tables by the use of difference curves. Let us suppose that it was desired to find the compressibility at 60°. Figure 6 represents graphically the operation which was actually performed by a computation. The curve of volume at 60° against pressure was displaced downwards (shown in the dotted line) so as to have the same origin as the curve for 40°. The difference between the curves was plotted on a large scale against pressure, and the graphically determined slope of the difference curve used as a correction to bring the compressibility from 40° to 60°. The process was performed for intervals of 20°, and the results were tabulated and plotted for the twelve liquids.

The compressibility is most likely to be in error at the lower pressures, as was the dilatation. In particular, the compressibility at atmospheric pressure can be found from the method outlined above only by a wide extrapolation, and therefore is not accurate. Another method was adopted, therefore, at atmospheric pressure. Of course, the compressibilities ought to be consistent with the tables of volumes, that is, it ought to be possible to compute from the compressibility to the change of volume given in the table. The compressibility at atmospheric pressure was accordingly computed so that when combined with the compressibility at 500 kgm. it should give the proper change of volume between atmospheric pressure and 500 kgm. It was assumed in the computation that the mean compressibility between 1 and 500 kgm. was the average of the compressibilities at 1 and 500 kgm. This is not quite true, because the compressibility varies rapidly with pressure at low pressures. The value computed in this way is likely to be somewhat low. The discrepancy cannot be large, however, and this method was accepted as the best under the circumstances. The compressibility at atmospheric pressure has also been determined in a number of instances by other observers. There is not always, however, very good agreement between other observers even at atmospheric pressure, so that the compressibility at atmospheric pressure might well be the subject for further experiment in some cases. The actual disagreement at atmospheric pressure and the probable accuracy of the value finally chosen is to be given in the detailed discussion for the separate liquids.

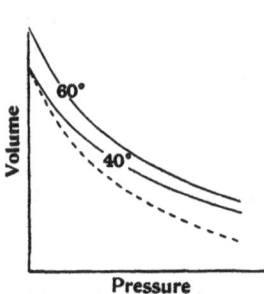

FIGURE 6. Illustrates the method for finding the temperature correction of the compressibility.

The Work of Compression. — The mechanical work of compression was the next quantity of thermodynamic interest to be computed. This is given by the formula $\left(\frac{\partial W}{\partial p}\right)_\tau = -p\left(\frac{\partial v}{\partial p}\right)_\tau$. To find the total quantity of work done from zero up to any given pressure it is evidently necessary to integrate the derivative. This integration was performed mechanically with the integraph of the mathematical department of Harvard University. In performing the integration there are two possible methods. We may either integrate $p\left(\frac{\partial v}{\partial p}\right)_\tau$ as given above, or we may integrate the equivalent expression $\left(\frac{\partial W}{\partial v}\right)_\tau = p$. The first involves an integration with pressure as the independent variable, and the second with volume. The first uses as the integrand the compressibility, which was obtained by computation from the experimental data, and the second uses as the integrand the pressure, which is one of the direct experimental data. It is well known that the derivative of an experimental quantity has considerably greater error than the experimental quantity itself. The second method was adopted, therefore, using the volume as the independent variable of integration. It was a fortunate accident that the method could be used without duplication of effort, because the volume had already been plotted against pressure for another purpose. However, the direct results of the integration were not immediately available because it was necessary to obtain the work of compression as a function of pressure instead of as a function of volume. The change of variable was made with the help of the curves of volume against pressure by reading off the pressures corresponding to the given volumes.

The work of compression was found by the method outlined above at 20° intervals of temperature. It differs only slightly for different temperatures, so slightly that the difference of the work at different temperatures could not be obtained directly from the curves with as great accuracy as was necessary for computing the specific heats. In order to obtain the differences of the work with greater accuracy, an independent integration of the quantity $\Delta p\left(\frac{\partial v}{\partial p}\right)_\tau dp$ was performed. The symbol Δ indicates the difference of the product $p\left(\frac{\partial v}{\partial p}\right)_\tau$ at a given pressure for the interval 20°–40° or 40°–60° or 60°–80°. These differences were taken from the tables of compressibility. The integration of the differences was performed with the integrating machine. Tests of the integrating machine showed that the accuracy of this part of the process alone was as high as $\frac{1}{10}$%. The differences found in this way were now used in finding a better mean value for the total work of isothermal compression. For it is evidently possible with these differences to correct the work of compression at 20° or 60° or 80° back to 40°. If there were no error all these values should agree, but of course complete agreement could not be expected. The agreement of the values corrected to 40° in this way was a few tenths of a per cent. The final value at 40° was taken as the mean of the four corrected values, and is shown in the diagrams. In the lower part of the diagrams the relation between pressure and the difference of the work of compression for 20° intervals is plotted on a larger scale.

The Heat of Compression.— The heat of compression was the next quantity to be computed. It is unfortunate that the expression, "heat of compression" is sometimes used in a sense which is not indicated by the words themselves; namely, as the rise of temperature when the pressure on a substance is increased adiabatically by the unit amount. But by no stretch of the imagination is it possible to identify a temperature with a "heat." A more descriptive name for this effect would seem to be the "temperature effect of compression." The effect was discussed under this name in the previous paper on water. By "heat of compression" we shall mean in this paper what is naturally suggested by the words, namely the heat, Q, which is given out by a unit quantity of a substance when it is compressed isothermally. It may be computed if the dilatation is known, by using the formula $\left(\frac{\partial Q}{\partial p}\right)_\tau = \tau \left(\frac{\partial v}{\partial \tau}\right)_p$. To find the total heat given out by the substance as it is compressed from the initial to the final state it is necessary to integrate this expression. The procedure was exactly analogous to that in computing the work of compression. The integration was performed first for the four temperatures. Then, in order to obtain the differences more accurately, a separate integration was made of the differences of $\tau \left(\frac{\partial v}{\partial \tau}\right)_p$ for intervals of temperature of 20°. With these differences the total heat at any temperature was corrected to 40°, thus giving four values for the heat of compression at 40°, of which the mean was taken for the final value at this temperature. From this final value, the values for the other temperatures were found by computing back again with the differences. The magnitude of the differences is much greater than the differences of the mechanical work, so that it was possible to plot the total heat for each temperature without confusing the diagram.

Change of Internal Energy.— From the heat of compression and the mechanical work of compression we may find at once the change of internal energy when the liquid is compressed isothermally. During compression the liquid receives work from the compressing force and delivers heat. The change of internal energy is the difference between the work received and the heat given out. It was computed in this way and is given in a set of diagrams (Folder 5) for the four regular temperature intervals.

Specific Heat at Constant Pressure.— Other thermodynamic quantities of a simple nature which are usually thought of as characteristic of a liquid are its two specific heats. They also may be found by thermodynamic computation from the data given, but the accuracy is not so great as the accuracy of the other quantities. There are two methods of attack open to us here, but both of them must assume as known the specific heats at atmospheric pressure as a function of temperature. In general, it may be shown that the characteristic equation of a substance is not sufficient in itself to determine the specific heats; we must know in addition the specific heats along some line not an isothermal. Unfortunately, the specific heat of very few of the liquids with which we are concerned is known with accuracy as a function of temperature at atmospheric pressure. The results of different observers are often in essential disagreement. But the characteristic equation can give us the *change* of specific heats along an isothermal. These are the results which will be tabulated in this paper, therefore, leaving for other experimenters the more accurate determination of the specific heats at atmospheric pressure. These future results may then be combined with the differences given here to determine the specific heat at any pressure.

The first method for calculating the specific heat at constant pressure is the method used in the paper on water. It makes use of the formula $\left(\frac{\partial C_p}{\partial p}\right)_\tau = -\tau\left(\frac{\partial^2 v}{\partial \tau^2}\right)_p$. Evidently in order to obtain the total change of specific heat at any pressure we must perform an integration. The weakness of the method is that it involves the use of a second derivative, which cannot be determined with great accuracy from measurements of volume. The method would be open to greater error if applied to these twelve liquids than in the case of water, because the dilatation varies more and more irregularly than for water.

The second method uses a cyclic process to determine the amount of heat absorbed in passing from one temperature to another at any constant pressure. Let us imagine a liquid in the condition represented by the point A on the diagram (see Figure 7). The liquid is now to be carried to the neighboring point D at the same pressure but at a higher temperature. The total change of internal energy when we arrive at D is independent of the path which we have traversed. One path from A to D may be described by raising the temperature from t_0 to t_1 at constant pressure. (Path AD in Figure 7.) In this case the liquid does a certain amount of mechanical work against external pressure and also absorbs a quantity of heat which we can compute immediately when we know the specific heat at constant pressure. The external work during this process is simply the product of the constant pressure and the change of volume, and may be computed directly from the table of volume as a function of pressure and temperature. Or we may pass from A to D by a more circuitous route, by lowering the pressure isothermally at t_0 from A to B, then raising the temperature at atmospheric pressure from B to C, and then increasing the pressure isothermally at the final temperature t_1 to the point D. Now the advantage of this longer route is that we know all the quantities of energy which enter the body on the way. The mechanical work of compression along the isothermals from A to B and from C to D we have already computed. We have also found the heat of compression along the lines A–B and C–B. No work is done in the expansion along the line B–C, and the heat absorbed along this line is known if we know the specific heat at atmospheric pressure. By comparing the inflow of energy along these different paths we are in a position to compute either the quantity of heat absorbed along the line A–D at constant pressure, or else the difference between this heat and the heat absorbed along the line B–C. This heat (or else the difference of heat) may be plotted against the difference of temperature between the points A and D. The same process may be performed at the same pressure for a number of temperature intervals, each with t_0 as the lower limit, giving a curve of the quantity of heat absorbed at constant pressure as a function of the temperature. The specific heat at any temperature is the slope of this heat curve at that temperature.

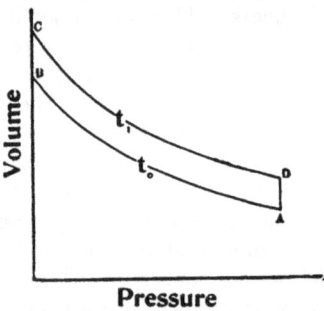

FIGURE 7. Shows the cycles described in finding the specific heat at constant pressure.

The slope was found by a method similar to that for computing the thermal dilatation at constant pressure. At any temperature the difference between the amount of heat actually absorbed and the amount which would have been absorbed if the relation between heat and temperature had been linear was computed and plotted against temperature. The slope of the difference curve was then found graphically, and applied as a correction to the value found on the assumption of a linear relation. The modification if C_p at atmospheric pressure is not known is obvious; a similar procedure, plotting now the difference of heats against temperature gives the difference of the specific heats at the pressure in question and atmospheric pressure. In a few cases, where the liquid boiled at a low temperature at atmospheric pressure so that it was not possible to prolong the curves to the origin of pressure, the difference between the specific heats at the pressure in question and 500 or 1000 kgm. has been given.

It is now obvious why it was necessary to know the differences of the mechanical work and the heat of compression at different temperatures with a greater accuracy than could be found from the curves obtained by a direct integration.

The units in which the specific heats are given should perhaps be mentioned because they are unusual. It is customary to give the specific heat in gm. cal. per gm. of the liquid. But this method of measuring specific heat makes no connection with the thermodynamic formulas, in which the heat is measured in mechanical units corresponding to the units of the other quantities. It was preferred here, therefore, to give the heat in units which are more unusual, but which are consistent with the other quantities, so that it is possible to substitute any of the quantities directly in the formulas without the troublesome work of changing the units. The unit of pressure is the kgm. per sq. cm., and the unit of volume the c.c. Therefore the unit of work which fits the formulas is the kgm. cm., and this is the unit in which the results have been tabulated. It is to be noticed that in making comparison with the usual values of the specific heats, it is not only necessary to change the unit of work, but the unit of quantity as well, because the amount of liquid to which this value of the specific heat is referred is not the gm., as is usual, but is the amount of liquid which at 0° C and atmospheric pressure occupies 1 c.c. In order to convert the usual value of the specific heat into these units, it is necessary to multiply by the density of the liquid at 0° and atmospheric pressure, and by 42.66, the number of kgm. cm. in 1 gm. cal.

As a check on the specific heat at constant pressure found in this way, the same quantity was computed for the first three alcohols, that is for methyl, ethyl, and propyl alcohol, by the alternative method involving the second temperature derivative of the volume which was used in the paper on water. The second derivatives were found graphically from the curves of dilatation against temperature at constant pressure, and were integrated mechanically. The results so found agree fairly well with the values found by the other more accurate method. The magnitude of the discrepancies might be as much as 10%, but all the essential characteristics of the curve as given by one method were reproduced by the other also, such as the maxima and the minima, and the points of inflection. Of course the pressure of maximum or minimum was sometimes displaced, as was to be expected.

Specific Heat at Constant Volume.—From the specific heat at constant pressure it is now possible to compute the specific heat at constant volume by the well known formula for the difference of the two specific heats, namely $C_p - C_v = -\dfrac{\tau \left(\dfrac{\partial v}{\partial \tau}\right)_p^2}{\left(\dfrac{\partial v}{\partial p}\right)_\tau}$. This formula involves only quantities which have already been determined, so that C_v may be found immediately. The values of $\left(\dfrac{\partial v}{\partial \tau}\right)_p$ and $\left(\dfrac{\partial v}{\partial p}\right)_\tau$ used in this computation were taken from the tables, not from the diagrams. Just as for the specific heat at constant pressure, the values found in this way are the differences between C_v at atmospheric pressure and the pressure in question. The differences are such that a positive value means that the specific heat is greater at atmospheric pressure than at the pressure in question. A decreasing curve indicates, therefore, that the specific heat is increasing with increasing pressure.

In the paper on water other quantities of thermodynamic interest were plotted. These are the temperature effect of compression and the adiabatic compressibility. They may be easily calculated from the data given in this paper. While they are of interest in themselves, they do not seem to be of such fundamental importance as the quantities already listed in suggesting the possible internal mechanism of the liquid. It was felt, therefore, that to give them would unduly increase the volume of this paper, and they have accordingly not been computed.

A word seems called for as to the general character of the curves. In many cases there are slight irregularities which may very well not correspond to the actual facts, the irregularities being beyond the limit of experimental accuracy of the work. It is true that if each of these quantities were being given for itself alone, without connection with other quantities, it would not have been justifiable to retain all the irregularities which some of the curves show. The reason for retaining the irregularities is that the attempt has been made to present a set of data which should be thermodynamically consistent. Let us suppose, for example, that the compressibility and the dilatation were both determined from the original tables of volumes and that they have been plotted against pressure. Both of these curves show irregularities which may be smoothed off by drawing smoother curves through the points, thus giving values of the compressibility and the dilatation which doubtless in themselves represent with greater probability the actual compressibility and dilatation. But each of these modified values for the compressibility and the dilatation will have a reflex effect on the table of volumes, which has now become inconsistent with the better values of the compressibility and the dilatation, and must therefore be altered slightly so as to be in accord with the new values. The alteration in the table necessary to accomplish this may be produced by changes less than the possible experimental error. But the point is this. Either the revised value of the compressibility or of the dilatation is sufficient of itself to completely revise the table of volume. If we are to adjust the compressibility or the dilatation we must do it so that both have the same reflex action on the table. Furthermore, all seven thermodynamic quantities must be adjusted in the same way. It is evident that this is a task of no small difficulty. To perform it, the only method seems to be a tedious one of trial and error. The labor of such an adjustment would be far beyond the labor of making the measurements with greater accuracy, and the labor had much better be so used in performing new experiments. It must furthermore be remembered that the values in the tables have been smoothed once with respect to both temperature and pressure. Any further changes would amount simply to slight changes in this smoothing; changes which were not justifiable by an examination of the data themselves but are rendered probable only by an examination of certain derived quantities. The choice has been made in this paper, therefore, to present results which may be slightly in error when taken by themselves, but which are nevertheless consistent thermodynamically.

IV. Numerical Details of Experiment and Computation.

In the detailed discussion and presentation of the results for the twelve liquids which is to follow, there will be given the experimental accuracy to be expected for each liquid (because for some liquids the accuracy is greater than for others), and also the sources and the numerical values of the results of other observers which have been used in computing the results given here. The results taken from other work are the density at atmospheric pressure, the thermal dilatation, the initial compressibility for low pressure ranges, and the specific heats at atmospheric pressure. Unless otherwise specified, the values for the density at atmospheric pressure have been taken from the recent tables of Kaye and Laby, and the values for the thermal dilatation have been deduced from the tables of Landolt and Börnstein. In these tables the volume at any temperature is given in terms of the volume at 0° by a power series of the form $V_t = V_0 (1 + at + bt^2 + ct^3)$. In reproducing this expression it will not be necessary to repeat the formula each time, but merely to give the values of the three constants a, b, and c.

It has been mentioned on page 22 that in computing the changes of volume with pressure at 40°, it was found that beyond 500 kgm. the shape of the curves was nearly the same for all twelve liquids, the only difference being in the numerical magnitudes. The constants used in the general pressure-volume formula of page 22 for the average of the twelve liquids were as follows; $\alpha = -0.0029$, $\beta = -0.0546$, $\gamma = +0.2969$, and $\delta = -0.1804$. To pass from this general formula to any one of the twelve liquids each of these four constants is to be multiplied by the same factor. This factor will be given in the following under the name of the "reduction factor."

The discussion is to be one of merely the numerical details of the measurements and the computations. The discussion of the general character of the results and their significance will be reserved until the data have all been presented.

Methyl Alcohol.— Three sets of measurements were made on this substance with three different fillings of the apparatus, the last being separated by nearly three months from the earlier two. The first measurement was of the thermal dilatation and compressibility at low pressures with the larger bulb adapted for low pressure work. The next set of measurements, made immediately afterwards, was of the compressibility and dilatation at the higher pressures with the smaller bulb for the high pressure work. The third measurement was with the high pressure bulb, and included the compressibility and dilatation over the entire pressure range, both high and low pressures. The measurements at low pressures were made, as already explained, before the piston had been upset by the higher pressures. The accuracy of the compressibility measurements may be estimated from the fact that the mean discrepancy of the piston displacements in the two sets of high pressure readings was 0.0035 inch, the maximum displacement being 2.07 inches. For the dilatation, the average discrepancy in the piston displacement for a rise of temperature of 20° was 0.0011 inch, the average displacement being 0.070 inch.

In computing the volumes at atmospheric pressure the density at 15° was taken to be 0.7960, from Kaye and Laby. The constants of the dilatation formula from Landolt and Börnstein are as follows; $a = 0.0_21186$, $b = 0.0_5156$, $c = 0.0_891^6$. This gives for the density at 0°, 0.8100. The quantity of methyl alcohol to which the tables and the diagrams refer weighs, therefore, 0.8100 gm. Since the boiling point of methyl alcohol is 64.7°, the volume listed in Table II for 80° and atmospheric pressure is, therefore, merely an extrapolation by means of the formula.

The "reduction factor" by means of which the transition was made from the mathematical formula for volume in terms of pressure at 40° to the experimental curve was 1.009.

The change of volume from 1 to 500 kgm. at 40° was taken as 0.0483, following Amagat. It should be noticed, however, that Amagat gives for the volume at 40° and atmospheric pressure 1.0438, against 1.0483 of the tables of Landolt and Börnstein. In this work the value 1.0483 was taken as the volume at 40°, but Amagat's value for the change of volume 1–500 kgm. was adopted without correction. At low pressures (20°) the present experimental values for the changes of volume were as follows: 1–500 atmos., 0.0530; 500–1000, 0.0294; 1000–1500, 0.0242; 1500–2000, 0.0199. The corresponding values of Amagat are 0.0480, 0.0300, 0.0239, 0.0194. The agreement is fairly good, except for the lowest pressure interval, where as has been pointed out, the present method can only indicate the probable result by an extrapolation. The newly published result of Richards is 0.0430 for the change of volume at 20° for an increase of 500 kgm. of pressure as against 0.0415 listed in the tables of volume.

The volume of methyl alcohol is shown as a function of pressure and temperature in Table II and in Figure 8.

The compressibility, β, of the first five alcohols has been measured by Pagliani and Palazzo,[7] who have collected their results into formulas of the type, $\beta_t = \beta_0 (1 + at + bt^2)$. Their pressure range was 1–4

---

[6] Pierre, Ann. chim. et,phys., **15**, 325 (1845).
[7] Pagliani and Palazzo, Mem. R. Acc. Lin., **19**, 279 (1883/84).

atmospheres. Within this range the change of compressibility with pressure is negligible. The values found from their formulas (reduc-

TABLE II.

VOLUME OF METHYL ALCOHOL.

Pressure. kgm. cm.²	Volume						
	20°.	30°.	40°.	50°.	60°.	70°.	80°.
1	1.0238	1.0361	1.0483	1.0610	1.0737	1.0869	1.1005
500	0.9823	0.9909	1.0000	1.0096	1.0197	1.0404	1.0416
1000	.9530	.9607	0.9684	0.9763	0.9844	0.9929	1.0023
1500	.9276	.9347	.9415	.9481	.9549	.9621	0.9697
2000	.9087	.9151	.9213	.9271	.9331	.9393	.9456
2500	.8930	.8988	.9044	.9098	.9151	.9205	.9260
3000	.8792	.8845	.8897	.8947	.8997	.9047	.9095
3500	.8663	.8712	.8761	.8808	.8854	.8899	.8944
4000	.8551	.8597	.8642	.8687	.8730	.8773	.8814
4500	.8449	.8492	.8535	.8577	.8618	.8657	.8695
5000	.8354	.8395	.8436	.8476	.8515	.8552	.8588
6000	.8192	.8232	.8271	.8307	.8344	.8379	.8412
7000	.8053	.8091	.8129	.8104	.8196	.8228	.8262
8000	.7936	.7972	.8008	.8040	.8070	.8108	.8134
9000	.7827	.7861	.7894	.7924	.7952	.7981	.8013
10000	.7725	.7757	.7788	.7818	.7847	.7876	.7905
11000	.7634	.7663	.7693	.7724	.7756	.7786	.7813
12000	.7559	.7586	.7614	.7647	.7682	.7712	.7738

tion being made from atmos. to kgm.) at 20°, 40°, 60°, and 80°, were 0.0_3113, 0.0_3124, 0.0_3142, and 0.0_3158 respectively. The values found from the present data by the method of computation outlined on page 29 are 0.0_3101, 0.0_3124, 0.0_3137, and 0.0_3147 respectively. The means adopted for this paper are 108, 124, 140, and 152 respectively. There are also values for the compressibility by other observers, but not under conditions so nearly comparable with those here. These values are: 0.0_3104 at $14.7°$ and 0.0_3221 at $100°$ between 8.7

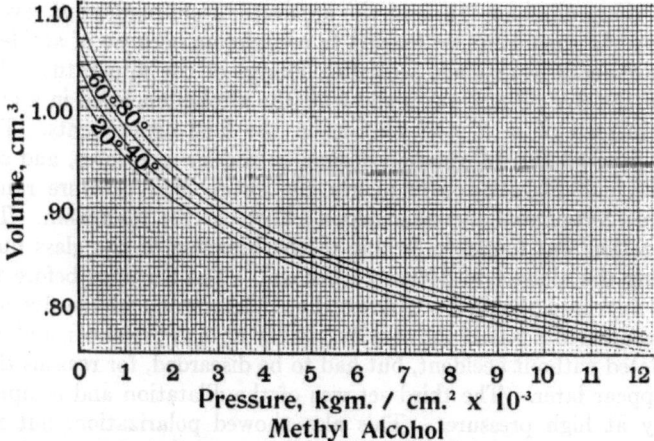

FIGURE 8. Methyl Alcohol. Volume at $20°$, $40°$, $60°$, and $80°$ plotted against pressure. The lower curve gives the volume at $20°$.

and 37 atmos. by Amagat[8]; 0.0_491 at $13.5°$ and 7.5 atmos. by Grassi[9]; 0.0_3108 at $2.7°$ and 8 atmos. and 0.0_3120 at $18°$ and 8 atmos. by Röntgen.[10] The newly published value of Richards for the compressibility at $20°$ is 0.0_3109 against 0.0_3108 adopted above.

There are a few measurements for the specific heat, C_p, at atmospheric pressure; 20.39 between $5°$ and $10°$, 20.28 between $10°$ and $15°$, and 20.76 between $15°$ and $20°$ by Regnault[11]; 22.29 between $23°$ and $43°$ by Kopp;[12] 21.56 between $5°$ and $13°$ by Lecher;[13] and 21.40 between $15.5°$ and $34.9°$, 21.82 between $19.6°$ and $45°$, 22.13 between $18.1°$ and $50.4°$, and 22.77 between $20.5°$ and $63.2°$ by von Reis.[14]

---

[8] Amagat, Ann. chim. et phys., **11**, 520–549 (1877).
[9] Grassi, Am. chim et phys., **31**, 437 (1851).
[10] Röntgen, Wied. Ann., **44**, 1 (1891).
[11] Regnault, Ann. chim. et phys., **9**, 322 (1843).
[12] Kopp, Pogg. Ann., **75**, 98 (1848).
[13] Lecher, Wien Ber., **76**, 937 (1877).
[14] von Reis, Wied. Ann., **13**, 447–465 (1881).

These values are in the units of this paper. It will be seen that the results of different observers do not agree within 5%. The results of von Reis, however, do justify us in assuming that C_p increases with rising temperature.

Ethyl Alcohol. More measurements were made on this than on most of the other substances, because it was the liquid with which the preliminary tests of the apparatus were made, but several of the early runs were not carried to completion because of accident. Measurements were made with five fillings of the apparatus. The first of the five fillings was made with the alcohol enclosed in a glass bulb, instead of in a steel one, as in the final experiments. This filling gave all the information desired at the low pressures, and also the thermal dilatation over nearly the entire high pressure range, but was terminated by polarization effects in the manganin. The polarization was found to be due to the breaking of the glass bulb, allowing the alcohol to diffuse to the coil. The readings before the break appeared should be trustworthy. The second of the five sets of measurements was of the compressibility at high pressures, and was completed without accident, but had to be discarded, for reasons that will appear later. The third set was of the dilatation and compressibility at high pressures. This also showed polarization, but not until the very end of the compressibility run. For the second and third runs a steel bulb was used, but the top was put on with soft solder. This soft solder gave way under pressure, allowing the kerosene to mix with the alcohol. The polarization probably did not occur as soon as the solder cracked, because it takes time for the alcohol to diffuse through the kerosene to the coil. The compressibility measurements of the third run are, therefore, more likely to be in error than the dilatation measurements, which were made a considerable time before the polarization appeared. Because the apparatus was the same, the second run is likely to be in error just as the compressibility measurements of the third, the polarization not having time to appear in the second run before the apparatus was taken apart. The early dilatation measurements were retained, therefore, and the early compressibility measurements discarded. The agreement of the early compressibility measurements, which presumably were made on a mixture of kerosene and ethyl alcohol, was good, 0.3%, but they were about 4% higher than the results of the final successful run. The last two of the five runs were carried through without accident, one being of the compressibility and dilatation at low pressures, and the other the corresponding measurements for high pressures. For these, and for all subsequent runs, the top of the bulb was put on with silver solder.

TABLE III.

Volume of Ethyl Alcohol.

Pressure. kgm. cm.²	Volume.						
	20°.	30°.	40°.	50°.	60°.	70°.	80°.
1	1.0212	1.0323	1.0438	1.0557	1.0679	1.0805	1.0934
500	0.9794	0.9873	0.9956	1.0044	1.0135	1.0233	1.0334
1000	.9506	.9570	.9636	0.9707	0.9781	0.9861	0.9944
1500	.9267	.9323	.9380	.9440	.9505	.9572	.9640
2000	.9081	.9131	.9182	.9235	.9291	.9349	.9407
2500	.8923	.8969	.9016	.9064	.9114	.9165	.9216
3000	.8786	.8830	.8874	.8919	.8964	.9010	.9055
3500	.8661	.8702	.8746	.8789	.8831	.8873	.8915
4000	.8545	.8586	.8628	.8668	.8708	.8747	.8787
4500	.8439	.8481	.8521	.8559	.8597	.8634	.8671
5000	.8343	.8383	.8424	.8461	.8498	.8533	.8568
6000	.8178	.8218	.8256	.8291	.8324	.8356	.8387
7000	.8038	.8075	.8110	.8142	.8171	.8200	.8229
8000	.7917	.7952	.7984	.8013	.8038	.8065	.8094
9000	.7807	.7840	.7868	.7893	.7917	.7954	.7973
10000	.7703	.7733	.7760	.7785	.7809	.7835	.7863
11000	.7606	.7633	.7659	.7693	.7713	.7741	.7765
12000	.7521	.7545	.7571	.7600	.7631	.7652	.7682

The average discrepancy of the piston displacement for a rise of temperature of 20° at constant pressure was 0.0016 inch, the mean displacement being about 0.070 inch.

The reduction factor from the mathematical formula for volume in terms of pressure at 40° was 0.9979.

The density at atmospheric pressure and 0° was taken as 0.8063. The constants of the dilatation formula of Landolt and Börnstein were $a = 0.0_21022$, $b = 0.0_5182$, $c = 0$. The values of the volume given

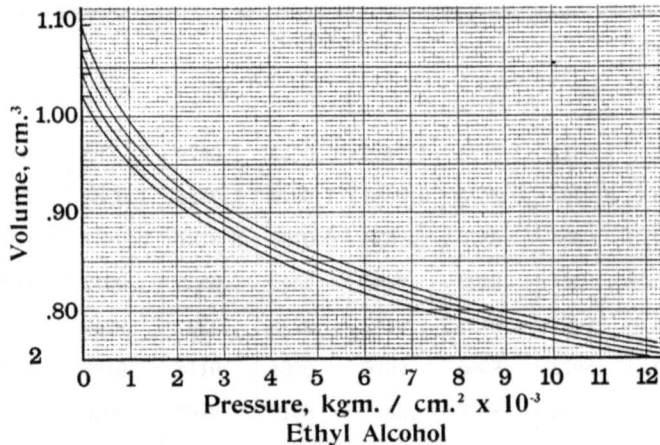

FIGURE 9. Ethyl Alcohol. Volume at 20°, 40°, 60°, and 80° plotted against pressure. The lower curve gives the volume at 20°.

by this formula are 1.0212, 1.0138, and 1.0679 at 20°, 40° and 60° respectively. We also have values of Pierre[15], which are 1.0216, 1.0448, and 1.0695 at the same temperatures respectively.

Amagat gives .0484 for the change of volume from 1 to 500 kgm. His value for the volume at 40° is 1.0442 against 1.0438 above. In the tables, 0.0484 was used as the change of volume 1–500, and 1.0438 as the volume at atmospheric pressure. At 20°, the change of volume between 1 and 500 atmos. was found to be 0.0477 against 0.0438 of Amagat. The numbers for the succeeding 500 atmos. intervals were 0.0287, 0.0236, 0.0193 against 0.0297, 0.0228, and 0.0188 of Amagat.

The volume as a function of pressure and temperature is given in Table III and in Figure 9.

The initial compressibilities at 20°, 40°, 60°, and 80°, computed as described, were found to be 0.0_3105, 0.0_3121, 0.0_3138, and 0.0_3151, respectively. The corresponding values of Pagliani and Palazzo are: 0.0_3102, 0.0_3114, 0.0_3130, and 0.0_3151. The agreement is as good as

---

[15] Pierre, Am. chim. et phys., **19**, 199 (1847).

could be expected for measurements for this nature. The means shown in the curves are: 0.0_3104, 0.0_3118, 0.0_3135, and 0.0_3151. Comparison may also be made with the values of Amagat,[16] which are

TABLE IV.

C_p for Ethyl Alcohol.

Observer.	Temp.	C_p (kgm. cm.).
Regnault [17]	—20°	17.37
	0°	18.81
	40°	22.63
	80°	26.44
Sutherland [18]	80°	24.49
	120°	31.27
Zetterman [19]	20°	31.23
De Heen and Deruyts [20]	40°	20.53
von Reis	15°.7–35°.1	19.93
	20°.7–45°.7	20.78
	18°.4–56°.0	21.32
	19°.8–62°.9	21.79
	20°.5–73°.4	22.44

0.0_4981 at 14° and 0.0_3196 at 99.4° at a mean pressure of 22 atmos. The agreement at the lower temperature is good; the upper temperature is beyond the range.

For C_p at atmospheric pressure we have a number of values which are shown in Table IV. The results are in very bad agreement, as may be seen by plotting them. It is however, perfectly certain that on the whole C_p for ethyl alcohol increases with rising temperature.

---

[16] Amagat, l. c. (1877).
[17] Regnault, Mém. Acad., **26**, 262 (1862).
[18] Sutherland, Phil. Mag., **26**, 298 (1888).
[19] Zetterman, Akad. Afh. Helsingfors (1880).
[20] De Heen and Deruyts. Bull de Belg., **15**, 168 (1888).

Propyl Alcohol.— Readings were made on this liquid with two fillings of the apparatus. The first was of the compressibility and the

TABLE V.

Volume of Propyl Alcohol.

Pressure. kgm. cm.²	Volume.						
	20°.	30°.	40°.	50°.	60°.	70°.	80°.
1	1.0173	1.0274	1.0380	1.0493	1.0612	1.0737	1.0865
500	0.9780	0.9864	0.9948	1.0034	1.0121	1.0213	1.0320
1000	.9498	.9571	.9641	0.9710	0.9779	0.9853	0.9934
1500	.9297	.9357	.9415	.9473	.9533	.9594	.9657
2000	.9142	.9192	.9242	.9293	.9344	.9396	.9448
2500	.9011	.9055	.9100	.9145	.9190	.9235	.9282
3000	.8897	.8937	.8979	.9021	.9062	.9103	.9145
3500	.8794	.8833	.8872	.8911	.8949	.8987	.9025
4000	.8700	.8738	.8776	.8813	.8849	.8884	.8919
4500	.8612	.8650	.8688	.8723	.8758	.8791	.8823
5000	.8529	.8567	.8604	.8639	.8671	.8702	.8732
6000	.8390	.8426	.8462	.8494	.8524	.8555	.8579
7000	.8200	.8000	.8333	.8363	.8391	.8416	.8442
8000	.8163	.8193	.8223	.8250	.8277	.8302	.8328
9000	.8069	.8098	.8124	.8150	.8175	.8201	.8230
10000	.7984	.8011	.8037	.8060	.8085	.8112	.8142
11000	.7909	.7934	.7958	.7980	.8004	.8031	.8061
12000	.7840	.7864	.7885	.7905	.7928	.7955	.7982

dilatation at the higher pressures, and the second was the complete set, both compressibility and dilatation at both high and low pressures. During the last set of readings, however, the moveable plug pinched off at a high pressure because of fatigue, so that there is only one reading for the dilatation at the two highest pressures. The agreement between the two sets at the lower pressures was good enough however, so that it did not seem necessary to set the apparatus up again merely to repeat these last two readings.

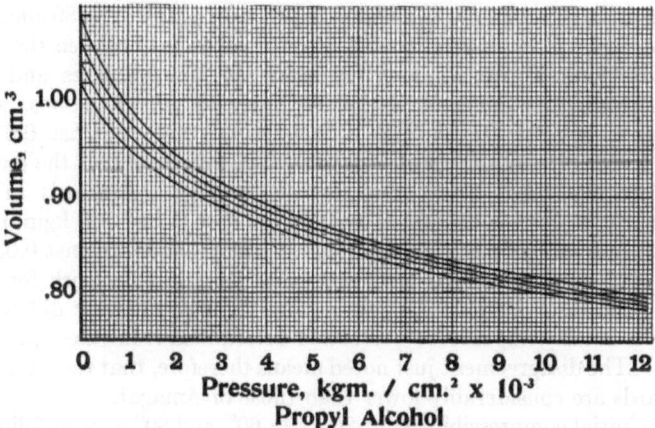

FIGURE 10. Propyl Alcohol. Volume at 20°, 40°, 60°, and 80° plotted against pressure. The lower curve gives the volume at 20°.

The average discrepancy between the piston displacement of the two determinations of the compressibility at high pressures was 0.0026 inch on a total stroke of 2 inches. The average discrepancy in the displacements at constant pressure corresponding to an increase of temperature of 20° was 0.0018 inch on a mean displacement of about 0.070 inch.

The reduction factor from the mathematical formula was 0.8726.

The density at 0° is 0.8179. The constants of the dilatation formula are: $a = 0.0_2 774$, $b = 0.0_5 497$, $c = - 0.0_7 141$. These values of a, b, and c are from results of Zahnder,[21] who gives for the density at 0° 0.8177, instead of 0.8179 above. The agreement is virtually perfect. In addition we have data by Naccari and Pagliani[22] who give for the density at 0° 0.8203, and for the volume at 20°, 40°, 60°, and 80°, 1.020, 1.042, 1.064, and 1.090 respectively, against 1.017, 1.038, 1.061, and 1.0865 adopted from Zahnder's formula above.

---

[21] Zahnder, Lieb. Ann., **225**, 114–193 (1882).
[22] Naccari and Pagliani. Att. R. Acc. dell. Sc., **16** (Sept. 1881).

Here again it seems as if the agreement between different observers should be better.

The volume of propyl alcohol as a function of pressure and temperature is given in Table V and in Figure 10.

At 40°, Amagat gives for the change of volume between 1 and 500 kgm. 0.0432, which is the value used in the table. He gives, however, 1.0406 for the volume at atmospheric pressure against 1.0380 adopted above. It will be noticed that Amagat's value lies between those of Zahnder and of Naccari and Pagliani. At low pressures and 20° the changes of volume for successive intervals of 500 atmos. were found to be: 0.0407, 0.0245, 0.0202, and 0.0170, against 0.0399, 0.0274, 0.0211, and 0.0176 of Amagat. The agreement for the lowest pressure interval is better than on the average. Richards in his recent paper gives a change of volume between 1 and 500 kgm. considerably smaller than that used here, namely 0.0355 against 0.0393. It should be remembered that the value used in this work for 20° is founded essentially on Amagat's value for 40°, the only difference being a small temperature correction determined from these present data. The disagreement just noted means therefore, that the values of Richards are considerably lower than those of Amagat.

The initial compressibilities at 20°, 40°, 60°, and 80° were as follows; to give the value of ΔV listed in the table 0.0_492, 0.0_3103, 0.0_3118, and 0.0_3130 respectively; the corresponding values of Pagliani and Palazzo are 0.0_490, 0.0_3101, 0.0_3115, and 0.0_3133. The agreement is good. The final values taken as a fair mean were: 0.0_491, 0.0_3102, 0.0_3117, 0.0_3131. Röntgen has also measured the compressibility at atmospheric pressure. His value for 20° would be 0.0_4955, judging from a linear extrapolation from his values at 4° and 18°. Richard's recent value at 20° is 0.0_4873, lower than any other of the values given above.

For C_p at atmospheric pressure we have the following values: — 21° to +12°, 18.02 by Nadejdine[23]; 21° to 23°, 22.99 by Pagliani;[24] 21° to 90°, 23.55 by Lougiunine[25]; and from 16.5° to 42.2°, 20.54, from 20.6° to 53.4°, 21.34, from 20.4°, to 65.2°, 21.99, from 19.5° to 78.5°, 22.63, and from 20.7° to 90.8°, 23.32 by von Reis. These results also indicate a considerable rise of C_p with rising temperature.

Isobutyl Alcohol.— Measurements on this were made with two fillings of the apparatus; the first gave the compressibility and the dilata-

---

[23] Nadejdine, Jour. Russ. Phys. Chem. Ges., **16**, 222 (1884).
[24] Pagliani, N. Cim., **11**, 229 (1882).
[25] Lougiunine, Am. chim. phys., **13**, 289 (1898).

tion over the high pressure range, and the second the compressibility and dilatation over both the high and low pressure ranges. The use of isobutyl instead of normal butyl alcohol was not intended. In ordering the chemicals, normal butyl was not specified, and it was not noticed that the substance sent was isobutyl until all the preparations had been made for a run. This substance has the disadvantage, of not being one of the same series as the four other alcohols. However, it makes little difference so far as the comparison of the results with those of Amagat is concerned, for Amagat did not work with either normal- or iso-butyl alcohol. Furthermore, the use of this substance has proved very instructive in showing that a change in the structural formula changes the properties even at high pressures. It might be expected that high pressures would wipe out variations due to structural differences, but such has not proved to be the case, at least to 12000 kgm.

The average discrepancy in the piston displacements of the two determinations of compressibility was 0.0024 inch on a total displacement of 2.0 inches. The mean discrepancy of the displacement for the thermal dilatation for 20° was 0.0008 inch on a mean of about 0.070 inch. The agreement between the two sets of readings for the highest temperature range, 60°-80°, was virtually perfect.

The reduction factor for the mathematical formula was 0.9342.

Landolt and Börnstein's tables do not contain the requisite data for the volume of isobutyl alcohol at atmospheric pressure. The values adopted here were obtained by Naccari and Pagliani, and are apparently the only data which have been published for this liquid. These authors have not expressed their results by a power series, but prefer instead to give the density for a considerable number of temperatures. By interpolation from their results the volumes at 20°, 40°, 60°, and 80° were found to be: 1.0195, 1.0406, 1.0625, and 1.0880. From these results the value of Kaye and Laby for the density at 18° is reduced to 0.8165 at 0°, against 0.8162 of Naccari and Pagliani, virtual agreement.

No measurements of the change of volume of isobutyl alcohol had been made beyond a few kgm. previous to these computations, so the value obtained from the low pressure determinations of the present work was adopted. This was 0.0484 at 40° between 1 and 500 kgm.; not at all an unlikely value, being the same as Amagat's for ethyl alcohol. Compressibility determinations of others at low pressures have shown that isobutyl alcohol has a compressibility considerably higher than that of normal butyl alcohol, so that we are to expect a value higher than we should predict from the behavior of propyl alcohol and a value as large as that of ethyl alcohol does not seem

TABLE VI.

Volume of Isobutyl Alcohol.

Pressure. kgm. cm.²	Volume.						
	20°	30°	40°	50°	60°	70°	80°
1	1.0195	1.0300	1.0406	1.0414	1.0625	1.0744	1.0880
500	0.9751	0.9838	0.9922	1.0006	1.0093	1.0184	1.0277
1000	.9486	.9560	.9632	0.9701	0.9768	0.9840	0.9918
1500	.9268	.9330	.9391	.9448	.9505	.9565	.9630
2000	.9097	.9150	.9202	.9252	.9303	.9355	.9410
2500	.8956	.9001	.9048	.9094	.9141	.9187	.9235
3000	.8822	.8867	.8905	.8949	.8994	.9038	.9080
3500	.8705	.8743	.8782	.8823	.8867	.8908	.8947
4000	.8601	.8637	.8673	.8712	.8755	.8794	.8830
4500	.8507	.8541	.8577	.8614	.8655	.8692	.8726
5000	.8409	.8443	.8477	.8513	.8552	.8587	.8619
6000	.8269	.8301	.8335	.8369	.8403	.8433	.8463
7000	.8130	.8163	.8196	.8228	.8260	.8289	.8317
8000	.8028	.8060	.8092	.8123	.8154	.8183	.8210
9000	.7927	.7959	.7990	.8021	.8050	.8079	.8105
10000	.7832	.7863	.7894	.7924	.7953	.7980	.8007
11000	.7742	.7772	.7803	.7833	.7862	.7888	.7913
12000	.7662	.7692	.7722	.7751	.7780	.7805	.7827

unlikely. The recent work of Richards gives for the change of volume at 20° between 1 and 500 kgm. 0.0355, against 0.0344 used in the tables. The agreement is as close as could be expected when the rough nature of the present determinations at the low pressures is considered; the agreement is better than the agreement of those values which have been taken directly from the work of Amagat.

The volume of isobutyl alcohol as a function of pressure and temperature is shown in Table VI and in Figure 11.

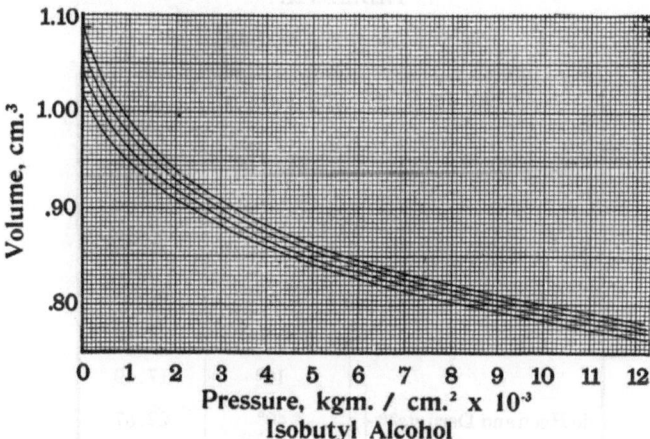

FIGURE 11. Isobutyl Alcohol. Volume at 20°, 40°, 60°, and 80° plotted against pressure. The lower curve gives the volume at 20°.

The compressibility determinations of Pagliani and Palazzo were fortunately made with isobutyl instead of normal butyl alcohol. They give for 20°, 40°, 60°, and 80° the values 0.0_492, 0.0_3103, 0.0_3118, and 0.0_3137. The values required to give the values of ΔV listed in the tables are 0.0_3122, 0.0_3133, 0.0_3144, and 0.0_3164. The discrepancy is large, too large. Instead, however, of taking the average of the discordant results, it was preferred to retain the values consistent with the table, it being understood that the initial values between 1 and 500 kgm., both for the total change of volume and for the compressibility are probably in error. Abnormal variations of compressibility, such as the rapid initial decrease with pressure, may possibly explain part of the discrepancy. We have also a value of Röntgen for the compressibility at 20°, 0.0_406, which is in very much better agreement with the value of Pagliani and Palazzo than the present value. The recent value of Richards is practically the same as Röntgen's.

For C_p we have a larger number of measurements than we should expect from the small number of dilatation measurements. These values are shown in Table VII. When plotted, they show considerable discrepancies. The value at $-5°$ of Nadejdine and that of Pagliani are almost certainly in error. The other points lie roughly

TABLE VII.

C_p for Isobutyl Alcohol.

Observer.	Temperature.	C_p (kgm./cm.)
Longuinine [26]	20°–114°	24.00
	21°–109°	24.94
Nadejdine [27]	$-21°$–$+10°$	17.70
	16°–70°	21.39
	18°–98°	23.24
de Heen and Deruyts [28]	10°	17.49
	40°	22.57
	85°	29.29
Pagliani [29]	26°–30°	23.90

on a straight line, such that C_p increases from about 15 at 0° to about 23 at 50°. This is a very considerable increase.

Amyl Alcohol.—Experiments were made on this with three fillings of the apparatus; the first for compressibility and dilatation at low pressures with the large bulb, the second for compressibility and dilatation over the high pressure range, and the third for compressibility and dilatation over both ranges, high and low. The runs were all accomplished without accident of any sort.

The average discrepancy of the piston displacement for compressibility at 40° was 0.0019 inch on about 2 inches. The corresponding discrepancy for changes of temperature of 20° at constant pressure was 0.00206 inch on an mean of about 0.070 inch.

---

[26] Louguinine, l. c.
[27] Nadejdine, l. c.
[28] De Heen and Deruyts, l. c.
[29] Pagliani, l. c.

The reduction factor from the mathematical formula was 0.8925, showing that at high pressures amyl alcohol is one of the most incompressible of the twelve liquids.

TABLE VIII.

VOLUME OF AMYL ALCOHOL.

Pressure. kgm. cm.²	Volume.						
	20°.	30°.	40°.	50°.	60°.	70°.	80°.
1	1.0181	1.0270	1.0374	1.0476	1.0583	1.0694	1.0814
500	0.9800	0.9880	0.9959	1.0039	1.0122	1.0210	1.0304
1000	.9526	.9593	.9660	0.9724	0.9792	0.9863	0.9936
1500	.9325	.9383	.9440	.9495	.9551	.9608	.9667
2000	.9158	.9210	.9259	.9307	.9354	.9402	.9452
2500	.9015	.9064	.9107	.9149	.9190	.9232	.9277
3000	.8892	.8938	.8979	.9018	.9055	.9094	.9136
3500	.8780	.8823	.8864	.8900	.8935	.8971	.9010
4000	.8682	.8725	.8764	.8800	.8832	.8867	.8903
4500	.8593	.8635	.8673	.8708	.8739	.8772	.8807
5000	.8508	.8548	.8585	.8618	.8648	.8679	.8713
6000	.8373	.8409	.8442	.8471	.8501	.8530	.8560
7000	.8251	.8281	.8310	.8337	.8363	.8390	.8418
8000	.8149	.8176	.8201	.8225	.8250	.8274	.8302
9000	.8044	.8068	.8092	.8114	.8138	.8163	.8190
10000	.7948	.7971	.7994	.8018	.8041	.8066	.8091
11000	.7860	.7886	.7904	.7932	.7954	.7976	.8001
12000	.7782	.7803	.7826	.7854	.7882	.7905	.7926

The density at 0° and atmospheric pressure was taken as 0.8266. The constants of the dilatation formula were: $a = 0.0_289$, $b = 0.0_657$,

$c = 0.0_7118$. This formula of Landolt and Börnstein seems to be taken from Pierre.[30] It gives for the volumes at 20°, 40°, 60°, and 80° the values 1.0181, 1.0374, 1.0583, and 1.0814 respectively. We have also the following values by Pierre and Puchot[31]; 1.0187, 1.0397, 1.0610, and 1.0864. These authors give for the density at 0°, 0.817.

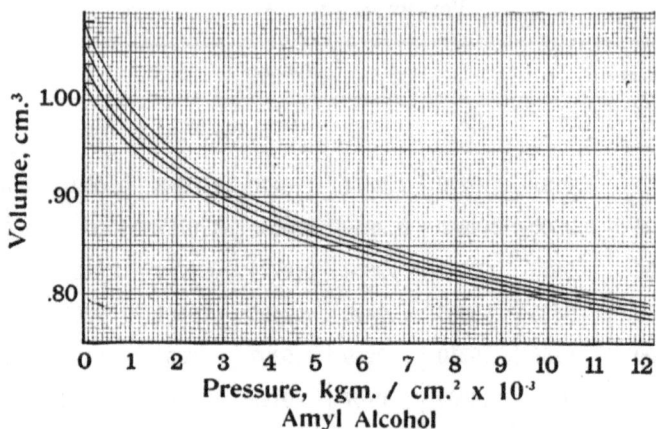

FIGURE 12. Amyl Alcohol. Volume at 20°, 40°, 60°, and 80° plotted against pressure. The lower curve gives the volume at 20°.

Zahnder[32] gives for the density at 0° 0.829, and for the constants of the dilatation formula, $a = 0.0_3919$, $b = -0.0_6461$, and $c = 0.0_7175$? Here again the discrepancies appear to be greater than they should in measurements of this character.

For the change of volume between 1 and 500 kgm. there seem to be no other data as a basis of comparison. Amagat used allyl instead of amyl alcohol for some unknown reason. The only course, therefore, was to accept the value given by this present work for the lower pressure interval, namely 0.0451. That this figure is about correct, however, is spoken for by the rather unusually close agreement of the two measurements of the piston displacement at 20°, 0.389 inch and 0.379 inch, a disagreement of 2.5%.

The volume of amyl alcohol as a function of pressure and temperature is shown in Table VIII and in Figure 12.

---

[30] Pierre, l. c. (1847).
[31] Pierre and Puchot, Ann. chim. et phys. (4), **22**, 306?
[32] Zahnder, l. c.

For C_p we have the following values: 19.89 by Kopp[33] for the temperature range 26°–44°; 24.43 by Regnault[34] for the range 10°–117°; 24.51 by Louguinine[35] between 21° and 130°; and 22.97 between 20.5° and 100.1°, 23.64 between 22.2° and 111.6°, and 24.23 between 22.2° and 124.5° by von Reis. These values, which are rather more consistent than usual, show a fairly rapid increase of C_p with temperature.

Ether.—Measurements were made on this liquid with four different fillings of the apparatus. The first two, made before the method had been perfected, were neither complete because of accidents, but between them they give completely the compressibility and the dilatation over the entire high pressure range. The third set of readings was over the low pressure range; this set was repeated without refilling the apparatus. The fourth set, made with the perfected apparatus, was over the high pressure range, and was completed successfully without accident.

There were three sets of piston displacements for the compressibility at 40°. The mean discrepancy of these was $\frac{1}{2}\%$ on the maximum displacement, which is below the average in accuracy. The mean discrepancy of the piston displacements for the thermal dilatation was 0.0022 inch on about 0.070 inch, which is nearly normal.

The reduction factor from the mathematical formula was 1.104, showing that over the entire range ether remains more compressible than any of the other liquids, except ethyl chloride.

The low boiling point of ether at atmospheric pressure, 34.6°, makes it impossible to tabulate the initial properties at the higher temperatures. For this reason many of the curves start at 500 kgm. as the zero instead of atmospheric pressure.

The density of ether at atmospheric pressure and 0° was 0.7382. The three constants of the dilatation formula had the values: $a = 0.0_2 1513$, $b = 0.0_5 236$, and $c = 0.0_7 400$.[36]

The change of volume at 40° between 1 and 500 kgm. was taken from Amagat as 0.0770. Amagat gives for the volume at 40° and atmospheric pressure 1.0672 against 1.0669 of the dilatation formula above. Amagat's value for the volume at 40° and 500 kgm. was corrected, therefore, in accordance with the above. The low pressure measurements of this present work at 20° are in unusually good agreement with Amagat; 0.0665 against 0.0656 for the interval 1–500

---

[33] Kopp, l. c.   [34] Regnault, l. c. (1862).
[35] Louguinine, l. c.   [36] Pierre, l. c. (1845).

atmos.; 0.0370 against 0.0379 between 500 and 1000; 0.0272 against 0.0275 between 1000 and 1500; and 0.0216 against 0.0215 between

TABLE IX.

VOLUME OF ETHER.

Pressure. kgm. cm.²	Volume.						
	20°.	30°.	40°.	50°.	60°.	70°.	80°.
1	1.0315	1.0492	1.0669				
500	0.9681	0.9790	0.9899	1.0011	1.0124	1.0247	1.0387
1000	.9363	.9445	.9530	0.9616	0.9707	0.9804	0.9906
1500	.9093	.9153	.9221	.9291	.9364	.9438	.9516
2000	.8871	.8980	.8980	.9038	.9099	.9164	.9223
2500	.8685	.8734	.8785	.8837	.8890	.8943	.8997
3000	.8530	.8576	.8623	.8670	.8718	.8765	.8912
3500	.8395	.8440	.8483	.8526	.8570	.8613	.8654
4000	.8275	.8318	.8359	.8400	.8439	.8478	.8515
4500	.8168	.8209	.8249	.8287	.8324	.8359	.8393
5000	.8071	.8111	.8149	.8186	.8220	.8253	.8284
6000	.7916	.7954	.7989	.8023	.8055	.8085	.8112
7000	.7773	.7806	.7838	.7869	.7899	.7927	.7953
8000	.7645	.7675	.7704	.7732	.7759	.7786	.7813
9000	.7525	.7554	.7580	.7606	.7632	.7658	.7687
10000	.7418	.7444	.7469	.7496	.7520	.7547	.7574
11000	.7312	.7335	.7360	.7388	.7418	.7445	.7469
12000	.7216	.7237	.7261	.7289	.7316	.7342	.7365

1500 and 2000. It may be expected, therefore, that the low pressure values of the various thermodynamic properties are rather more than usually accurate for ether.

The volume of ether as a function of pressure and temperature is shown in Table IX and in Figure 13.

For the initial compressibility at 20° and 40° we have values of Amagat[37]; 0.0_3184, and 0.0_2218 respectively. There are also measurements by Avenarius[38] at 20° and 40°; 0.0_3191, and 0.0_3232, respec-

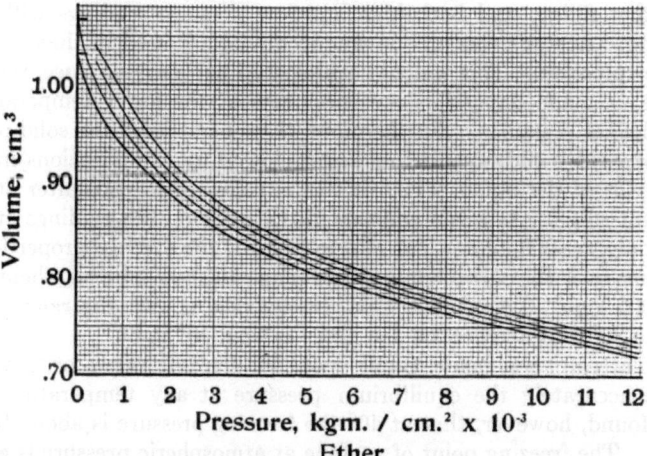

FIGURE 13. Ether. Volume at 20°, 40°, 60°, and 80° plotted against pressure. The lower curve gives volume at 20°. The curves for the higher temperatures could not be extended to the origin because of the low boiling point.

tively. The values needed to give the changes of volume listed in the table are 0.0_3170, and 0.0_3215. The mean values finally adopted were 0.0_3184, and 0.0_3220. There are also other measurements by Grimaldi[39] and Amagat at temperatures considerably above the normal boiling point, and at accordingly increased pressures, 20 kgm. or so. It was not attempted to make connections with these values.

C_p has been measured by Regnault[40], who gives 20.97 at 0°, and 21.69 at 30°; by Sutherland[41], whose values are 27.35 at 80°, and 31.86 at 120°; and by de Heen[42], who found 32.59 at 140°, and 41.27 at

---

[37] Amagat, l. c., (1877).
[38] Avenarius, Bull. Acc. St. Pet., 10 (1877).
[39] Grimaldi, N. Cim., 19, 7 (1886).
[40] Regnault, l. c. (1862).
[41] Sutherland, l. c.
[42] de Heen, Bull de Belg., 15, 522 (1888).

180°. These values all lie roughly on a curve passing through 20.0 at 0°, 23.5 at 40°, 27.0 at 80°, 31.0 at 120°, and 36.5 at 160°. The increase of C_p with temperature is fairly rapid, and becomes more rapid at the higher temperatures.

Acetone.— Two fillings of the apparatus with this substance were made. The high pressure measurements only were made with the first filling, and both high and low pressure readings with the second. Acetone was unique among the liquids used in that it froze under pressure. This was not anticipated nor desired, since, for one thing, it made impossible measurements at the lower temperatures and higher pressures. Furthermore, the separation of the solid phase is apparently accompanied or foreshadowed by complications in the behavior of the liquid, which it was not desired to encounter at the present stage. As a consequence of the freezing, the readings at 20° run only to 8000 kgm. The curves showing the average properties of acetone over the entire temperature range all show a break, therefore, at 8000 kgm. Below 8000 kgm. the average is over the range from 20° to 80°, but above 8000 the range is from 40° to 80°.

No attempt was made to follow out the freezing curve, or to determine accurately the equilibrium pressure at any temperature. It was found, however, that at 40° the freezing pressure is about 10000 kgm. The freezing point of acetone at atmospheric pressure is given by Kaye and Laby at − 95°. This raising of the freezing point by 135° seems to be larger than any previously recorded.

Acetone also showed one other peculiarity. When the liquid was examined at the close of the second run, it was found to be of a slight rose color, and there was a small amount of a fine white precipitate. The rose color deepened in the course of several days to a dirty brown, and the precipitate appeared to increase slightly in quantity. It was thought at the time that this was a chemical reaction brought about by pressure alone, but subsequent investigation showed that the effect was doubtless due to the presence of a slight impurity of phosphorus trichloride, left from the previous run. Phosphorus trichloride when mixed with acetone and allowed to stand at atmospheric pressure was found to produce very slowly the same discoloration and precipitate as observed after exposure to pressure. The effect of pressure apparently is merely to hasten the reaction. In a subsequent experiment, in which every trace of phosphorus trichloride had been carefully removed by prolonged heating, acetone was submitted to the pressures and temperatures of the regular experiment for a day, with absolutely no trace of discoloration. Unfortunately, no examination was made of the condition of the liquid after the end of the first run, so it cannot be told whether the effect was present then or not; probably not. In any event the error so introduced is probably very small, because not more than a very small impurity of PCl_3 could have escaped attention in the weighing. There is, however, a slight possibility that the reaction was catalytic, in which event the error might be greater. The close agreement of the two sets of readings makes this unlikely, however.

The mean discrepancy in the piston readings for compressibility was about 0.005 inch on a total stroke of 2.0 inches. The first compressibility readings were made at 40° The liquid does not freeze until it has been considerably subcooled, so that it was possible to cover the entire pressure range at 40°. But in order to avoid the possibility of the liquid freezing the second time at a less degree of subcooling than at first, the second run was made at 60°, and then reduced to 40° for comparison with the first run. The average discrepancy of the displacements for thermal dilatation was 0.0013 inch on an average of 0.070 inch.

The reduction factor from the mathematical formula was 1.049, showing that acetone is more compressible than the average.

The boiling point of acetone is 56.5°, so that for this reason the initial point of the 80° curve is taken as 1000 kgm. The initial values at 60° were obtained by extrapolation, disregarding the boiling, and strictly apply only to a zero of a few kgm.

The density of acetone at atmospheric pressure and 0° was assumed to be 0.8136. The constants of the dilatation formula were: $a = 0.0_2 1324$, $b = 0.0_5 380$, and $c = -0.0_8 88$. These constants are taken from data of Zahnder, who gives for the density at 0° 0.8125, and for the boiling point 56.3°, values slightly different from those given above.

The change of volume at 40° between 1 and 500 kgm. was taken as 0.0541 from Amagat. His value for the volume at 40° and 1 kgm., however, is 1.0575 against 1.0585 given by the formula above. The probable accuracy of the low pressure measurements of acetone may be judged from a comparison of the values at 20° with those of Amagat. For the successive pressure intervals 1–500, 500–1000, 1000–1500, and 1500–2000 atmos. the present work gave the following changes of volume; 0.0526, 0.0308, 0.0251, 0.0208, while Amagat gives 0.0483, 0.0325, 0.0245, and 0.0196.

The volume of acetone as a function of temperature and pressure is shown in Table X and in Figure 14.

The initial compressibility at 20°, 40°, and 60° may be computed from values of Amagat[43] at 14° and 99°, between 15 and 22 atmos.

TABLE X.

VOLUME OF ACETONE.

Pressure. kgm. cm.²	Volume.						
	20°.	30°.	40°.	50°.	60°.	70°.	80°.
1	1.0279	1.0426	1.0585	1.0752	1.0929		
500	0.9829	0.9931	1.0044	1.0165	1.0297		
1000	.9553	.9638	0.9728	0.9821	0.9924	1.0015	1.0107
1500	.9307	.9385	.9463	.9541	.9619	0.9694	0.9764
2000	.9100	.9173	.9243	.9309	.9374	.9436	.9497
2500	.8927	.8997	.9058	.9116	.9173	.9229	.9285
3000	.8775	.8841	.8897	.8948	.8999	.9051	.9105
3500	.8646	.8707	.8759	.8806	.8853	.8902	.8953
4000	.8532	.8586	.8636	.8681	.8725	.8771	.8819
4500	.8430	.8482	.8528	.8572	.8614	.8657	.8699
5000	.8334	.8380	.8425	.8469	.8510	.8549	.8586
6000	.8175	.8216	.8257	.8299	.8339	.8374	.8403
7000	.8028	.8064	.8103	.8145	.8182	.8215	.8243
8000	.7898	.7933	.7969	.8005	.8039	.8071	.8101
9000			.7847	.7879	.7910	.7942	.7974
10000			.7737	.7767	.7797	.7827	.7857
11000			.7634	.7667	.7697	.7725	.7750
12000			.7546	.7583	.7614	.7638	.7657

by a linear interpolation for temperature and by assuming the variation of compressibility with pressure found here. We obtain in this

[43] Amagat, l. c. (1877).

way from Amagat's data the following values: 0.0_3121, 0.0_3143, and 0.0_3194, respectively. The corresponding values from the present work to give the correct changes of volume are 0.0_3120, 0.0_3146, and 0.0_3167; rather good agreement except at 60°, where for one thing the linear interpolation from Amagat's data would be accountable for

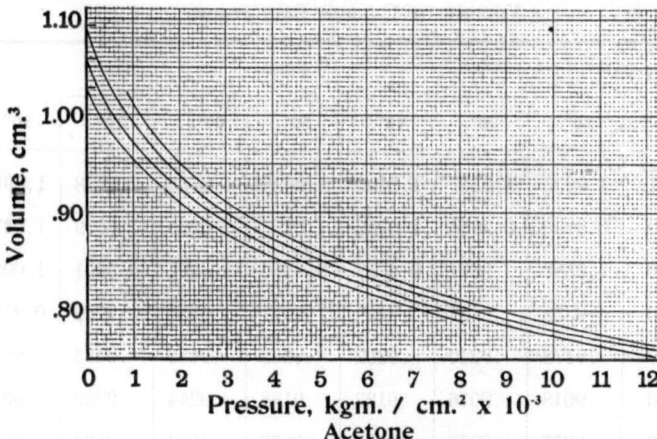

FIGURE 14. Acetone. Volume at 20°, 40°, 60°, and 80°, against pressure. The lower curve is for 20°. The curve for 20° is terminated at 8000 kgm. because acetone freezes at this point. The curve for 80° starts from 1000 kgm. because acetone boils at atmospheric pressure below 80°.

some of the divergence in the direction shown. The values taken as the average were; 0.0_3120, 0.0_3145, and 0.0_3167.

The only values we have for C_p are by von Reis, who gives 19.43 between 16.4° and 52.6°, 19.65 between 17.6° and 60.3°, 19.68 between 18.9° and 70.2°, and 19.72 between 18.7° and 79.1°. These data show an unusually slight increase of C_p with the temperature.

Carbon Bisulphide.— Three sets of measurements were made on this substance, the first of compressibility and dilatation at low pressures with the larger low pressure bulb, the second of compressibility and dilatation over the high pressure range with the smaller high pressure bulb, and the third of compressibility and dilatation with the smaller bulb over the entire pressure range. All of these runs were made without accident of any sort.

The average discrepancy of the two sets of piston displacements for the isothermal compressibility at 40° was 0.002 inch, on a total stroke of about 2.0 inches. The corresponding discrepancy for the thermal expansion due to a change of temperature of 20° was 0.0013 inch for a mean displacement of about 0.070 inch. As far as self consistency goes, the measurements on carbon bisulphide are among the best of the series.

TABLE XI.

VOLUME OF CARBON BISULPHIDE.

Pressure. kgm. cm.²	Volume.						
	20°.	30°.	40°.	50°.	60°.	70°.	80°.
1	1.0235	1.0357	1.0490	1.0630	1.0775	1.0928	1.0992
500	0.9865	0.9964	1.0063	1.0158	1.0256	1.0359	1.0473
1000	.9586	.9671	0.9752	0.9829	0.9907	0.9991	1.0083
1500	.9358	.9432	.9504	.9571	.9639	.9709	0.9787
2000	.9173	.9240	.9302	.9362	.9423	.9485	.9552
2500	.9018	.9076	.9133	.9188	.9244	.9299	.9357
3000	.8877	.8928	.8981	.9033	.9084	.9134	.9185
3500	.8756	.8801	.8849	.8897	.8946	.8991	.9035
4000	.8647	.8688	.8732	.8770	.8823	.8855	.8902
4500	.8548	.8586	.8627	.8672	.8714	.8752	.8786
5000	.8453	.8489	.8528	.8570	.8610	.8645	.8676
0000	.8205	.8329	.8367	.8406	.8442	.8472	.8501
7000	.8147	.8184	.8222	.8257	.8290	.8319	.8347
8000	.8022	.8061	.8100	.8131	.8162	.8191	.8220
9000	.7911	.7954	.7989	.8020	.8049	.8078	.8107
10000	.7805	.7844	.7879	.7910	.7940	.7969	.7997
11000	.7715	.7745	.7777	.7809	.7839	.7867	.7894
12000	.7638	.7658	.7682	.7710	.7743	.7772	.7795

The reduction factor from the mathematical formula was 0.9947, showing pretty nearly average compressibility.

The density at 0° and atmospheric pressure was assumed to be 1.292. The three constants of the dilatation formula were as follows: $a = 0.0_21140$, $b = 0.0_5137$, and $c = 0.0_7191$.[44]

Amagat did not measure the volume of CS_2 at 40° at less than 600 atmos. The change of volume between 1 and 500 kgm. was found,

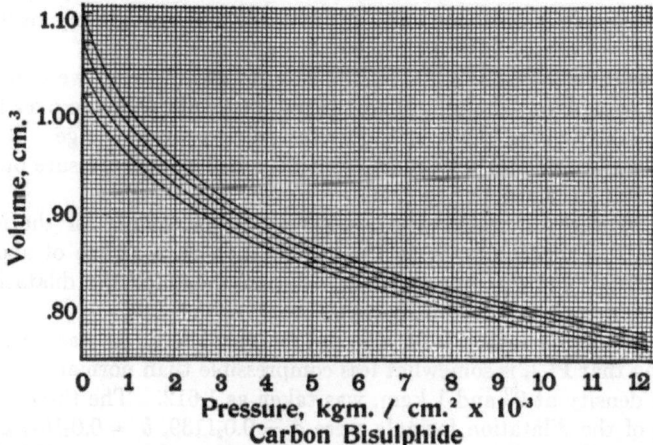

FIGURE 15. Carbon Bisulphide. Volume at 20°, 40°, 60°, and 80° plotted against pressure. The lower curve is for 20°.

therefore, by subtracting the value found here for 500–1000 kgm. from Amagat's value for the change 1–1000 kgm., giving the result 0.0427 between 1 and 500 kgm. Amagat's value for the volume at 40° and 1 kgm. is 1.0484 against 1.0490 given by the formula above. The present low pressure determinations gave results consistently higher than those of Amagat: These values were as follows: 0.0476 (not accurate) for 1–500 atmos., 0.0295 between 500 and 1000, 0.0227 between 1000 and 1500, 0.0198 between 1500 and 2000. For the same pressure intervals Amagat has 0.0387, 0.0277, 0.0222, and 0.0183.

The volume of carbon bisulphide as a function of pressure and temperature is shown in Table XI and in Figure 15.

The initial compressibilities at 20°, 40°, 60°, and 80° may be found from Amagat to have the values 0.0_490, 0.0_3107, 0.0_3128, and 0.0_3149 respectively. The corresponding values required to give the change of volume listed in the tables are 0.0_492, 0.0_3107, 0.0_3133, and 0.0_3150; good agreement. Röntgen also gives 0.0_487 at 20°. The values

---

[44] Pierre, l. c. (1845).

shown in the curves are the values to give the correct change of volume, except at 20°, where 0.0₄91 was adopted.

There are a number of measurements of C_p for CS_2. Regnault[45] gives 12.68 at $-30°$, 12.95 at 0°, 13.23 at 30°; Hirn[46] gives 13.13 at 30°; Sutherland [47] 14.33 at 80°, and 15.22 at 120°; and Forch [48] 13.34 at 18°. These results are more consistent than usual, lying on a smooth curve within about 1%. C_p increases with rising temperature, the rate of increase also increasing.

Phosphorus Trichloride.— Two sets of measurements were made on this substance; the first of compressibility and dilatation over the high pressure range, the second over the entire pressure range. Both sets of measurements were made with the smaller high pressure bulb. There was no accident.

The average discrepancy in the piston displacements for the isothermal compressibility at 40° was 0.006 inch on a stroke of about 2.0 inches. The discrepancy in the displacement for the dilatation was 0.0011 inch on 0.070 inch, mean.

The reduction factor from the mathematical formula was 0.9335, showing that PCl_3 is somewhat less compressible than normal.

The density at 0° and 1 kgm. was taken as 1.612. The three constants of the dilatation formula were $a = 0.0_21139$, $b = 0.0_5167$, and $c = 0.0_840$.[49] There are also values for the volume of PCl_3 by Pierre, who gives for 20°, 40°, and 60°: 1.0231, 1.0477, and 1.0747. The corresponding values computed with the above values for the constants are 1.0234, 1.0485, and 1.0752, rather better agreement than we have come to expect.

The change of volume at atmospheric pressure between 1 and 500 kgm. was taken from Amagat as 0.0445. Amagat gives for the atmospheric volume at 40°, 1.0483, in substantial agreement with 1.0485, given by the formula. At 20° and low pressures, the values found for the change of volume for successive intervals of 500 atmos. were, 0.0451, 0.0263, 0.0219, and 0.0187, against the values of Amagat; 0.0396, 0.0282, 0.0224, and 0.0186.

The volume of phosphorus trichloride as a function of pressure and temperature is given in Table XII and in Figure 16.

There seem to be no other determinations of the initial compressibility at atmospheric pressure. Accordingly, the values given in the

---

[45] Regnault, l. c. (1862).
[46] Hirn, Am. d. chim., **10**, 32 (1867).
[47] Sutherland, l. c.
[48] Forch, Ann. d. Phys., **12**, 202 (1903).
[49] Thorpe, Jour. Chem. oc., **63**, 273 (1893).

tables are the values computed in the manner described on page 29 to give the correct changes of volume.

TABLE XII.

VOLUME OF PHOSPHORUS TRICHLORIDE.

Pressure. kgm. cm.²	Volume.						
	20°.	30°.	40°.	50°.	60°.	70°.	80°.
1	1.0234	1.0358	1.0485	1.0616	1.0752	1.0893	1.1039
500	0.9862	0.9949	1.0040	1.0136	1.0238	1.0346	1.0459
1000	.9593	.9666	0.9739	0.9816	0.9896	0.9980	1.0065
1500	.9382	.9445	.9509	.9575	.9643	.9712	0.9783
2000	.9205	.9262	.9318	.9377	.9437	.9498	.9557
2500	.9057	.9107	.9159	.9212	.9268	.9322	.9375
3000	.8926	.8973	.9022	.9072	.9123	.9171	.9220
3500	.8809	.8853	.8899	.8946	.8994	.9038	.9082
4000	.8705	.8747	.8790	.8836	.8880	.8922	.8962
4500	.8611	.8652	.8693	.8736	.8779	.8819	.8851
5000	.8521	.8560	.8600	.8641	.8682	.8720	.8757
6000	.8375	.8411	.8448	.8486	.8524	.8561	.8596
7000	.8245	.8279	.8313	.8347	.8382	.8416	.8450
8000	.8133	.8165	.8196	.8228	.8260	.8292	.8323
9000	.8029	.8059	.8089	.8120	.8150	.8180	.8210
10000	.7929	.7959	.7989	.8020	.8050	.8080	.8109
11000	.7838	.7866	.7897	.7927	.7957	.7985	.8014
12000	.7761	.7789	.7818	.7847	.7875	.7902	.7928

For C_p there seems to be only one determination, due to Regnault,[50] who finds the mean value 13.67 between 10° and 15°.

[50] Regnault, l. c. (1843).

Ethyl Chloride.— Two sets of measurements were made on this substance, separated by thirty-six days in time, both with the smaller bulb, and both complete for compressibility and dilatation over the entire pressure range. Both runs were entirely without accident of any kind. The very low boiling point of this substance, 12.5°, and

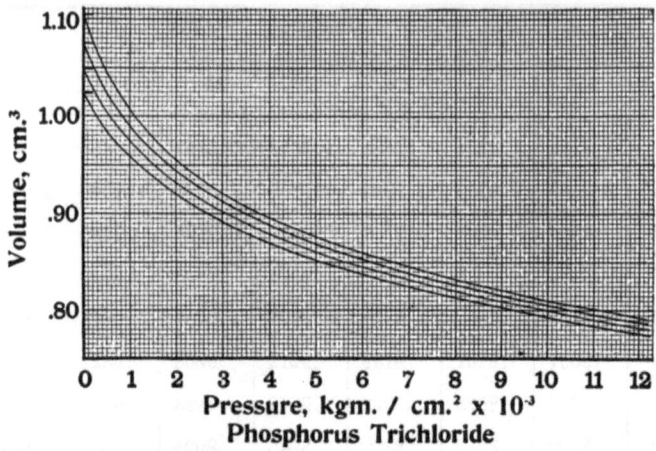

FIGURE 16. Phosphorus Trichloride. Volume at 20°, 40°, 60°, and 80° plotted against pressure. The lower curve is for 20°.

its abnormally high compressibility, made slight changes necessary in the details of the manipulation. The ethyl chloride was furnished by Kahlbaum, in sealed glass bulbs, which were accordingly exposed to an internal pressure greater than atmospheric. The steel compressibility bulb was filled after the ethyl chloride and the bulb had been brought to 0° in an ice bath. The steel bulb was then allowed to warm sufficiently to boil away a slight quantity of the ethyl chloride, when the capillary stem of the bulb was closed by forcing into it a small rubber stopper, considerably too large for it. The friction of the stopper was sufficient to hold it in place against the vapor pressure of the ethyl chloride at room temperature. The first application of a very moderate pressure by the pump was sufficient to drive the stopper into the bulb, where it remained during the rest of the measurements. In this way the filling could be accomplished without the troublesome necessity of cooling the large cylinder below 12° and maintaining it there until pressure could be applied. The small rubber stopper was weighed, and its weight applied as a correction to the weight of the bulb full of the chloride. The weight was about 0.03 gm.

The piston displacement for isothermal compressibility at 40° showed a mean discrepancy of 0.0016 inch on a total of about 2.1 inches. The discrepancy in the displacements for dilatation averaged 0.002 inch on a mean of about 0.070 inch. The larger discrepancies were at the lower pressures; the mean above 2000 kgm. was 0.001 inch, half as much.

It has already been mentioned that ethyl chloride is so abnormally compressible that it was not possible to use the same formula as for the other eleven liquids to smooth the changes of volume at 40° for the tables. A formula of the same type was used, but with different coefficients (see page 22). The best values of the coefficients for ethyl chloride were found to be, $\alpha = 0.06723$, $\beta = 0.17139$, $\gamma = 0.04030$, and $\delta = -0.06261$. The maximum differences between the observed and the calculated change of volume were $+0.0007$ at 3000 kgm. and -0.0024 at 9000 kgm. The four constants were determined so that the curve passed through the experimental points at 500, 2000, 5000, and 12000 kgm. It should perhaps be mentioned that this formula is merely an empirical expression for the change of volume over the pressure range of the experiment. It has no theoretical significance whatever, and should not be used for purposes of extrapolation. For instance, it is seen immediately that it predicts an impossible behavior at infinite pressure.

The density at 0° was taken as 0.9120. The liquid boils at atmospheric pressure for every temperature within the range of the table. The ordinary dilatation formula would have been valueless, therefore, to fix the volume at any one point of the table. The fiducial point was taken at 40° and 500 kgm. from the data of Amagat,[51] who gives 0.9951 for the volume. The fact that the liquid boils at all temperatures of the table at atmospheric pressure has necessitated starting from 500 or 1000 kgm. as the initial point from which most of the thermodynamic properties have been computed.

The initial compressibilities given in the diagrams for 20° and 40° were taken from Amagat by interpolation and extrapolation from 22 atmos. The values are $0.0_3 163$, and $0.0_3 211$; they correspond to pressures somewhat higher than atmospheric.

The volume of ethyl chloride as a function of pressure and temperature is given in Table XIII and in Figure 17.

---

[51] Amagat, l. c. (1877).

For C_p we have apparently only one value, again due to Regnault,[52] who gives 16.80 at $-28.4°$, a temperature beyond the range of this work.

TABLE XIII.

VOLUME OF ETHYL CHLORIDE.

Pressure. kgm. cm.²	Volume.						
	20°.	30°.	40°.	50°.	60°.	70°.	80°.
1							
500	0.9714	0.9831	0.9951	1.0075	1.0105	1.0339	1.0379
1000	.9276	0.9358	.9446	0.9544	0.9647	0.9741	0.9827
1500	.8988	.9059	.9132	.9211	.9296	.9373	.9444
2000	.8774	.8836	.8900	.8967	.9039	.9104	.9264
2500	.8596	.8652	.8709	.8768	.8831	.8888	.8938
3000	.8442	.8492	.8544	.8599	.8654	.8703	.8749
3500	.8311	.8358	.8405	.8456	.8506	.8551	.8591
4000	.8200	.8245	.8289	.8337	.8384	.8426	.8462
4500	.8087	.8129	.8172	.8217	.8262	.8301	.8335
5000	.7994	.8035	.8076	.8118	.8161	.8199	.8230
6000	.7821	.7860	.7900	.7938	.7976	.8010	.8040
7000	.7680	.7718	.7756	.7791	.7825	.7856	.7887
8000	.7561	.7597	.7633	.7666	.7699	.7730	.7762
9000	.7454	.7490	.7522	.7553	.7581	.7611	.7644
10000	.7352	.7385	.7415	.7444	.7473	.7502	.7533
11000	.7259	.7288	.7317	.7347	.7376	.7405	.7432
12000	.7176	.7199	.7225	.7254	.7286	.7314	.7336

Ethyl Bromide.— Two sets of measurements were made on this, both being with the smaller bulb and over the entire pressure range.

[52] Regnault, l. c. (1862).

An interval of forty days separated the two sets. Both were completed without accident.

The average discrepancy of the piston displacements for isothermal compressibility at 40° was 0.0025 inch on a stroke of 2.05 inches. The mean discrepancy in the displacements for dilatation was 0.0013 inch on a mean of about 0.070 inch.

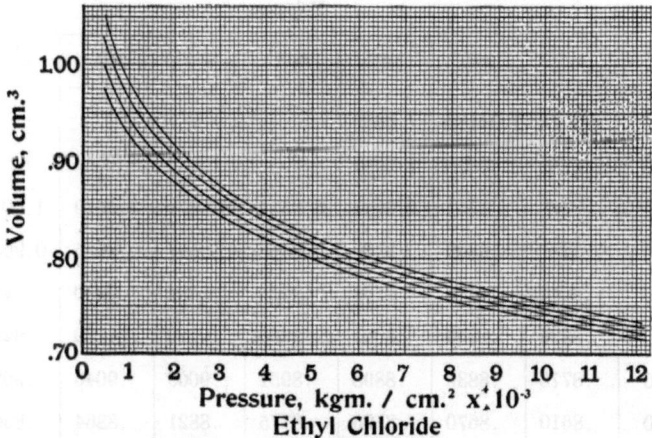

FIGURE 17. Ethyl Chloride. Volume at 20°, 40°, 60°, and 80° plotted against pressure. The lower curve is for 20°. The boiling point at atmospheric pressure is at 12°.5, so that it was necessary to take the origin of pressure as 500 kgm.

The reduction factor for passing from the mathematical formula was 1.032, indicating a compressibility somewhat more than normal.

The density at 0° and atmospheric pressure was assumed to be 1.483. The three constants of the dilatation formula were taken as $a_2 = 0.012275$, $b = 0.0_64437$, and $c = 0.0_70258$.[53] The boiling point of ethyl bromide is 38.4°. The formula gives for the volume at 20° 1.0249, and for the extrapolated value at 40°, 1.0515. Pierre also gives the volumes at 20° and 40°, 1.0275 and 1.0578 respectively. The discrepancies are large, 0.6% at 40°. In this case the preference has been given to the values of Pierre against those of Landolt and Börnstein, because Pierre actually measured the volumes at the temperatures in question, where as the formula of Landolt and Börnstein is directly applicable only at lower temperatures. Furthermore,

---

[53] Pierre, l. c. (1845).

Amagat's value at 40°, 1.0583, agrees much more closely with Pierre's than with that given by the formula.

TABLE XIV.

VOLUME OF ETHYL BROMIDE.

Pressure. kgm./cm.²	Volume.						
	20°.	30°.	40°.	50°.	60°.	70°.	80°.
1	1.0275	1.0418(?)	1.0578				
500	0.9788	0.9890(?)	1.0004				
1000	.9478	.9557	0.9644	0.9534	0.9824	0.9919	1.0018
1500	.9237	.9309	.9380	.9448	.9517	.9585	0.9654
2000	.9044	.9110	.9175	.9235	.9294	.9350	.9407
2500	.8885	.8950	.9011	.9066	.9120	.9170	.9218
3000	.8776	.8839	.8898	.8951	.9000	.9046	.9090
3500	.8610	.8670	.8725	.8775	.8821	.8864	.8904
4000	.8505	.8556	.8606	.8652	.8696	.8735	.8772
4500	.8410	.8455	.8500	.8543	.8585	.8622	.8657
5000	.8317	.8358	.8399	.8439	.8478	.8514	.8546
6000	.8163	.8201	.8237	.8273	.8307	.8340	.8371
7000	.8020	.8056	.8092	.8125	.8156	.8187	.8220
8000	.7900	.7935	.7968	.7999	.8028	.8059	.8091
9000	.7787	.7821	.7852	.7881	.7911	.7939	.7968
10000	.7686	.7717	.7747	.7777	.7807	.7834	.7858
11000	.7598	.7623	.7653	.7684	.7715	.7741	.7762
12000	.7521	.7546	.7572	.7601	.7633	.7659	.7677

The change of volume at 40° between 1 and 500 kgm. was taken as 0.0573 from Amagat. The values found here at 20° for successive intervals of 500 kgm., beginning at 1 kgm. were 0.0536, 0.0320, 0.0255, and 0.0202, against 0.0492, 0.0322, 0.0248, and 0.0199 of Amagat. The agreement is rather good, except at the lowest pressure, where agreement is not to be expected. The just published value of Richards for the change of volume at 500 kgm. and 20° is 0.0446, against 0.0487 given in the table, which is essentially that of Amagat.

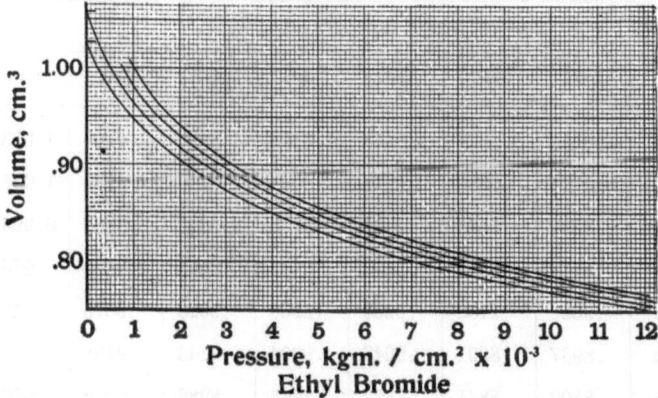

FIGURE 18. Ethyl Bromide. Volume at 20°, 40°, 60°, and 80° plotted against pressure. The lower curve is for 20°. The curves for 60° and 80° start from 1000 kgm. because the boiling point at atmospheric pressure is below 60°.

The volume of ethyl bromide as a function of pressure and temperature is given in Table XIV and in Figure 18.

When the computations of this paper were made the only value for the compressibility at low pressures was that of Amagat at 99° and a mean pressure of 20 atmos. This was too far removed from the range of this paper to justify any correction. The initial compressibilities at 20° and 40° were taken, therefore, so as to give the values of the changes of volume listed in the tables; 0.0_31248 and 0.0_31476 respectively. At 60° and 80° the compressibility is not given for pressures lower than 1000 kgm. The recently published data of Richards give for the compressibility at atmospheric pressure and 20° the value 0.0_3106, considerably lower than the value given above.

Regnault[54] gives a few values for C_p at atmospheric pressure; 14.62 between 5° and 10°, 14.42 between 10° and 15°, 14.54 between 15° and 20°. The temperature range is too small and the variations

---

[54] Regnault, l. c. (1843).

too great to enable us to decide whether C_p really increases with temperature or not.

TABLE XV.
VOLUME OF ETHYL IODIDE.

Pressure. kgm. cm.²	Volume.						
	20°.	30°.	40°.	50°.	60°.	70°.	80°.
1	1.0214	1.0324	1.0438	1.0555	1.0677	1.0803	1.0935
500	0.9785	0.9880	0.9979	1.0081	1.0180	1.0276	1.0366
1000	.9502	.9584	.9665	0.9746	0.9825	0.9900	0.9969
1500	.9277	.9345	.9412	.9479	.9544	.9605	.9663
2000	.9092	.9150	.9209	.9266	.9323	.9375	.9425
2500	.8937	.8991	.9043	.9094	.9143	.9188	.9231
3000	.8802	.8851	.8899	.8945	.8988	.9028	.9065
3500	.8684	.8728	.8770	.8811	.8848	.8883	.8917
4000	.8583	.8621	.8659	.8694	.8728	.8759	.8790
4500	.8487	.8522	.8558	.8592	.8624	.8653	.8681
5000	.8394	.8429	.8463	.8496	.8529	.8557	.8581
6000	.8236	.8271	.8306	.8340	.8371	.8398	.8418
7000	.8093	.8129	.8164	.8193	.8220	.8243	.8264
8000	.7968	.8006	.8038	.8065	.8090	.8113	.8134
9000	.7856	.7902	.7922	.7945	.7967	.7989	.8013
10000	.7755	.7789	.7817	.7841	.7862	.7885	.7909
11000	.7665	.7694	.7722	.7747	.7771	.7794	.7817
12000	.7588	.7611	.7638	.7667	.7693	.7717	.7737

Ethyl Iodide.— Two runs were made on this with the smaller high pressure bulb over the entire pressure range, without accident. The mean variation of the displacement readings for compressibility at 40° was 0.0020 inch on a total of 2.05 inches. The agreement is quite perceptibly better than the average. The thermal dilatation measurements show however, by far greater disagreement than any other of the twelve liquids. 0.0030 inch on a mean of 0.070 inch. The discrepancy was greater at the higher temperatures;

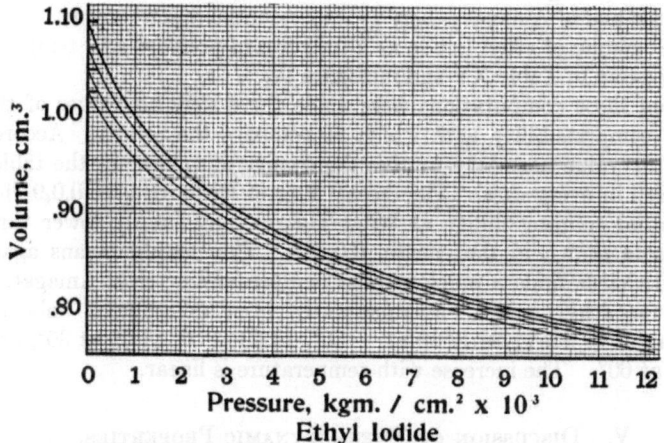

FIGURE 19. Ethyl Iodide. Volume at 20°, 40°, 60°, and 80° plotted against pressure. The lower curve is for 20°.

0.0012 inch from 20° to 40°, 0.0026 inch from 40° to 60°, and 0.0052 inch from 60° to 80°.

The reduction factor from the mathematical formula was 0.9817, showing slightly less than normal compressibility.

The density at 0° was assumed to be 1.973. The three constants of the dilatation formula were as follows: $a = 0.0_21054$, $b = 0.0_6636$, and $c = 0.0_71004$.[55] Pierre also gives values for the volume at 20°, 40°, and 60° respectively; 1.0232, 1.0484, 1.0749, against 1.0214, 1.0438, and 1.0677 given by the formula. The agreement should be better. Probably Pierre's values are better, as is shown by the agreement of Amagat's value at 40°, but Pierre does not give the volume at 80°; so the value of the formula was accordingly selected.

The change of volume between 1 and 500 kgm. was taken from Amagat as 0.0459. Amagat gives for the initial volume at 40° 1.0486 against 1.0438 of the formula. The low pressure determina-

---

[55] Dobriner, Lieb. Ann., **243**, 1–23 (1888).

tions of the change of volume at 20° gave for successive intervals of 500 atmos. 0.0509, 0.0278, 0.0233, and 0.0195 respectively, against 0.0404, 0.0289, 0.0227, and 0.0184 of Amagat. The agreement is fair, except of course for the first interval. The recent work of Richards gives 0.0370 for the change of volume at 20° and 500 kgm. against 0.0429 given in the tables. It is evident that the work of Richards and of Amagat is here in very essential disagreement.

The volume of ethyl iodide as a function of pressure and temperature is given in Table XV and in Figure 19.

When these computations were made there were no values of the initial compressibility with which to compare the results. Accordingly the value necessary to give the change of volume in the tables was used in every case. The recent data of Richards give 0.0_491 for the initial compressibility at 20°. This is considerably lower than the value shown in the curves, 0.0_4108. This simply means again that Richards finds a much smaller compressibility than Amagat.

C_p for ethyl iodide has apparently been determined only by Regnault[56]. He gives 13.19 at $-30°$, 13.60 at 0°, 14.03 at 30°, and 14.44 at 60°. The increase with temperature is linear.

V. Discussion of Thermodynamic Properties.

In the following sections the general characteristics of the several thermodynamic functions will be discussed. The discussion will include suggestions as to what modifications it may possibly be necessary to make in our conceptions of a liquid, or what features that we have neglected at low pressures it may be necessary to emphasize at high pressures. Incidentally in the course of the discussion, suggestions will be made bearing on the theory of liquids, but any detailed examination of the problems that confront us in trying to frame a theory of liquids valid for high pressures will be reserved for section VI.

Volume.— The tables and diagrams of volume as a function of pressure and temperature have already been given, but with little comment.

One of the significant facts about the change of volume is in regard to the volume at infinite pressures, that is, the so-called volume of the molecules themselves, which is one of the quantities entering into nearly every theory of liquids. In particular, Tumlirz[57] and

---

[56] Regnault, l. c. (1862).
[57] Tumlirz, Sitz. k. Akd. Wiss. Wien, **118**, 1–39 (1909).

Tammann[58], in their recent theories, give values for the volume at infinite pressure. The values are listed in Table XVI for four of the liquids here investigated, and compared with the volumes found experimentally at 12000 kgm. and 20°. The observed value for ether at 12000 kgm. is actually less than the value predicted by either of

TABLE XVI.

Substance.	Volume		
	Calculated, $p = \infty$		Observed. $p = 12000$.
	Tumlirz.	Tammann.	$t = 20°$
Methyl Alcohol	0.6970	0.7255	0.7559
Ethyl Alcohol	0.7037	0.7380	0.7521
Ether	0.7274	0.7246	0.7216
Carbon Bisulphide	0.6881	0.7246	0.7638

these theories for an infinite pressure; and for the other liquids the observed value is close to the predicted minimum. This result serves to emphasize more strikingly a point made in the preceding paper on water; namely, that at high pressures a liquid is more compressible than we might expect from its behavior at low pressures.

One line of inquiry is worth mentioning which seemed promising before the experiments were performed. The question, suggested by such properties of the atom as the atomic re-fraction, was this; is it possible at the higher pressures to assign to each atom its own specific volume as a function of the pressure, and so compute the volume of a compound at any pressure from its chemical constitution? But an examination of the changes of volumes of the two isomers, ether and isobutyl alcohol, shows that the supposed relation does not hold. For if we compare the volumes of equal weights, that is the volume occupied by the same number of atoms, we shall find that at atmospheric pressure the ratio of the volume of ether to that of isobutyl alcohol is 1.102, and that at 12000 kgm. it has dropped to 1.038. If the above relation were true, this ratio would be unity.

---

[58] Tammann, l. c., see also Korber.

The fact that the ratio is approaching unity shows that the atoms are approaching the behavior suggested above, but if they ever reach it, it can only be at pressures considerably beyond those reached here.

In Table XVII are given the average volumes of the twelve liquids between 20° and 80°. This table corresponds to the diagrams for the average between 20° and 80° of the other thermodynamic properties; it will prove useful in plotting any of the average properties against volume, which may in some cases give more significant results than when pressure is used as the independent variable, as here.

Thermal Expansion. — The mechanism ordinarily assumed in explanation of thermal expansion is as follows. Any liquid is continually striving to expand, because of the thermal agitation of its molecules. The tendency to expand is resisted by two forces, the external pressure, and the forces of attraction between the molecules. An increase in temperature means an increase in the expanding force, which results in an increase of volume. This increase of volume would be expected to be greater if the force preventing expansion were less. Now the force preventing expansion becomes less as the volume becomes greater, because the cohesional forces decrease as the volume, or the distance apart of the molecules, becomes greater. The result is that the thermal dilatation increases with increasing temperature, that is with increasing volume. In other words, $\left(\frac{\partial^2 v}{\partial \tau^2}\right)_p$ is positive. Furthermore, as pressure increases, the force resisting expansion increases because of the decreased distance apart of the molecules, so that we are to expect a decreased dilatation at the higher pressures.

An examination of the curves for dilatation against pressure shows that these expectations are much more nearly fulfilled as regards the behavior of the dilatation with respect to pressure than with respect to temperature.

The general tendency of the dilatation of the separate liquids (Folder I, Figures 20 to 31) is to decrease with rising pressure. The decrease is very much more rapid at the lower than at the higher pressures. But beyond this general fact the curves give only an impression of bewildering complexity, crossing and recrossing in apparent disorder at the higher pressures. It is possible to find many instances where the dilatation increases with rising pressure over a range of several thousand kilograms, ultimately, however, to decrease again. One of the most striking examples of this is the 20° curve for carbon bisulphide; other well marked examples are afforded by ace-

TABLE XVII.

Average Volume between 20° and 80°.

Pressure. kgm./cm.²	Methyl Alcohol.	Ethyl Alcohol.	Propyl Alcohol.	Isobutyl Alcohol.	Amyl Alcohol.	Ether.	Acetone.	Carbon Bisulphide.	Phosphorus Trichloride.	Ethyl Chloride.	Ethyl Bromide.	Ethyl Iodide.
1	1.0621	1.0573	1.0519	1.0538	1.0498	1.085†	1.075†	1.0664	1.0637		1.0755†	1.0575
500	1.0119	1.0064	1.0050	1.0014	1.0052	1.0034	1.0165	1.0169	1.0161	1.0097	1.0120†	1.0076
1000	0.9776	0.9725	0.9716	0.9712	0.9731	0.9635	0.9830	0.9835	0.9329	0.9552	0.9748	0.9736
1500	.9486	.9453	.9477	.9449	.9496	.9305	.9536	.9573	.9583	.9216	.9446	.9470
2000	.9271	.9244	.9295	.9254	.9335	.9047	.9299	.9363	.9381	.8969	.9226	.9259
2500	.9095	.9069	.9147	.9096	.9146	.8841	.9106	.9188	.9216	.8767	.9052	.9084
3000	.8943	.8920	.9021	.8951	.9014	.8671	.8943	.9031	.9073	.8596	.8933	.8934
3500	.8803	.8788	.8910	.8826	.8895	.8525	.8800	.8896	.8946	.8451	.8757	.8801
4000	.8682	.8666	.8810	.8716	.8793	.8395	.8676	.8775	.8834	.8331	.8639	.8687
4500	.8572	.8555	.8718	.8617	.8700	.8281	.8565	.8667	.8734	.8211	.8534	.8584
5000	.8471	.8455	.8631	.8514	.8611	.8178	.8460	.8565	.8639	.8112	.8432	.8488
6000	.8302	.8282	.8485	.8366	.8467	.8014	.8289	.8398	.8486	.7931	.8267	.8327
7000	.8157	.8134	.8354	.8224	.8335	.7863	.8136	.8247	.8348	.7784	.8120	.8179
8000	.8035	.8006	.8246	.8119	.8226	.7729	.8000	.8121	.8228	.7662	.7996	.8051
9000	.7920	.7890	.8150	.8016	.8117	.7606	.7911°	.8009	.8120	.7549	.7878	.7935
10000	.7815	.7783	.8063	.7920	.8020	.7496	.7797°	.7901	.8019	.7443	.7772	.7832
11000	.7723	.7695	.7985	.7828	.7931	.7391	.7692°	.7805	.7926	.7346	.7680	.7741
12000	.7668	.7602	.7911	.7745	.7854	.7291	.7602°	.7717	.7845	.7256	.7599	.7663

† Extrapolated to 50°.   ° Average between 40° and 80°.

Folder I. The thermal expansion, $\left(\dfrac{\partial v}{\partial \tau}\right)_p$, against pressure. In Figures 20 to 31 the thermal expansion for the separate liquids at 20°, 40°, 60°, and 80° is shown against pressure. In Figure 32 the average expansion between 20° and 80° is plotted against pressure for all twelve liquids in a single diagram. In the curves for the separate liquids, the curves cross and recross so many times that it would be confusing to try to determine from the diagrams alone what curve belongs to each temperature. The following detailed description is given, therefore, of the order of the curves at intervals of 2000 kgm., reading from below up.

FIGURE 20. Methyl Alcohol. The order of the curves, reading from below up, is as follows; at 1 kgm., 20°–40°–60°–80°; at 2000 kgm., 40°–60°–80°–20°; at 4000 kgm., 80°–60°–40°–20°; at 6000 kgm., 80°–60°–40°–20°; at 8000 kgm., 80°–60°–40°–20°; at 10000 kgm., 60°–40°–20°–80°; and at 12000 kgm., 80°–20°–40°–60°.

FIGURE 21. Ethyl Alcohol. The order of the curves, reading from below up, is as follows. At 1 kgm., 20°–40°–60°–80°; at 2000 kgm., 20°–40°–60°–80°; at 4000 kgm., 80°–60°–40°–20°; at 6000 kgm., 80°–60°–40°–20°; at 8000 kgm., 60°–80°–40°–20°; at 10000 kgm., 60°–40°–80°–20°; and at 12000 kgm., 20°–80°–40°–60°.

FIGURE 22. Propyl Alcohol. The order of the curves, reading from below up, is as follows; at 1 kgm., 20°–40°–60°–80°; at 2000 kgm., 20°–40°–60°–80°; at 4000 kgm., 80°–60°–40°–20°; at 6000 kgm., 80°–60°–40°–20°; at 8000 kgm., 80°–60°–40°–20°; at 10000 kgm., 40°–60°–20°–80°; and at 12000 kgm., 40°–20°–60°–80°.

FIGURE 23. Isobutyl Alcohol. The order of the curves, reading from below up, is as follows. At 1 kgm., 20°–40°–60°–80°; at 2000 kgm., 60°–40°–20°–80°; at 4000 kgm., 80°–20°–40°–60°; at 6000 kgm., 80°–60°–20°–40°; at 8000 kgm., 80°–60°–40°–20°; at 10000 kgm., 80°–60°–40°–20°; and at 12000 kgm., 80°–60°–40°–20°.

FIGURE 24. Amyl Alcohol. The order of the curves, reading from below up, is as follows. At 1 kgm., 20°–40°–60°–80°; at 2000 kgm., 60°–40°–80°–20°; at 4000 kgm., 60°–80°–40°–20°; at 6000 kgm., 60°–40°–80°–20°; at 8000 kgm., 60°–40°–20°–80°; at 10000 kgm., 20°–40°–60°–80°; and at 12000 kgm., 80°–20°–40°–60°.

FIGURE 25. Ether. The order of the curves, reading from below up is as follows. At 1 kgm., 20°; at 2000 kgm., 20°–40°–60°–80°; at 4000 kgm., 80°–60°–40°–20°; at 6000 kgm., 80°–60°–40°–20°; at 8000 kgm., 60°–80°–40°–20°; at 10000 kgm., 20°–40°–60°–80°; and at 12000 kgm., 20°–80°–40°–60°.

FIGURE 26. Acetone. The order of the curves, reading from below up, is as follows. At 1 kgm., 20°–40°–60°; at 2000 kgm., 80°–60°–40°–20°; at 4000 kgm., 60°–80°–40°–20°; at 6000 kgm., 80°–60°–20°–40°; at 8000 kgm., 80°–60°–20°–40°; at 10000 kgm., all equal; and at 12000 kgm., 80°–60°–40°.

FIGURE 27. Carbon Bisulphide. The order of the curves, reading from below up, is as follows. At 1 kgm., 20°–40°–60°–80°; at 2000 kgm., 60°–40°–20°–80°; at 4000 kgm., 80°–20°–40°–60°; at 6000 kgm., 80°–20°–60°–40°; at 8000 kgm., 80°–60°–40°–20°; at 10000 kgm., 80°–60°–40°–20°; and at 12000 kgm., 20°–80°–40°–60°.

FIGURE 28. Phosphorus Trichloride. The order of the curves, reading from below up, is as follows. At 1 kgm., 20°–40°–60°–80°; at 2000 kgm., 20°–40°–80°–60°; at 4000 kgm., 80°–20°–60°–40°; at 6000 kgm., 80°–20°–40°–60°; at 8000 kgm., all equal, at 10000 kgm., 80°–[20° and 60°]–40°; and at 12000 kgm., 80°–[20° and 60°]–40°.

FIGURE 29. Ethyl Chloride. The order of the curves, reading from below up, is as follows. At 1 kgm., 20°; at 2000 kgm., 20°–40°–60°; at 4000 kgm., 80°–[20° and 60°]–40°; at 6000 kgm., 80°–60°–40°–20°; at 8000 kgm., 80°–60°–40°–20°; at 10000 kgm., 60°–40°–80°–20°; and at 12000 kgm., 80°–20°–40°–60°.

FIGURE 30. Ethyl Bromide. The order of the curves, reading from below up, is as follows. At 1 kgm., 20°; at 2000 kgm., 80°–60°–40°–20°; at 4000 kgm., 80°–60°–40°–20°; at 6000 kgm., 80°–60°–40°–20°; at 8000 kgm., 60°–40°–80°–20°; at 10000 kgm., 80°–60°–40°–20°; and at 12000 kgm., 80°–20°–40°–60°.

FIGURE 31. Ethyl Iodide. The order of the curves, reading from below up, is as follows. At 1 kgm., 20°–40°–60°–80°; at 2000 kgm., 80°–60°–40°–20°; at 4000 kgm., 80°–60°–40°–20°; at 6000 kgm., 80°–60°–40°–20°; at 8000 kgm., 80°–60°–40°–20°; at 10000 kgm., 60°–80°–40°–20°; and at 12000 kgm., 80°–20°–60°–40°.

FIGURE 32. The average dilatation between 20° and 80° for all twelve liquids. The numbers on the curves indicate the liquids in the same order as the diagrams for the separate liquids. In order to prevent overlapping the origin of each curve has been displaced one square with respect to the one next to it. The origin is so situated that the dilatation for all the liquids at 12000 kgm. is between 0.0002 and 0.0003. The scale is shown at the right hand side. As an example of the use of the diagrams, the initial dilatation of amyl alcohol is 0.001056.

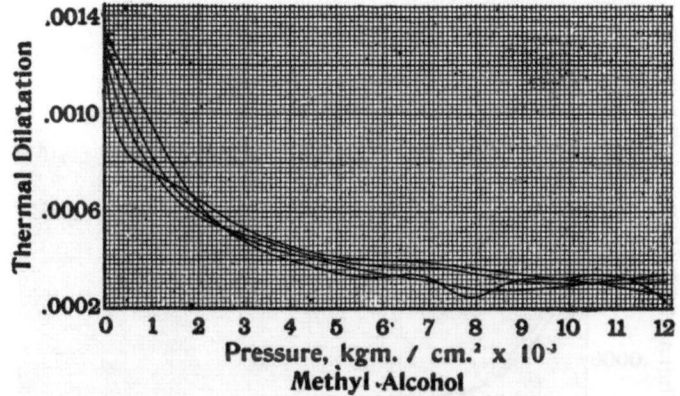

FIGURE 20.

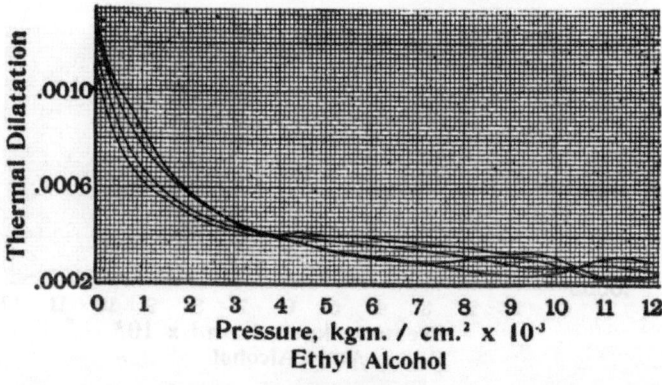

FIGURE 21.

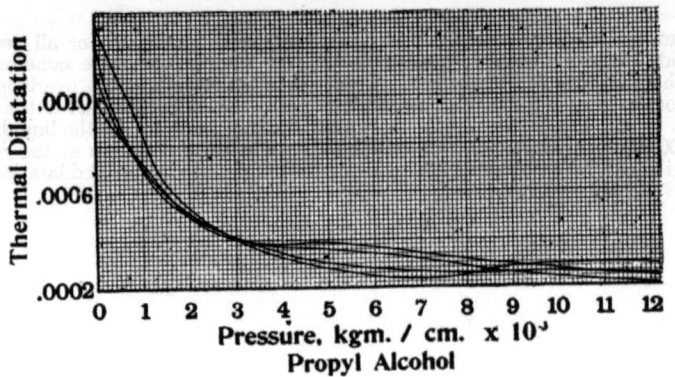

FIGURE 22.

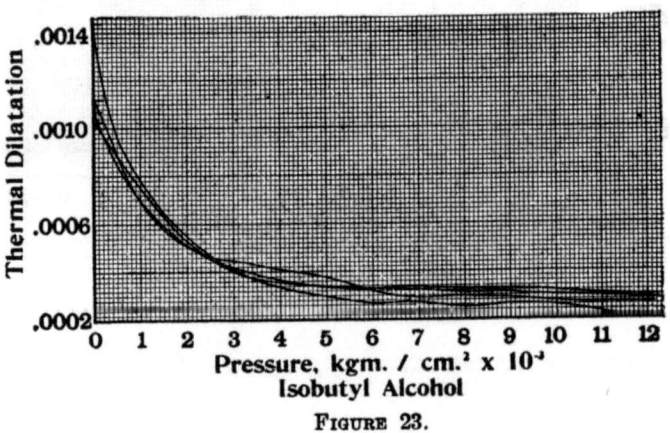

FIGURE 23.

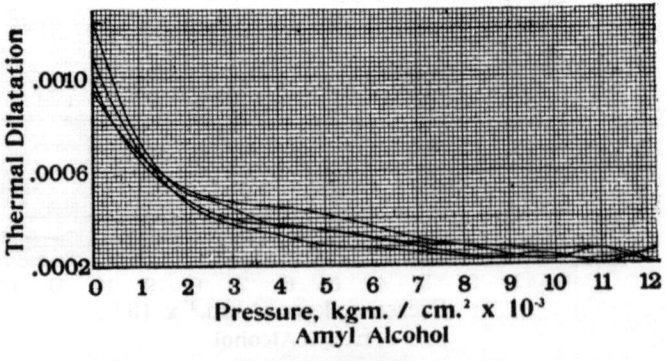

FIGURE 24.

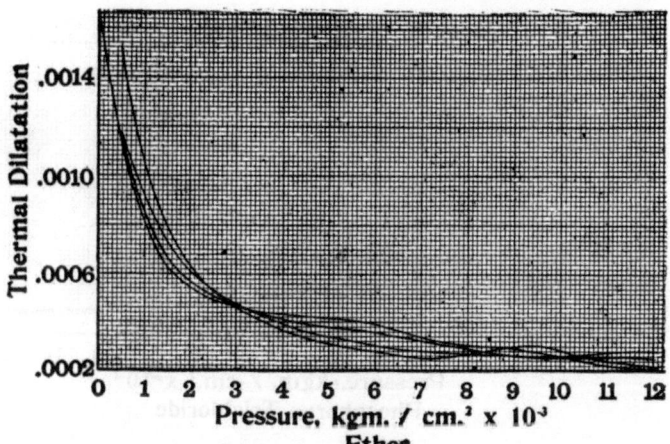

FIGURE 25. Ether

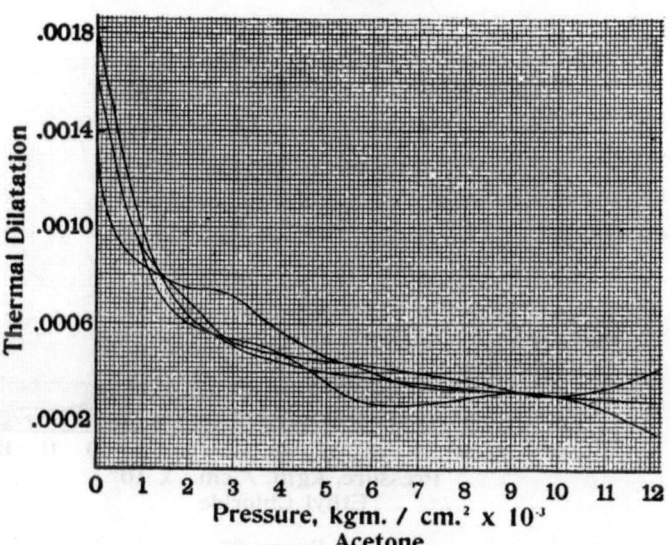

FIGURE 26. Acetone

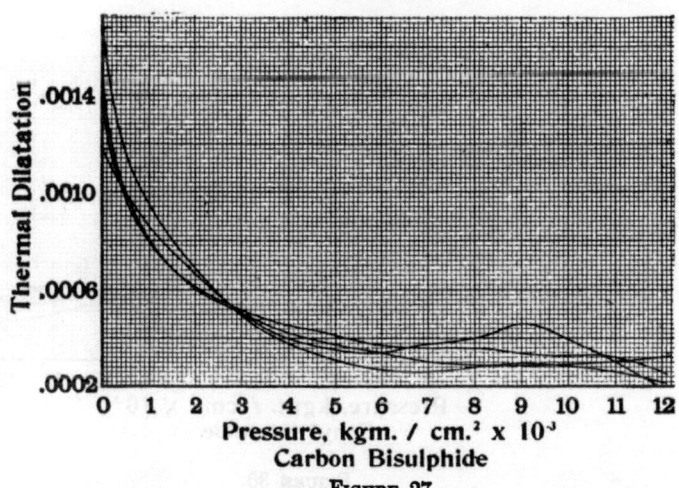

FIGURE 27. Carbon Bisulphide

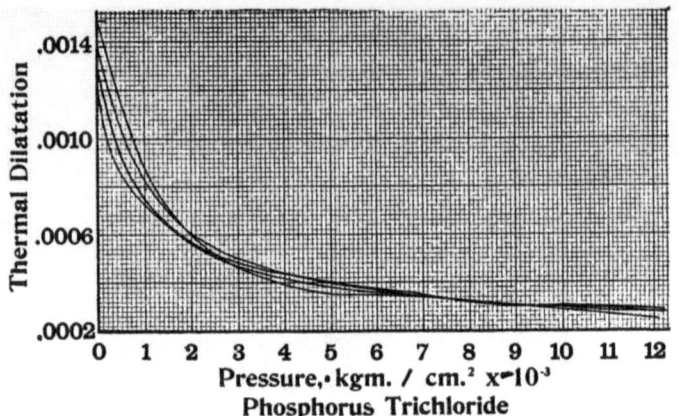

FIGURE 28.

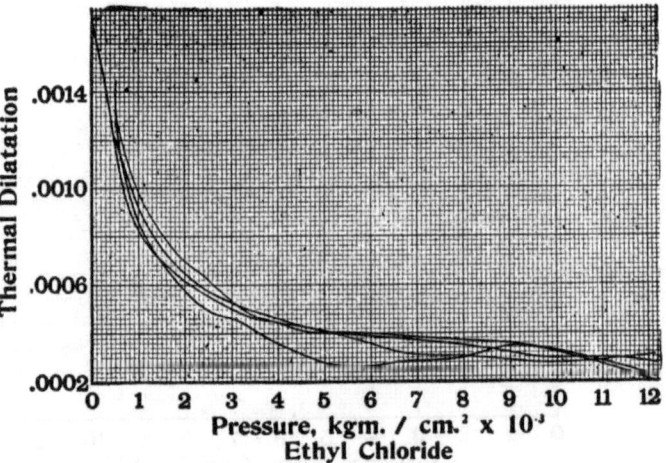

FIGURE 29.

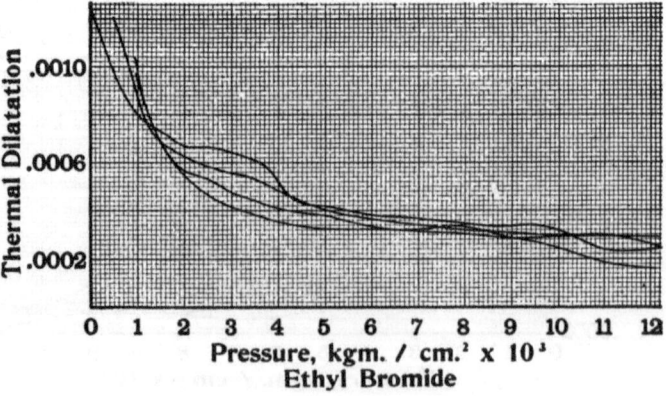

FIGURE 30.

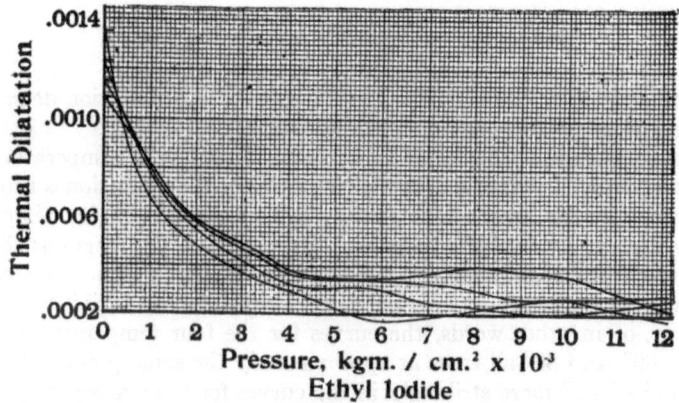

FIGURE 31.

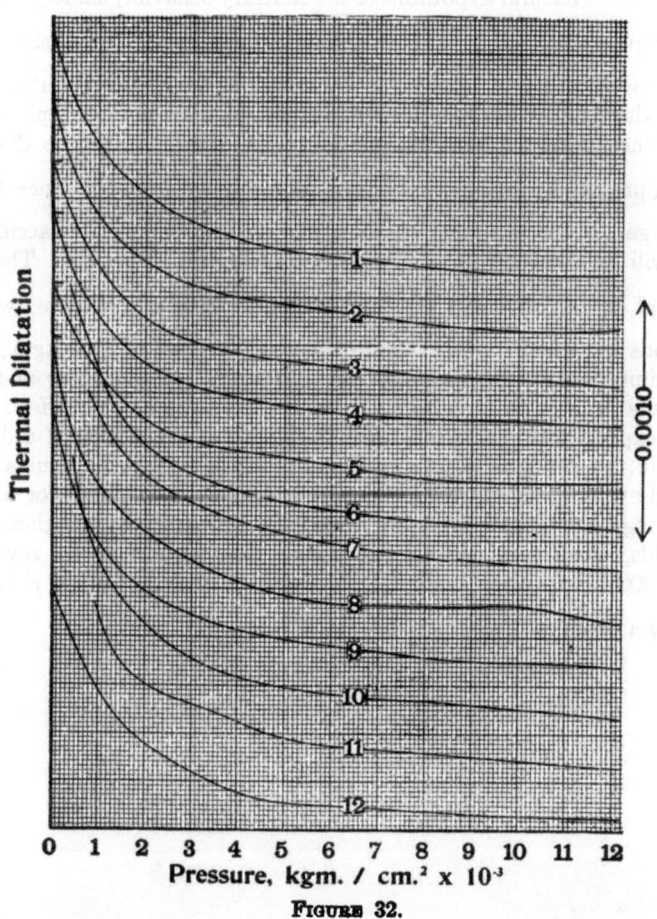

FIGURE 32.

tone and ethyl iodide. The rule, therefore, that dilatation decreases with rising pressure has many exceptions.

As regards the behavior of dilatation with respect to temperature, it is a striking fact that at the higher pressures the dilatation is usually greatest at the lower temperatures, instead of at the higher temperatures, as at atmospheric pressure. Every one of the diagrams shows this. What is more, the reversal of the effect in almost all cases takes place sharply at a definite pressure, the same for all temperatures; or in other words, the curves for the four temperatures, 20°, 40°, 60°, and 80° all cross at approximately the same point. This is exhibited still more strikingly in the curves for the average C_p of the twelve liquids (Figure 99). We have the thermodynamic formula $\left(\frac{\partial C_p}{\partial p}\right)_\tau = -\tau\left(\frac{\partial^2 v}{\partial \tau^2}\right)_p$, so that when the average value of $\left(\frac{\partial^2 v}{\partial \tau^2}\right)_p$ over the temperature range vanishes C_p will have a maximum. All of the curves show this maximum at the same pressure.

This universal reversal in the sign of $\left(\frac{\partial^2 v}{\partial \tau^2}\right)_p$ is a fact of no little interest and importance, and seems not to have been anticipated. In fact, the natural hypothesis of the contrary behavior, namely that at high pressures $\left(\frac{\partial^2 v}{\partial \tau^2}\right)_p = 0$, has recently been made the basis of an empirical theory of liquids by Tammann as was pointed out in the introduction. This hypothesis of Tammann is based on very plausible evidence from the data of Amagat, which seem to indicate that at high pressures $\left(\frac{\partial^2 v}{\partial \tau^2}\right)_p$ does vanish. But this apparent evidence from Amagat is founded on an accident, and a rather remarkable accident, as will be evident from an inspection of Figure 99 for C_p. The reversal in the sign of $\left(\frac{\partial^2 v}{\partial \tau^2}\right)_p$ takes place for nearly all the twelve liquids at pressures which are in the neighborhood of 3000 kgm., the maximum pressure reached by Amagat. As a matter of fact, Amagat's data do show in some cases the reversal of the effect, but the experimental error was fairly high, and Amagat himself did not credit the reversal as genuine. Only six of Amagat's liquids can yield evidence on this point, because they are the only ones for which readings were made at more then two temperatures; of these six liquids, ether, ethyl alcohol and carbon bisulphide show the reversal at 3000 kgm., while methyl and propyl alcohol and ethyl chloride show a positive $\left(\frac{\partial^2 v}{\partial \tau^2}\right)_p$ over the entire range.

At higher pressures there is in many cases, though not in all, a tendency for the effect to reverse again, that is, for the dilatation to again become greater at the higher temperature. The pressure of this second reversal is in the vicinity of 9000 or 10000 kgm. This fact also is indicated by the C_p curves.

A comparison of the curves for the different liquids shows a few features of interest. In the group of the five alcohols, isobutyl stands out as being the simplest, there being none of the crossing and recrossing which the others exhibit at the higher pressures. In this respect the four normal alcohols are much alike in their high pressure complications. That isobutyl alcohol should be different from the other alcohols was not anticipated before these experiments were made, since it seemed probable that at high pressure the effect of structural differences in the molecule would be eliminated. The effect of structural difference is also shown by a comparison of isobutyl alcohol and ether, since these two have the same formula, $C_4H_{10}O$. The high pressure effects are more complicated for ether. Acetone is peculiar in the wide divergence of the curves at the maximum pressure. It does not show any unusual effects in the neighborhood of the freezing point. It will be seen later that the specific heats are the quantities most susceptible to irregularities at the freezing point. Carbon bisulphide is remarkable for the curve at 20°, which shows a large increase of dilatation between 6000 and 9000 kgm., and also in this region shows a much greater dilatation than the curves for the higher temperatures. Phosphorus trichloride is the only one of the twelve liquids which behaves approximately as had been expected, since it shows little irregularity at the high pressures, and $\left(\frac{\partial^2 v}{\partial \tau^2}\right)_p$ nearly vanishes. The curves for the three ethyl halogen compounds do not show any particular progressive change of character such as one might expect, except with regard to the pressure at which $\left(\frac{\partial^2 v}{\partial \tau^2}\right)_p$ reverses in sign. This is shown best on the diagram for C_p. The pressure for reversal is lower than for most of the other liquids, and becomes less as the molecular weight of the compound increases. This may be because the increased molecular weight produces an increase in the cohesional force of attraction, and a consequent increase in the internal pressure, with the same effect as an increase of the external pressure. Ethyl iodide is remarkable for the low value of the dilatation at 80° at 6000 kgm. The minimum as 0.00017, and is lower than for any of the other liquids at the highest pressures.

It is evident that in order to explain these complicated facts we shall have to give up the simple picture of things that led us to expect the dilatation always to increase with increasing temperature and to decrease with increasing pressure. A natural way of modifying our conceptions so as to make room for these effects would seem to be as follows. We are to think of the molecules as having complicated shapes, or what for our purposes would amount to the same thing, of being surrounded by fields of force different in different directions. This concept does not seem to be a forced or an unlikely one; it must certainly be the fact for liquids which eventually crystallize under pressure, and for liquids which do not crystallize it is hard to conceive how an assemblage of atoms with unlike chemical affinities can coalesce into a molecule identical in all aspects. The effect of pressure on such an assemblage of molecules may be somewhat as follows. As the molecules are crowded closer together, the localized centers of force on the corners and edges play an increasingly individualized part, so that for small volumes we can no longer regard the molecules as centers of force and the cohesional force as a function of the mean distance apart of the molecules, but the orientation of the molecules with respect to each other begins to have its effect. Now we suppose that the natural tendency of the molecules in a liquid under normal conditions is to arrange themselves at haphazard with relation to each other. But as the constraints increase with decreasing volume, the characteristic shape compels an arrangement not entirely at haphazard, but such that the molecules fit into each other to some extent. Now it is evident that with the increasing orderliness of the arrangement of the molecules the relative positions of the localized centers of force will change. What is more, it is conceivable that the change in position of the force centers will be such that on the average the mean attraction between the molecules shall decrease with decreasing volume instead of increasing as is usual. Still more would this be true if there are centers of both attraction and repulsion in the molecule, as seems very likely from our present views of the electrical nature of matter. In this case we should expect the greater dilatation to be at the smaller volumes, because the cohesional force is less. But since the smaller volumes correspond to the lower temperatures, we have the effect already met, namely, that the dilatation is greater at the lower temperatures. The explanation of the sometimes increasing dilatation with rising pressure is similar. With increasing pressure the molecules may be forced to assume positions more regular in arrangement, of less volume, and of less average attractional force also. The number of possible positions which the molecules may assume in this attempt to adapt themselves to the diminished space at their disposal may evidently be very great with molecules of at all complicated shapes, so that there is here the possibility of such very complicated dilatation curves as we actually have.

Another possible explanation which in the end amounts to very much the same thing for some purposes, is that of association. If we suppose that association takes place with decrease of volume, and that the amount of association is increased with increasing pressure, and is decreased with rising temperature, then we have also the possibility of decreasing dilatation with rising temperature and of increasing dilatation with rising pressure. In this case the phenomena are essentially similar to those of solidification under pressure of a mixture of different liquids. Such a case has already been investigated for kerosene, which shows the same general features as above. The phenomena for kerosene are much simpler than for these liquids, however. It is evident that one simple association (such as single to double molecules) is not sufficient of itself to explain the facts, but if we assume several associated molecules of varying degrees of complexity, and that the relative numbers of these change with pressure and temperature, the explanation would account for several of the facts actually found. But reasons will be given in the next section, for supposing that association cannot have a very large part in the phenomena at high pressures.

We now turn our attention to Figure 32, which gives in one diagram the average dilatation between 20° and 80° for all twelve liquids. The origin of each of these curves has been displaced downwards one unit with respect to the one above it. The origin is so located that the value of the thermal dilatation for each of the liquids at 12000 kgm. is between 0.0002 and 0.0003. The scale of the drawing is indicated at the side. The immediately striking feature is that the curves are nearly equi-spaced at the higher pressures. Approximate equal spacing of the curves would of course be a consequence of their all approaching zero, but this is not the entire effect by any means. For instance, the ratio of the dilatation of ether to that of amyl alcohol at low pressure is 1.50 while at 12000 kgm. pressure it has dropped to 1.03. (In the computation for ether, the initial dilatation at 20° was used.) This is only the first example of many we shall meet tending to show that at high pressures liquids lose the individual differences which characterized them at low pressures, and become more alike.

The very gradual change of dilatation with pressure at the high pressures was also a surprise. The grand average for the dilatation at 12000 kgm. for the twelve liquids is 0.00025, about 40% more than that of mercury at atmospheric pressure and much higher than for any solid except possibly gutta-percha. It is ten times as great as that of aluminum, for example. The very slow change of dilatation

TABLE XVIII.

PRESSURE AT WHICH THE THERMAL EXPANSION IS TWICE AS LARGE AS IT IS AT 12000 KGM.

Substance.	Pressure. $\frac{kgm.}{cm.^2}$	Substance.	Pressure. $\frac{kgm.}{cm.^2}$
Methyl Alcohol	2150	Acetone	2900
Ethyl "	2000	Carbon Bisulphide	2900
Propyl "	2250	Phosphorus Trichloride	2200
Isobutyl "	1850	Ethyl Chloride	2750
Amyl "	2080	Ethyl Bromide	3100
Ether	2750	Ethyl Iodide	2400

with pressure is shown in Table XVIII, giving the pressure at which the dilatation has twice its value at 12000 kgm. The table shows that between atmospheric pressure and 2400 kgm. (on the average) the dilatation falls to about 0.4 its initial value, while over a further pressure range four times as large it falls off only an additional 50%. From these data it would appear that the dilatation must remain very considerable for pressures far in excess of those reached here. The approximate equality of the pressures for the twelve liquids at which the dilatation is double its value at 12000 kgm. indicates that the liquids behave similarly over much the greater part of the pressure range.

The curves in general fall with increasing pressure, but all the alcohols show a tendency to become stationary or to rise; while carbon bisulphide shows an extended stationary region between 6000 and 10000 kgm. but beyond 10000 its curve drops with unusual steepness.

The four normal alcohols show a dilatation continually decreasing with increasing molecular weight. But isobutyl alcohol shows the effect of its different structure by a higher dilatation than its position in the series would indicate. A comparison of isobutyl alcohol with ether, its isomer, shows that although initially the dilatation of ether is greater, at 12000 kgm. it has become less. This emphasizes again that the structure of the molecule continues to play a part even at high pressures. It is not to be wondered at in view of the suggested explanation of the complicated nature of the dilatation curves. For molecules of the same atomic formula, but of different structure, are to be thought of as possessing different shapes, and it is at high pressures that the effect of shape is greatest.

Isothermal Compressibility.— The isothermal compressibility is shown in Folder II; the values for the liquids separately at 20° intervals in Figures 33 to 44, and in Figure 45 the average results over the entire temperature range are collected into a single diagram for the twelve liquids.

The curves require a word of explanation. Up to 4000 kgm. the curves for each liquid are drawn for the four different temperatures, but at pressures higher than 4000 kgm. the curves would be so close together, sometimes crossing each other, that it would have been very confusing to draw them on the same scale. Therefore, at higher pressures, the only complete curve given is for 40°, while in the upper part of the diagram are shown on a larger scale the differences between the compressibilities for 20° intervals. The zero of these difference curves is drawn as a heavy line. Negative ordinates of the difference curve 20°–40° indicate that the compressibility is less at 20° than at 40°; positive values for the difference curve 40°–60° indicate a greater compressibility at 60° than at 40°, and similarly positive differences 60°–80° mean a greater compressibility at 80° than at 60°. To find the compressibility at 20°, one adds to the value obtained from the 40° curve the ordinates of the difference curve 20°–40° (this ordinate is usually negative); the compressibility at 60° is found by adding to the ordinate of the 40° curve that of the 40°–60° curve, and the compressibility at 80° by adding to that at 60° (obtained as above) the ordinate of the 60°–80° curve. A larger compressibility at 20° than at 40° is indicated by the difference curve 20°–40° rising from below and crossing the axis, while a smaller compressibility at 60° than at 40° is similarly shown by the curve 40°–60° crossing the axis from above. The mutual crossing of the difference curves, from this point of view, is not especially significant; it is the crossing of the axis that counts. The meaning of the mutual crossing of the curves 40°–60° and 60°–80°,

Folder II. Isothermal compressibility, $\left(\dfrac{\partial v}{\partial p}\right)_T$, against pressure. Figures 33 to 44 give the compressibility for the liquids separately at intervals of 20°, and Figure 45 shows the average between 20° and 80° for all twelve liquids. In order to avoid confusion in the diagrams the following course is adopted. The compressibility at 40° is given over the entire pressure range. Up to 4000 kgm. the compressibilities at 20°, 60° and 80° are also shown on the same scale, but at the higher pressures, where the difference between the compressibilities becomes smaller, the difference for intervals of 20° is plotted in the upper right hand of the diagram on a larger scale. The scale of these difference curves is shown at the right hand side. A negative ordinate for the curve 20°–40° indicates that the compressibility is less at 20° than at 40°. The appearance of the difference curves is the same as would be presented if the curves of the lower part of the diagram were magnified twenty times and the curve for 40° straightened out into the heavy fiducial line. The order of the curves in the lower left hand part of the diagram is the same for all liquids, 20°–40°–60°–80°, reading from below up.

Figure 45 shows the average for the twelve liquids. In order to avoid overlapping, the origin of each curve is displaced one square with respect to the next one. The scale of the curves is shown at the right hand side. The origin is so situated that the compressibility for all twelve liquids is between 0.0_51 and 0 at 12000 kgm. The numbers on the curves indicate the liquid in the same order as the curves for the separate liquids (Figures 33 to 44).

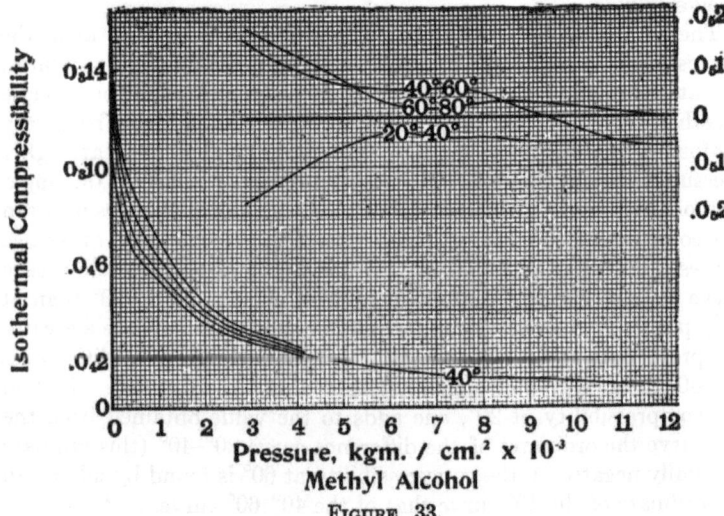

Methyl Alcohol
FIGURE 33.

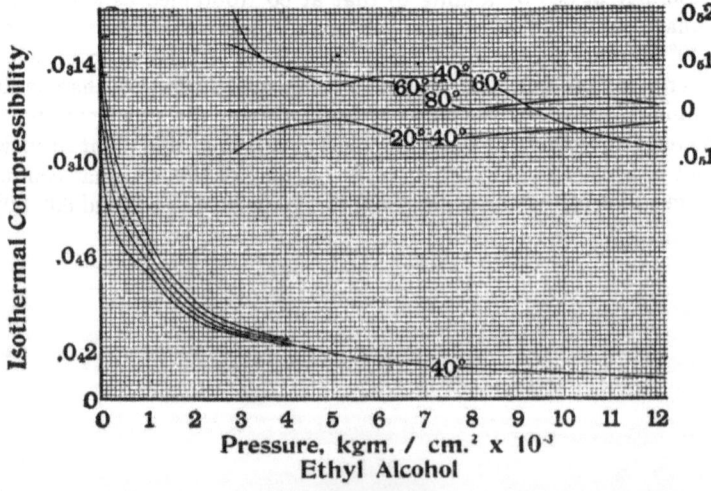

Ethyl Alcohol
FIGURE 34.

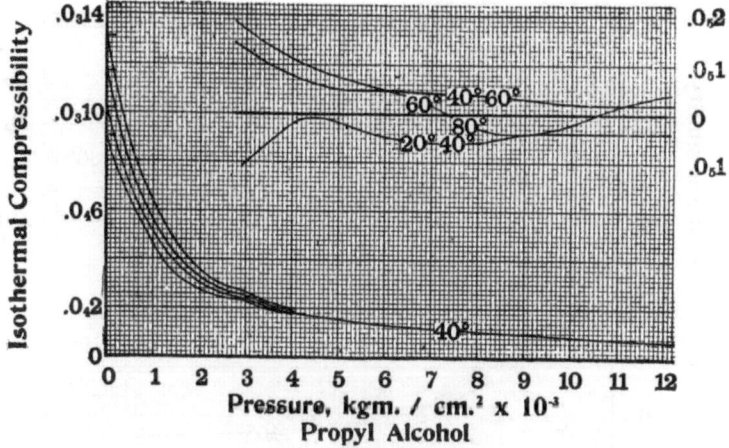

FIGURE 35.

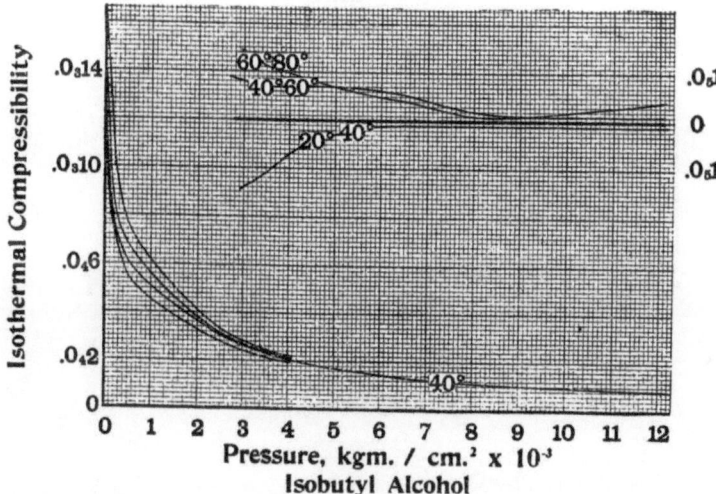

FIGURE 36.

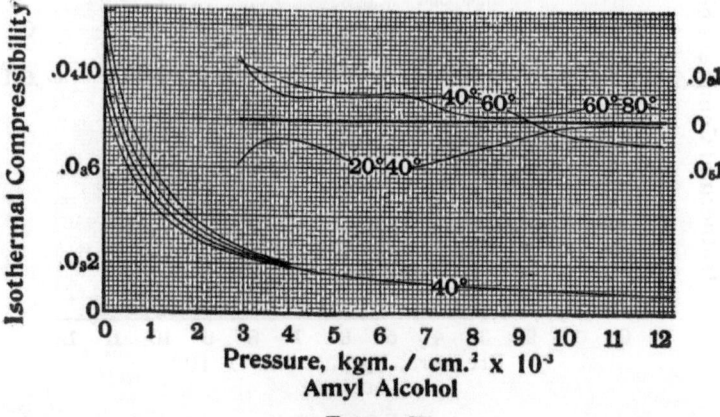

FIGURE 37.

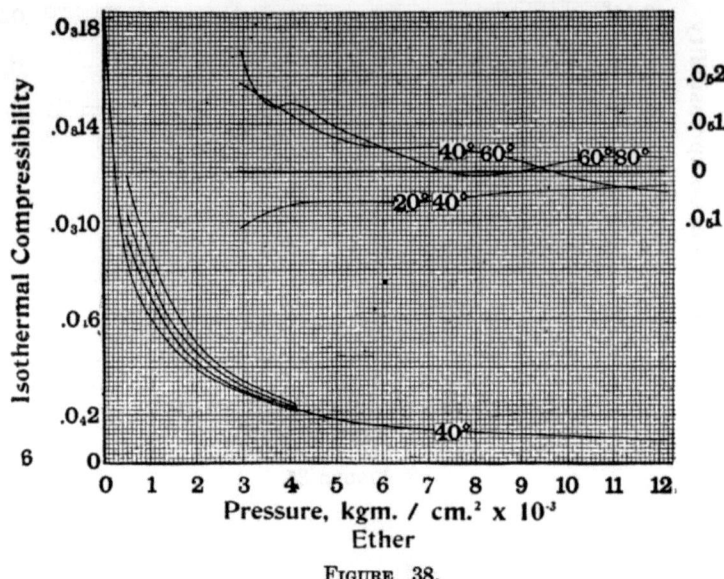

FIGURE 38.

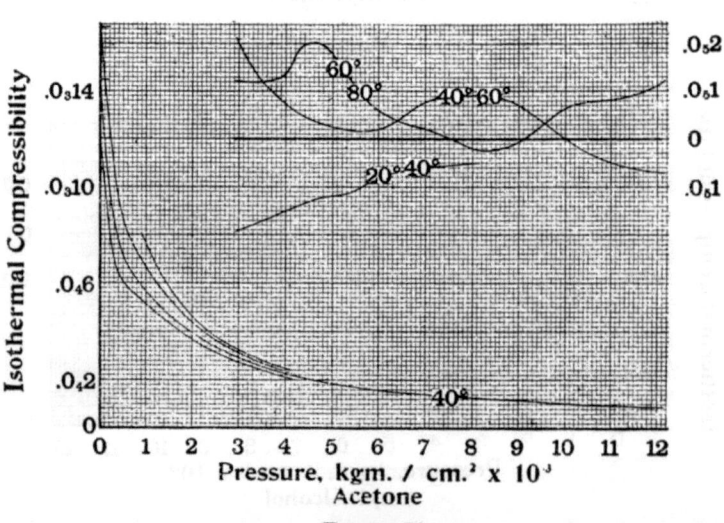

FIGURE 39.

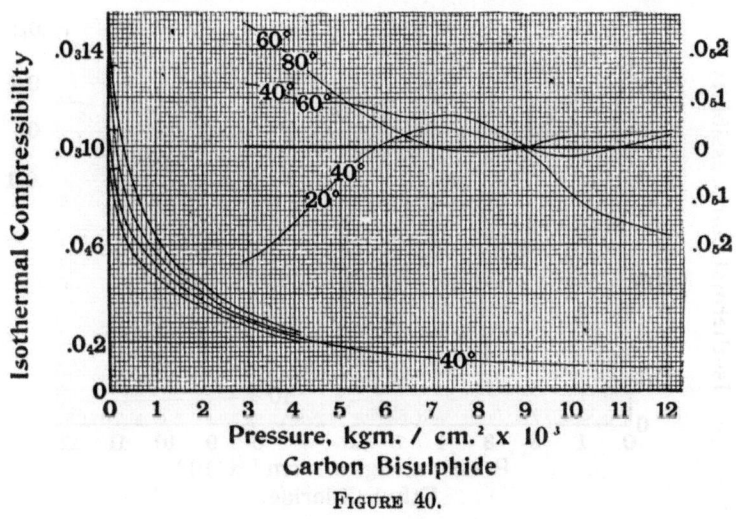

Carbon Bisulphide
FIGURE 40.

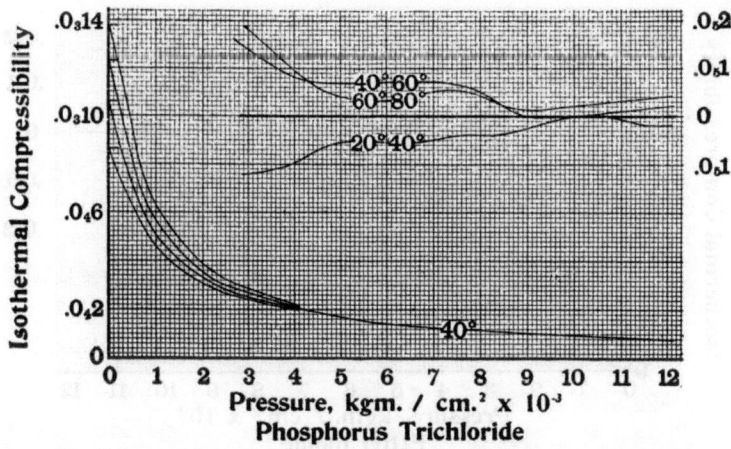

Phosphorus Trichloride
FIGURE 41.

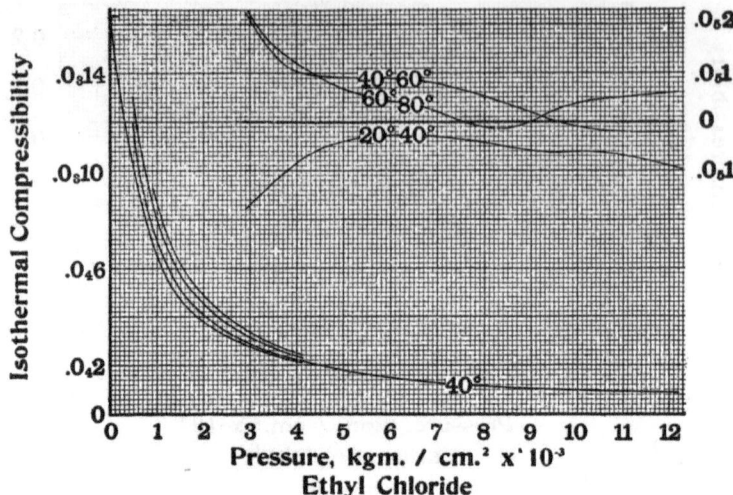

FIGURE 42.

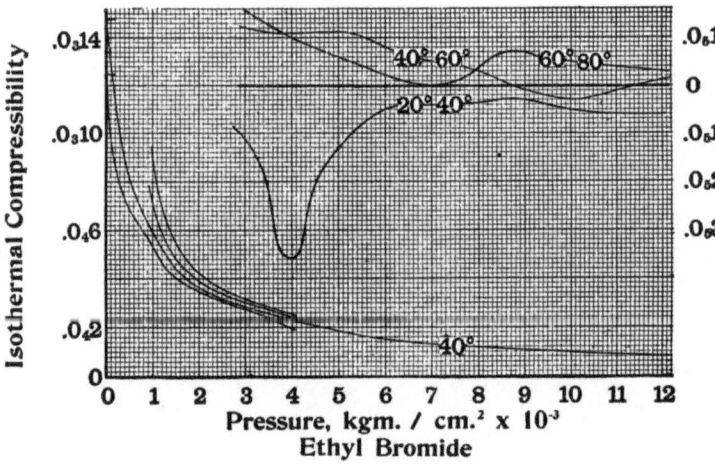

FIGURE 43.

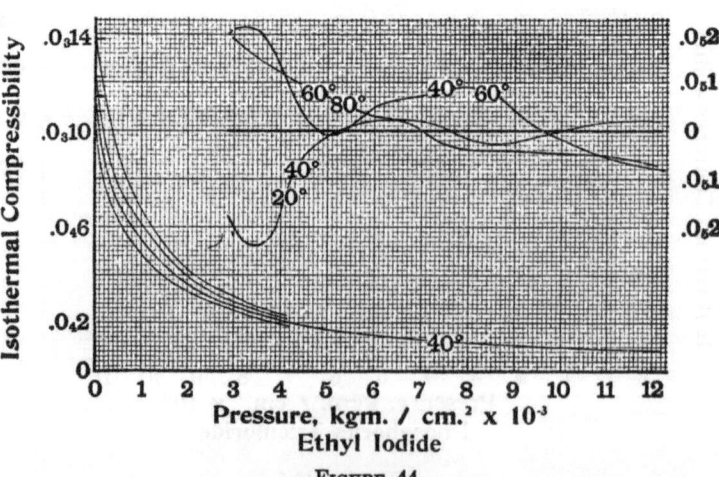

FIGURE 44.

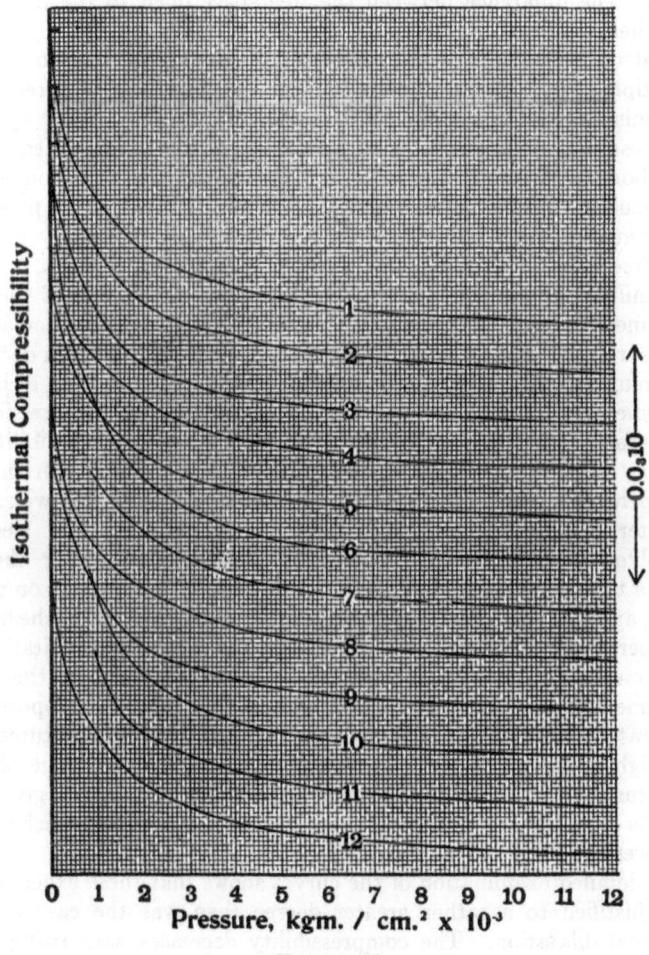

FIGURE 45.

for example, would be that over the temperature range 40°–80°,
$\frac{\partial^2}{\partial \tau^2}\left[\left(\frac{\partial v}{\partial p}\right)_\tau\right]_p = 0$.

The general behavior of the compressibility that one is prepared to expect from experiments at low pressures is a decrease with increasing pressure, and an increase with increasing temperature. The familiar conception of a liquid due to van der Waals is competent to explain this. The difference between the free space open to the molecules for their temperature vibrations and the total volume of the liquid is equal to the volume of the molecules themselves, or else to a small multiple of it. When pressure is increased, therefore, the free space diminishes much more rapidly than the total volume. Now we may suppose the pressure exerted by a liquid to be due in large measure to the bombardment of the walls by the temperature agitation of the molecules. The more frequent the collisions, the greater the pressure. Now at constant temperature, the number of collisions is inversely as the free space. At higher pressures, that is smaller volumes, a given diminution of total volume implies a greater diminution of the free volume than at low pressures, and therefore a greater increase of pressure. So that at small volumes (high pressures) a given decrease of volume carries with it a greater increase of pressure than at larger volumes (lower pressures), or in other words, the compressibility decreases with rising pressure. The increase of compressibility at higher temperatures is to be explained in the same way. At higher temperatures (constant pressure) the volume is greater and we expect greater compressibility. This however, is not the only element involved in the change of compressibility with temperature; there is also a temperature effect as such. It was found in the paper on water that, at equal volumes, the compressibility was always less at the higher temperatures. The reason is evidently the more rapid agitation of the molecules. At equal volumes, a given decrease in the total volume, and so of the free volume, will produce the same proportional increase in the number of impacts at high and low temperatures, but at high temperatures each impact involves a greater change of momentum, with the result that at the higher temperatures a given decrease of volume produces a greater increase in pressure, which means a lower compressibility.

A detailed examination of the curves shows that these expectations are justified to a rather greater degree than was the case for the thermal dilatation. The compressibility decreases with rising pressure for all twelve liquids at 40°. A careful analysis of the difference curves will show that this is also the case for the other three temperatures, with a single exception. For carbon bisulphide, between 11000 and 12000 kgm. there is an increase of compressibility from 0.0_3101 to 0.0_3102 at 60°, and from 0.0_3103 to 0.0_3105 at 80°. The change is very small and may well be due to experimental error. The difference curves show as bewildering small variations with temperature as the thermal dilatation curves, so that it is hopeless to try to explain them in detail at this stage of our knowledge. It would seem, however, that an actual reversal of the effect, that is, a smaller compressibility at a higher temperature, does not occur so often as was the case for the dilatation. The reversal of the dilatation was universal for all liquids and took place between 1000 and 3000 kgm. But the reversal of the compressibility, indicated by the difference curve crossing the axis, is sporadic in occurrence, and only once occurs for all temperatures simultaneously.

Comparison of the curves shows some points of interest. The first three alcohols show a slight kink; propyl alcohol at 3000 kgm., ethyl at 1100, and methyl possibly at a somewhat lower pressure. Amyl alcohol apparently has lost the kink. These four alcohols show in rather desultory fashion for at least one temperature the temperature reversal at high pressures. Isobutyl alcohol, as we expect, shows the effect of structural variation by curves of different character from the other four alcohols, in that they show no reversal, but decrease with fair regularity under increasing pressure. Isobutyl alcohol is also remarkable for the extremely rapid initial drop of compressibility with rising pressure. Ether, the isomer of isobutyl alcohol, still further shows the importance of the structure of the molecule; the difference curves for ether show reversal at high pressures, and are also more widely separated than the curves for isobutyl alcohol. The last six liquids show a somewhat greater temperature effect on compressibility than the first six. Acetone shows a strange maximum in the difference curve between 4000 and 5000. Carbon bisulphide shows greater variations at the highest pressures than any other liquid (it will be remembered that the dilatation curves also show abnormally great variations), and a reversal of the effect at 9000 kgm. simultaneous for all temperatures. This simultaneous reversal is shown by no other liquid, and reminds one of the dilatation. Phosphorus trichloride shows no marked features; the compressibility curves are not so simple as the dilatation curves. The halogen compounds do not show any marked similarities. Ethyl chloride is without particular features. Ethyl bromide has a very pronounced separation of the curves for 20° and 40° at 4000 kgm., while ethyl iodide shows a similar, but smaller separation between the 20°–40° and also the 40°–60° curves. The halogens all show reversals at the high pressures.

The occasional reversal of compressibility with temperature, that is, a smaller compressibility at a higher temperature, has ready explanation if we adopt the hypothesis of molecules with shape. We have seen that in this event, because of the varying completeness with which the molecules interlock, it may sometimes happen that the increase of volume with temperature results in a decrease of the free space open to the thermal agitation of the molecules. In this case the compressibility is less at high temperatures than at low. The same argument would admit also the possibility of compressibility increasing with increasing pressure if the effect of pressure is to produce a better fitting together of the molecules and so greater free space. It may be, therefore, that the effect found for carbon bisulphide is genuine, and not to be explained away by experimental error.

As a general rule, the result found for water applies to these twelve liquids also, namely that compressibility when plotted against volume is less at the higher temperatures. There are, however, a few exceptions. Carbon bisulphide and ethyl chloride, for example, show a reverse effect at the higher pressures. A possible explanation of this is to be found in the opposition of two effects. In general we think of the effects of lowering the temperature and of increasing the pressure as the same, namely to increase the degree of interlocking of the molecules. If now we compare two states of the liquid, each occupying the same volume, but one at a higher temperature than the other, we see that the high temperature condition differs from the low in that the molecules are less interlocked so that they have less free space at their disposal. An increase of pressure will tend to produce a smaller change of volume at the high temperature therefore, because of the smaller free space which the molecules possess. But on the other hand, an increase of pressure will tend to produce a greater change of volume because the effect of increased pressure is to increase the amount of interlocking, and so to decrease the volume. As one or the other of these effects predominates, we shall have smaller or greater compressibility at the higher temperatures with constant volume. We have seen that the compressibility is usually smaller.

We turn now to the diagram (Figure 45) in which are collected the average of the results for all twelve liquids. As for the curves of thermal dilatation, the zeros of the successive curves are displaced

TABLE XIX.

Changes of Compressibility and Thermal Expansion produced by Pressure.

Liquid.	Compressibility, κ.				Dilatation, δ.			
	$\dfrac{\kappa_1}{\kappa_{12000}}$	$\dfrac{\kappa_{1000}}{\kappa_{12000}}$	$\dfrac{\kappa_{6000}}{\kappa_{12000}}$	κ_{12000}	$\dfrac{\delta_1}{\delta_{12000}}$	$\dfrac{\delta_{1000}}{\delta_{12000}}$	$\dfrac{\delta_{6000}}{\delta_{12000}}$	δ_{12000}
1	18.4	8.2	2.20	0.0_574	4.29	2.76	1.23	0.0_3298
2	13.7	7.4	2.02	81	4.50	2.73	1.30	268
3	15.8	7.8	1.94	70	4.80	3.06	1.33	237
4	16.6	6.3	1.68	86	4.15	2.62	1.17	275
5	14.4	7.1	1.88	74	4.40	2.84	1.30	240
6		7.7	1.62	96		3.65	1.32	248
7		7.3	1.85	87		3.27	1.35	282
8	13.8	6.3	1.82	87	5.47	3.16	1.31	262
9	14.2	7.1	1.81	80	4.84	2.83	1.31	278
10		8.4	1.78	90		3.44	1.37	267
11	14.9	8.3	1.87	82		3.46	1.33	260
12	14.9	7.2	1.89	81	4.86	3.15	1.22	248
Water	4.9	3.7	1.64	89	1.00	1.00	1.00	400
Kerosene			1.82	87			1.14	280

one square with respect to each other. The origin of each curve is so situated that at 12000 kgm. the compressibility has approached to within less than one unit of zero. Thus, the compressibility of ethyl alcohol at 12000 is 0.0_581.

The most striking feature is that the curves become nearly equispaced at the higher pressures. This appearance of increasing equality of behavior is not an illusion due to the approach of all the curves to zero, but it is real, as shown by the fact that the variation in the ratios of the initial compressibilities of the liquids to each other is greater than the variation in the ratios at 12000 kgm. The increasing equality at high pressures is also made more strikingly visible to the eye by noting on the diagram at what pressure the compressibility is equal to 0.0_420. For all the twelve liquids except propyl, isobutyl, and amyl alcohol this pressure is between 4000 and 5000 kgm., and for the alcohols it is not far removed. This is the same sort of thing that we have seen to hold for the dilatation curves.

The compressibility curves are, however, quite different from the dilatation curves in several respects. The effect of pressure is very much greater in decreasing the compressibility than in decreasing the dilatation. Furthermore, the decrease of compressibility at the higher pressures continues to be more rapid than that of the dilatation; both the initial and the final relative rates of change are more rapid for the compressibility than for the dilatation. This is shown in Table XIX. In the second and sixth columns of this table the relative changes of compressibility and dilatation are given over the entire pressure range. The change is about four times greater for the compressibility than for the dilatation. The fourth and eighth columns show that the change between 6000 and 12000 is greater for the compressibility than for the dilatation. In other words, the dilatation comes much nearer to approaching a finite asymptote than the compressibility. This is not what was expected at first. It was thought that at high pressures the molecules would be squeezed into virtually perfect contact, that the compressibility would be provided for by the compression of the molecules, but that under these conditions the dilatation would practically vanish. The exact reverse has turned out to be the case. The data previously obtained for water and kerosene have also been included in the table. At high pressures kerosene behaves much like the other liquids. Water is nearly normal at high pressures as regards compressibility, but is abnormal over the entire range with respect to temperature effects.

The compressibility of these liquids at 12000 kgm. may be compared with the compressibility of metals under ordinary conditions. For mercury the value is about 0.0_539, and for iron 0.0_658. The average compressibility of these liquids at 12000 kgm. is therefore, about twice the initial compressibility of mercury and about fourteen times that of iron. The average dilatation at 12000 approaches more nearly to that of mercury, being about 40% greater, but is farther removed from the dilatation of iron, being about 20 times greater. The difference between compressibility and dilatation is further accentuated by the fact that the compressibility of iron is nearly as low as that of any solid, while there are a number of solids with a smaller dilatation. There seems, therefore, to be more difference between a solid and a liquid with respect to thermal expansion than with respect to compressibility.

Pressure Coefficient.— No diagrams have been given for this quantity, but it is nevertheless worth some discussion, because of the part it has played in previous theoretical discussions. This so-called "pressure coefficient" is the thermodynamic quantity $\left(\frac{\partial p}{\partial \tau}\right)_v$, the change of pressure when the temperature is raised one degree at constant volume, and is mathematically equivalent to the ratio of dilatation to compressibility $\left(\frac{\partial p}{\partial \tau}\right)_v = -\left(\frac{\partial v}{\partial \tau}\right)_p / \left(\frac{\partial v}{\partial p}\right)_\tau$. It has been proposed as an empirical law by Ramsay and Young that the pressure coefficient is a function of the volume only. That is, if the pressure coefficient is plotted against volume, the curves for different temperatures will fall together. The experiments of Ramsay and Young covered a wide temperature range, but a comparatively low pressure range, since their chief concern was with the relations between a liquid and its vapor, and their pressures seldom exceeded the critical pressure, a matter of a few hundred atmospheres. Amagat, in his discussion of his own results for liquids up to 3000 atmos., has devoted considerable attention to the pressure coefficient. One of his results was that the pressure coefficient is approximately independent of temperature at constant volume, but does nevertheless show small consistent variations, which Amagat was unwilling to ascribe to experimental errors. The coefficient of different substances may increase or decrease with rising temperature, or show still more complicated variations. The coefficient increases with decreasing volume, that is with increasing pressure. Tammann, however, in his recent empirical theory of liquids for high pressures, has concluded from an examination of Amagat's work that the variations which Amagat found in the pressure coefficient do not exceed the possible experimental errors. Tammann has accordingly taken as one of the fundamental hypotheses of his theory the assumption that the pressure coefficient is a function of the volume only.

The discussion of the pressure coefficient to be given here has for its only purpose to show that at high pressures, whatever the facts may be at low pressures, there is absolutely no ground for the assumption that the coefficient remains independent of the temperature. This was shown to be the fact in the previous paper on water, but the argument lost force because water is abnormal. However, for none of the twelve liquids of this paper is the relation even approximately

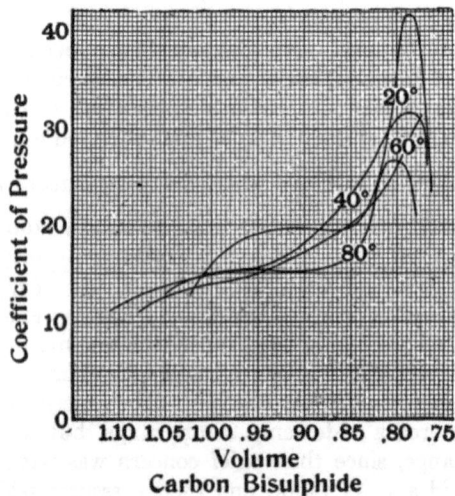

FIGURE 46. The pressure coefficient of carbon bisulphide plotted against volume. The diagram shows that at high pressures the pressure coefficient is a function of temperature as well as of volume.

satisfied. It is not necessary to show curves for all twelve liquids in order to disprove this one point. The data are at hand so that any one may make a complete test for himself. A single diagram, chosen at random, is sufficient to show that the proposed relation breaks down completely, as indeed one would expect it to in view of the complications of compressibility and dilatation. Figure 46 shows this for carbon bisulphide. It speaks for itself.

Work of Compression.— The mechanical work of compression is shown in Folder III; Figures 47 to 58 are for the liquids separately, and Figure 59 shows the average between 20° and 80° of all twelve liquids together. The difference of the work for different temperatures is so nearly the same that the differences could not have been read accurately directly from the curves. The course was adopted, therefore, of plotting the work at 40° only, and in the lower

Folder III. The work of compression $\left(W = \int p\, dv\right)$ against pressure. In Figures 47 to 58 W is shown for the liquids separately, and in Figure 59 the average between 20° and 80° for all twelve liquids. The work of compression varies very little with the temperature, so in order to avoid confusion, the complete curve for only 40° is given, and in addition, in the lower part of the diagram, on a scale one hundred times as large, the difference of the work at 20° intervals. Each of these difference curves starts from a new origin. Thus for example, the work of compression of ethyl iodide at 6000 kgm. and 80° is found to be 4.30 (the work of compression at 40°) plus 0.23 (the difference between the work at 40° and 60°), plus 0.30 (the difference between the work at 60° and 80°), or 4.83 kgm. m. in all. In Figure 59, giving the average work of compression over the entire temperature range, the numbers indicate the liquids in the same order as the immediately preceding figures for the separate liquids. In order to avoid confusion, the origin of each curve is displaced one square with respect to the next one. The break in the curve for acetone (7) is because of the freezing, which made it possible to take the average only from 40° to 80° over the upper part of the pressure range, instead of between 20° and 80° as at the lower pressures. The origin of the curve for ethyl chloride (10) is at 500 kgm. instead of atmospheric pressure as for the other liquids, because of the low boiling point of this substance at atmospheric pressure.

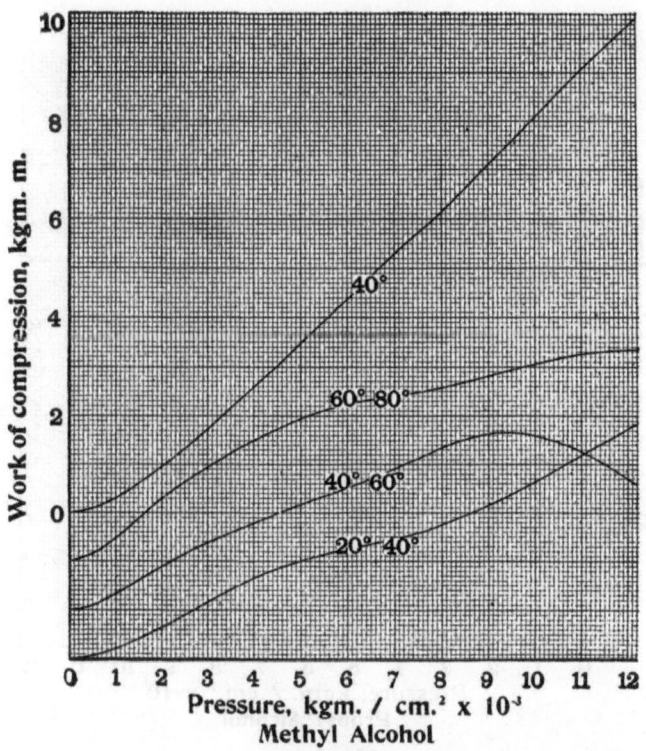

Methyl Alcohol

FIGURE 47.

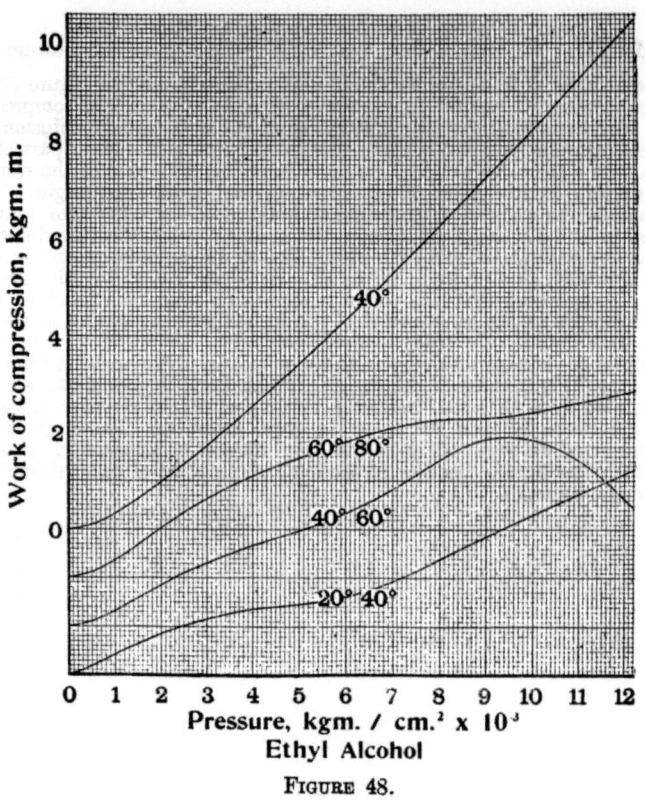

Ethyl Alcohol

FIGURE 48.

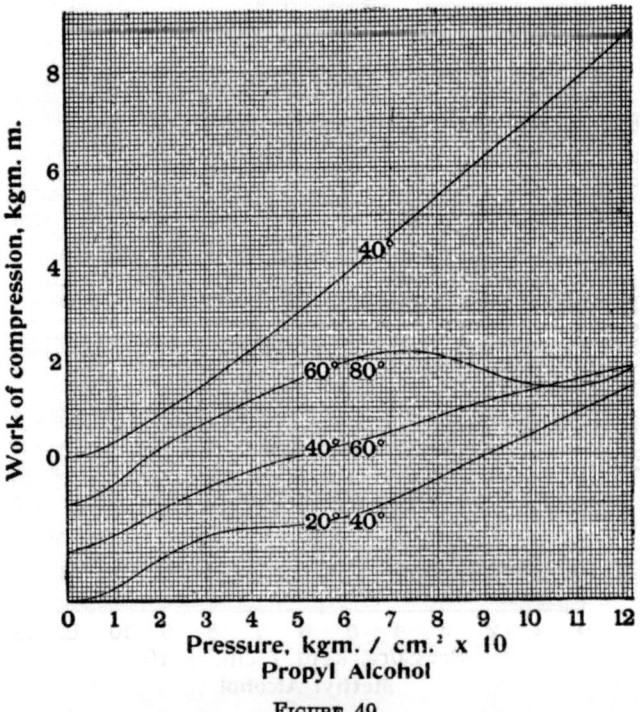

Propyl Alcohol

FIGURE 49.

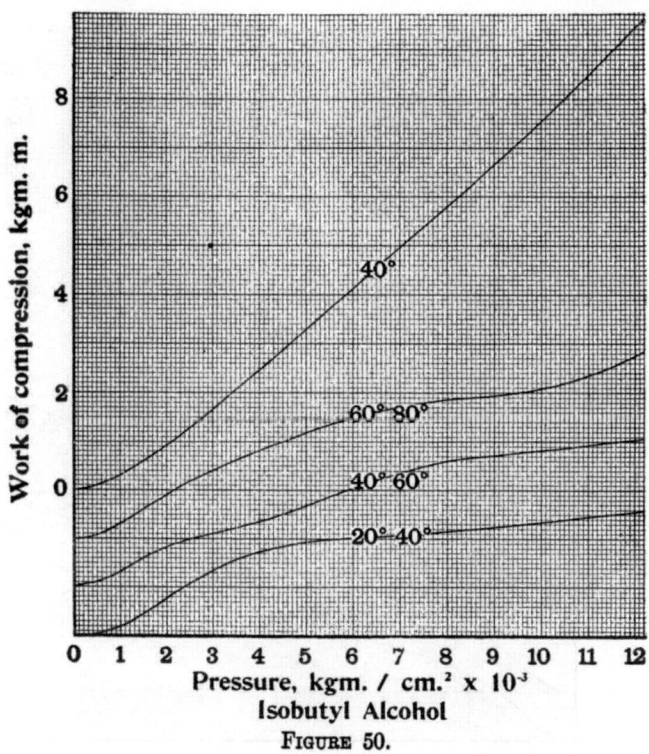

FIGURE 50.

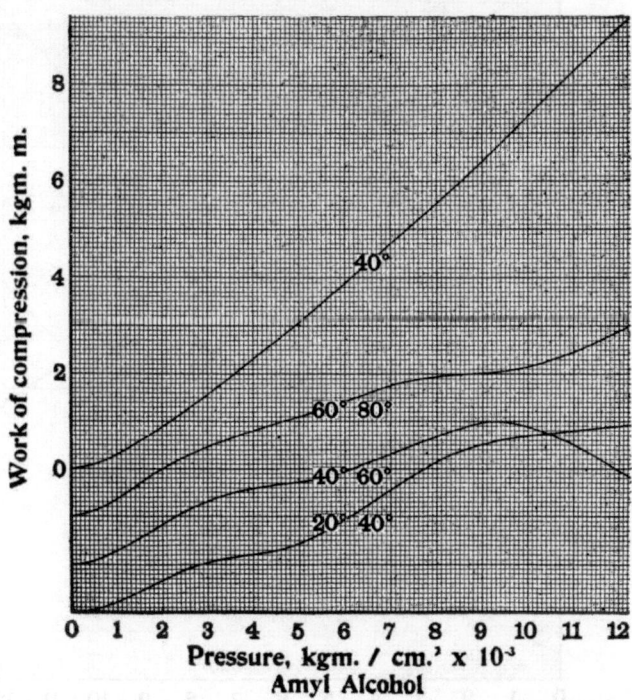

FIGURE 51.

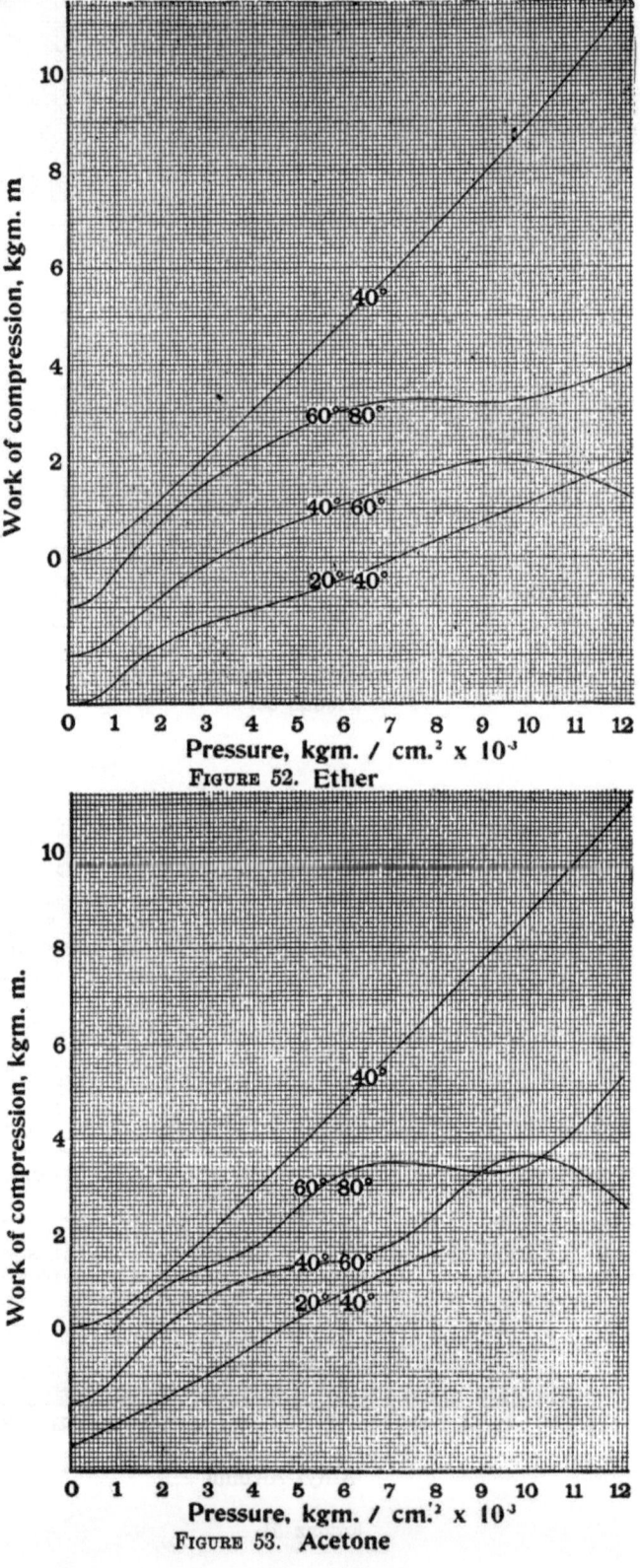

FIGURE 52. Ether

FIGURE 53. Acetone

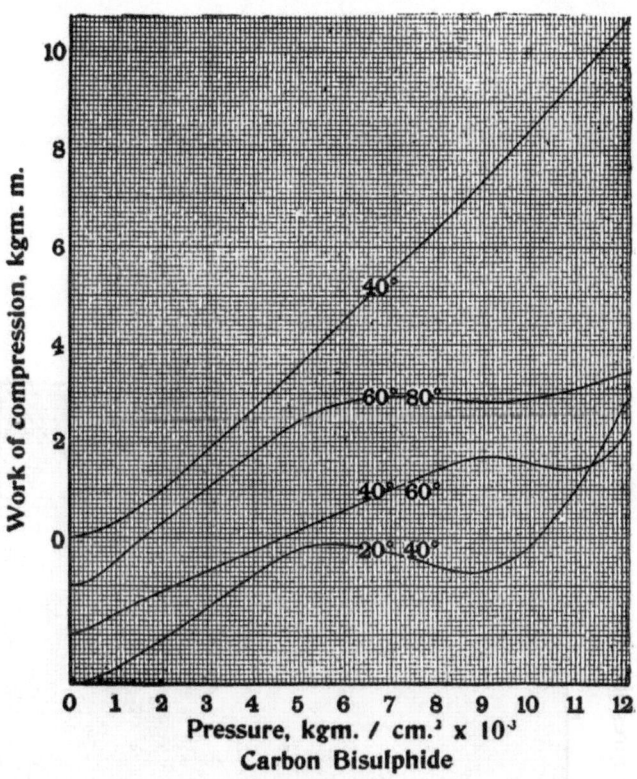

FIGURE 54.

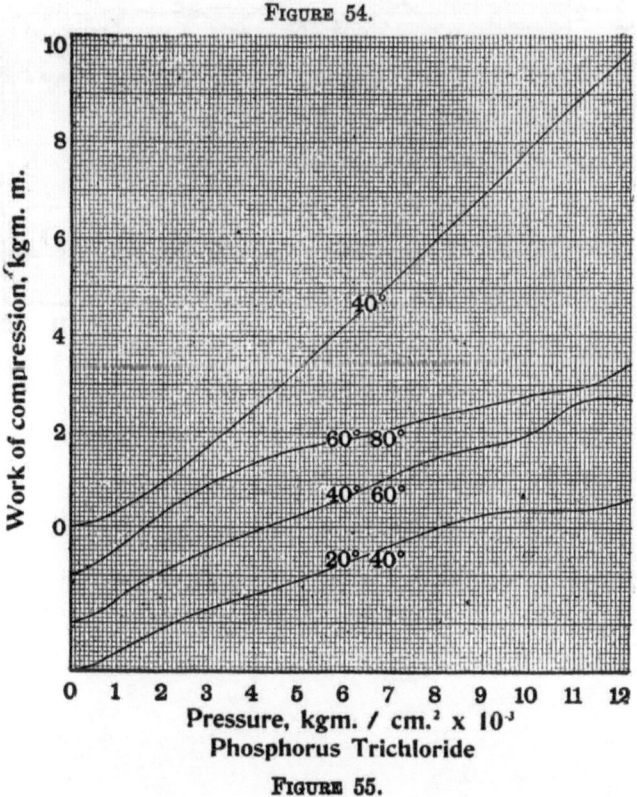

FIGURE 55.

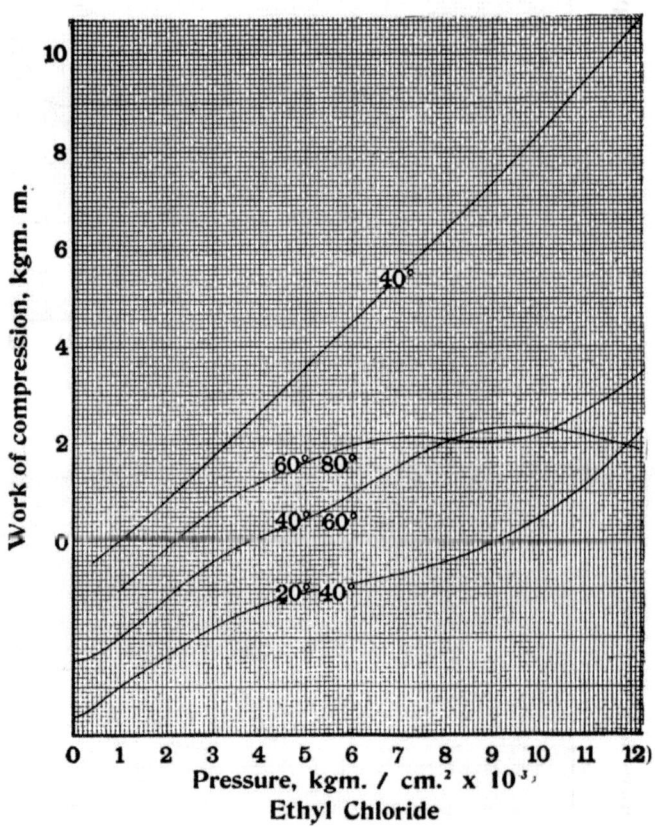

FIGURE 56.

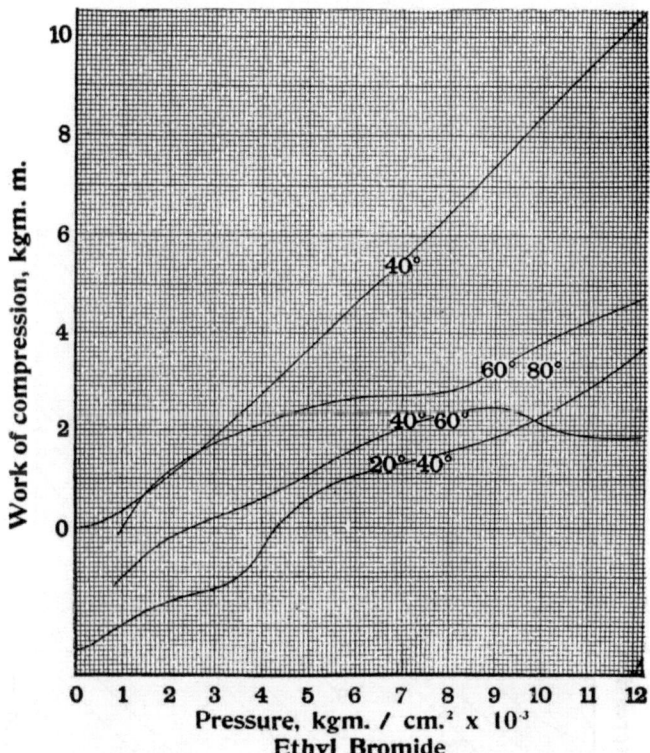

FIGURE 57.

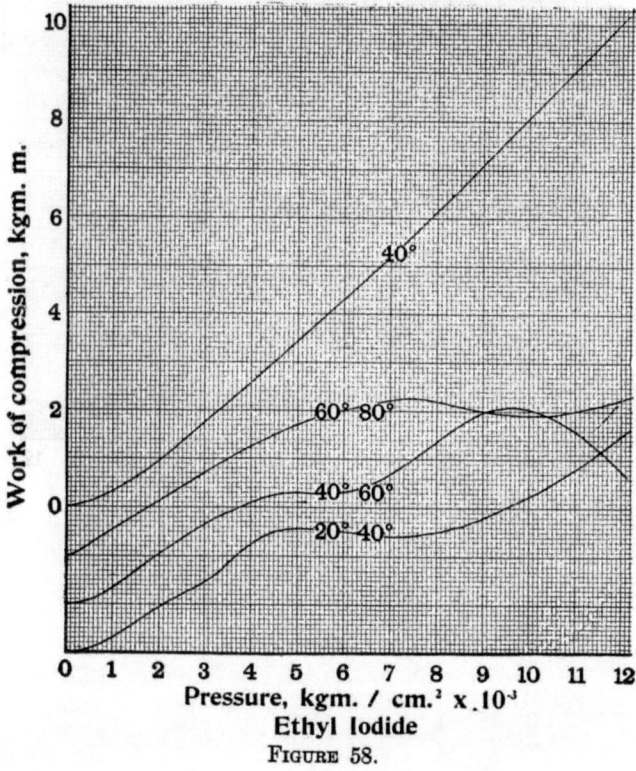

FIGURE 58.

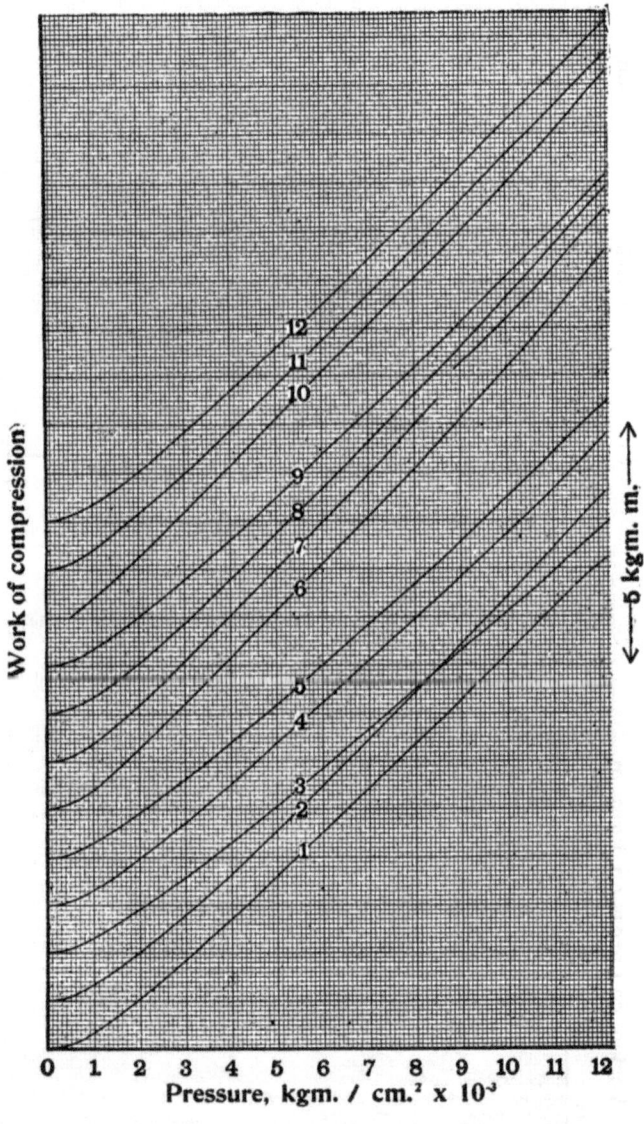

FIGURE 59.

part of the diagram giving on an enlarged scale the difference of the work for intervals of 20°.

The curves for the separate liquids do not require much comment. The difference curves are universally positive; that is, it is always true that more mechanical work is expended when a liquid is compressed to a given pressure at a high temperature than at a lower one. Beyond this, however, there do not seem to be many common features. The curves show irregular and apparently unrelated variations, but the irregularities are not so great as for the dilatation or the compressibility. Of course this was to be expected, because the curves are essentially integral curves.

Three of the alcohols, methyl, ethyl, and amyl, show similar difference curves. Ether and isobutyl alcohol, isomers, are unlike, as we have found them before. The difference curve 20°–40° for carbon bisulphide shows the effect of the abnormally high compressibility at 20°, which we saw previously, and the three ethyl halogens show similarities. Other variations in the difference curves are not particularly illuminating.

Figure 59, combining the average results for the twelve liquids, is of more interest. The similarity in general shape of all the curves is perhaps the most interesting feature. The curves become nearly linear at the higher pressures. This of course is not the usual relation between stress and work for a body like a steel spring, which maintains a stiffness independent of stress. For such bodies the work stored up as potential energy of strain varies as the square of the stress. This is true for liquids also over a pressure range so small that the compressibility may be regarded as constant, and is shown in the initial stages by all the curves, which are tangent to the axis at the origin. The fact that at high pressures the curves tend to become linear, still remaining slightly concave upwards, means that at high pressures the compressibility is becoming less, so that the change of volume, and therefore the work, is less for a given increment of pressure.

If we assume tentatively that the work of compression is linear at high pressures, we have a means of finding the compressibility and the volume at high pressures. For;

$$W = -\int_1^p p \left(\frac{\partial v}{\partial p}\right) dp = a + bp \quad (a \text{ is negative, } b \text{ positive}).$$

Differentiating this equation, we obtain,

$$\left(\frac{\partial v}{\partial p}\right)_t = -\frac{b}{p}$$

whence,
$$\Delta V = \int^p \left(\frac{\partial v}{\partial p}\right)_t dp = -b \log p + C.$$

This equation cannot be expected to hold for low pressures. The constants may be determined as follows. From the curves for W above 5000 kgm. we find that an approximate value for b is 0.1. If furthermore we assume as a fair average of all twelve liquids that $\Delta V = -0.25$ when $p = 5000$, the equation becomes,

$$\Delta V = -0.1 \log_e p + 0.6017.$$

The equation evidently cannot hold for infinite pressures, because it demands that ΔV decrease indefinitely, and it is physically impossible that ΔV should become less than -1. But the pressure at which this would happen according to the above formula is 9,060,000 kgm. This is so very far beyond the range of these experiments, that it would probably be safe to apply a similar equation as an approximate expression for any experimental pressures above 5000 kgm.

The mathematical analysis gives us, furthermore, information as to the ultimate behavior of the work of compression. For the actual work must eventually be less than that given by the above formula, which corresponds to a smaller change of volume. It must be, therefore, that at higher pressures the curvature reverses, and the curve becomes convex upwards. Methyl and amyl alcohol show the beginning of this effect, but in view of the extreme remoteness of the vanishing of the volume predicted above, it may well be that this slight change of curvature is merely one of the many local variations.

The maximum amount of work stored up at 12000 kgm. is nearly the same for all twelve liquids; much more nearly equal than the initial differences in the compressibilities would lead us to expect. The variations in the maximum are about 25%, whereas there are initial variations in the compressibility of 100%. If we admit water to our family of curves, as we may because the final work of compression is over 9 kgm. m., the variation in the initial compressibility may be 400%. Initial differences of compressibility have little effect on the work at the maximum pressure, because of the comparatively small amount of work done at low pressures.

The total amount of mechanical work stored up in a liquid is quite considerable. For example the work of compression of ether at 12000 kgm. would suffice to raise it through about 45000 ft., or to give it a velocity of 1700 ft. sec.

Heat of Compression.— The heat of compression, that is the quantity of heat in kgm. m. which flows out of a substance as it is compressed isothermally, is shown in Folder 4; for the twelve liquids separately in Figures 60 to 71, and the average for the twelve liquids collected into a single diagram in Figure 72. The differences of the curves for different temperatures are sufficiently great so that the total heat of compression for each temperature could be plotted and the difference found with sufficient accuracy directly from the curves, without the necessity of drawing difference curves as was the case for the work of compression. The zero of each curve has been displaced upwards one square for the successive temperatures. In the case of those liquids which boil at low temperatures, the zero of the curves for all temperatures has been taken at the same pressure, 500 or 1000 kgm., although it would have been possible to extend the curves to atmospheric pressure for the lower temperatures. In the case of acetone and ethyl chloride the curves for the lower temperatures have been extended backwards from the origin (1000 kgm.) to atmospheric pressure.

The heat of compression is positive, that is, as a substance is compressed isothermally, heat flows out to the surroundings. Examination of the curves in detail shows also that the total heat always increases with increasing pressure. This is a direct consequence of the fact that the thermal expansion is always positive. The curves therefore, show less pronounced irregularities than some, such as those for compressibility for example.

In general, the concavity is toward the pressure axis, that is, the increase of the heat of compression becomes less rapid at the higher pressures. The curves for the different liquids at different temperatures show that the heat is not universally greater at the higher temperatures, although such is generally the case. This is shown by the curves drawing together at high pressures in some cases to within less than the one square which separated them at the origin. An example of this is afforded by amyl alcohol between 20° and 40°, and by ethyl chloride and ethyl iodide between 60° and 80°. In general, a drawing together of the curves with increasing pressure means that the expansion is less at the higher temperature. There are many instances of this, although it is not usual that the drawing together is great enough to bring the curves to within less than the original arbitrary distance of separation.

Figure 72, in which are collected the average results for all twelve liquids, shows again that the twelve liquids are alike in character.

Folder IV. The heat of compression $\left(Q = \int \tau \left(\frac{\partial v}{\partial \tau}\right)_p dv\right)$ against pressure. In Figures 60 to 71 are shown the curves for the separate liquids at intervals of temperature of 20°. In order to avoid overlapping, the origin for different temperatures is not the same. The scale of the diagram is shown at the right hand side. Because of the low boiling point, some of the curves do not start from an origin at atmospheric pressure. Figure 72 shows the average heat of compression between 20° and 80° for all twelve liquids. The origin of each curve is displaced one square with respect to the next. Ethyl chloride (10) and ether (6) start from an origin at 500 kgm. because of the low boiling point. The curve for acetone (7) shows a break at 8000 kgm. because above 8000 the average had to be taken between 40° and 80° by reason of the freezing under pressure. The numbers in Figure 72 indicate the liquids in the same order as that of the separate diagrams.

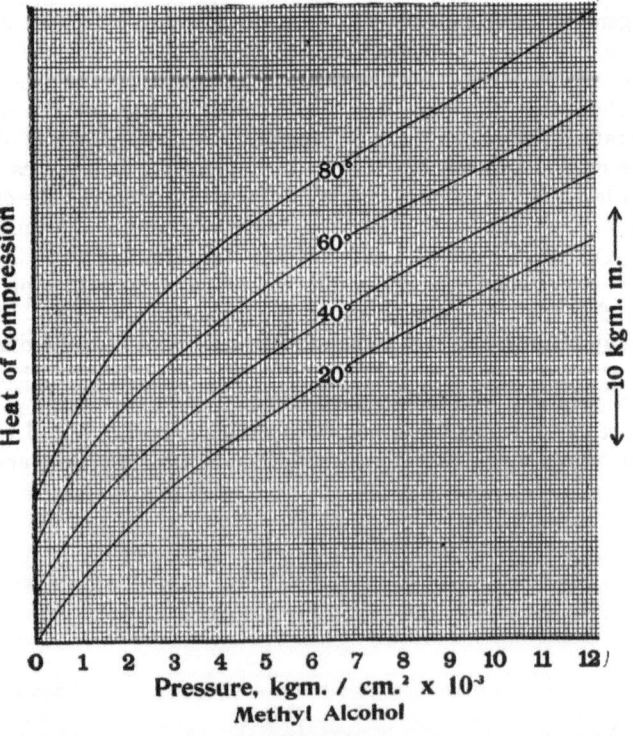

Figure 60.

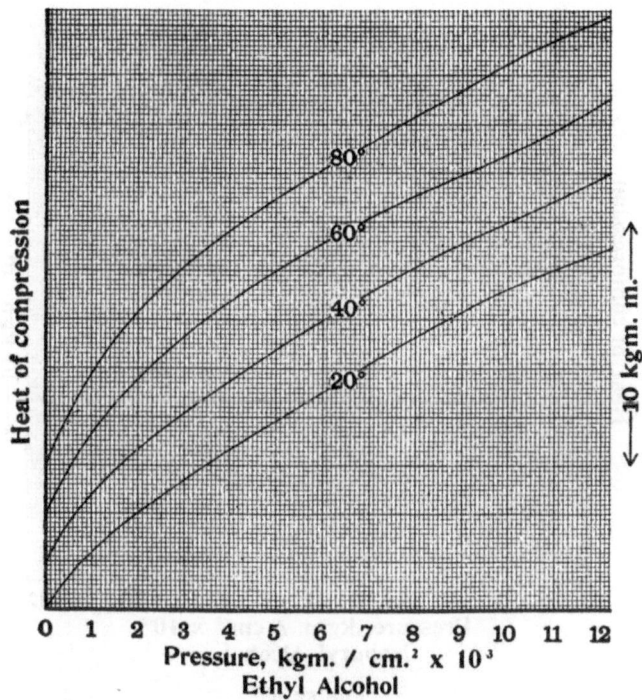

FIGURE 61.

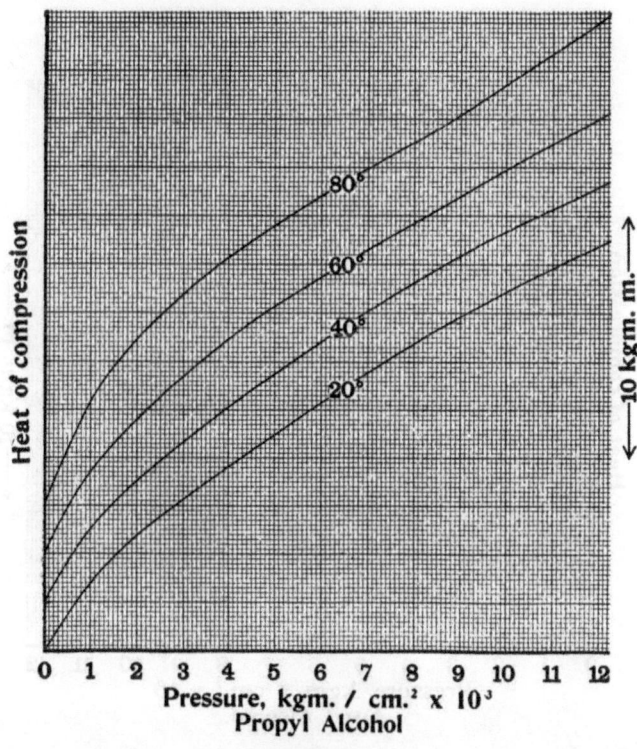

FIGURE 62.

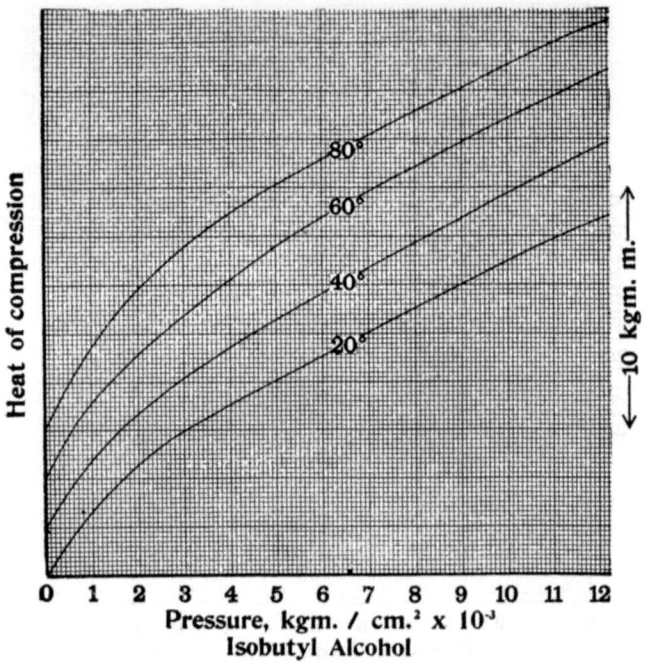

FIGURE 63.

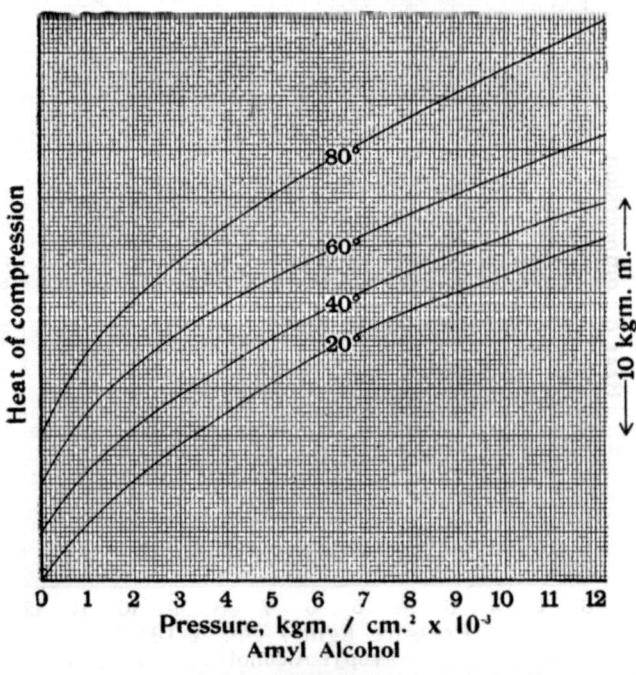

FIGURE 64.

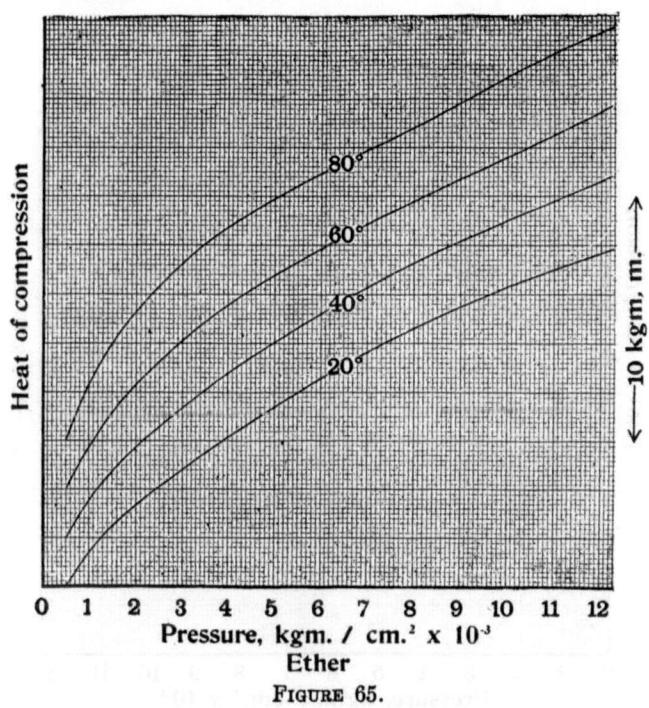

Ether

FIGURE 65.

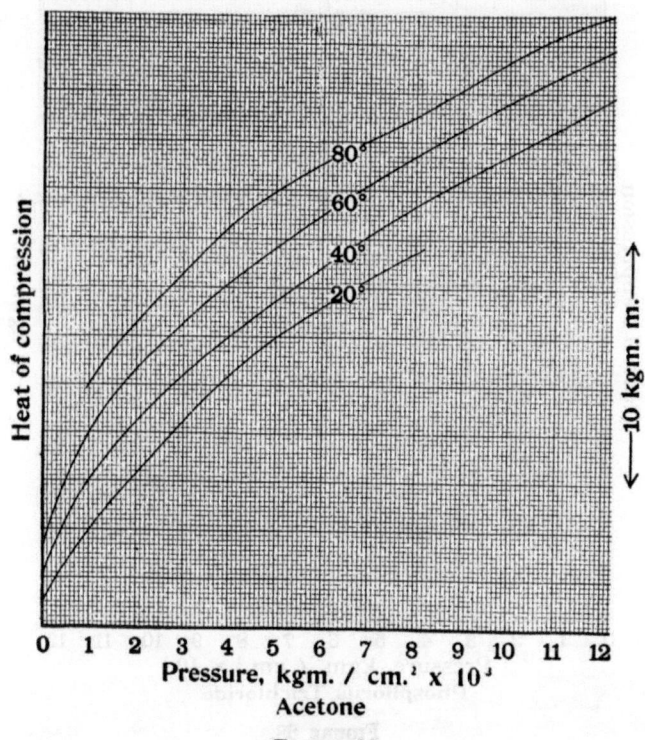

Acetone

FIGURE 66.

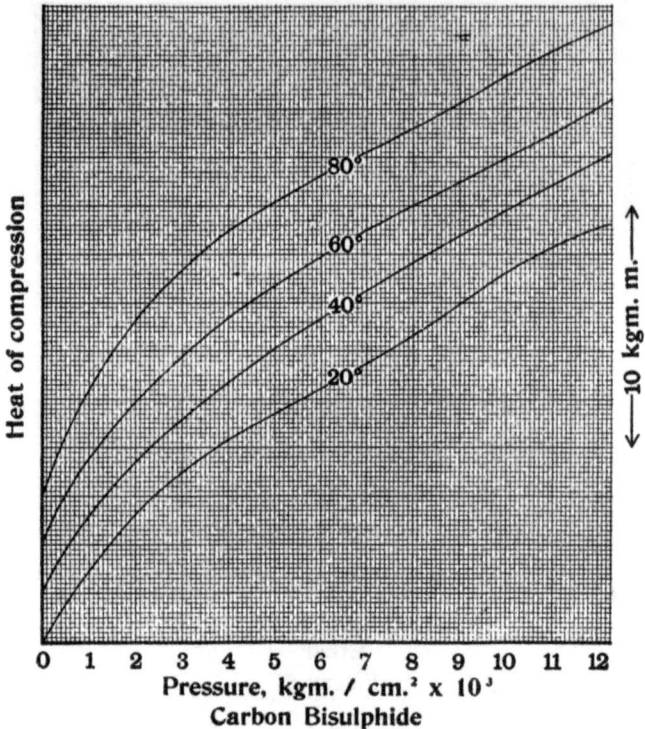

FIGURE 67.

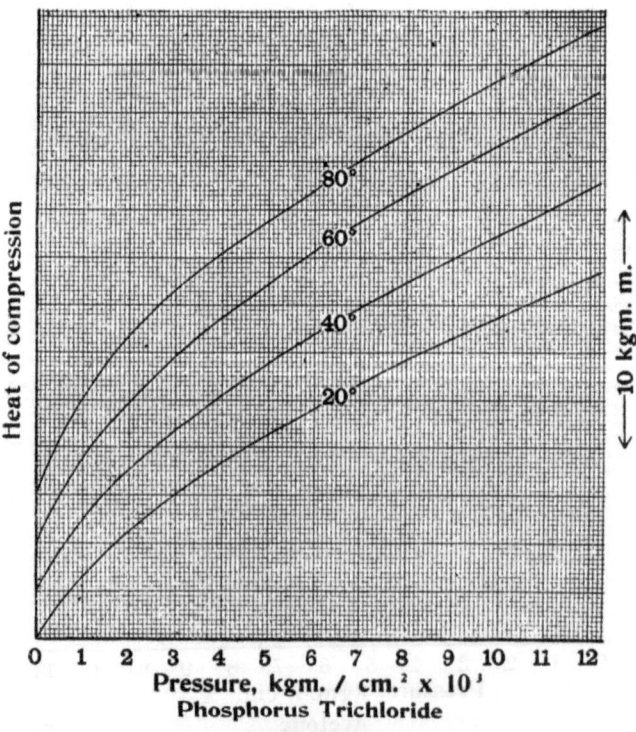

FIGURE 68.

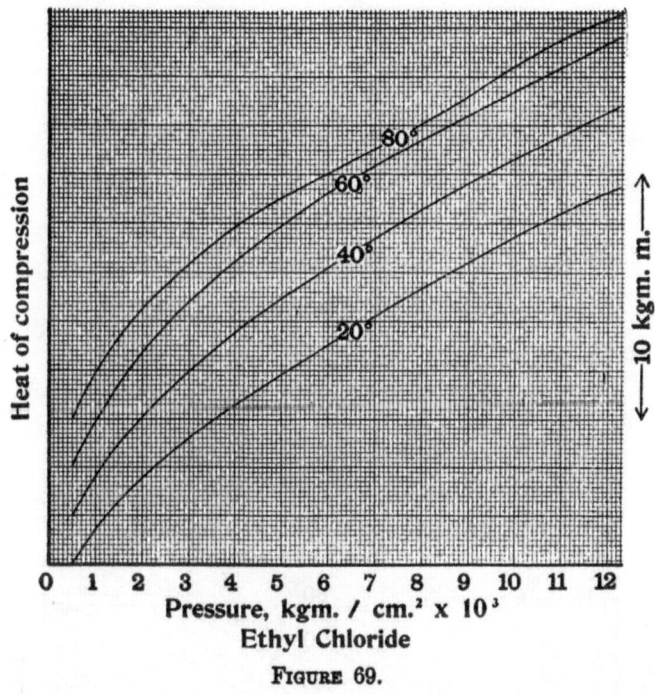

FIGURE 69.

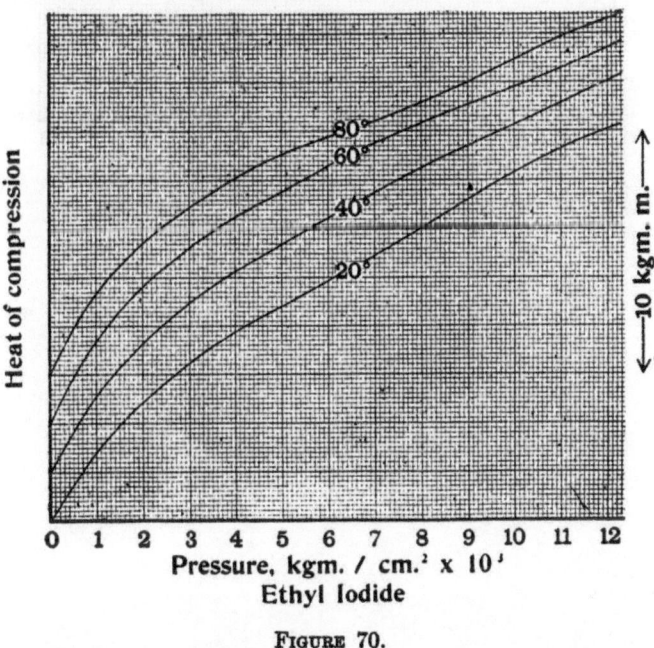

FIGURE 70.

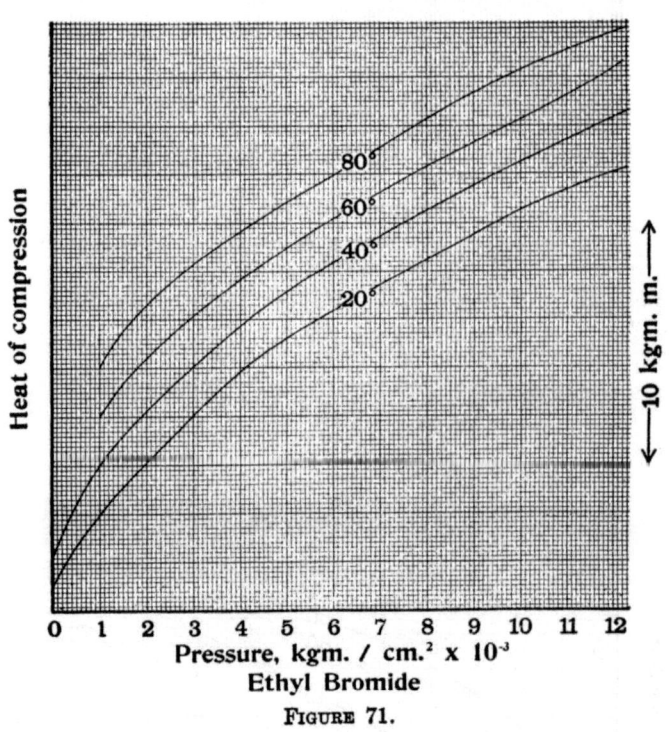

FIGURE 71.

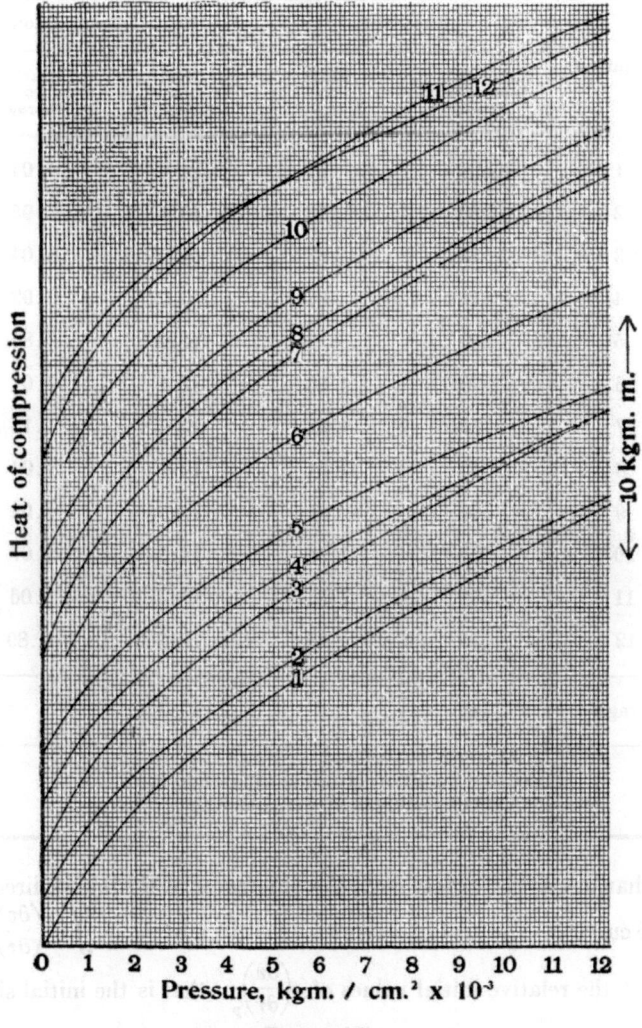

FIGURE 72.

It is true that the curves cross in one or two instances, and are unevenly spaced at the high pressures, but the differences are very much

TABLE XX.

Comparison of Thermal Expansion with Heat of Compression.

Liquid.	Expansion.		Heat of Compression	
	Expansion at Atmospheric Pressure.	Ratio to Average.	Heat at 12000 kgm.	Ratio to Average.
1	.00128	.91	18.0	1.04
2	120	.85	16.4	.95
3	115	.82	18.0	1.04
4	114	.81	16.0	.92
5	106	.75	14.9	.86
6	170(?)	1.20	17.4	1.01
7	172(?)	1.22	19.6	1.13
8	143	1.01	18.0	1.04
9	134	.95	17.5	1.01
10	190(?)	1.35	18.5	1.07
11	185(?)	1.31	18.3	1.06
12	120	.85	15.5	.89
Average	141		17.3	

Maximum variation of initial expansion, 1:1.8
Maximum variation of heat at 12000 kgm. 1:1.3

less than one would expect from the differences in the initial direction of the curves. The heat of compression is given by $Q = \int \tau \left(\frac{\partial v}{\partial \tau}\right)_p dp$, so that the relative initial values of $\left(\frac{\partial v}{\partial \tau}\right)_p$, that is the initial slopes of the curves for Q, give us an idea of what we might expect to be the relative values of Q if the liquids preserved over the entire pressure range their relative initial behavior. Table XX shows this. It gives the initial dilatation and its ratio to the average, the total heat given out at 12000 kgm. and its ratio to the average. The dilatation tabulated for the low boiling liquids was obtained by a linear extrapolation, which gives values too small. Larger values for these liquids would only increase the force of the argument. It is obvious that the two ratio columns show only a general rough agreement; large values of one corresponding to large values of the other. There seems to be no correspondence in the small variations in the ratio column. This shows that the initial behavior of a liquid with respect to the heat of compression does not fix its behavior at the higher pressures. Furthermore, the magnitude of the variations in the ratios of the dilatation are greater than the variations of the heat ratios; a maximum variation in the one column of 1:1.8 against 1:1.3 in the other. This shows again that the liquids become more alike at the high pressures than we should expect from their initial behavior at atmospheric pressure.

The magnitude of the heat of compression is of interest. For the average liquid this is about 17 kgm. m. at 12000 kgm. If we take as a fair average for C_p 25 kgm. cm. this means that the amount of heat flowing out of the average liquid as it is compressed isothermally to 12000 kgm. would raise it through 68° at atmospheric pressure.

Change of Internal Energy. — This quantity is shown on Folder V; for the twelve liquids separately at four temperatures in Figures 73 to 84, and the average results for the twelve liquids plotted against pressure in Figure 85 and against volume in Figure 86. The change of energy plotted in these figures is the internal energy at atmospheric pressure minus the internal energy at the pressure in question. A positive value for the change means, therefore, that the internal energy is less at the higher pressure than it is at atmospheric. The origins of the curves for the separate temperatures have been displaced with respect to each other, so as not to confuse by over lapping. In the cases where the liquid boils at low temperatures the origin of the curve for the higher temperatures has been taken at 500 or 1000 kgm.

The change of internal energy was found by taking the difference between the heat and the work of compression. The curves show nothing, therefore, not already given.

The different substances show irregularities which cannot be dwelt

Folder V. Change of internal energy, $\Delta E = Q - W$, against pressure. Figures 73 to 84 show the change of internal energy at intervals of 20° for the separate liquids. The change of energy plotted is the difference between the energy at atmospheric pressure and the pressure in question, so that a positive ordinate means that the internal energy has decreased with increasing pressure. In order to avoid confusion, the origin of the separate curves is displaced one square with respect to the next. The scale of the diagram is shown at the right hand side. The origin of some of the curves is not at atmospheric pressure, but at 500 or 1000 kgm. This is because the low boiling point makes it impossible to carry the liquid to atmospheric pressure at the temperature in question. In these cases, the change of internal energy shown is the difference between the internal energy at 500 or 1000 kgm. and the pressure in question. Figure 85 shows the average change of energy between 20° and 80° for the twelve liquids plotted in a single diagram against pressure, and in Figure 86, the same quantity against volume. The origin of these curves has been displaced in the usual way, and the liquids have been shown by numbers as is usual, (the order being the same as that of the separate diagrams.) The scale is shown at the right hand side.

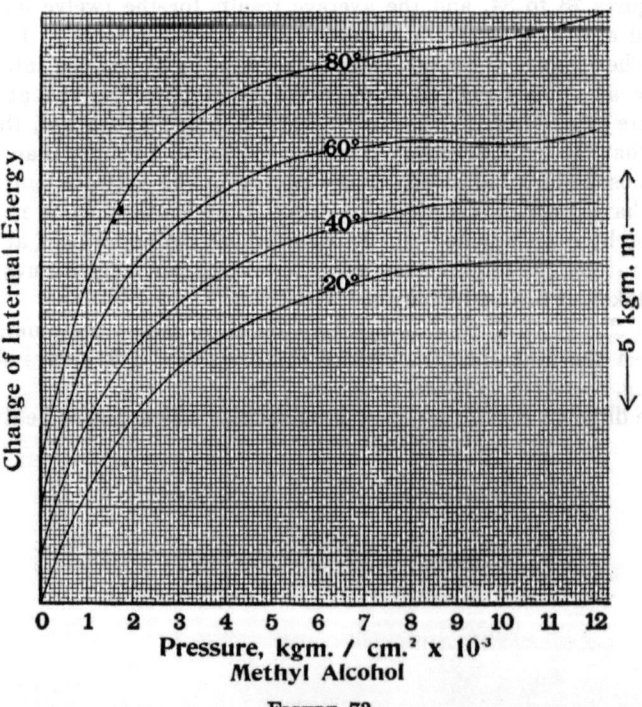

FIGURE 73.

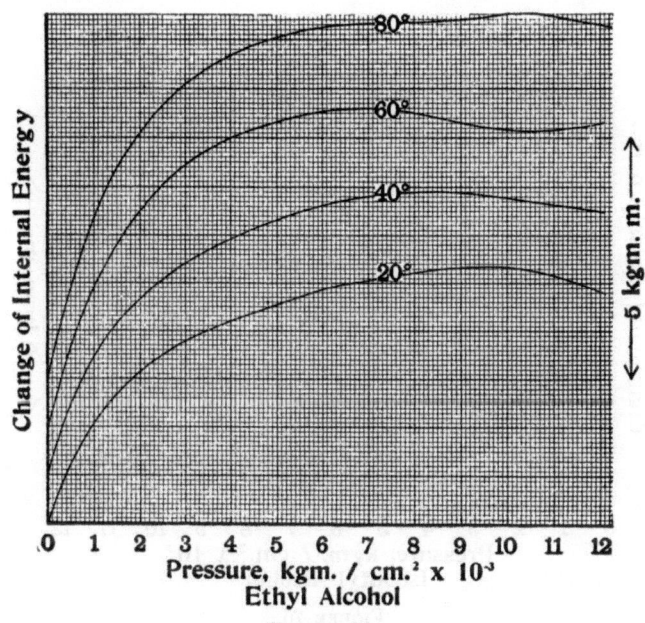

FIGURE 74

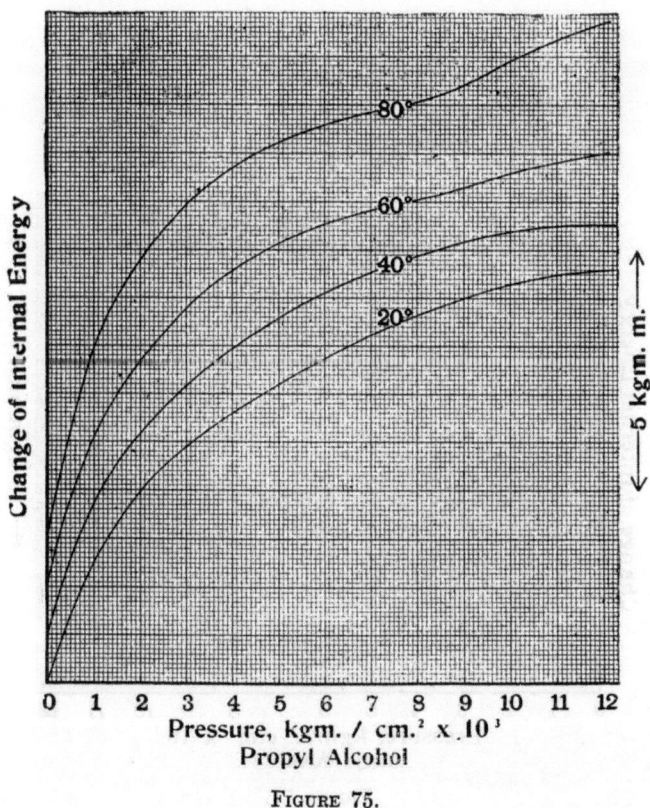

FIGURE 75.

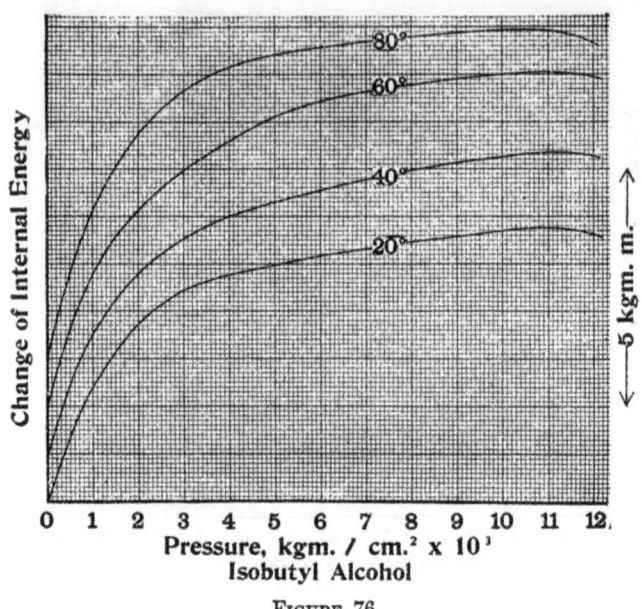

FIGURE 76.

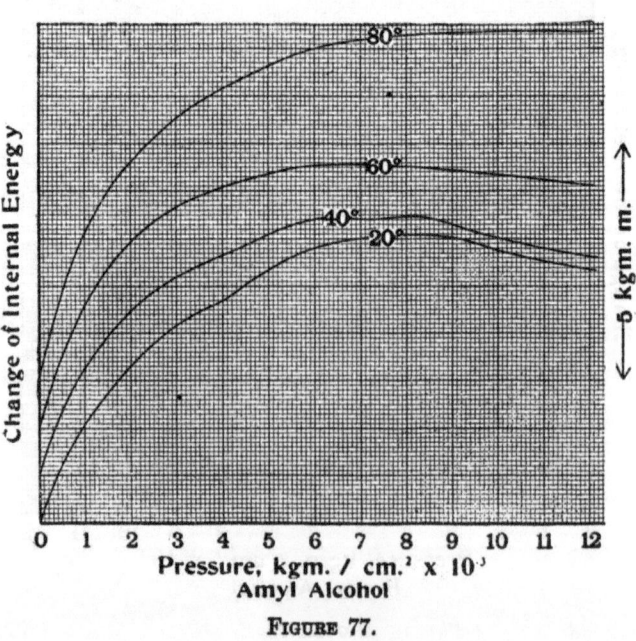

FIGURE 77.

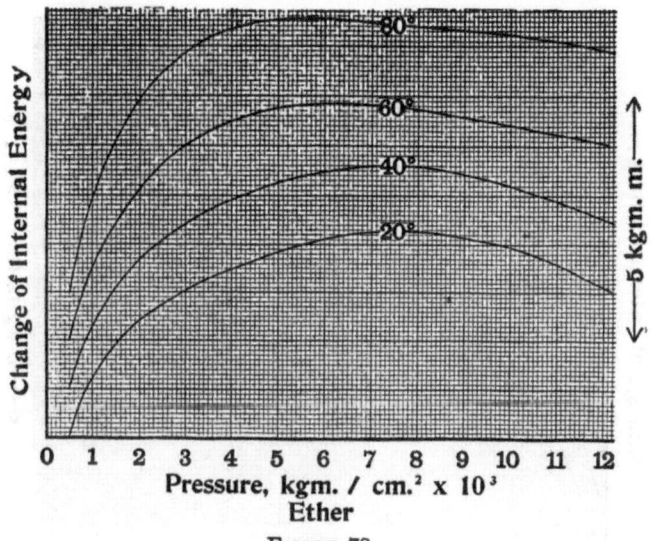

Ether

FIGURE 78.

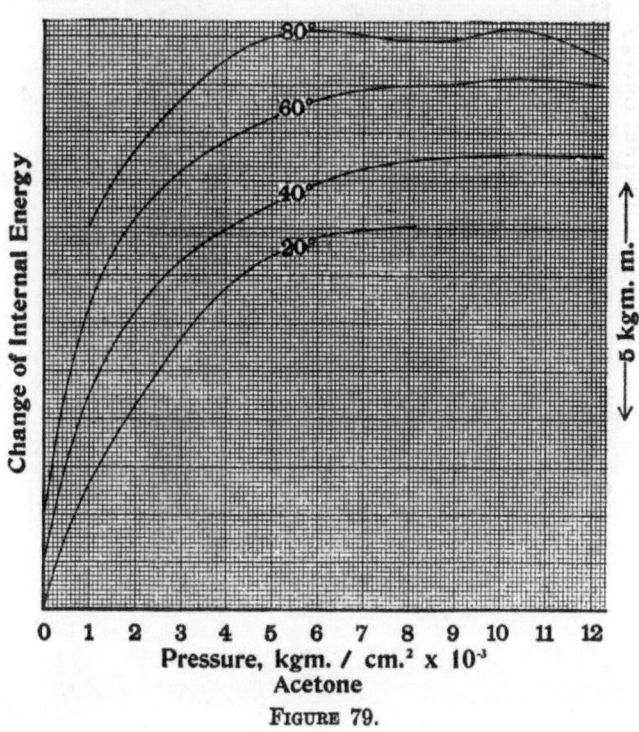

Acetone

FIGURE 79.

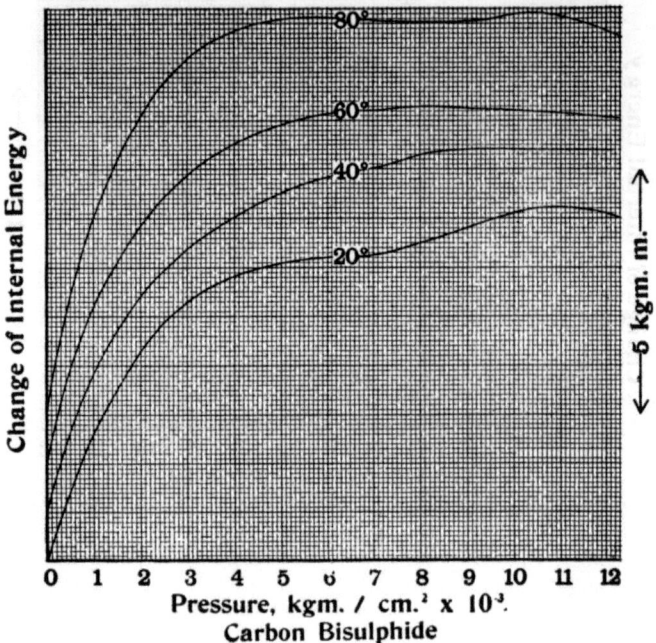

FIGURE 80.

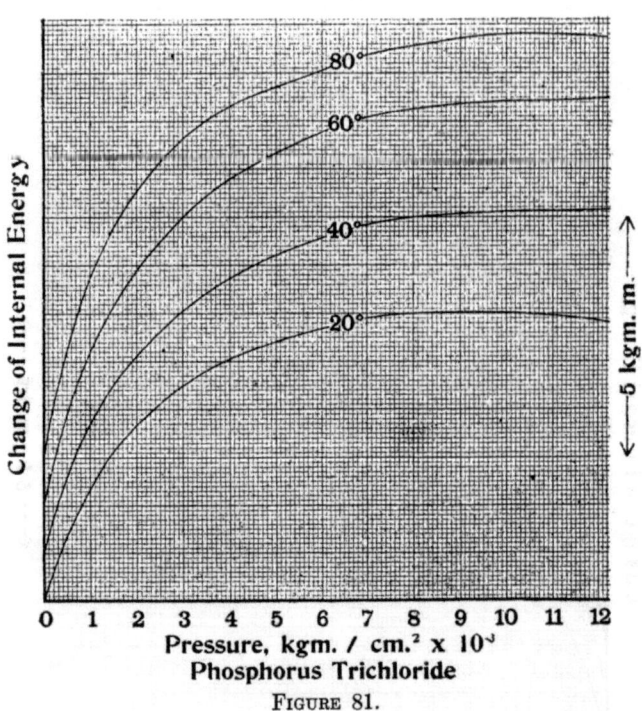

FIGURE 81.

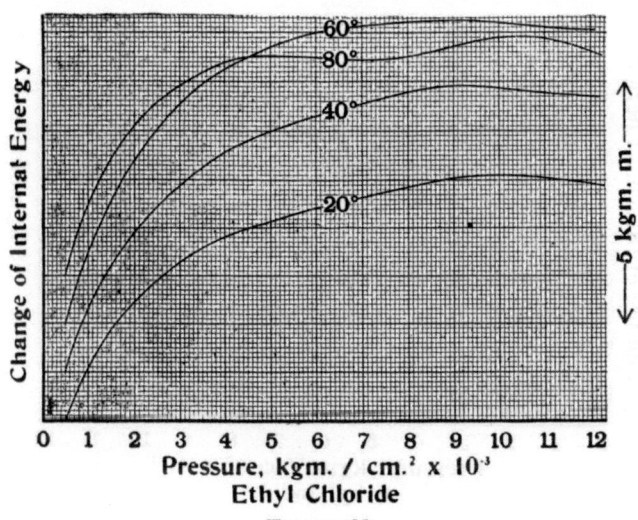

FIGURE 82.

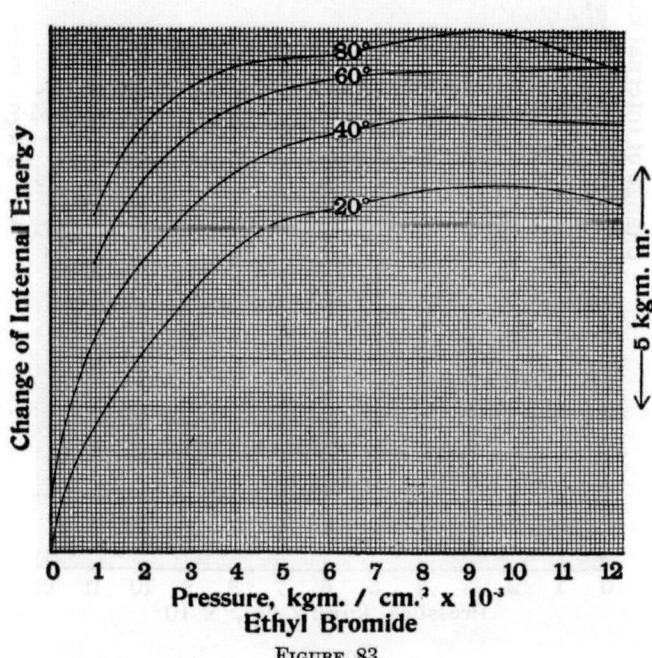

FIGURE 83.

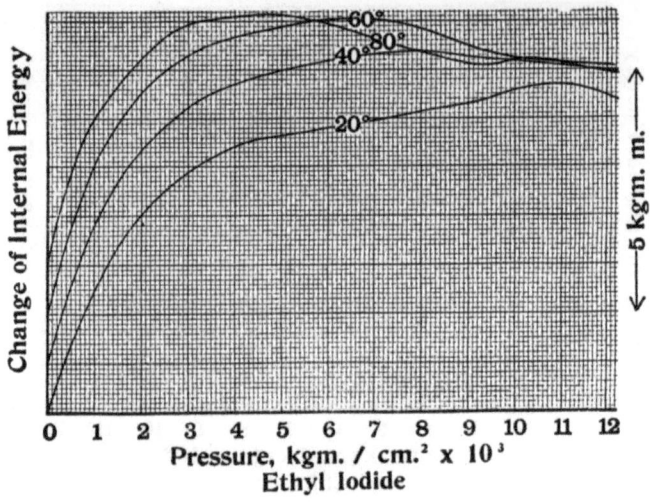

FIGURE 84.

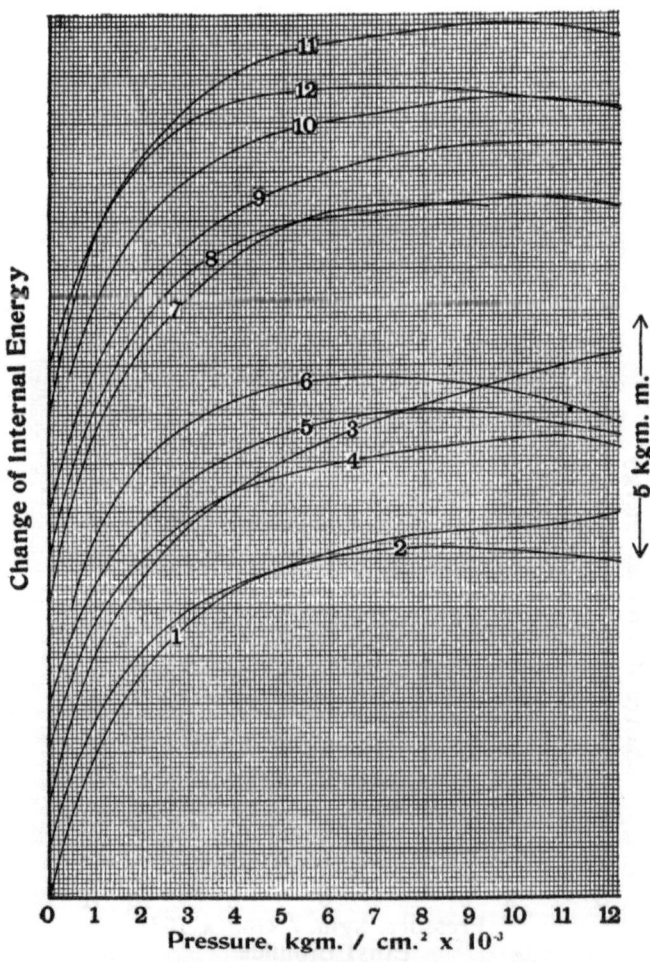

FIGURE 85.

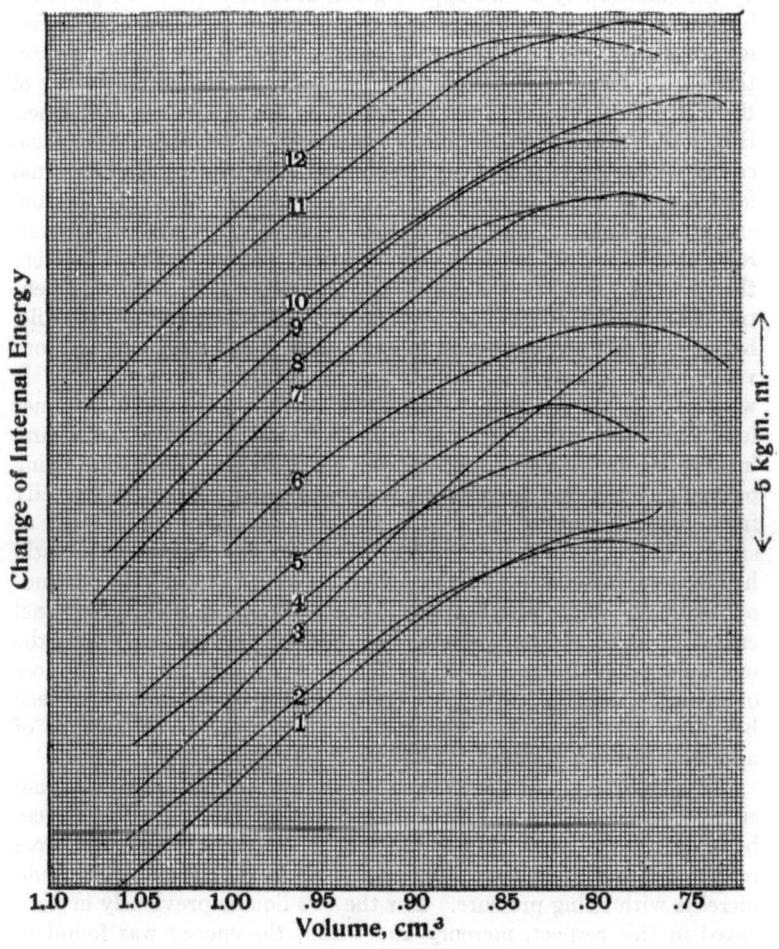

FIGURE 86.

on here. In general, the curves for the higher temperatures tend to draw apart, that is, the decrease of internal energy is greater at the higher temperature, but there are several well pronounced exceptions to this rule. Amyl alcohol, ethyl chloride, and ethyl iodide, for example, are exceptional at the same temperatures where we found exceptional behavior with respect to the work of compression.

The change of internal energy is to be thought of as brought about by the counterplay of two opposing sets of forces, and it is significant because of what it can tell us about these forces. When a substance is compressed, the molecules are brought closer together, the attractive forces between the molecules do work, the potential energy of the attractive forces decreases, and the internal energy decreases. But at the same time, the molecules become compressed by mutual contact, energy is stored up inside the molecule by the external forces in the form of potential energy of strain, and the internal energy increases. Now these two sets of forces play very different roles at different stages of the compression. At low pressures, where the molecules may be thought of as possessing a free path, no potential energy of strain can be permanently stored up in the molecule, because during the motion in the free path it has been entirely converted into translational kinetic or temperature energy. This is what takes place in a gas. But as the volume of the liquid becomes less, the length of the free path rapidly becomes smaller, and at any instant an increasingly large number of molecules is not describing part of any free path at all, but is merely being handed on directly from one collision to the next. At this stage, potential energy of strain can be permanently stored up within the molecule. At still higher compressions, when the molecules are practically in continuous contact, there is still greater possibility of storing up internal energy of strain. The possibility is limitless, provided only that the molecule never becomes incapable of further compression. The loss of internal potential energy by the attractive forces cannot proceed beyond a certain limit, however, imposed by the least distance of approach of the molecules.

We should expect, therefore, that at low pressures the internal energy would decrease with rising pressure, the attractive forces being in the ascendant, but at the higher pressures, where the forces resisting compression have become dominant, that the energy would increase with rising pressure. For the two liquids previously investigated in this respect, mercury and water, the energy was found to continue to decrease over a pressure range of 12000 kgm. The direction of curvature was in each case such as to suggest that this decrease might not continue indefinitely, and it was suggested then that there must ultimately be a reversal of the effect. It seemed surprising that at pressures as high as 12000 kgm. the attractive forces should still do more work than could be stored up as strain by the external forces.

Figure 85, giving the average for the twelve liquids, shows that the anticipation of a reversal in the change of energy is justified for nearly all the liquids; that is, the internal energy, after decreasing for a while, passes through a minimum (on the curves a maximum) and from here increases with rising pressure. The necessary existence of this maximum could of course have been predicted from the curves for the heat and the work of compression, since the one is either linear or concave upwards, while the other is concave downwards.

The change of energy is markedly different for the different liquids, whereas the other thermodynamic properties are similar. Of course the reason is that we are here concerned with the difference of two effects. The position of the maximum of the difference of two functions is very sensitive to slight changes in the functions themselves. Under these circumstances, the mere existence of a maximum is evidence of similarity. The only curves which do not show the maximum are methyl and propyl alcohol. It will be remembered that the work of compression curve for methyl alcohol had a reversed curvature at the upper end, and that the work of compression of propyl alcohol was abnormally low.

It is of interest to plot the change of energy against volume, because it may give information about the attractive forces. If the attractive forces are central forces, functions only of the distance from the centers of the molecules, then the potential energy of the attractive forces will be a function of the volume. Thus if the attractive forces are proportional to the inverse fifth power, as has often been supposed, then the potential energy is inversely as the fourth power of the distance apart of the molecules, or as the inverse four thirds power of the volume. This relation was tried for four of the twelve liquids; for amyl alcohol, ether, phosphorus trichloride, and ethyl iodide. The change of internal energy of these liquids was plotted against $V^{-\frac{4}{3}}$. The diagram was the same in character as the diagram plotting the change of energy against V, except of course that small values of $V^{-\frac{4}{3}}$ correspond to large values of V. Now if the change of internal energy is proportional to $V_0^{-\frac{4}{3}} - V^{-\frac{4}{3}}$, this curve plotted against $V^{-\frac{4}{3}}$ should be linear. The curves were very nearly linear in the initial stages, but of course ultimately diverged greatly from linearity, passing through a maximum. If the straight portion of the curve at the origin represents a region in which the inverse four thirds power law is satisfied, then the tangent to the curve at the origin represents over the entire range of the experiment what the change of energy would have been if the four thirds law had held throughout. The difference between the actual curve and the tangent at the origin is then, according to the above view, equal to the energy which has been stored up as strain inside the molecule. This difference was determined and plotted against volume, in order to find what function of the volume the strain energy might be. For these four liquids, it turned out that the strain energy varies over the entire range approximately as the cube of the change of volume reckoned from a suitable origin. The greatest discrepancy is for ether at low pressures. For PCl_3 a variation of only 0.01 in the arbitrary zero of volume from which the change is reckoned would wipe out the discrepancy; for amyl alcohol a variation in the zero of 0.005 would give perfect agreement; for ethyl iodide a change of 0.01, and for ether a change of 0.04. The variation for ether is all below 3000 kgm.; above this the energy of strain is almost exactly proportional to the cube of the change of volume, taking 1.06 as the origin. The zero of volume for phosphorus trichloride is 1.11, for amyl alcohol 1.07, and for ethyl iodide 1.045. All the curves showed slight consistent variations from the cube law; at low pressures the strain energy varies more rapidly than the cube and less rapidly at the high pressures.

The fact that the internal energy of strain varies as the cube of the change of volume probably does not have very much significance in showing us what the elastic mechanism of the molecule is. If the entire change of volume of the liquid were due to change of volume of the molecules, then we should expect the strain energy to vary as the square of the change of volume, provided the elastic constants of the molecule were unaffected by pressure. If, as is likely, the molecule becomes less compressible at high pressure, then the strain energy would vary less rapidly than the square. But the strain energy was found to vary as the cube. The reason for this is probably that at low pressures strain energy is stored up in only a few of the molecules; those molecules which are describing a free path and are not in contact with other molecules have no strain energy of compression. With increasing pressure the number of molecules in which strain energy is stored up increases rapidly, and the strain energy of each molecule increases at the same time as the square of the strain, so that the total energy of strain would be expected to increase more rapidly than the square of the change of volume, as we found it to. But any more detailed speculation as to the precise way in which the number of molecules taking part in the strain, and the way in which the strain energy of the average molecule varies with the total pressure, would probably be useless because the argument as to the attractive forces breaks down at small volumes. It is probable that the molecules are not really homogeneous spheres, but that there are localities in which the attractive forces are more or less concentrated. Therefore, although we may regard the attraction exerted by the molecule as toward its center, and inversely as the fifth power of the distance when the molecules are separated by wide intervals, we cannot conceive of this law continuing to hold when the molecules are so close as to be in contact. Under these circumstances the force may increase more rapidly than as the inverse fifth power. What is more, the potential energy of the attractive forces will not under these circumstances be a function of the volume only, that is of the mean distance apart of the centers of the molecules, but will also vary with the orientation. We saw that the average orderliness of orientation may be expected to vary with temperature, being on the whole more haphazard for equal volumes at the high temperatures. Even with this picture of what is happening, it would be difficult to say whether the potential energy of attraction should be expected to be greater at the higher or lower temperature. We have seen that the lower temperature usually means a greater space open for occupation, but it may still be that because of the greater approach to order at the low temperatures the localities of intense force are brought closer together, so that the potential of the attractive forces may be less.

The main conclusion to be drawn from the fact that the strain energy of the molecules varies as the cube of the change of volume is, therefore, that at low pressures the greater part of the change of volume is due to the decrease in the distance apart of the molecules, but that at high pressures an increasingly large part of the change of volume is occasioned by the actual change of volume of the molecules themselves.

It is interesting that the initial slopes of the curves of change of energy against volume are very nearly the same for all twelve liquids. This is a little unusual. Previously we have found the twelve liquids to become similar at high pressure, but in respect to the change of energy they appear to be more alike at low pressures.

Specific Heat at Constant Pressure.—The specific heat at constant pressure is shown on Folder VI; the curves for the twelve liquids at four temperatures in Figures 87 to 98, and the collection of the average results in Figure 99. The quantity listed as change of C_p is the specific heat at atmospheric pressure minus the specific heat at the pressure in question. A positive change means, therefore, that the specific heat is less at the pressure in question than at atmospheric pressure. In order not to confuse the curves, the origin for each temperature has been displaced with respect to the neighboring curves. The scale of the drawing is shown at the right hand side.

The twelve liquids show a bewildering variety, so bewildering that speculation as to the cause of all the variations is hopeless. It is to be pointed out nevertheless, that such great variety is to be expected if the molecules take up different positions more or less symmetrical in arrangement with increasing pressure. The process is similar in many ways to a process of association, which is accompanied by much greater changes in the specific heats than in the volume, or compressibility, or dilatation. The curves show some points of similarity, however. It is an almost universal rule that the initial change of C_p at any temperature is a decrease. For the majority of liquids the specific heat on the whole decreases at the high temperature and increases or does not increase so much at the low temperatures. We have seen that as a rule the specific heat at atmospheric pressure is higher at the higher temperatures. For some liquids the temperature effect may be very marked. The change under pressure is in such a direction as to bring the specific heats at high pressures more nearly to equality for all the temperatures. The three halogen compounds are an exception to the rule, however. The very large increase of C_p for ethyl chloride at 80° is very much like that already found in the case of water. It may mean an abnormally high rate of dissociation at the higher pressures. The four normal alcohols show similarities in the abnormally large decrease of C_p at 80°. The decrease evidently cannot go on indefinitely. For methyl and ethyl alcohol, the maximum pressure is sufficient to change the decrease into an increase, as it must eventually, but for propyl and amyl alcohols, the reversal in direction must be at higher pressures than reached here. It seems to be at hand for propyl alcohol.

There is one very rough check which may be applied to the values given here for the specific heats; namely, in no case must the specific heat decrease by an amount more than its original value, for a nega-

Folder VI. The change of specific heat at constant pressure ($\dot{C_p}$) against pressure. Note that the units in which C_p are measured are the number of kgm. cm. of work necessary to raise through one degree the quantity of liquid which at 0° C. and atmospheric pressure occupies one c.c. Figures 87 to 98 show the change of C_p at intervals of 20° for the liquids separately. A positive ordinate means that C_p is greater at atmospheric pressure than at the pressure in question. The origin of each of the curves has been displaced with respect to the next in the usual way, and the scale is shown at the right hand side. Because of the low boiling point, the origin for some of the curves is at 500 or 1000 kgm. In these cases, the difference shown is the difference between C_p at 500 or 1000 kgm. and the pressure in question. Figure 99 shows the average of C_p from 20° to 80° for all twelve liquids. The liquids are indicated by numbers, the origin of the separate curves has been displaced, and the scale is shown at the right hand side, in the usual way.

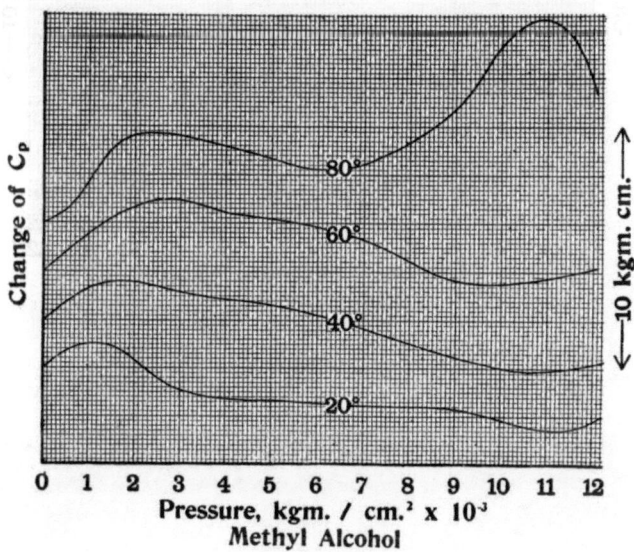

FIGURE 87.

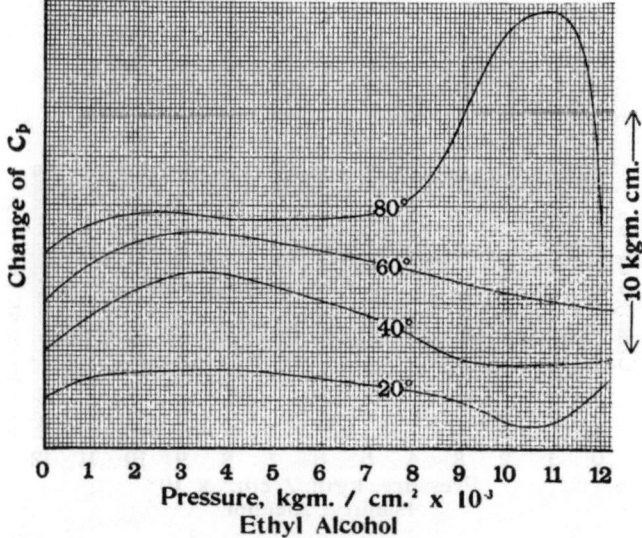

FIGURE 88.

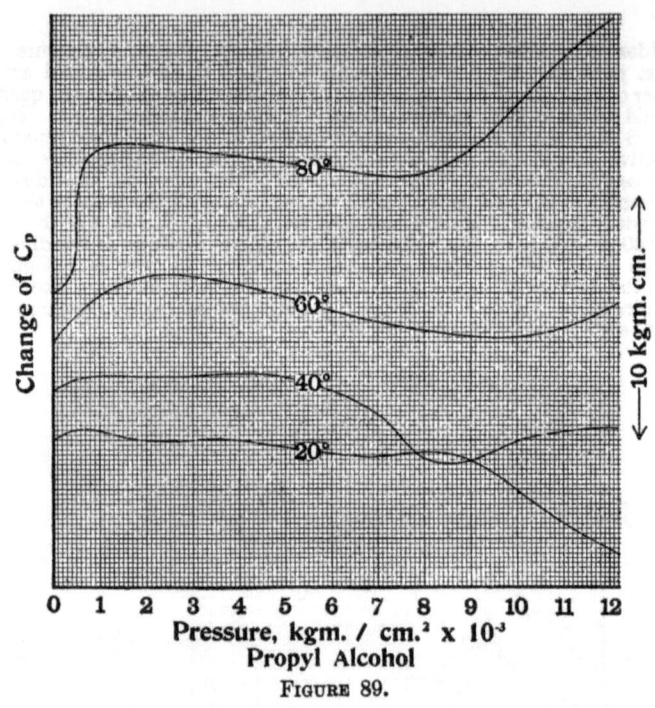

FIGURE 89.

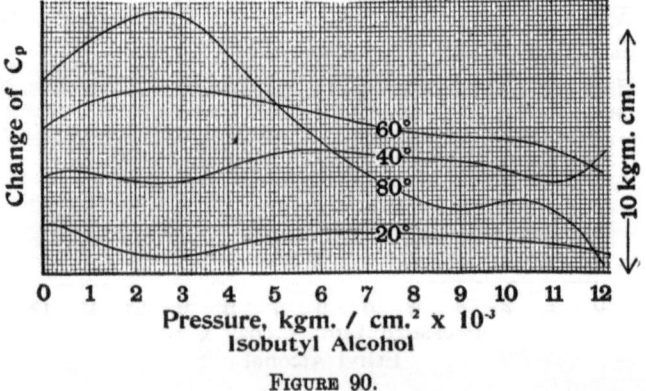

FIGURE 90.

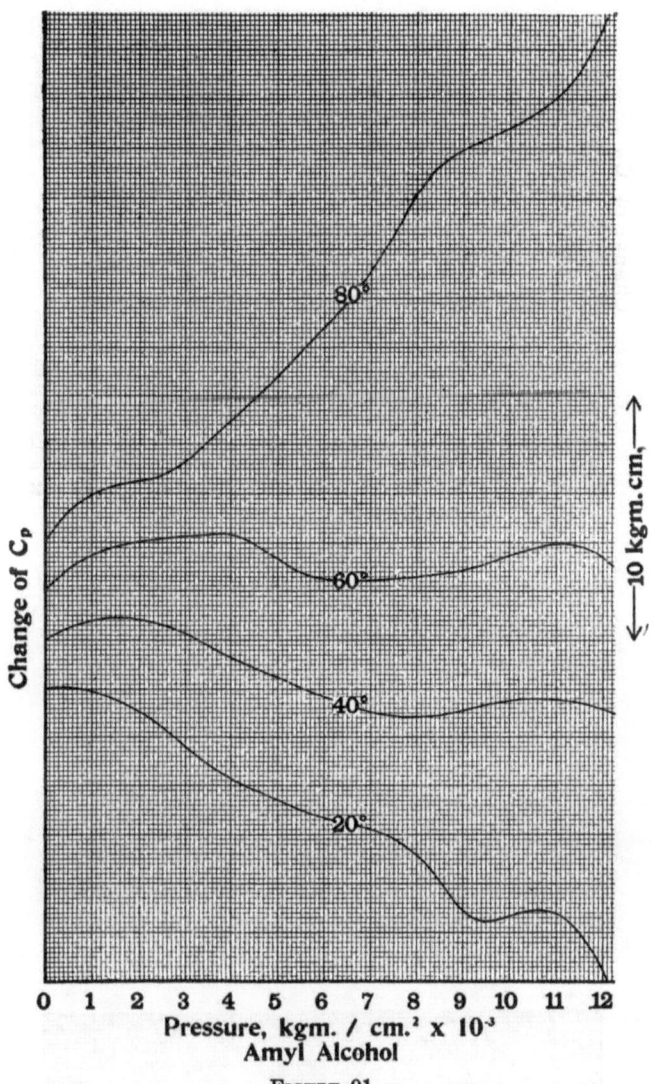

Amyl Alcohol

FIGURE 91.

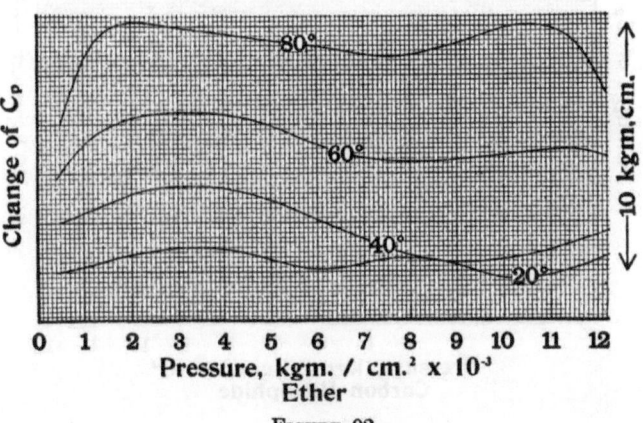

Ether

FIGURE 92.

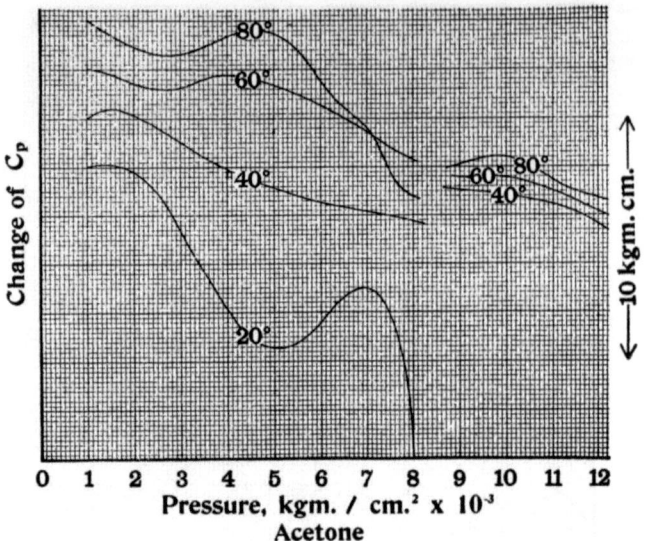

Figure 93.

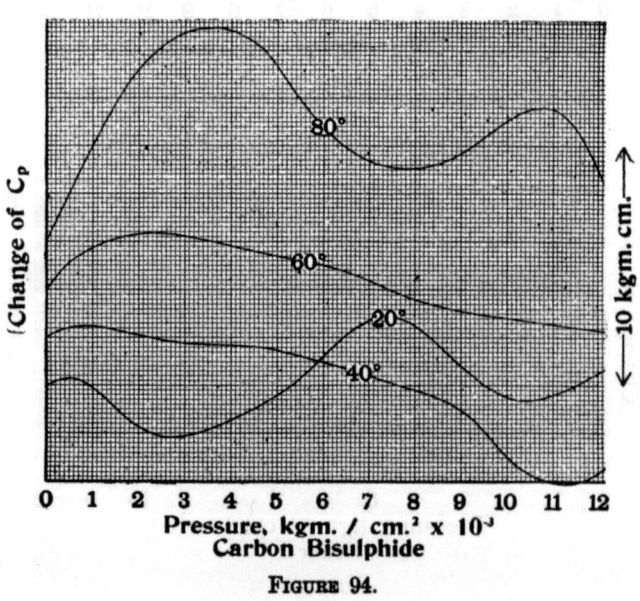

Figure 94.

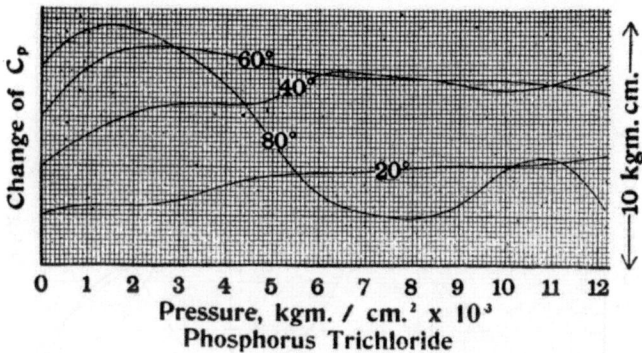

FIGURE 95.

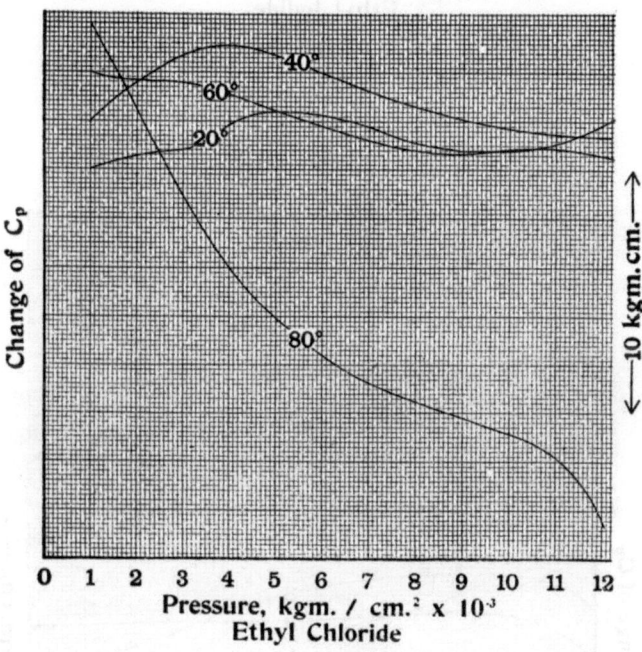

FIGURE 96.

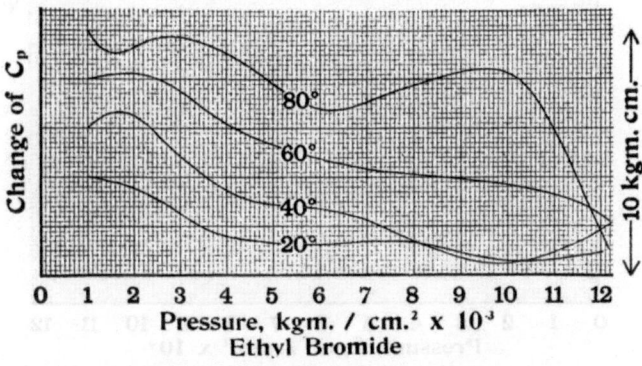

FIGURE 97.

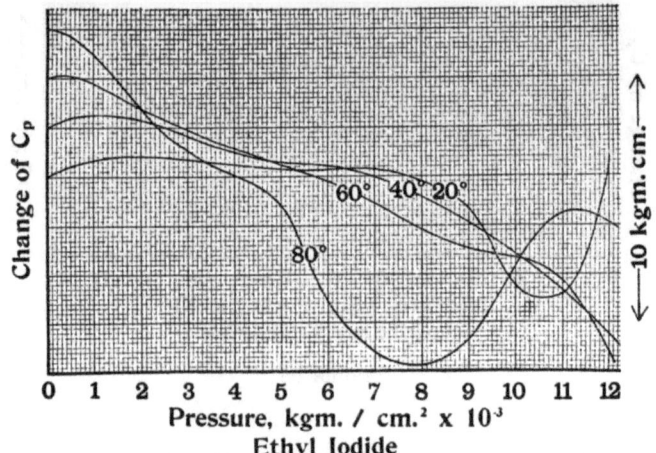

FIGURE 98.

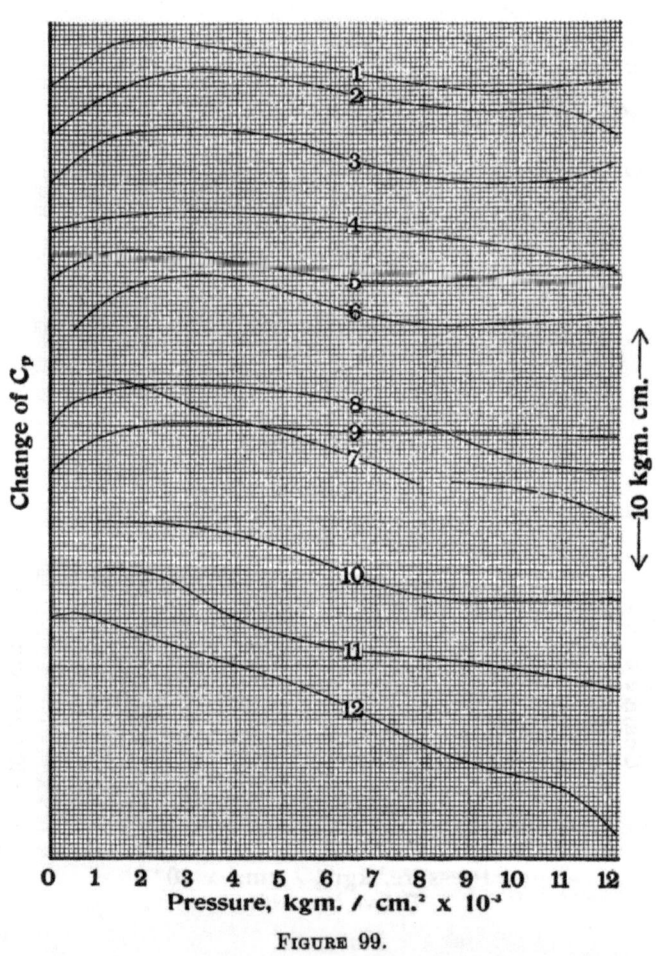

FIGURE 99.

tive specific heat is impossible. The only case where this condition comes anywhere near making trouble is for amyl alcohol. The curve for 80° shows a decrease of C_p of 21 kgm. cm. The data for atmospheric pressure are discordant, but 25 kgm. cm. seems to be a fair average value at 80°. This would mean that the specific heat at 80° and 12000 kgm. is only 4 kgm. cm., about one sixth its initial value. It is evident that the reversal in the effect must come speedily.

Figure 99 for the average change of C_p over the entire temperature range for the twelve liquids shows much less variation from liquid to liquid than one might expect from the irregular variations at the different temperatures. The curves in general all show the same characteristics; at first a decrease of specific heat and then an increase at higher pressures. The first minimum has already been commented on in another connection. Beyond this minimum in C_p (maximum on the curves) there is in general a continuous increase with rising pressure, although there are several cases where C_p decreases again slightly at the highest pressures. The three halogen compounds are exceptional (and also acetone) for the rather large increase of C_p with rising pressure. The increase is greater for the heavier members of the series.

The data of the previous paper on water show the same behavior, a minimum in C_p for all temperatures (except at 0° where the curve is broken off by the freezing) and then an increase with still further increasing pressure. The pressure for the minimum is higher than for these twelve liquids because of the abnormal behavior of water at low pressures.

So far as is known to the author, these are the only measurements from which an attempt has been made to find the specific heat at high pressures. The probable behavior of the specific heats does not seem to have been suspected. Thus Tumlirz deduces from his empirical equation for liquids that both the specific heats decrease with rising pressure. For water, Tumlirz finds the limiting value of C_p for infinite pressure to be about 0.5 gm. cal. He does not compute it for the other liquids. We see from the above that we are to expect for all liquids an ultimate increase instead of a decrease with rising pressure.

Specific Heat at Constant Volume. — The specific heat at constant volume is shown in Folder VII, the curves of the different liquids at different temperatures in Figures 100 to 111, and the collected averages for all the liquids in Figure 112. The treatment of these curves is the same as for C_p.

Folder VII. The change of specific heat at constant volume (C_v) against pressure. Note that the units in which C_v is measured are the number of kgm. cm. of work necessary to raise through one degree the quantity of liquid which at 0° C. and atmospheric pressure occupies one c.c. Figures 100 to 111 show the change of C_v for intervals of 20° for the liquids separately. A positive ordinate means that C_v is greater at atmospheric pressure than at the pressure in question. The origin of each curve has been displaced with respect to the next in the usual way, and the scale is shown at the right hand side. Because of the low boiling point, the origin of some of the curves is at 500 or 1000 kgm. In these cases, the difference shown is the difference between C_v at 500 or 1000 kgm. and the pressure in question. Figure 112 shows the average of C_p from 20° to 80° for all twelve liquids. The liquids are indicated by numbers, the origin of the separate curves has been displaced, and the scale is shown at the right hand side, in the usual way.

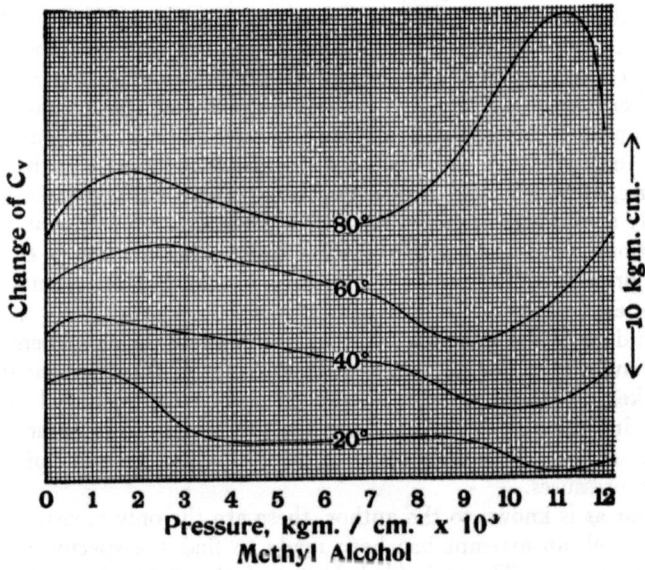

FIGURE 100.

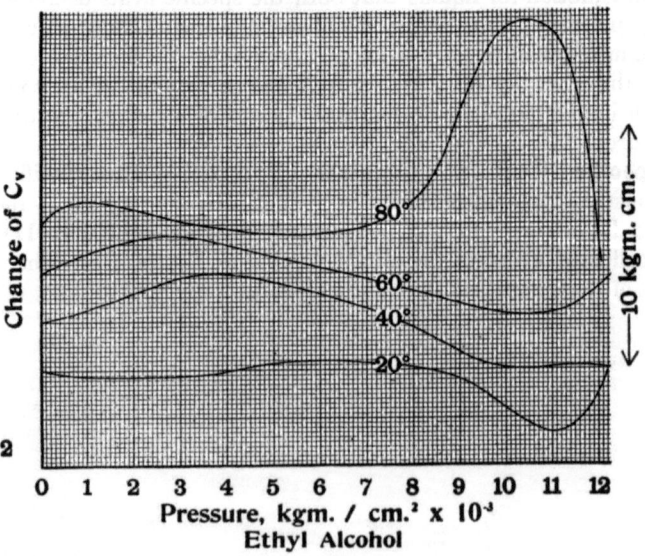

FIGURE 101.

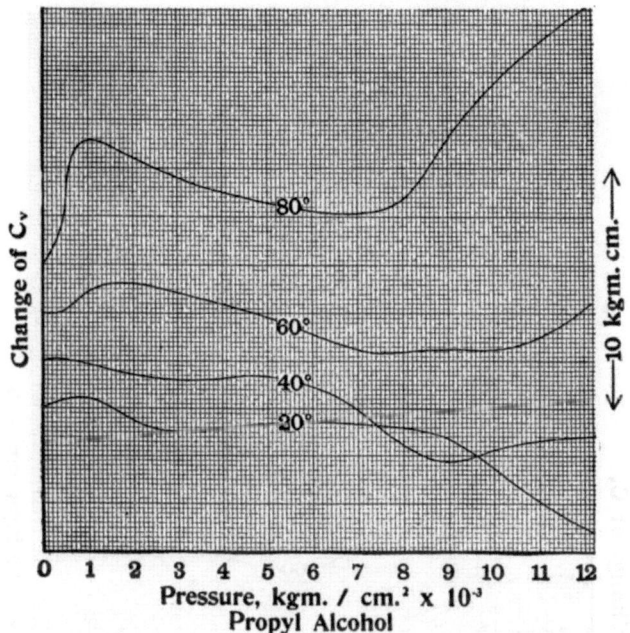

FIGURE 102.

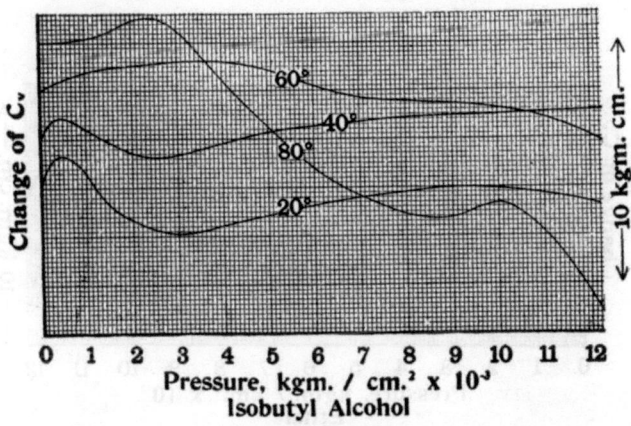

FIGURE 103.

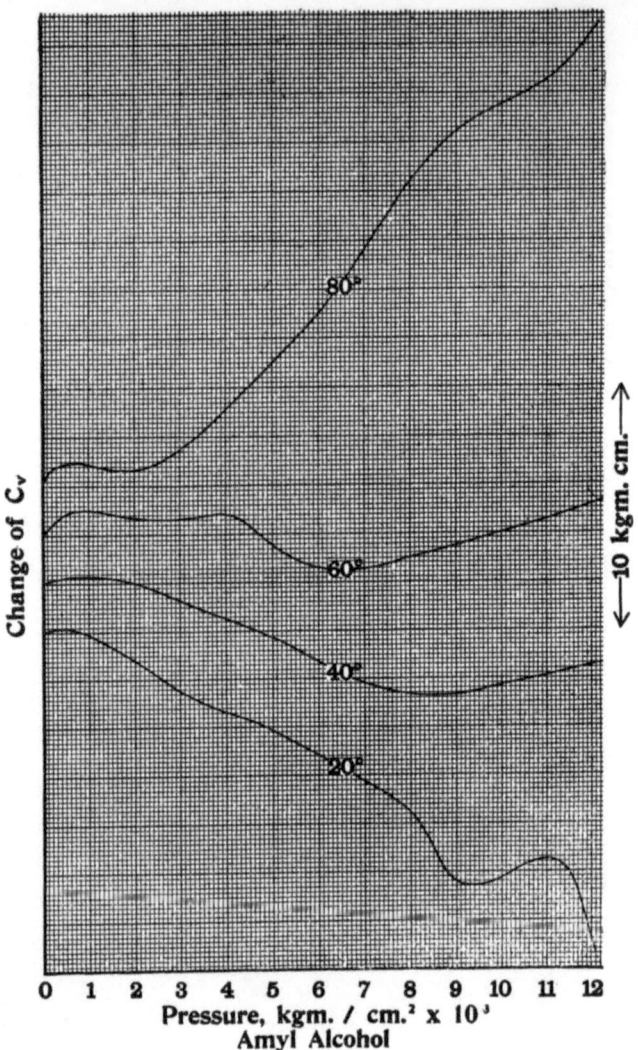

FIGURE 104.

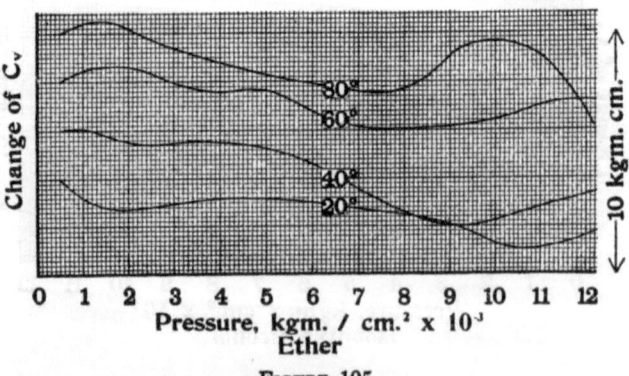

FIGURE 105.

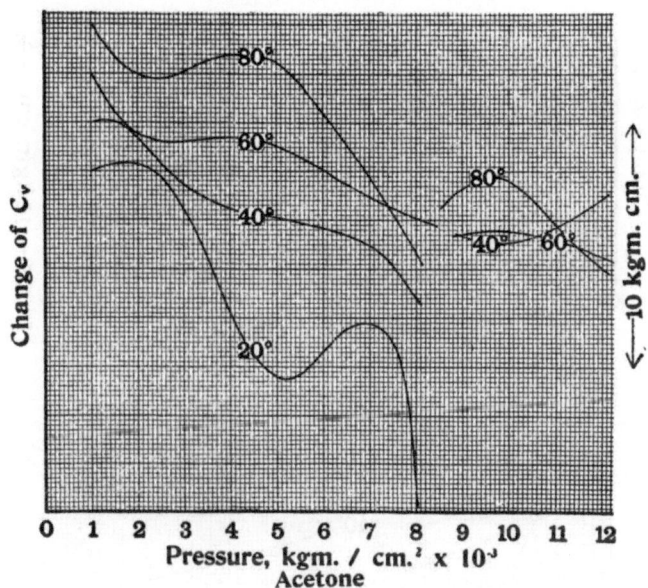

FIGURE 106.

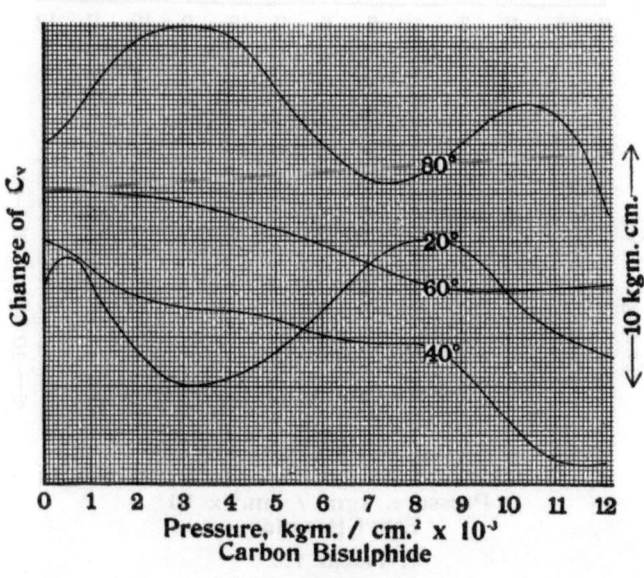

FIGURE 107.

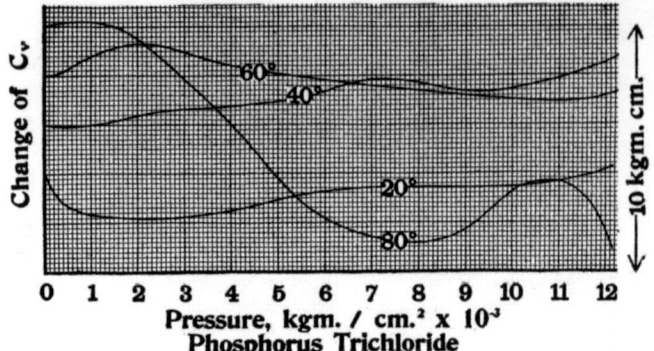

FIGURE 108.

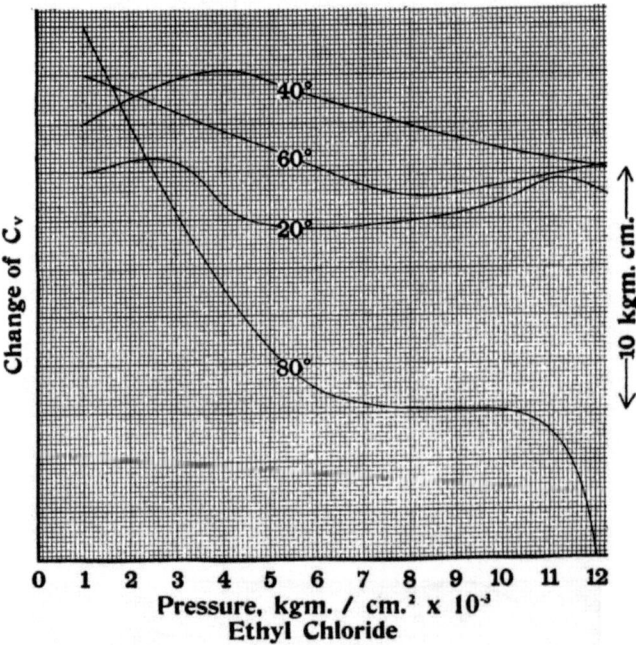

FIGURE 109.

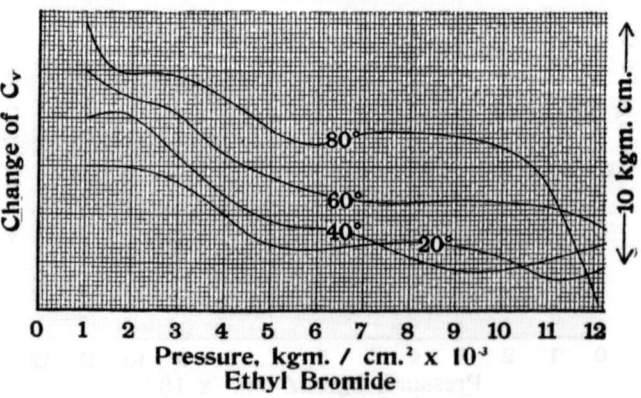

FIGURE 110.

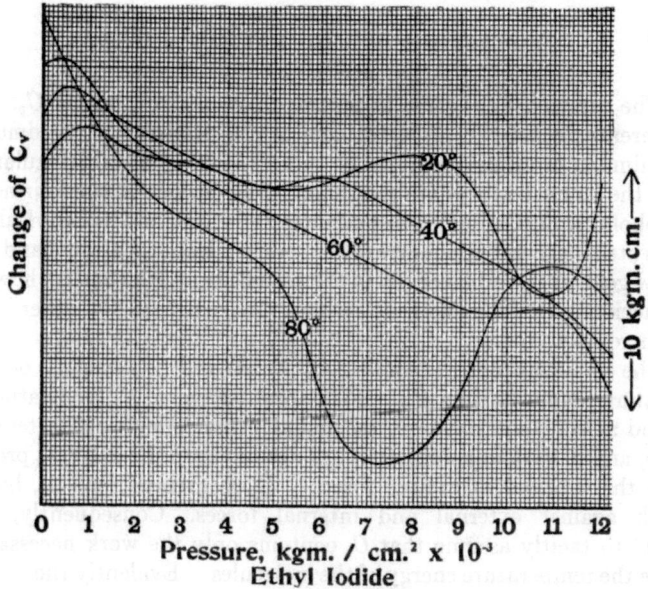

FIGURE 111.

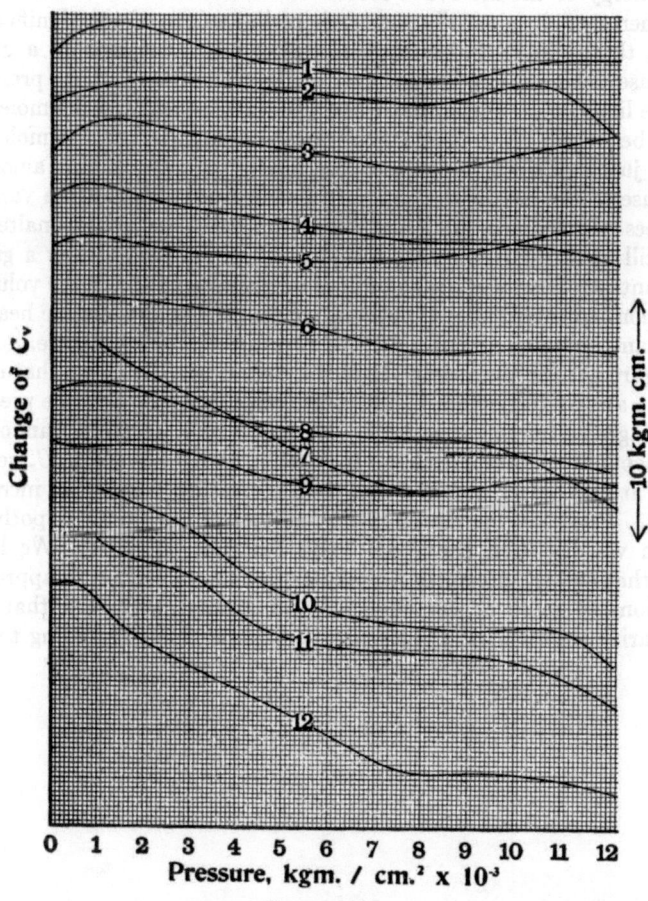

FIGURE 112.

The curves are the same in general character as those for C_p. The differences consist in displacements of the pressures of maximum or mimimum, or occasionally in the suppression of small irregularities; but the larger features are the same. The curves for the four normal alcohols retain the same resemblances as before, and isobutyl alcohol is as strikingly not a member of the series. Isobutyl alcohol and ether show general resemblances as to the specific heat, which may, however, be accidental. The halogens are also different from the other liquids with respect to C_v.

The importance of tabulating C_v is that it is a quantity of much greater simplicity than C_p. When we determine C_p by heating the liquid at constant pressure, the liquid expands with rising temperature, and in so doing performs work against both the external pressure and the internal attractive forces. C_p is greater than C_v by this work against external and internal forces. Consequently, it is usual to tacitly assume that C_v contains only the work necessary to raise the temperature energy of the molecules. Evidently the assumption is merely another way of stating the assumption that the potential energy of the attractive forces is a function of the volume only. Furthermore, it is usually assumed in discussions of the significance of C_v that a given increase of temperature corresponds to a given increase of kinetic energy of the molecule, no matter what the pressure of the liquid. It is true that when the kinetic energy of the molecule is to be raised, more total energy must be imparted to the molecule than just sufficient to increase the kinetic energy by this amount, because of the law of the equipartition of energy among the various degrees of freedom, but if the degrees of freedom remain unaltered, we still have the result that to increase the temperature by a given amount requires the same amount of work independent of the volume. The consequence of the hypothesis would be that the specific heat at constant volume is independent of pressure and temperature.

Figure 112 for the average C_v of the twelve liquids, shows that even on the average C_v cannot be independent of pressure, and the preceding diagrams for the separate liquid show that it certainly cannot be independent of temperature. In general the behavior of C_v seems to be at first a decrease, and then with increasing pressure an increase again. Probably the reason is that neither of the simple hypotheses which we discussed above are valid for high pressures. We have seen that at high pressures we may expect the molecules to approach positions of more or less regularity of arrangement, and that the regularity is greater at the low temperatures. Now according to the particular arrangement which the molecules adopt, and the relation of the local centers of force to this arrangement, it is conceivable that the potential of the attractive forces should be either greater or less at the lower temperatures and equal volumes. The likelihood is, however, that the potential will usually be less at the lower temperatures. Similarly, it is usually more likely that the potential of the attractive forces should be less at the higher pressures. As a rule, then, an increase of temperature of one degree at a higher pressure or a lower temperature must provide the energy to do more work against the attractive forces, so that the specific heat at constant volume will be greater at higher pressure and lower temperature. But in those more infrequent cases where the potential energy of position is less at the higher temperature or lower pressure, the specific heat will be less at higher pressures and lower temperatures. It is as a rule true, as we have seen from the curves, that C_v does become greater at the higher pressures and lower temperatures.

The considerations just discussed are somewhat similar to considerations regarding the association of the molecules, but do not in all cases lead to the same results. For instance, if we suppose a liquid of single molecules to associate to one of double molecules, the specific heat of the associated liquid would be one half that of the simple one, if we neglect the effect of the altered number of internal degrees of freedom.

The second hypothesis made above, that a given increase of temperature always corresponds to the same increase of molecular energy, probably breaks down also at high pressures. The difficulty of determining what happens in this case is increased by uncertainty as to what the definition of temperature shall be at high pressures. We may perhaps, however, think of temperature at low pressures as being roughly proportional to the average translational energy of the molecule during its free flight. Now we have seen that as pressure increases, the time of free flight decreases rapidly, and an increasing fraction of the time is spent in collision. During collision the kinetic energy of translation has become potential within the molecule. The result is that as pressure increases, the potential strain energy of the molecules becomes a greater part of the total energy, leaving a smaller residue to become kinetic. Now if temperature corresponds to translational kinetic energy, it is evident that at high pressures more total energy must be imparted to the substance to increase the translational energy a given amount, or in other words, the specific heat will increase with increasing pressure.

The initial decrease of C_v may very possibly be an association effect. A word should be said about this association effect. The reason that an association from single to double molecules, for example, reduces the specific heat to one half, neglecting the effect of the internal degrees of freedom, is that a rise of temperature of one degree corresponds to a definite increase in the kinetic energy of each molecule, and when there are half as many molecules, half the additional energy is needed to increase the energy by an amount corresponding to one degree. This means that association has an effect on specific heat as long as temperature remains a molecular affair. This is true for a gas. But the law of atomic heats for solids suggests very strongly that in solids temperature is no longer an affair of the molecule, but has become a matter of the atom. We naturally expect somewhere a transition from the one state to the other, and the liquid is the natural place to look for it. That is, if a liquid temperature is on its way from being an affair of the molecule to becoming an affair of the atom. It is not likely, therefore, that the existence of large molecular complexes in a liquid, as in the case of association, will modify very much the behavior with respect to temperature, and in particular to the specific heats. Such effect as there is should be looked for at low pressures.

A word should be said about the curves for the average of C_p and C_v over the entire temperature range. The average C_p is the total heat absorbed at constant pressure between 20° and 80° divided by 60. C_v is the average of the four values at 20°, 40°, 60°, and 80°. These two averages are not always equivalent when there are large variations with temperature, but they are always approximately equal. In cases of question, the average C_p corresponding to C_v may be found from the curves for C_p at the four temperatures. To find the average C_v corresponding to the average C_p would involve a more complicated procedure.

General Discussion of the Bearing of the Results on a Theory of Liquids for High Pressures.

It is proposed to discuss here the nature of the problems which confront one at high pressures. No attempt is to be made to develop a new theory, but merely to indicate some of the directions in which the data of this paper suggest modifications of conceptions which have hitherto worked at low pressures.

Perhaps the most far reaching modification is in regard to our idea of the kinetic origin of pressure. We suppose that the pressure of a gas, for instance, is produced by the impact of the molecules against the walls of the vessel; and we compute the magnitude of the pressure from the total change in one second of the momentum of the molecules striking on unit area of the wall. The velocity with which the molecules strike the walls decreases with falling temperature and vanishes at the absolute zero. The pressure exerted by a liquid also is thought of in most attempts at a theory of liquids as exerted by the same mechanism and is computed in the same way. But it is entirely obvious that the molecules may exert pressure on the walls in another way. It is inconceivable that at the absolute zero a substance should not resist an attempt to compress it; yet at this temperature there can be no kinetic effect exerted by the molecules as a whole. Under these conditions the molecules transmit pressure in the same way that a compressed spring does; that is, they behave like elastic solids. It does not concern us here to inquire what the ultimate origin of this elasticity is; it may be kinetic within the atom. The point is simply this; from our point of view, which regards the molecule as a whole, we must recognize two possible functions or modes of action; the molecule may behave like a moving centre of mass with kinetic energy and momentum, or it may behave like an elastic solid. The molecule may exert pressure in virtue of either one of these two modes of action. Under ordinary conditions the momentum effect of its motion as a whole is almost the entire effect. But if we examine the mathematical analysis which justifies us in putting the pressure equal to the total change of momentum in unit time, we shall find that we made certain simplifying assumptions. We assumed that each collision with the wall is unimpeded; that is, the molecule approaches the wall during free flight, has a single encounter and makes a clean get-away. And this assumption was necessary; as soon as the molecule is interfered with during its collision, as it may be by a collision from another molecule from behind, the simple change of momentum relation ceases to give the pressure, and we must treat our molecule during collision as an elastic solid. Now it is evident that as the volume becomes less, that is, as the pressure increases, the total number of collisions with interference will increase, and our kinetic conception of pressure becomes less and less useful.

A very simple model, which may or may not correspond to the physical facts, is instructive in showing how under different conditions we may compute the pressure from the momentum effect alone, or must consider also the elasticity effects. The substance we are to consider consists of a single molecule. The molecule consists of a heavy particle of mass m, with two weightless springs projecting on either side of length l_0, giving a total diameter of the molecule of $2l_0$. (Figure 113.) This molecule travels back and forth in a horizontal line between the two opposite vertical walls of the enclosure.

The distance apart of the walls is the volume of the enclosure, the force exerted by the springs on the walls during an encounter gives the pressure exerted by the substance, and the kinetic energy of the particle at its maximum velocity represents temperature.

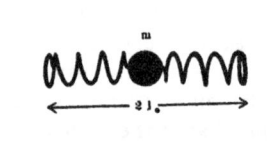

FIGURE 113. Model of a substance consisting of one molecule. The model is to show that at high compressions the pressure is not given by the change of momentum of the molecules striking the walls of the vessel in unit time.

This model is so simple that the entire discussion may be carried through with rigorous mathematical analysis. The pressure exerted on the walls is evidently to be found by taking the time average (over a long interval of time) of the force exerted on the walls by the springs during the encounters. Now in the solution of this problem there are three different cases.

I. The first case is when the distance apart of the walls of the vessel is greater than $2l_0$, the diameter of the molecule. Under these conditions we have collision without interference, which are the only conditions to which the usual analysis applies. In this case it may be proved by a detailed mathematical solution which will not be given here that the time average of the force is exactly equal to the change of momentum in unit time with respect to one of the walls, as it should be. The results of the mathematical analysis for this case may be given in the form of a distinctive equation of the substance,

$$P[(v-b) + a\, T^{\frac{1}{2}}] = 2\, T.$$

Here P is pressure, v = volume (distance apart of walls), b = diameter of molecule ($2l_0$), $a = \pi\sqrt{\dfrac{2}{k}}$, where k is elasticity of the springs, and T = absolute temperature (kinetic energy of particle). The equation bears a resemblance to van der Waal's equation without the attractive forces. In van der Waal's equation $a = 0$, or $\kappa = \infty$, which means that the time of collision of the particles with the wall may be neglected in comparison with the time of free flight.

II. The distance apart of the walls is less than $2l_0$, the diameter of the molecule, and the kinetic energy of agitation is so small that at no time is the molecule out of contact with *both* walls simultaneously. The detailed mathematical solution shows that under these conditions the momentum effect has no influence whatever, and the pressure is determined simply by the relative magnitude of the volume and the unstressed diameter of the molecule. Under these conditions we obtain as the characteristic equation

$$P = \frac{k}{2}(b-v).$$

It is remarkable that the temperature has disappeared from the characteristic equation, or, in other words, the thermal dilatation is zero. The substance still remains compressible, however. Something like this was expected for the liquids of these experiments; that is, it was expected that the dilatation would tend to vanish more rapidly then the compressibility.

III. Case II passes into this case when the violence of the temperature vibration becomes so great that during part of the vibration the molecule is in contact with only one wall. The critical temperature at which this occurs is when $T = k\left(\frac{b-v}{2}\right)^2$. The mathematical analysis is more complicated for this case, because the motion must be split up into two stages, during which the restoring forces are different functions of the displacement. But just as in Case II, the change of momentum of the molecule in unit time does not give the mean pressure exerted on the wall. The complete equation of state for Case III is complicated, involving antisines, so that it is hardly worth giving. It reduces to one or the other of the other two cases, however, under proper critical conditions. In this case, the pressure computed by the change of momentum is too low, as we should expect it would be, because we have neglected an elasticity term which modifies conditions when the molecule is in contact with both walls.

The sequence of events when we compress a substance at a given temperature from a large volume is first Case I, then Case III, when $v = b$, and then Case II, when the volume has been still further reduced by an amount depending on the temperature. Case I passes smoothly into Case III without discontinuity in either compressibility or dilatation. The difference between Cases I and III is that a higher pressure corresponds to a given volume in Case III than we should expect from the formula of Case I. The pressure at a given volume in Cases I and III depends on temperature, but in Case II, the relation between volume and pressure is the same for every temperature. The only way in which temperature has an effect in Case II is with regard to the critical temperature which determines when Case II passes into Case III. The pressure for a given volume in Case III ultimately becomes less than we would have expected it to be if we extrapolated by the formula of Case I. In other words, the substance is more compressible than we should expect it to be from its behavior at low pressures. This reminds us of the formulas of Tumlirz and Tammann. It should be said however, that the mathematical analysis applied to Case III cannot continue to have physical significance indefinitely, for it was assumed that the springs obey Hooke's law, which is true only for small displacements. The characteristic equation given above for Case III predicts the vanishing of the volume at a finite pressure.

Something similar to the action of the model must take place in a liquid. At any instant there are collisions taking place, some free collisions similar to those of Case I, some collisions with interference like Case III, and some contacts between molecules like Case II, which are not properly called collisions at all. The momentum computation of pressure applies only to Case I. At low pressures, by far the greater number of collisions is of type I, but as the volume decreases and pressure increases, collisions of type II and III become increasingly predominant, until at infinite pressure we may suppose type II only to be present. Under these conditions, the momentum effect has absolutely no connection with pressure.

The ordinary conception of internal pressure must obviously be modified in a similar way. There are many different meanings attached to "internal pressure." One way of defining internal pressure is by constructing an imaginary surface in the interior of a liquid, and finding the momentum of all the molecules which cross this surface in unit time. This process evidently fails to have physical meaning when there are molecules in the liquid in contact with each other for any length of time. The modification of the definition necessary to meet these new conditions would be very complicated. It seems better under the circumstances to give up this conception of internal pressure altogether. Other conceptions may still be useful at high pressures; such, for example, as to regard the internal pressure as the external pressure plus the unbalanced attractive effect of the molecules at the surface of the liquid. But it is a difficult matter to define internal pressure in such a way as to have a physical meaning for one of Maxwell's demons inhabiting the interior of a liquid.

Accordingly, in the previous discussion no use has been made of this "internal pressure."

The usual kinetic conception of temperature must undergo modification at high pressures just as our kinetic conceptions of pressure. We think of the temperature of a gas as proportional to the average kinetic energy of translation of its molecules, the translational energy being the energy of the center of mass during free flight. Now it is inconceivable that at infinite pressure there should be any free flight, that is, there can be no kinetic energy of the molecule as a whole, but it is also inconceivable that at infinite pressures a substance should not possess temperature and be capable of temperature equilibrium with surrounding objects. Our model of the molecule may be helpful to us again. When the volume decreases beyond a certain limit, we saw that the boundaries of the molecule become fixed in position, and that the temperature is represented by the energy of internal agitation. This suggests that temperature changes in character from an affair of the molecule as a whole at low pressures to an affair of agitation within the molecule at high pressures. The behavior of the specific heats of gases and solids also strongly suggests the same thing. We compute the specific heat at constant volume of a gas by supposing that the kinetic energy of translation of each molecule must be increased by a fixed amount to produce a rise of temperature of one degree. But for most solid elements the law of Dulong and Petit holds, which is equivalent to the statement that to increase the temperature of a solid by one degree we must increase the energy of each atom by a fixed amount. Now a microscopic analysis of a solid like iron discloses a crystalline structure of great complexity; we find it hard to think that there are not groups of associated molecules and that the molecules are monatomic. It appears, then, that temperature has become connected somehow with what is going on in the atom. It view of this it would seem that another conception of temperature is desirable. It must be such a conception as not to be at variance with what we suppose to happen in a gas or a solid.

The material for such a conception is at hand. It is a property common to the temperature energy of a gas molecule and to the internal energy of our suppositious molecule, that it is constantly in a state of flux, changing during collision from kinetic to potential and back to kinetic again. A natural generalization, then, is as follows: Temperature is proportional to that part of the energy associated with a representative molecule which undergoes periodic changes from kinetic to potential and back to kinetic. In complicated cases where the transformation from kinetic to potential does not take place simultaneously in all parts of the atom, an equivalent generalization is: Temperature is the difference between the maximum and minimum potential energy of a representative molecule. This evidently applies to both the extreme cases above; the one for which temperature is proportional to kinetic translational energy, and the one for which it is proportional to internal energy. In one respect these two extreme cases are not entirely unlike; in accordance with the law of equipartition, only a certain proportion of the total energy communicated to a molecule of a gas becomes kinetic; the rest goes to the internal degrees of freedom. So that an increase of temperature always carried with it an increase of the internal energy of the molecule. Whether this internal energy also oscillates in character between kinetic and potential is not obvious.

Of course the justification of this proposed general definition of temperature must be furnished by experiment. It does have the advantage, however, of being applicable at high pressures where the ordinary definition breaks down completely, and it does agree with our physical feeling of what temperature must be in the case of one very simple model of a substance under high pressure.

The effect of this conception of temperature on our conception of the equipartition of energy between the different degrees of freedom is interesting. At high pressures, where the molecules press on each other from all sides, one degree of freedom has been lost (or more properly three) namely, the possibility of motion of the molecule as a whole. We may suppose, if we like, that under these circumstances the total energy is equally divided between the remaining degrees of freedom. Now at any instant in a liquid, there are molecules with varying degrees of freedom, according to the kind of collision in which they are entangled. Furthermore, the same molecule at different stages of its career may enjoy a different number of degrees of freedom. When the degrees of freedom change, there must follow a redistribution of the total energy among the remaining degrees of freedom, and this process takes time. The result is that we cannot ascribe to the average molecule of the substance any definite number of degrees of freedom. The number cannot be an integer in the first place, and in the second place must vary continuously as pressure varies.

The idea that the temperature of a substance need not be proportional to the kinetic energy of translation of the molecules is not new. For instance, there is a recent paper by Brillouin,[59] in which he discusses at length the possible necessity of a change in the usual definition of temperature as the volume changes.

One other very important consideration which we shall probably be obliged to introduce into a theory valid for high pressures has already been mentioned several times in the discussion of the various thermodynamic properties. It is this; in gases we think of the molecules as perfect spheres, but at high pressures we shall undoubtedly have to recognize that they possess more complicated shapes. It is inconceivable that a molecule should not have a characteristic shape; we cannot well imagine the possibility of so fitting together the atoms of a complicated organic compound as not to produce some irregularities in the molecule. Shape becomes increasingly important at high pressures where the molecules are forced together and constrained to adapt themselves as best they can to each others irregularities. Beside the possibility of shape, there is the possibility that there are local centers of force in the molecule; that we cannot regard the molecule as a homogeneous sphere exerting a force towards its own center, but that when we approach too close to the molecule, there are individualized centers of force that begin to act of their own account. This again, seems by no means a forced conception.

Along with the idea of molecules with shape goes the conception that at high pressures these shapes must be forced to more or less adapt themselves to each other; in other words, the molecules must begin to show traces of regular arrangement. The regularity is by no means the thorough going regularity of a crystal in which the molecules are permanently moored to certain mean positions: the molecules of the liquid still circulate about among each other, but as they slide past each other there may be a growing tendency at higher pressures to point the long axes in the direction of relative motion, for example. Just as at a crowded ball room, there is a tendency for the throng of young men making their way to and from the refreshment room to hold their plates out from them in the direction of motion. This increasing order of arrangement seems not only natural, but inevitable at high pressures. It may ultimately terminate in crystallization. We should expect furthermore, at equal volumes, there should be nearer approach to order at lower temperatures where the violence of temperature agitation is less.

---

[59] Brillouin, Ann. Chim. et phys., **18**, 387–400 (1909).

Now the combination of these two effects, namely that when the molecules come very close to each other the attractive forces depend on the orientation as well as on the mean distance apart, and that the molecules may assume a greater uniformity of arrangement, has far reaching consequences that provide the possible explanation of all the complicated effects which we have found to exist. Thus, if we consider two possible configurations of the liquid, each having the same volume, but one with a more orderly arrangement of the molecules, we see that the more orderly arrangement involves a greater effective space open to occupation. One consequence is that the more orderly arrangement has the greater compressibility. One striking example of this has been found in the case of mercury. The compressibility of solid mercury has been found to be less than the compressibility which the liquid would have at the same temperature if it could be compressed without freezing to the same volume as the solid.

In the detailed discussion of the thermodynamic properties it has been shown that we have here a possible explanation of many complicated effects. It explains increasing compressibility with rising pressure, decreasing compressibility with rising temperature, increasing thermal expansion with increasing pressure, and decreasing expansion with rising temperature. It is not necessary to go into the details of the argument again. It is to be emphasized, however, that we have here a mechanism capable of explaining a bewildering array of experimental facts. There must be at least some validity in the point of view.

Not much has been said in the explanation above of the results of possible association, because under high pressures, when the molecules find difficulty in adapting themselves to the space at their disposal, it seems unlikely for groups of molecules to unite themselves into very close knit units. The molecules, on the other hand, do apparently preserve their individuality under these high pressures, and do not break up into simpler compounds. It might be expected, for instance, that pressure would transform ether into isobutyl alcohol, a substance of the same atomic constitution, but with a smaller volume. Such was not the case, however. But it may be that association does play an important part at low pressures. In this case it would be capable of explaining various irregularities in very much the same way as suggested above. For instance, if association takes place with decrease of volume, the thermal expansion or compressibility may be greater at lower than at high temperatures. Association has been discussed in greater detail on page 104.

The results have exhibited one striking feature which has been frequently emphasized, namely that at high pressures all twelve liquids become more nearly like each other. This suggests that it might be useful in developing a theory of liquids to arbitrarily construct a "perfect liquid" and to discuss its properties. Certainly the conception of a "perfect gas" has been of great service in the kinetic theory of gases; and the reason is that all actual gases approximate closely to the "perfect gas." In the same way, at high pressures all liquids approximate to one and the same thing, which may be called by analogy the "perfect liquid." It seems to offer at least a promising line of attack to discuss the properties of this "perfect liquid," and then to invent the simplest possible mechanism to explain them.

Summary of Results.

These measurements have disclosed an unexpectedly complicated state of affairs at high pressures, in many respects the exact opposite of what we would expect from the behavior at low pressures. The compressibility may decrease with increasing temperature, or in a few cases may increase with increasing pressure. The thermal expansion also may decrease with increasing temperature or increase with increasing pressure. The peculiarity of thermal expansion with respect to temperature is possessed by all liquids above 3000 kgm. This has been shown to have a bearing on previous theories. Of the other thermodynamic properties, perhaps the most important is the internal energy. This passes through a minimum and then increases again with increasing pressure. The reason is that beyond a certain pressure the attractive forces do less work than is done by the external forces in compressing the molecule.

Among considerations which would seem to be of importance for a theory of liquids at high pressures, that of the shape of the molecules is worthy of attention. It is inconceivable that the molecules should not have shape, and it is natural to suppose that the shape will play an important part when the molecules are forced into close contact. It is shown in detail that considerations of this sort offer possible explanations of the complicated effects actually found. Other modifications of the ordinary conceptions of liquids that may be necessary have to do with our ideas of the kinetic origin of pressure and temperature. There can be no doubt that at high pressures there are other than kinetic effects involved at least in pressure.

It is a pleasure to acknowledge generous assistance from the Rumford Fund of the American Academy of Arts and Sciences for the purchase of apparatus and supplies, and from the Bache Fund of the National Academy for an assistant in some of the experimental work.

JEFFERSON PHYSICAL LABORATORY,
HARVARD UNIVERSITY, CAMBRIDGE, MASS.

[On page 74, line 15, *for* $0.0_4 108$ *read* $0.0_3 108$.]

Bei Fragen zur Produktsicherheit wenden Sie sich bitte an:
If you have any questions regarding product safety,
please contact:

Walter de Gruyter GmbH
Genthiner Straße 13
10785 Berlin
productsafety@degruyterbrill.com